Introductory
Plant Physiology

INTRODUCTORY PLANT PHYSIOLOGY

Second Edition

G. Ray Noggle
Department of Botany
North Carolina State University

George J. Fritz
Department of Agronomy
University of Florida

PRENTICE-HALL, INC., Englewood Cliffs, New Jersey   07632

Library of Congress Cataloging in Publication Data

Noggle, G. Ray (Glen Ray), (date)
   Introductory plant physiology.

   Includes bibliographies and index.
   1. Plant physiology. I. Fritz, George J.
(George John), (date). II. Title.
QK711.2.N63 1983   582'.01   82-24143
ISBN 0-13-502096-4

Editorial/production supervision:
  Zita de Schauensee
Cover design: Photo Plus Art
Cover photograph by Roy Saigo
  and Barbara Saigo
Manufacturing buyer: John Hall

Printed in the United States of America

10 9 8 7 6 5 4 3 2 1

ISBN 0-13-502096-4

Prentice-Hall International, Inc., *London*
Prentice-Hall of Australia Pty. Limited, *Sydney*
Editora Prentice-Hall do Brasil, Ltda., *Rio de Janeiro*
Prentice-Hall Canada Inc., *Toronto*
Prentice-Hall of India Private Limited, *New Delhi*
Prentice-Hall of Japan, Inc., *Tokyo*
Prentice-Hall of Southeast Asia Pte. Ltd., *Singapore*
Whitehall Books Limited, *Wellington, New Zealand*

Contents

Chapters 2, 3, 5, 6, 14, 15, 16, 17, and 18 were contributed by G. Ray Noggle. Chapters 1, 4, 7, 8, 9, 10, 11, 12, and 13 were contributed by George J. Fritz.

Preface

This book originated as a course in plant physiology for undergraduate students interested in higher plants and how they grow. Many of the students were majors in forestry, crop science, soil science, horticulture, or pest management and had a variety of backgrounds and training. The course also served as an introduction to botanical science for students who had had a one-semester course in general biology. Majors in botany took the course after a course in elementary botany and plant anatomy. All the students had completed one year of college chemistry, but most had had limited exposure to biochemistry or physics. With these students in mind, we have attempted to write a book suitable for use in a one-semester course in plant physiology emphasizing higher green plants and the concept of "whole plant" physiology.

The table of contents indicates our views of the important major topics of contemporary plant physiology. We have tried to make each chapter a reasonably complete and unified account of a major topic or concept. Complete documentation is not included in the text, but a reading list is provided at the end of each chapter. The references are of two kinds: general writings—usually textbooks, monographs, or reviews—and research papers bearing on the material discussed in the chapter. These references will give the interested student access to the current literature of plant physiology.

Since the first edition in 1976, the science of plant physiology has undergone several major changes. The energy crisis called attention to the dependence of modern agriculture on fossil fuels and dimmed the promise of the Green Revolution for providing food to a growing world population. Plant physiologists have responded by initiating new research on biological nitrogen fixation, mechanisms of carbon dioxide assimilation, plant responses to a wide variety of stress factors (drought, temperature, salinity, pests, inorganic nutrient imbalance, pollutants), and the roles of plant growth substances in growth and development. Accompanying these studies have been new studies on the uses of plants for fuel and chemicals, referred to as biomass conversion. Research in these areas has provided new information which is incorporated in this edition.

Plant physiology also has entered the field of molecular biology. Biochemistry, biophysics, microbiology, and genetics have given new tools for investigating problems of plant growth and development. Many scientists believe that plants offer many advantages for moving theoretical concepts of molecular biology into practical applications which can be used in improving production of food, fuel, and fiber. Recent advances in these fields have been brought into this new edition.

As with the first edition, each of the authors has been responsible for certain chapters; these are identified in a footnote on the first page of the table of contents. We read and criticized each other's work but did not settle on a single style of writing or arrive at complete agreement on all the material covered. In all cases we reminded ourselves continually that we were writing an introductory book for undergraduate students with the background described in the first paragraph.

Over a number of printings we made corrections in the original edition. In this second edition we attempt to bring the material up to date and incorporate useful suggestions made by many individuals. Special thanks are given to B. E. Michel, Ed. J. Stadelmann, and J. B. Shive, Jr. Their comments have been extremely helpful.

Raleigh, North Carolina   G. Ray Noggle
Gainesville, Florida   George J. Fritz

1

Introduction

The higher vascular plants that cover the land areas of the earth and inhabit the seas have a crucial role in mankind's existence and survival. They furnish us with food. They provide our livestock with forage. They supply the air we breathe with O_2. From them we obtain fibers for clothing, wood for shelter and furnishings, and medicines that we use to alleviate our ailments.

Not only mankind's primary biological needs but also many of the items used in everyday life are obtained from higher vascular plants. [The term *higher vascular plants* refers to those plants that produce seeds (i.e., gymnosperms and angiosperms).] Included among the items obtained are paper; rubber; spices; nonalcoholic beverages, such as tea, cocoa, and coffee; and alcoholic beverages, such as wine, beer, whiskey, gin, and vodka. Moreover, higher vascular plants minister to the aesthetic needs of humans by beautifying our physical environment. In fact, these plants are the most prominent feature of the natural green landscape.

The fact that we are completely and absolutely dependent on higher vascular plants for the necessities of life makes it imperative that we gain as thorough a knowledge as possible of the science of plant physiology. Moreover, our insatiable curiosity about the world in which we live provides ample justification for studying plant physiology.

THE SCIENCE OF PLANT PHYSIOLOGY

Plant physiology is a study of vital phenomena in plants. It is the science concerned with processes and functions, the responses of plants to changes in the environment, and the growth and development that result from the responses. *Process* means a natural continuing sequence of events. Examples of processes that occur in living plants are photosynthesis, respiration, ion absorption, translocation, stomatal opening and closing, assimilation, transpiration, flowering, and seed formation. To describe and explain plant processes is one of the tasks of plant physiology. *Function* refers to the natural activity of a thing, whether cell, tissue, organ, chemical substance, or whatever. A second task of plant physiology is to describe and explain the function of each kind of organ, tissue, cell, and cellular organelle in plants and also the function of each chemical constituent, whether ion, molecule, or macromolecule. But because processes and functions are dependent on and modified by such external factors as light and temperature, a third task of plant physiology is to describe and explain how processes and functions respond to changes in the environment. Essentially the overall goal of plant physiology is to evolve a detailed and comprehensive knowledge of all the natural phenomena that occur in living plants and thus to understand the nature of plant growth, development, and movement.

The methods used in plant physiology are derived mainly from chemistry and physics. In fact, an important feature in the development of modern plant physiology has been the increasing role of chemistry and physics in furnishing new ideas and new tools, not only for the solution of old problems but also for the recognition of new ones. Of course it will be realized that processes and functions cannot be understood properly unless something is known about the structures with which they are associated. Therefore some knowledge of plant anatomy, in addition to chemistry and physics, is needed in the study of plant physiology.

The tools used in the study of plant physiology are varied and include all those traditionally associated with chemistry, physics, and plant anatomy as well as many of the more sophisticated ones developed in recent years. Examples of the latter are chromatography, electron scanning microscopy, electrophoresis, freeze etching, mass spectrometry, and radioactive tracer techniques. In addition, many of the tools distinctively associated with cell biology are used in plant physiology. Maceration techniques, for example, have made it possible to extract soluble and particulate fractions from plant cells and to study metabolic reactions carried on by individual enzymes and by cellular organelles, such as mitochondria, chloroplasts, ribosomes, and nuclei. As a result, a great deal of information concerning cellular activities has been gathered during the past several decades. Thus our understanding of the biochemical reactions that occur in plants and the functional significance of cellular organelles has been broadened considerably. It should be kept in mind, however, that individual components of cells operate under more subtle conditions in vivo (i.e., in the living condition) than in vitro (i.e., when separated from living cells). When organized cellular structure is destroyed and the various parts of the cell are separated, the unique capabilities conferred on a cell by the organized arrangement of interacting components is impaired.

The science of plant physiology is never static but always changing as new facts are discovered and fresh concepts developed. It is only natural that plant physiologists may not always have the same opinion regarding the mechanism of certain plant processes or the function of certain plant constituents. Just as two cameras placed at two different points and both trained on the same scene can result in an overall picture that is slightly out of focus, so it is possible for two plant physiologists to reach different conclusions on the basis of a given set of experimental data. The theory proposed by one is challenged by others. As more and more data are collected, however, old interpretations are sifted, weighed, and reexamined in the light of these new data and new interpretations more in harmony with the known facts of the physiology of plants arise. The close interweaving of experimentation and interpretation is a guarantee that progress in understanding will be made. Plant physiology is a self-correcting body of knowledge. Through continuing research efforts and the unfolding of new evidence, new ideas evolve and replace the old.

PRACTICAL ASPECTS OF PLANT PHYSIOLOGY

Quite apart from the considerable theoretical significance of plant physiology in helping us understand the world in which we live, a knowledge of plant physiology is essential to all fields of applied botany—agronomy, floriculture, forestry, horticulture, landscape gardening, plant breeding, plant pathology, pharmacognosy, and so forth. All these applied sciences depend on plant physiology for information about how plants grow and develop.

Plant physiology will probably also assume an increasingly important role in agricultural research programs. As world population increases, mankind faces enormously complex problems. Their solutions will require input from many sources: social, economic, technological, and agricultural. One of the primary tasks of the future will be to increase food, forage, fiber, and wood production substantially throughout the world. Future agricultural research programs will continue, as in the present, to have as their major goals the production of new and better varieties and strains of crop plants; the improvement of plant protection against insects, diseases, and weeds; the control of soil fertility; and an increase in mechanization efficiency. But in addition there will be a sharp intensification of demands on plant physiologists not only to supply basic information regarding how plants grow and develop but also to undertake research programs designed specifically to increase yields of plant products.

Many aspects of practical agriculture can benefit from more intensive research in plant physiology. Only a few examples will be mentioned here. The efficiency of photosynthetic conversion of solar radiation in the production of food crops acceptable to human diets can be increased by one or more of several means, including a decrease in the rate of photorespiration and breeding genetic changes in plants with the goal of increased display of leaves so as to create better light-capturing systems. Improved biological N_2 fixation will offset the enormously expensive chemical synthesis of commercial nitrogen fertilizers. Techniques of tissue culture and cell fusion developed by plant physiologists during the past several years may be used to breed desirable strains of crop plants. Means of avoiding or reducing environmental stresses, such as drought, frost, and pests (insects, weeds, disease, etc.), can be developed (e.g., leaf sprays designed to decrease transpiration and thereby increase efficiency of water use in drought resistance). Crop yields can be increased by learning how and when the application of plant growth regulators to plants is most effective. Now useless weeds and jungle plants can be converted to high-quality fodder by the addition of fats and proteins produced through large-scale culture of algal and yeast cells. Greater efficiency of nutrient uptake from soils can be realized by obtaining superior strains of microorganisms for formation of mycorrhizal symbioses with roots. These and other potential applications of research in plant physiology to the solution of practical problems in agronomy, forestry, horticulture, and other fields give an added dimension to the already recognized importance of research in basic plant physiology.

HEREDITARY AND ENVIRONMENTAL INFLUENCES ON PLANT BEHAVIOR

It is a basic principle of plant physiology that two sets of factors, hereditary and environmental, regulate the internal processes and conditions of the plant and thereby determine plant growth and development (see Figure 1-1). Thus the ultimate shape, size, form, and degree of complexity of a plant are the result of the interaction between its genetic composition and the environment in which it grew. Just as the genetic composition of petunia seeds will ensure that they always produce petunia plants, not zinnias or roses, so the environmental factors will determine whether the petunia plants are vigorous or stunted, bright green or yellowish, or turgid or wilted. Modifications caused by variations in environmental factors normally are not inherited.

The heredity information that "tells" a plant how to behave is determined by the nucleic acids present in all cells of the plant body. Deoxyribonucleic acid (DNA) is the primary genetic substance that conveys hereditary information from generation to generation. How the genetic "blueprint" is transmitted is described in Chapter 3.

Because much of the science of plant physiology concerns plant responses to the physical environment, it will be useful, for purposes of orientation, to make a few comments regarding each of the major factors of the physical environment.

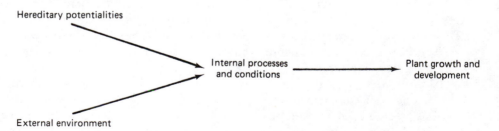

Figure 1-1   Two sets of factors, hereditary and environmental, determine internal processes and conditions in the plant (i.e., the physiology of the plant). Internal processes and conditions, in turn, are expressed in plant growth and development.

The quality, intensity, and duration of radiation that impinges on plants have profound effects on many physiological processes. Light, for example, affects chlorophyll formation (Chapter 5), photosynthesis (Chapter 6), and photorespiration (Chapter 7). Moreover, light induces cylindrical plant organs (e.g., stems, petioles, coleoptiles) to grow at unequal rates when they are differentially illuminated on the two sides; this phenomenon is referred to as phototropism (Chapter 17). Also alternations in light and dark periods from day and night (photoperiodicity) control many aspects of plant growth and development [e.g., flower formation (Chapter 18)].

The critical importance of soil moisture supply and atmospheric humidity on plant growth and development is recognized by everyone. When a plant does not receive sufficient water or when transpiration rates are excessive, its leaves wilt, its growth slows down or ceases altogether, and it may even die (Chapter 13).

Not only water but also a host of chemicals naturally present in the soil and atmosphere affect the growth and development of plants. Mineral ions are supplied by the soil to roots (Chapter 8). Certain substances excreted by plants may inhibit the germination of seeds of other plants; this phenomenon is called allelopathy. Also, the gases O_2 and CO_2 are vital components of the environment. All plant cells require O_2 for respiration (Chapter 4). CO_2 is assimilated in photosynthesis by green leaves in the light (Chapter 6). O_2 inhibits the magnitude of net photosynthesis in many plants by stimulating a process known as photorespiration (Chapter 7). In addition to naturally occurring chemicals, certain industrial and automobile waste products, both gaseous and particulate, pollute the atmosphere in many geographic areas and pose a special hazard to plants.

The temperature of the soil and air not only affects the rates of physiological processes (e.g., photosynthesis, ion absorption, growth) but may also determine the course of development in plants. Only three examples will be given here.

1. Seeds of certain species will not germinate unless exposed to low temperatures for several weeks or months; during this period certain biochemical changes occur that result in the breaking of dormancy (Chapter 16).
2. In a number of species exposing the seed to low temperatures for a prolonged period of time results in a hastening of the subsequent flowering of the plants. In fact, some plants will not produce flower buds unless their seeds are subjected to low temperatures; this phenomenon is referred to as vernalization (Chapter 18).
3. Optimal growth of many plants is possible only when high day temperatures alternate with lower night temperatures. This requirement for a diurnal thermo-periodicity is considered in Chapter 17.

A final category of external environmental factors that influence plant growth and development includes a few physical forces. In response to the natural field of gravity on earth, roots grow down toward the center of the earth and stems grow up, away from the center of the earth (Chapter 17). Other physical influences that could also be included in this last category would be electrical and magnetic fields of force artificially applied to plants, and winds of extreme velocity.

THE NATURE OF LIVING THINGS

Inasmuch as plant physiology deals with life processes, it will be useful to consider the nature of living things. No fully precise and completely satisfying definition that includes all living things and excludes all nonliving can be formulated. Nevertheless, living matter possesses a number of attributes that, taken together, characterize it even though each attribute individually may be shared with nonliving objects. Growth, for example, is one of the properties of life, although it closely resembles the growth of crystals in saturated solutions.

Among the attributes of a living organism (i.e., an individual plant or animal) are its specific form and unique structure. The former is manifested in its distinctive external appearance and the latter in the arrangement of its parts in three-dimensional space. Another attribute of a living organism is its capacity for development—that is, its ability to change its form and structure through its lifetime. Also, a living organism preserves its hereditary traits and transmits them to its progeny. A living organism is characterized by its unique chemical composition, consisting mostly of organic matter (i.e, carbonaceous compounds) of six major classes: carbohydrates, amino acids, nucleotides, porphyrins, lipids, and enzymes. And, added to these, a living organism has the ability to absorb substances from its environment (nutrition); to carry on literally hundreds of chemical reactions at the same time, each catalyzed by an enzyme with a high degree of specificity and all closely and precisely coordinated through appropriate interactions (metabolism); to increase its size and mass (growth); to produce others like itself (reproduction); to undergo from time to time permanent changes in its genetic material and to transmit these changes to its progeny (mutation); and to respond and adjust continually to changes in the environment (adaptation). Finally, by utilizing an input of energy (chemical energy in the form of organic matter or light energy trapped by chloroplasts), a living organism is able to synthesize more of itself from materials in its environment.

Some biologists have suggested that the *essence* of life is its *organization*. A living organism represents an organized arrangement with interacting components. Although complexity abounds within living organisms, it is a highly developed interconnected complexity that can be said to represent a mutual adaptation of parts. This organization extends through all levels, from organ, tissue, and cell to the subcellular level and the domain of macromolecular assemblies. The organization of its components confers capabilities on a living organism which are completely lacking in any of the isolated components. Other biologists suggest that the single most important difference between biological systems and other complex inanimate systems is that the former contains an internal representation in the form of *genetic apparatus*. To understand how this genetic apparatus participates in the elaboration and specification of patterns and structures present in biological systems appears to be one of the major unsolved problems of biology; this second group of biologists hopes that its solution will give us a greater understanding of the behavior (physiology) of biological systems. To sum up, it is clear from these two separate beliefs that unanimity of opinion has not yet been reached concerning the *essence* of life.

LIMITATION OF SUBJECT MATTER

Even though living matter is fundamentally similar in all living things and shares a set of common properties, differences exist. The most obvious evidence of protoplasmic differences among living organisms is reflected in their external morphology. The plant kingdom alone accounts for about half a million species. Included are more than 300,000 species of flowering plants. To these must be added conifers and their relatives (about 1000 species), ferns (10,000), club mosses and lycopods (1000), mosses and liverworts (25,000), and bacteria, fungi, and algae (100,000).[1] (It is impossible to count hybrids, cultivars, and varieties; there are so many.) Clearly the differences in morphology in this vast array of plants are accompanied by differences in physiological behavior.

This book is concerned primarily with the physiology of the seed plants (flowering plants and conifers), especially those that produce flowers (i.e., those that belong to the group known as angiosperms). Specialized physiological problems associated with such plant groups as bacteria and fungi (including problems such as spore formation and factors affecting pathogenicity) are not considered.

The plants known as *seed plants* produce seed during the reproductive phase of their life cycle. They have vascular systems consisting of xylem and phloem tissues. Individual cells are characterized by a cellulosic-pectinaceous wall and a central vacuole; the latter constitutes a large portion of the cell volume at maturity. (This last characteristic is shared by many nonvascular plants, which also have cellulosic walls and large vacuoles.) Also, seed plants are characterized by the independence and dominance of the sporophyte, which consists of roots, stems, and leaves (see Chapter 2).

Sometimes the seed plants are called *higher* vascular plants [in contradistinction to the *lower* vascular plants (i.e., club mosses, lycopods, and ferns)]. Moreover, they are called *higher green plants* (to distinguish them from the algae, which have chloroplasts and carry on photosynthesis but do not have vascular systems). Generally, however, the seed plants are referred to simply as *higher plants*. Almost all live in terrestrial rather than aquatic habitats.

Unlike the relatively simple plant cells in bacteria and blue-green algae (which are not nucleated), cells of higher plants have elaborate internal protoplasmic membranes that not only compartmentalize the cell into nuclei, chloroplasts, mitochondria, and other organelles, but also function in transport of substances into and out of cells and from one organelle to another within a cell; a three-dimensional skeletal framework of long, thin, hollow cylinders of proteinaceous nature called microtubules (see pp. 24–25) which function in maintenance of the shape of the cell, in cell division, and in movement of newly synthesized materials from place to place within the cell; and nuclear genomes that are many times richer in genes (than those in bacteria and blue-green algae) for the specialized functions of higher plant cells.

---

[1] In contrast, 5 million chemical substances are catalogued. The number is increasing at the rate of about 350,000 per year.

Although the following chapters are primarily concerned with the physiology of higher plants, it is often instructive to compare similar physiological processes in plants of different taxonomic groups. This comparative approach yields information of general physiological interest and also often contributes to the solution of problems specific to higher plants. Photosynthesis is an example of a process that occurs in green leaves of higher plants as well as in certain unicellular plants, such as algae and a few bacteria. These unicellular organisms are relatively simple living objects whereas leaves of higher plants consist of a number of tissues, some of which are chlorophyllous and carry on photosynthesis (e.g., mesophyll tissue) and others that are non-chlorophyllous with functions other than photosynthesis (e.g., phloem, xylem, and epidermis). Certain kinds of experiments in photosynthesis are often easier to carry out with unicellular organisms than with multicellular leaves. In fact, studies carried out with these single cells have led to a greatly increased understanding of some of the fundamental aspects of photosynthesis (see Chapter 6).

REFERENCES

ANN. REV. PLANT PHYSIOL. 1950–. Vols. 1–. Annual Reviews, Inc., Stanford, Calif.

BERNAL, J. D. 1967. Definitions of life. *New Scient*. 33:12–14.

BLACK, S. 1973. A theory of the origin of life. *Adv. Enzymol.* 38:193–234.

FRITZ, G. J., ed. 1969–. *What's New in Plant Physiol.*, vols. 1–. WNPP, Agronomy Dept/Bldg 857/Museum Road, University of Florida, Gainesville.

NORTHRUP, J. H. 1961. Biochemists, biologists and William of Occam. *Ann. Rev. Biochem.* 30:1–10.

PARKER, J. 1953. Criteria of life: some methods of measuring viability. *Am. Scient.* 41:614–618.

PIRSON, A., AND M. H. ZIMMERMANN, eds. 1975–. *Encyclopedia of Plant Physiology*, new series, vols. 1–. Springer-Verlag, Berlin.

RUHLAND, W., ed. 1955–1967. *Handbuch der pflanzenphysiologie*, vols. 1–18. Springer-Verlag, Berlin.

SCHERY, R.W. 1952. *Plants for Man*. Prentice-Hall, Englewood Cliffs, N.J.

SPRAGUE, G. F., D. E. ALEXANDER, and J. W. DUDLEY. 1980. Plant breeding and genetic engineering: a perspective. *BioScience* 30:17–21.

STEWARD, F. C., ed. 1960–1972. *Plant Physiology, a Treatise*, vols. 1–6. Academic Press, New York.

STUMPF, P. K., and E. E. CONN, eds. 1980–1981. *The Biochemistry of Plants, a Comprehensive Treatise*, vols. 1–8. Academic Press, New York.

WENT, F. W. 1974. Reflections and speculations. *Ann. Rev. Plant Physiol.* 25:1–26.

WITTWER, S. H. 1979. Future technological advances in agriculture and their impact on the regulatory environment. *BioScience* 29:603–610.

2

The Organization of Plants

The seed plants exhibit the greatest diversity in form and occupy the most variable habitats of all members of the plant kingdom. Numbering approximately 300,000 species, they range in size from the small aquatic duckweeds with diameters of 2 to 5 mm to the giant Sierra redwoods (*Sequoiadendron giganteum*) with trunk diameters approaching 20 m and the coast redwoods (*Sequoia sempervirens*) with heights in excess of 100 m. Life spans are also quite variable. Some annuals, such as certain *Arabidopsis* species, complete their life cycle in 20 to 28 days. Some tree species, such as the Sierra redwood and bristlecone pine (*Pinus aristata*), are still alive at ages of about 4000 and 4600 years, respectively. Many species of seed plants grow in very restricted areas because of narrow tolerances to environmental factors, such as temperature, moisture, and chemical composition of the soil, whereas other species are quite adaptable to the environment and have a worldwide distribution. The fact that seed plants are found growing in all but the most inhospitable regions of the earth attests to the success of the plant "way of life."

Despite their great diversity in form, size, and habitat, seed plants have a similar external structural plan. The above-ground portion of the plant, the shoot, consists of leaves and stems. The leaves are attached at regular intervals to the stem at the nodes. In the axil of each leaf is a bud that may develop into either a vegetative shoot or a floral shoot. The roots lie below ground and may consist of a taproot without extensive branching or of a profusely branched fibrous root system. Unlike the stem branches, the lateral or branch roots are not attached at definite regions (nodes). Nor do buds exist on the roots. In addition to these vegetative organs—roots, stems, and leaves— seed plants also have flowers and eventually fruits. For any given species, the vegetative structures may be quite variable in size and form and may be modified by environmental factors. Flowers and floral structures do not display such variability. The numbers of floral parts, their position, and their arrangement on the plant are relatively constant in any given plant species, a fact useful in plant identification and classification.

INTERNAL STRUCTURE OF PLANTS

Plant organs are composed of specialized groups of cells and tissues that carry out the function of protection, support, translocation, storage, metabolism, and reproduction (Figure 2-1). In seedlings and short-lived annuals the plant body arises from the division, growth, and development of meristematic cells located at the tips of the shoots and roots. From these apical cells come primary tissues, such as the epidermis,

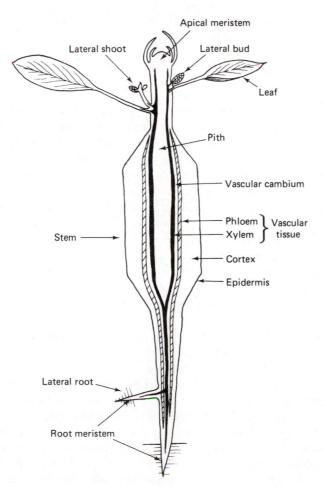

Figure 2-1 Diagrammatic representation of tissue systems and organs of seedling plant (dicotyledon).

vascular elements, pith, and cortex. In longer-lived plants, however, secondary tissues are formed through the activity of a vascular cambium. Secondary xylem and phloem cells, ray parenchyma cells, and others develop from this cambium. As a result of this secondary growth, the girth of the stem and root is increased. This increase in size is generally accompanied by the rupture of the epidermis and the development of a new cambium, the cork cambium, beneath the epidermal cells. From this cork cambium there develops the periderm, or cork, which replaces the epidermis as a protective layer on the roots and stems.

As noted, groups of cells at the tips of roots and stems remain in an embryonic condition and maintain the capacity to divide. Such cell groups form what are known as meristematic regions—stem meristem or root meristem. As long as conditions are favorable, stems and roots can continue to grow and develop throughout the life of the plant. Leaves of most plants, on the other hand, are organs of limited growth. The leaf initials, formed at the stem apex, initially are meristematic and the cells so laid down undergo division and differentiation, resulting in the final leaf form. No groups of cells within the leaf, however, remain embryonic and continue to divide. When the cells of the leaf have undergone expansion and differentiation, the final leaf form is achieved.

The leaf is covered by a layer of epidermal cells, which are covered with a layer of cutin. The epidermis contains stomata, thereby allowing for the exchange of carbon dioxide, oxygen, water vapor, and other gases. Chloroplast-containing mesophyll cells are found between the upper and lower epidermal cells. These cells are the major centers of photosynthesis. The leaf is well supplied with vascular tissue, which provides for the translocation of water and solutes between the leaf and other organs and tissues of the plant.

A generalized plant cell, or parenchymatous cell, is shown in Figure 2-2. The cellular fraction—protoplasm—is surrounded by the primary cell wall. Protoplasm is divided into two components, nucleus and cytoplasm. Such a separation emphasizes the crucial importance of the nucleus in cellular activity. The cytoplasm contains organelles (plastids, mitochondria, Golgi bodies, microbodies, lysosomes, vacuole, ribosomes, spherosomes, endoplasmic reticulum), fibrillar structures (microfibers, microtubules), and numerous inclusions (oil and fat droplets, spherosomes, protein bodies, starch grains, crystals). The cytoplasmic fraction exterior to membrane-bound organelles is referred to as the *cytosol*. This fraction contains proteins, amino acids, saccharides, organic acids, and inorganic ions, as well as cytoplasmic inclusions.

Two major membranes are shown in Figure 2-2: The tonoplast (vacuolar membrane) separating vacuolar contents from the cytoplasm and the plasmalemma (plasma membrane) separating protoplasm from the primary cell wall. Later it will be shown that many organelles are also surrounded by membranes that separate their cellular constituents from other cytoplasmic material. In addition, there are protoplasmic connections, or plasmodesmata, connecting the protoplasm of adjoining cells. This continuous network of protoplasm is referred to as *symplasm*. The fraction of the cell outside the protoplasm also forms a continuous network and is known as *apoplasm*. Symplasm and apoplasm will be mentioned later (pp. 305–306) in discussing pathways of transport in the plant.

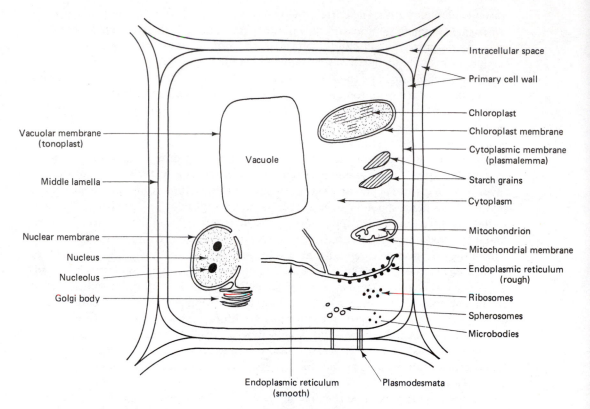

Figure 2-2   Generalized plant cell.

CELLS

Organ and tissue systems are multicellular structures and their complexity reflects the diversity and complexity of the individual cells of which they are composed. The cell is the basic unit of structure and function, the smallest biological unit having those attributes characteristic of living matter—unique chemical composition, metabolism, growth, reproduction, and organization. Although each cell has its own individual properties, the plant or the organ or the tissue is more than simply a loose aggregation of cells. The cells interact with each other and modify their environment, giving rise to the multicellular organism with its characteristic structure and function.

The concept of the generalized plant cell was developed from studies by cytologists on cells especially prepared for microscopic examination. In these studies the plant material is killed and fixed in a mixture of acid and alcohol to preserve the internal organization; afterward it is sliced into thin sections stained with dyes and then examined under the microscope. Using appropriate staining techniques, the various inclusions and organelles can be seen. A typical leaf mesophyll parenchyma cell measures about 200 μm[1] long, 75 μm wide, and 75 μm thick. The cytoplasmic matrix also appears to have additional structures, but they cannot be resolved by the ordinary light microscope. The resolving power of a light microscope depends on the wavelength of light used to illuminate the objects and objects less than one-half wavelength in diameter cannot be resolved. Thus if white light with an average wavelength of 550 nm is the source of illumination, objects less than 275 nm cannot be resolved. With the presently available light microscopic equipment, it is not possible to see any subcellular structures much smaller than 275 nm. The mitochondria are just at the lower limit of visibility.

The cell in Figure 2-2 represents a composite of information derived from numerous sources. The role of the chloroplast as the center of photosynthesis was suggested well over a hundred years ago, but only within past years has definite experimental proof been available. The recognition of the importance of the nucleus in the life of the cell came first from cytological studies. They were followed by chemical studies of isolated nuclei, which led ultimately to the elucidation of the structure of the deoxyribonucleic acid molecule. Cytological studies indicated that the structures known as mitochondria were able to reduce certain dyes, and it was suggested that the mitochondria were probably involved in respiratory activities. Biochemical studies on isolated cells and macerated cells showed that the preparations were capable of carrying out a great number of enzyme-catalyzed reactions, including dehydrogenations, hydrolyses or phosphorolyses, anabolic and catabolic reactions, and oxidation-reductions, to name a few. Not all these reactions proceed simultaneously within the cell. These diverse reactions must be compartmentalized and organized so as to provide for an orderly development of the plant.

Subcellular Organization

Undoubtedly the cell is the basic unit of structure and function in plants. With the acquisition of more information regarding the physiological capabilities of cells, however, it has become difficult to encompass these findings in the framework of the generalized plant cell shown in Figure 2-2. It is recognized that some unique features of a cell are related to its position in a tissue or organ. A cell influences its neighbors; in turn, neighboring cells influence it. Equally important is the subcellular organization of the cell. One exciting development in cell biology during the past several decades has been the discovery that the organelles of the cell are highly structured, containing internally organized membranes.

Progress in any science depends on the development of new techniques and methods. This fact is particularly clear in biology, where the above-mentioned

---

[1]The micron (abbreviated μm) is a unit of length commonly used in dealing with very small objects. The relationship of the micron to other units of length is as follows:

1 meter (m) = 39.37 inches	1 micron (μm) = 10^{-3} mm or 10^{-6} m
1 centimeter (cm) = 10^{-2} m	1 nanometer (nm) = 10^{-3} μm or 10^{-9} m
1 millimeter (mm) = 10^{-3} m	

subcellular organization of cells was revealed as the result of two newly developed experimental techniques—transmission electron microscopy and the centrifugal separation of subcellular organelles. It was pointed out earlier that the resolving power of a microscope is a function of the wavelength of electromagnetic radiation used to illuminate the object. Because of technical limitations inherent in optical systems, the best light microscopes cannot resolve objects smaller than 275 nm in diameter or less than 275 nm apart. By using a quartz optical system and a photoelectric cell rather than the eye to observe the image, a resolution of about 200 nm is possible with the light microscope. The electron microscope enables the observer to visualize objects of much smaller dimensions. Instead of electromagnetic (light) waves to illuminate the object, the electron microscope uses a beam of electrons. These electron beams behave as light waves and can be propagated with wavelengths of about 0.005 nm. Theoretically objects 0.0025 nm in diameter should be resolved with such a beam, but technical difficulties at present place the practical limits of resolution of the electron microscope at between 1.0 and 2.0 nm. The electron beam is not visible to the eye; so the microscopic image is recorded on a photographic film. The film image may be further magnified by photographic procedures giving total magnifications between 20,000 and 200,000.

When viewing an object on the stage of an optical microscope, it is possible to get an idea of its three-dimensional nature by focusing the lens up and down. Of course, the depth-of-focus feature is lost if the object is photographed as it is then fixed in space. The same difficulty arises when using the transmission electron microscope. Extremely thin sections are necessary for examination and the microscopic image must be recorded on a photographic film. A recent development in electron microscopy, however, now enables the viewer to obtain a depth-of-field view of the object under examination.

This technique, known as scanning electron microscopy (SEM), utilizes a moving beam of electrons across the object being viewed. It is not necessary to use harsh fixatives and embedding methods to prepare thin sections, as in transmission electron microscopy. Nonliving specimens are coated with a 10-nm-thick film of gold-palladium for viewing. Figure 2-3 is an SEM micrograph of the cross section of a tobacco leaf. Beneath the upper epidermis are palisade parenchyma cells containing chloroplasts. The chloroplasts are oriented against the outer walls of the palisade parenchyma cells by the central vacuole. Spongy mesophyll cells lie between the palisade parenchyma cells and the lower epidermis. One spongy mesophyll cell has retained its cellular contents and chloroplasts; the others have been cut during preparation. Note the connections between the palisade parenchyma cells and the spongy mesophyll cells as well as the connections between the mesophyll cells. Also note the amount of space between the mesophyll cells, indicating the ready access of gaseous materials to the cells. A single stomata with guard cells is seen in cross section on the lower epidermis. The micrograph clearly illustrates the three-dimensional character of SEM preparations.

Although transmission and scanning electron microscopy revealed the wealth of structural organization within cells, another technique, differential centrifugation,

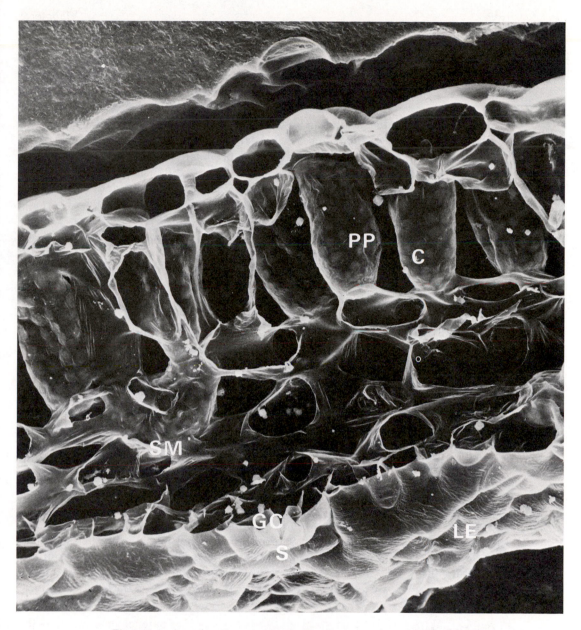

Figure 2-3 Scanning electron micrograph of cross section of tobacco leaf. The palisade parenchyma cells (PP) and spongy mesophyll cells (SM) contain chloroplasts (C). On the lower epidermis (LE) a stoma (S) and guard cells (GC) are visible. Note the amount of air space in the leaf. ×1000. (Courtesy Dr. C. G. Van Dyke, North Carolina State University.)

furnished the means of isolating these subcellular organelles. The centrifugation technique depends on the familiar observation that objects in solution settle out at rates proportional to their size and density. If a soil sample is suspended in water and then allowed to stand quietly, for example, the heavy pebbles and sand quickly fall to the bottom while the smaller particles settle out more slowly. The very fine particles may remain suspended in solution for a long time. The subcellular components of the cell are also of different sizes and densities and this fact has been used in devising a separation method. Instead of allowing the subcellular components to settle out by gravity, the suspension of cell organelles is centrifuged at different speeds. By adjusting the density of the suspending solution, as well as the speed and duration of centrifugation, various kinds of organelles can be separated.

To liberate the organelles, the cell wall and plasma membrane must first be disrupted by grinding or tearing the tissue. This step is done in a solution whose osmotic potential approximates that of the cytoplasm of the cell; if the osmotic potential is too low, the isolated organelles will absorb an excessive amount of water and burst, thereby losing their structural integrity. The grinding medium usually contains sucrose or a buffered salt solution, but their use depends on the nature of the tissue being studied and the ultimate use of the isolated organelles. No hard and fast rules apply to all tissues; the proper conditions must be determined for each tissue or cell type. All operations are generally carried out at temperatures just above freezing so as to preserve the biological activities of the isolated organelles.

Early centrifugation studies with plant tissues, such as leaves and meristems, led to the successful isolation of chloroplasts, nuclei, mitochondria, and ribosomes. The preparations were used to study the chemical composition and physiological activity of the isolated subcellular organelles. Electron micrographs of the organelles showed that the preparations were not pure but contained populations of other kinds of subcellular components whose function was unknown. Further refinements in the centrifugation techniques have enabled workers to isolate some of these new particles.

The techniques of electron microscopy and organelle isolation complement one another very well. Structures seen in electron photomicrographs have been isolated from disrupted cells and their structure compared with that of the intact cell. The structural details of the isolated organelles have been correlated with their chemical composition and biological activity, thereby giving a more complete picture of cell structure and function. Some aspects of the structure and function of the subcellular organelles now known to be present in plant cells are discussed in the following pages.

Cell Membranes

In describing the generalized plant cell (Figure 2-2) earlier, attention was drawn to two cell membranes, the plasma membrane that separates the cytoplasm from the external environment and the vacuolar membrane surrounding the vacuole. The importance of these membranes in water and solute movement, cell and organ turgor, and metabolic compartmentation had been recognized by cell physiologists for years (see Chapter 12). With the revelation by electron microscopy (Figure 2-4) that the cytoplasm

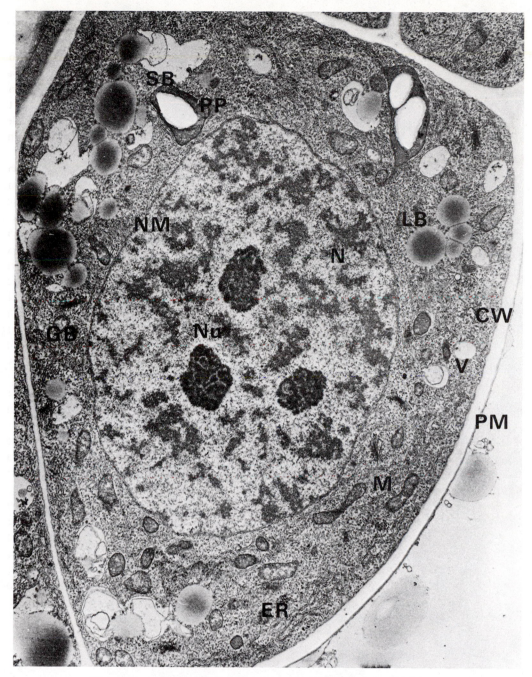

Figure 2-4 Transmission electron micrograph of meristematic cell of *Pinus taeda*. The large nucleus (N) contains three nucleoli (Nu) and is surrounded by a nuclear envelope (NM). A primary cell wall (CW) and plasma membrane (PM) are also shown. Within the cell are numerous subcellular organelles: Golgi bodies (GB), proplastid (PP) with starch body (SB), mitochondrion (M), lipid body (LB), endoplasmic reticulum (ER), and small vacuoles (V). ×2500. (Courtesy Dr. Henry Amerson and Dennis Gray, North Carolina State University.)

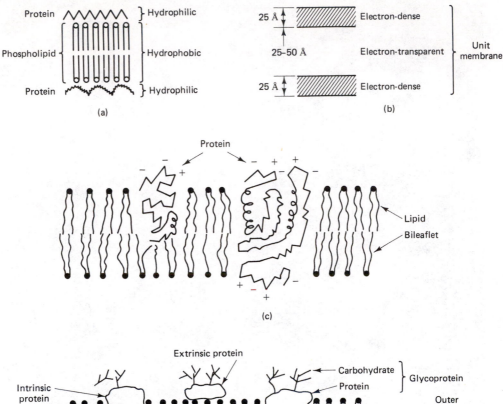

Figure 2-5 Interpretations of membrane structure and organization. (a) Bimolecular leaflets of phospholipid in association with protein molecules. (b) "Triple structure" of unit membrane as seen under electron microscope [see Figure 2-7(a)]. (c) Lipid globular protein mosaic model of membrane structure of Singer and Nicolson (1972) (d) Modified fluid mosaic model showing glycoproteins on outer surface.

contains subcellular organelles bounded by membranes, as well as interconnecting membrane surfaces between cells, new interest developed in membrane composition and organization.

From studies on the chemical composition of isolated membranes and their permeability to a variety of solutes (see Chapter 12) it was concluded independently by Davson and Danielli in 1935 that cell membranes are organized as shown in Figure 2-5(a). The central core of the membrane is composed of a double layer of phospholipid molecules (see Figure 3-21 for chemical structure of phospholipids). The hydrophobic fatty acid chains are oriented within a lipid-rich region whereas the hydrophilic ends of the phospholipids are oriented to an outer aqueous phase where they are in contact with the hydrophilic segments of protein molecules. Support for this model came from ultrastructural studies of cells by electron microscopy where cell membranes were observed [Figure 2-7(a)] as tripartite "sandwichlike" structures. Figure 2-5(b) compares the structure of membranes as seen by electron microscopy with the Davson–Danielli model in Figure 2-5(a). A proposal was made by Robertson in the 1960s that the Davison–Danielli model be considered the unit of structure for cell membranes (the unit membranes).

The Davson–Danielli–Robertson model accounts for some of the observed properties of membranes. The outer and inner surface proteins may be different, for instance, thereby allowing for a differential passage of solutes between the two cell surfaces. Also, the hydrophilic "head" groups of the phospholipids may be different at the two membrane surfaces, which might contribute to the "sidedness" of the membranes. With improved isolation and analytical techniques, however, it was found that the membrane proteins are globular in structure and quite heterogeneous as to size. Moreover, the proteins appear to be distributed within the membrane itself as well as located at the membrane surfaces. In addition, investigations of membrane organization by various physical methods suggest a more heterogeneous structure than depicted by the Davson–Danielli–Robertson model. From these later studies of membrane composition and organization a new model was put forward by Singer and Nicolson in 1972. This model, known as the *fluid-mosaic* or lipid-globular protein mosaic, is illustrated in Figure 2-5(c).

The major difference between membrane models 2-5(a) and 2-5(c) lies in the arrangement and organization of the protein molecules. The proteins may be entirely embedded within the lipid bilayer (intrinsic proteins), as shown schematically in Figure 2-5(d); or organized at or on the cell surface (extrinsic proteins). Another feature of the membrane proteins shown in Figure 2-5(d) is that those located at the outer membrane surface are likely to be glycoproteins with the carbohydrate moiety exposed to the aqueous external environment. The arrangement of the hydrophobic and hydrophilic amino acid subunits (see Figure 3-1) of the protein facilitates the organization of the proteins within the membrane. Many membrane proteins are not firmly bound within the membrane and may migrate in a lateral direction (the phospholipids also migrate laterally within the membrane). Extrinsic membrane proteins frequently can be removed by water or mild detergents without destroying membrane integrity. Intrinsic proteins, however, cannot be removed without disrupting membrane organization.

Some intrinsic proteins completely span the lipid bilayer and present protein surfaces to the outside and inside of the cell. Such proteins may be channels (or pores) for the movement of solutes through the membrane. The intrinsic and extrinsic proteins also have enzymic properties. The fluid-mosaic model of membrane structure appears to accommodate quite well the reactions and processes known to occur at membranes.

As stated earlier, cell cytoplasm contains many subcellular structures bounded by, or enclosed in, membranes. The nature of the membrane allows two broad classes of subcellular structures to be recognized. In the first group are those structures with single unit membranes—for example, vacuoles, endoplasmic reticulum, plasma membrane, and dictyosomes. The second group of subcellular structures—nucleus, chloroplasts, and mitochondria—are enclosed in double unit membranes. The outer unit membranes of chloroplasts and mitochondria seem similar in structure to those of vacuoles and other single unit membrane organelles. Their inner membranes, however, appear to consist of subunits. More will be said of these subunits when the chloroplasts and mitochondria are described more fully later on.

Organelles enclosed by single unit membranes are considered by some investigators to be related in composition, function, and assembly. Before discussing these relationships, a brief description of the organelles is appropriate.

Endoplasmic Reticulum (ER)

A striking feature of electron micrographs of cells is the ubiquitous occurrence of interconnecting membrane surfaces within the cytoplasm known as endoplasmic reticulum (ER). The ER ramifies throughout the cytoplasm, forming channels between various organelles. In some areas within the cell the ER has a smooth appearance (Figure 2-6) whereas in others it appears covered with small particles. Consequently, the ER is referred to as smooth ER (SER) or rough ER (RER). The particles on rough ER are ribosomes, which are involved in protein synthesis (see Chapter 3).

Golgi Apparatus (Dictyosomes)

The Golgi apparatus in many plant cells has the appearance of a series of stacked, flattened vesicles [Figure 2-7(b)]. The flattened vesicles (also referred to as dictyosomes) are small vesicular sacs that appear to bud off from the central vesicles.

Dictyosomes are especially numerous in the vicinity of the developing cell plate during cytokinesis (cell division). It is thought that they are centers of polysaccharide synthesis and that during cell division these materials are secreted into small vesicles in contact with the developing cell plate where the primary cell wall is laid down. Dictyosomes are also believed to be involved in the synthesis of plasma membrane constituents, such as phospholipids, polysaccharides, and glycoproteins.

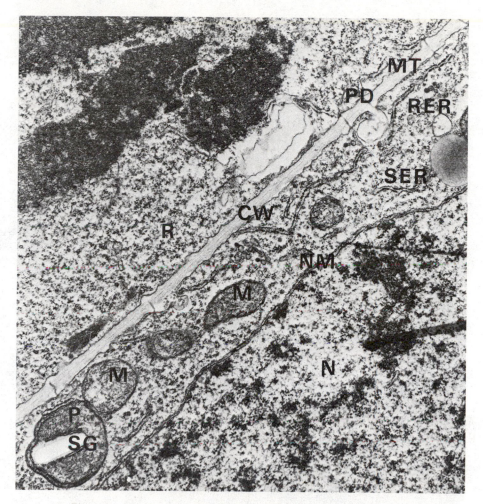

Figure 2-6 Electron micrograph of cell organelles in stem meristem (*Pinus taeda*). Cell wall (CW), plasmodesmata (PD), microtubules (MT), free ribosomes (R), nuclear membrane (NM), nucleus (N), mitochondria (M), proplastid (P), containing starch grain (SG), smooth endoplasmic reticulum (SER), rough endoplasmic reticulum (RER). ×10,000. (Courtesy Dr. H. V. Amerson and Dennis Gray, North Carolina State University.)

Vacuoles

Actively dividing meristematic cells seem devoid of vacuoles or they are too small to be seen. As cell development proceeds, small vesicles appear and coalesce to form a single vacuole that increases in volume as water is taken up by the cell. The vacuolar membrane is composed of a single unit membrane [Figure 2-7(a)], as is the plasma membrane. Vacuoles contain dissolved solutes, such as inorganic ions, amino acids,

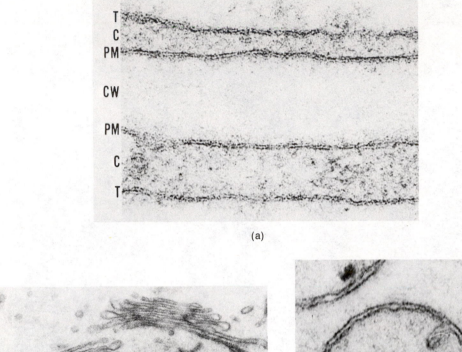

(a)

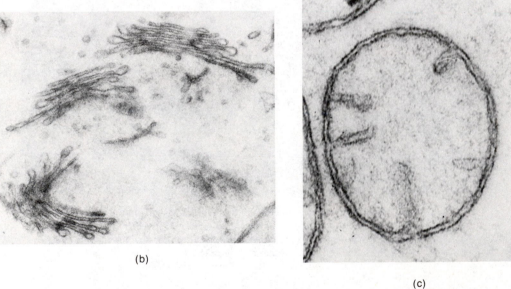

(b)

(c)

Figure 2-7 Electron micrographs of cell organelles. (a) Cell wall (CW) with plasma membrane (PM), and tonoplasts (T) of contiguous cells in *Nicotiana tabacum* leaf mesophyll. ×200,000. (b) Golgi bodies in root tip meristem of *Haplopapus gracilis*. ×50,000. (c) Mitochondrion in taproot secondary phloem of carrot (*Daucus carota*). ×66,000. (Courtesy Dr. H. W. Israel, Cornell University.)

organic acids, sugars, water-soluble pigments (anthocyanins), and insoluble materials in the form of crystals and needles. In addition, the vacuole contains proteins with enzymic activity as hydrolases, catalase, phosphatases, and so forth.

Vacuoles are not inert entities, serving only as receptacles for waste metabolic products or secondary plant substances. Certainly vacuoles are important in maintaining plant form through turgidity of cells, tissues, and organs. Vacuoles are very important in developmental processes, however, because they serve as storage compartments for metabolically active materials, such as sugars, amino acids, amides, organic acids, and inorganic ions. Depending on the specific stage of development of the cell, the vacuole may contain high concentrations of soluble sugars or amides, such as asparagine and glutamine. Under still different conditions the vacuoles may be rich in the potassium salts of organic acids. To mediate such metabolic activities, vacuoles also contain an array of enzymes. The vacuole should be viewed as an important subcellular organelle that participates in a variety of metabolic processes.

Lysosomes

Lysosomes were first described in animal cells as a cytoplasmic compartment containing hydrolytic enzymes capable of digesting macromolecules, such as proteins, phospholipids, nucleic acids, and polysaccharides. As long as the lysosomal membrane is intact, these hydrolytic enzymes are prevented from digesting cytoplasmic contents. With aging, disease, or cellular damage, the lysosomal membranes break down, releasing hydrolytic enzymes into the cytoplasm and thus initiating the digestion of cellular constitutents. Lysosomes do not have a characteristic internal organization, such as is shown by mitochondria and chloroplasts. Instead they are identified by cytochemical tests that identify the hydrolytic enzymes, such as proteinases, nucleases, and lipases. The enzyme catalase is also generally present in lysosomes.

Organelles containing catalase and hydrolytic enzymes are present in many plant cells. Based on their chemical composition, these organelles are called lysosomes. Their function is similar to that noted for animal lysosomes. Under conditions of wounding, senescence, and stress the hydrolytic enzymes of plant lysosomes are released to the cytoplasm and cytoplasmic macromolecules are digested.

It should be noted that other subcellular compartments in plants contain hydrolytic enzymes. Vacuoles, for example, frequently contain hydrolytic enzymes that are released into the cytoplasm following cell damage or senescence. Also, vesicles often contain hydrolytic enzymes. The point to keep in mind is that the various hydrolytic enzymes are contained within a membrane-bound organelle and only released under conditions of cell damage or at specific stages of cell damage.

 ## Spherosomes

Many plant cells, particularly those that are metabolically active, contain small spherical bodies (0.7–0.9 μm in diameter) known as spherosomes. These bodies are visible under the light microscope and are found to be rich in fats. Under the electron microscope it was discovered that spherosomes were not bounded by the usual tripartite unit membrane but rather by what appears to be a single layer of phospholipids and proteins. It has been interpreted as a half of a unit membrane. The significance of such a bounding membrane is not clear.

Although spherosomes are numerous in the endosperm cells of oil-bearing seeds, they have also been found in other parts of the plant. In sieve tube cells, for example, they are quite numerous but much smaller than those present in endosperm cells. Furthermore, the chemical composition of spherosomes is more complex than originally believed. Newer techniques of microchemistry and histochemistry reveal the presence of phospholipids and considerable protein. Moreover, much of the protein is enzymically active as phosphatases, lipases, and other hydrolytic enzymes. In this sense, spherosomes resemble lysosomes in that they are active in the degradation of macromolecules.

Aleurone Bodies

The endosperm of the seeds of many plants, particularly the grasses (maize, wheat, rice, etc.), contain specialized structures rich in protein known as aleurone bodies. These organelles are surrounded by membranes similar in structure to the unit membranes of endoplasmic reticulum, plasma membrane, and dictyosomes. During seed germination the aleurone bodies release hydrolytic enzymes involved in the breakdown of starch, lipid, and protein, the major storage compounds of seeds. Aleurone bodies are also rich in phytin, the potassium, magnesium, and calcium salt of phytic acid (phosphorylated phytin). When phytin is broken down by the enzyme phytase—a constituent of aleurone cells—potassium, magnesium, calcium, and phosphate ions are released and transported to the growing embryonic axis (see Chapter 16, Seed Physiology). Phytase also releases inositol, a polyhydric alcohol, from phytin.

Microtubules and Microfilaments

Early studies on cells showed cytoplasm to be a viscous fluid, but light microscopy did not reveal any structured entities other than such organelles as the nucleus, mitochondrion, and plastid. Nevertheless, cell physiologists believed that structured elements probably were present and used the term *cytoskeleton* to denote such a condition. With electron microscopy, two kinds cytoplasmic structures have been observed—microtubules and microfilaments. As the names suggest, microtubules are hollow whereas microfilaments are solid rods.

Microtubules consist of subunits of a protein known as tubulin. The formation of microtubules from the subunits appears to depend on the particular phase of cell development, calcium ion concentration, and metabolic activity. It is thought that microtubules are involved in cell division, particularly cytokinesis, when the cell walls of daughter cells are being laid down. Others have suggested that microtubules are involved in the orientation of cellulosic microfibers during secondary cell wall formation. Figure 2-6 shows microtubules oriented beneath the plasma membrane.

Microfilaments are also structures composed of protein, but the protein is different in nature from that found in microtubules. Microfilament protein appears to resemble actin, a protein widely present in animal cells. Actin or actinlike microfilaments are believed to play a role in different kinds of cell movements. Actin, plus myosin, is responsible for muscle contraction and extension. Such observations have led plant cell physiologists to suggest that microfilaments are responsible for cytoplasmic streaming and other cell movements (amoeboid movements in fungi).

Although there is reason to believe that microtubules and microfilaments play an important role in the life of the cell, much additional work is needed on a variety of plant cell types before their precise function is known. Of particular interest is the relationship between calcium ions and the protein fractions of microtubules and microfilaments. Studies with animal tissues have revealed a calcium-protein known as calmodulin. Calmodulin is involved not only in cell motility (as with microtubules and microfilaments) but also in regulating many enzyme systems. It has also been isolated from plants, where it appears to act similarly to animal calmodulin. Calmodulin seems to play an important role in cellular metabolism and further studies with plants undoubtedly will open up new areas of research.

Microbodies: Peroxisomes, Glyoxysomes

Electron micrographs of plant cells show that the cytoplasm contains a large number of structureless microbodies. Many microbodies fall in a size class between mitochondria and ribosomes. With refinements in the techniques of isolating organelles, two distinct classes of microbodies have been isolated from plant tissues. They are bounded by a single unit membrane, similar to the tonoplast, but lack internal organization. The two classes of microbodies are separated on the basis of their enzyme composition and their specific metabolic reactions.

Microbodies found in cells of green leaves are called peroxisomes. They are localized within the cell in close association with chloroplasts (Figure 7-6), where they participate in the metabolism of products secreted by the chloroplasts in the process of photorespiration, a topic discussed in Chapter 7.

A microbody, originally isolated from the cells of oil-and fat-rich endosperm tissue, is known as a glyoxysome. The name is derived from the presence of enzymes involved in the glyoxylate cycle pathway of carbohydrate breakdown (Chapter 16, Figure 16-3). Although similar in appearance to peroxisomes, glyoxysomes are found in different cells and tissues and differ markedly in their enzyme composition. Glyoxysomes play an important role in the germination of fat-rich seeds.

Microbodies can be visualized as subcellular organelles containing specialized complements of enzymes with specific metabolic functions. It is not clear whether there are distinctly different microbodies with a specialized structural organization or if all microbodies are similar but develop unique enzyme profiles, depending on their location—peroxisomes in green leaves, glyoxysomes in fatty endosperm tissue. The packaging of enzymes into compartments capable of performing specialized chemical reactions is an excellent mechanism for regulating metabolism.

Ribosomes

During the separation of cell homogenates by centrifugation small particles rich in protein and ribonucleic acid (RNA) can be observed. These particles, called ribosomes, play an active role in protein synthesis. Ribosomes are composed of two subunits—one small, the other large. In the absence of magnesium ions, the subunits do not aggregate into a functional ribosome active in protein synthesis.

Two populations of ribosomes are present in cells (Figure 2-6). One group is found attached to endoplasmic reticulum, the so-called rough endoplasmic reticulum. Other ribosomes occur within the cytoplasm, apparently unattached to membrane surfaces. They are not bounded by a unit membrane.

Because of their role in protein synthesis (Chapter 3), the distribution of ribosomes within plant cells is of interest. Not only are they found free and attached within the cytoplasm but they also occur in chloroplasts and mitochondria. Thus protein synthesis seems to be compartmented in several locations within the cell.

The Endomembrane Concept

With the exception of ribosomes and spherosomes, the subcellular organelles just described—dictyosomes, lysosomes, vacuoles, endoplasmic reticulum, aleurone bodies, peroxisomes, glyoxysomes, and plasma membrane—are composed of or bounded by single unit membranes. Ribosomes lack a membrane whereas the

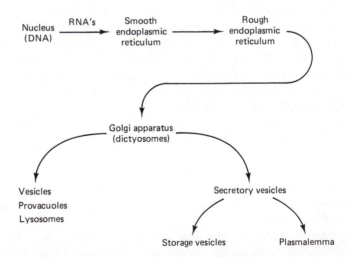

Figure 2-8   The endomembrane concept. Genetic information (DNA) in the nucleus is passed via several RNAs into a membrane system composed of smooth and rough endoplasmic reticulum and dictyosomes. Vesicles are budded off the membrane system, leading to the formation of vacuoles, lysosomes, plasma membrane, and secretory materials. The secreted material may be passed out of the cell through the plasma membrane or sequestered in vacuoles or storage vesicles. (After D. J. Morré and H. H. Mollenhauer, 1973; P. Matile, 1975.)

spherosome membrane appears to consist of a single layer of phospholipid molecules (half of a unit membrane?). The other major organelles to be described later—nucleus, chloroplast, and mitochondrion—are enclosed in two unit membranes.

Because of the similarity in membrane structure of the organelles bounded by a single unit membrane, it has been suggested that they can be visualized as being related in their formation and metabolic activity. This relationship, schematically shown in Figure 2-8, is referred to as the endomembrane concept. It is proposed that information coded in the deoxyribonucleic acid (DNA) of the nucleus is passed, in the form of RNA, to the endoplasmic reticulum by membrane connections with the outer membrane of the nucleus. Specific enzymes involved in the synthesis of membrane components—proteins, phospholipids, glycolipids, glycoproteins—are synthesized at various stages within the smooth and rough ER and dictyosomes. The dictyosomes, in turn, are the centers of formation of vesicles, vacuoles, lysosomes, and perhaps spherosomes. The formation of these last organelles is not completely understood. The vesicles pinched off from the dictyosomes are thought to contribute to the formation of the chemical components of the plasma membrane.

The purpose of the endomembrane concept is to provide a model for understanding the interrelationships of the subcellular components and their integration into a functional unit. Much remains to be discovered about the organization of the cell and its organelles, but the endomembrane concept does provide a point of view for considering how such an organization might be achieved.

Plasmodesmata

There is an active exchange of substances between plant cells. The cells in meristems, for instance, obtain water and inorganic and organic solutes from neighboring cells. Connecting the cytoplasm of contiguous cells are specialized structures known as plasmodesmata. The structural nature of the plasmodesmata is shown in Figure 2-9. It is difficult to make a general statement about the subcellular organization of plasmodesmata, but they appear to be composed of a tube that is continuous with the plasma membranes of two cells. Within the tube is a microtubulelike structure consisting of endoplasmic reticulum. Plasmodesmata may be quite numerous between metabolically active cells, such as in apical meristems, or between leaf mesophyll cells and leaf bundle sheath cells. A single cell may have 1000 to 10,000 plasmodesmata connecting the cytoplasms of adjacent cells, providing a pathway for water and solute exchange.

Another feature of the electromicrograph in Figure 2-9 is noteworthy. Two leaf cells of maize are shown. The one at the left is a leaf bundle sheath cell; the cell at the right is a leaf mesophyll cell. It will be shown later (Chapter 6, Photosynthesis) that these two cells—bundle sheath and mesophyll—carry out quite different biochemical reactions during photosynthesis. A major feature of photosynthesis in maize is that metabolites formed in the mesophyll cells are transported to the bundle sheath cells. From Figure 2-9 it can be seen that plasmodesmata provide for such transport through the symplasm, or intercellular continuum.

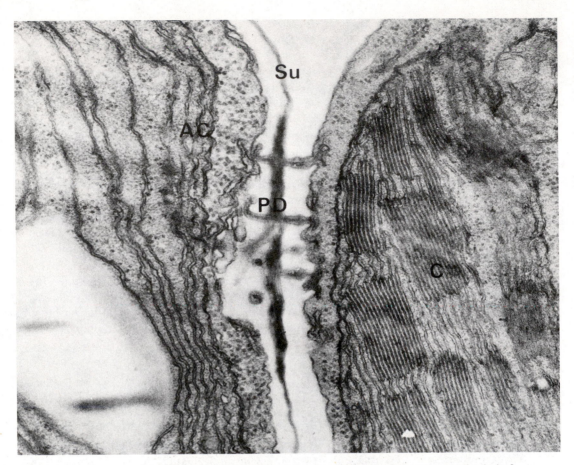

Figure 2-9 Electron micrograph showing plasmodesmata and organelles in leaf of *Zea mays*. The plasmodesmata (PD) provide cytoplasmic continuity between contiguous cells. Various solutes, both organic and inorganic, can be exchanged between cells in this cytoplasmic compartment (symplast). The cell on the left is a bundle sheath cell containing agranal chloroplasts (AC). The cell on the right is a mesophyll cell with granal chloroplasts (C). Note the layer of suberin (SU) around the bundle sheath cell. Uranyl acetate-lead citrate stained. ×50,000. (Courtesy Dr. C. G. Van Dyke, North Carolina State University.)

Also shown in Figure 2-9 is the presence of a suberin layer around the bundle sheath cell. It might be imagined that this relatively impermeable barrier around the bundle sheath cell isolates the cell from its surroundings and any exchange of solutes or water must occur through the plasmodesmata or other intercellular connections.

Nucleus

The nucleus is the center of heredity and is involved in the control of cellular metabolic processes. The genetic information passed on to daughter cells at nuclear division is present in the nucleus as a macromolecule of deoxyribonucleic acid (DNA). Proteins are also associated with DNA to form a nucleoprotein complex known as chromatin. In addition, the nucleus contains ribonucleic acid (RNA). It should be noted that DNA and RNA are also present in several other subcellular organelles—mitochondria and chloroplasts.

In Figure 2-4 the nucleus is shown surrounded by an envelope composed of two unit membranes. Enclosed within the envelope is nuclear material as unorganized chromatin (DNA-protein) and nucleoli. The nuclear envelope is interrupted by pores, thereby providing a continuum between the nuclear material and the cytoplasm. Precursors of RNA (nitrogen bases, ribose, phosphate) from the cytoplasm are organized into specific RNAs, such as messenger RNA, ribosomal RNA, and transfer RNA within the nucleus, which then move into the cytoplasm, where proteins are synthesized on ribosomes. Moreover, DNA precursors (nitrogen bases, deoxyribose, phosphate) move from the cytoplasm into the nucleus during DNA replication.

The outer nuclear membrane seems connected to the endomembrane complex as described earlier. Within this complex, protein is synthesized on ribosomes attached to the endoplasmic reticulum, the rough ER. Thus two centers of protein synthesis are present in cells. That occurring in the cytoplasm is believed to concern the synthesis of enzymes active in metabolic reactions, such as amino acid synthesis, organic acid interconversions, and carbohydrate breakdown (glycolysis). Proteins synthesized in the endomembrane complex are active as enzymes in the formation of such macromolecules as secretory proteins, phospholipids, complex carbohydrates, glyco-lipids, and glycoproteins, molecules involved in the formation of membranes and other structural components (cell walls).

The micrograph in Figure 2-4 showed the DNA of the nucleus dispersed as chromatin within the nuclear envelope. As nuclear division proceeds, the nuclear envelope disappears and the chromatin is organized into discrete chromosomes (Figure 2-10). After the chromosomes separate, two daughter nuclei form and new nuclear envelopes organize around each nucleus. In addition, during this period new organelles and primary cell wall materials are formed.

Mitochondria

Mitochondria are centers of respiratory activity within the cell. They are also active in protein synthesis. Because the mitochondria contain different DNA from that found in the nucleus, it is believed that they replicate independently of the nuclear DNA.

The structural organization of mitochondria as revealed by electron microscopy [Figure 2-7(c)] shows them to be enclosed in a continuous outer unit membrane with an inner unit membrane that is infolded to form structures referred to as cristae. The interior of the mitochondrion, as well as the space between the outer and inner unit membranes, is filled with a protein-rich matrix.

An interpretation of mitochondrial ultrastructure is shown in Figure 2-11. The outer unit membrane is continuous and similar in structure to that noted in the plasma and vacuolar membranes. The inner unit membrane consists of structured subunits

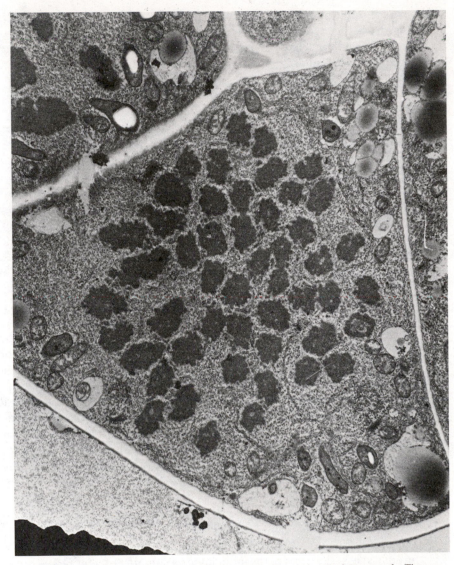

Figure 2-10 Electron micrograph of nucleus in meristematic cell of *Pinus taeda*. The nuclear membrane has disappeared and the chromatin is now organized into chromosomes. ×3000. (Courtesy Dr. Henry Amerson and Dennis Gray, North Carolina State University.)

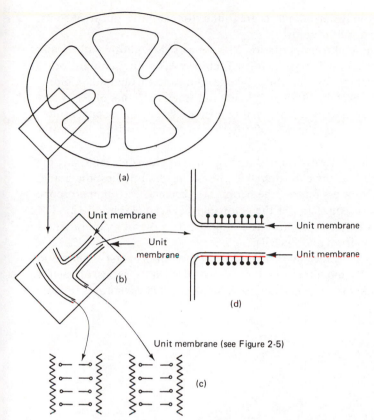

(a)

Unit membrane

Unit
membrane

Unit membrane

(b)

Unit membrane

Unit membrane

(d)

Unit membrane (see Figure 2-5)

(c)

Figure 2-11 Interpretation of fine structure of a mitochondrion.

that may assume several different forms. The subunit depicted in Figure 2-11(d) is composed of stalked knobs. These membrane subunits are concerned with some aspects of energy transformation that accompany oxidative respiration. It has been noted that when mitochondria are actively engaged in respiratory activity, the subunits of the inner membrane undergo changes in shape or conformation. These conformation changes are related to changes in the shape of the protein molecules present in the membranes. Similar conformational changes have been noted in the inner membranes of chloroplasts during photosynthesis.

Plastids: Chloroplasts

Plant cells contain several different kinds of plastids: chromoplasts, which are pigmented, and leucoplasts, which are colorless. Leucoplasts function as storage bodies, storing such products as starch (starch grains), oil, and protein.

Chromoplasts are of two general types: Those lacking chlorophyll and those containing chlorophyll. The chromoplasts lacking chlorophyll usually contain oils and a variety of fat-soluble carotenoid pigments, generally yellow but sometimes red or orange. Carotenoid-containing chromoplasts are found, for example, in carrot roots, where they give the tissue a yellow color.

Chlorophyll-containing chromoplasts, or chloroplasts, are photosynthetically active. In addition to chlorophyll (several forms), chloroplasts may contain other pigments, such as phycocyanin, phycoerythrin, fucoxanthin, and carotenoids. Chloroplasts consist of 45 to 50% protein, 50 to 55% lipid, and small amounts of RNA and DNA. The protein fraction has associated with it manganese, iron, and copper atoms; these metals are considered components of specific enzymes involved in photosynthetic reactions.

Chlorophyll is not uniformly distributed throughout the chloroplasts. Using the light microscope, it is seen that the chlorophyll is present in small bodies, the grana, which are embedded in a colorless matrix, the stroma. Under the electron microscope an even greater degree of organization of the chloroplast is seen (Figure 2-12). The chloroplast is bounded by a double membrane composed of two unit membranes, as was noted for the mitochondrion and nucleus.

An interpretation of the subcellular organization of a chloroplast is shown in Figure 2-13. The inner unit membrane system is organized as a series of saclike structures known as thylakoids. An aggregation of small stacked thylakoids forms a

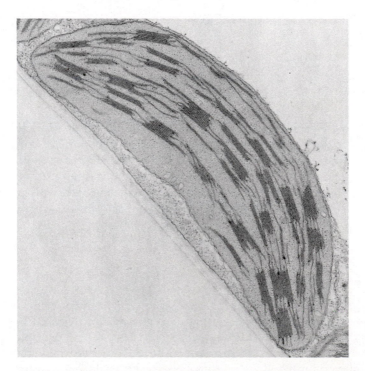

Figure 2-12   Electron micrograph of palisade parenchyma mesophyll chloroplast of *Nicotiana tabacum*. ×20,100. (Courtesy Dr. H. W. Israel. Cornell University.)

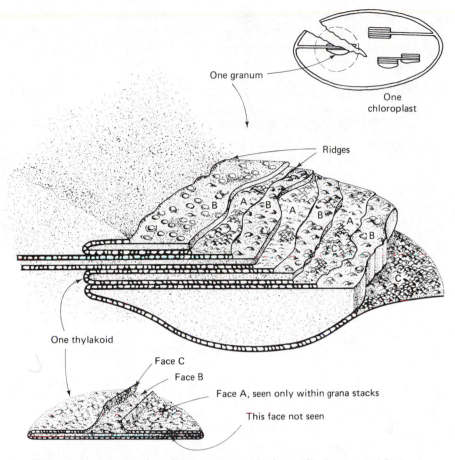

One granum

One chloroplast

Ridges

B A B A B A B

C

C

One thylakoid

Face C

Face B

Face A, seen only within grana stacks

This face not seen

Figure 2-13 Interpretation of chloroplast organization. (After Branton and Park, 1967; Figure 19.)

granum. The grana are not found free within the chloroplast membrane but are interconnected through stroma thylakoids.

Cell Wall

The cell wall usually has been considered an inert structure providing protection and support to the enclosed cytoplasmic and nuclear material. It is now clear, however, that the cell wall, particularly the primary cell wall, is a metabolically active organelle whose chemical composition changes under the influence of external environmental factors and internal stimuli. The primary cell wall should be regarded as an organelle interacting with other subcellular organelles during growth and differentiation.

The primary cell walls of meristematic cells are plastic and extensible. As cells mature and differentiate, a secondary wall is deposited on or within the primary cell wall and further cell expansion ceases. The factors responsible for the initiation of secondary cell wall formation and the differentiation of meristematic cells into vessel elements, tracheids, fibers, sieve tube cells, and other cell types are important in understanding plant development.

The change in our view of the primary cell wall from that of an inert, static envelope to one of a dynamic, changing organelle in intimate contact with the rest of the cell has largely occurred because of recent information concerning its chemical composition. Cellulose has always been recognized as a major cell wall constituent, but the chemical nature of the other constituents was poorly understood. These constituents, referred to as pectic polysaccharides and hemicelluloses, have been studied with new analytical procedures revealing the identity of the individual building blocks and the arrangement of these building blocks within the primary cell wall. In addition, a protein fraction has been identified as a component of the primary cell wall. Thus the primary cell wall can be viewed as an aggregation of macromolecules (polysaccharides and protein) whose synthesis and deposition are closely tied in with the activities of the rest of the cell. The chemical nature of the building blocks of the primary cell wall is described further in Chapter 3.

At the time of cell division a cell plate forms between the daughter nuclei because of metabolic activities of the endoplasmic reticulum and Golgi bodies (dictyosomes). Polysaccharides and proteins, synthesized within these organelles, are transported to the developing cell plate. When the cell plate is fully formed between the daughter cells, it is termed the middle lamella. At this time the daughter cells have also formed new plasma membranes. Through further metabolic activities within the daughter cells, a primary cell wall composed of cellulose microfibrils, pectic polysaccharides, hemicellulose, and protein is deposited on and within the middle lamella. The cellulose microfibrils can be viewed as the structural component of the primary wall whereas the pectic polysaccharides and hemicelluloses form a matrix surrounding the microfibrils. Within this matrix, the protein fraction is arrranged.

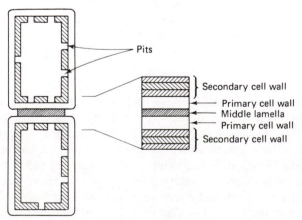

Figure 2-14 Structure of secondary cell wall.

A secondary cell wall usually develops after the cell has fully expanded. In addition to the constituents present in primary cell walls, secondary cell walls often contain lignin. Lignin is not a polysaccharide but rather a macromolecule composed of several closely related phenolic monomers. The synthesis of these phenolics (Chapter 3) depends on intermediates of carbohydrate metabolism. Indirectly, therefore, lignin is related to cell wall polysaccharides.

Mention should also be made of another structural feature of the mature cell wall, the pits. These are thin places in the secondary cell wall that are aligned with corresponding areas in the secondary cell walls of adjoining cells, as shown diagrammatically in Figure 2-14. Pits allow the exchange of water and solutes between contiguous cells. Protoplasmic strands, or plasmodesmata, may also extend through the pit areas and connect the cytoplasm of adjoining cells. Plasmodesmata are not confined to pit areas of secondary cell walls but also connect meristematic cells that lack secondary cell walls, as discussed earlier.

Cuticle

The outer surfaces of roots, shoots, and leaves are covered by a layer of cells known as the epidermis. Depending on the location, the epidermal cells are covered by a protective layer of varying thickness composed of lipid material in which either cutin or suberin is embedded. If located on the aerial parts of the plant—stem, petioles, leaves, flower parts, fruits, seed coats—the protective layer is known as the cuticle. Substomatal cavities are also lined by a thin layer of cutin. The epidermal cells of roots and tubers, on the other hand, are protected by a waxy layer containing suberin. The suberin layer also occurs on any organ during the formation of wound periderm in response to wounding. Suberin is present in the endodermis (Casparian band) of roots, the bundle sheaths of grasses (see Figure 2-9), and other specialized regions within the plant body.

As shown in Figure 2-15, the cuticle may have an amorphous appearance or present a laminar structure. The cuticular material is synthesized within the cell cytoplasm and then extruded through the cell wall. In some instances, the cuticle is laid down in patterns on the surface of the organ. Depending on environmental conditions during growth and on the age of the organ, the cuticle appears as a loose layer of plates or threads, relatively permeable to gases and water, or the cuticle may be compact and closely appressed to the epidermal cells so as to form a rather impervious layer.

In general, the epidermis and cuticle function to decrease the evaporative loss of water and dissolved solutes from the aerial parts of the plant. The cuticle also acts as a barrier to the entrance of various substances into the plant. Insects may find it impossible to chew or ingest plant parts because of the presence of the cuticle. The entrance of fungal hyphae into plants may be retarded by the cuticle. In some instances, it has been found that the fungus excretes an enzyme, cutinase, which degrades the cuticle and allows the fungal hyphae to enter the plant. The effectiveness of applied insecticides, herbicides, and fungicides on plants depends on their passing through the cuticular barrier.

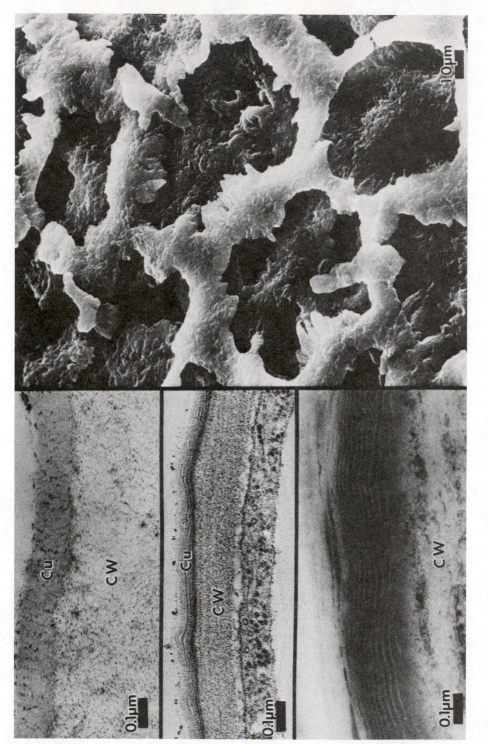

Figure 2-15 Electron micrographs illustrating organization of cutin and suberin. (a) Amorphous cutin layer on *Tropaeolum majus*. (b) Lamellar cutin layer on *Atriplex semibaccata*. (c) Lamellar suberin layer of potato. (d) Scanning electron micrograph of tomato fruit cutin. Cu, cuticle; CW, cell wall. (Courtesy Dr. P. E. Kolattukudy, Washington State University. *Science* 208:990–1000, Figure 1. Copyright 1980 by the American Association for the Advancement of Science.)

Suberin is also an extracellular layer on the epidermal cells of underground plant parts. It differs in chemical composition from cutin (see Chapter 3) but appears to serve similar functions—to control water and solute transport between the plant and the external environment.

REFERENCES

BRANTON, D., and R. B. PARK. 1967. Subunits in chloroplast lamellae. *J. Ultrastructural Res*. 19:283–303.

FINEAN, J. B., R. COLEMAN, and R. H. MICHELL. 1974. *Membranes and Their Cellular Functions*. Halsted Press, John Wiley & Sons, New York.

HALL, J. L., T. J. FLOWERS, and R. M. ROBERTS. 1974. *Plant Cell Structure and Metabolism*. Longmans, Green & Co. Ltd. London.

HARRISON, R., and G. G. LUNT. 1975. *Biological Membranes, Their Structure and Function*. Halsted Press, John Wiley & Sons, New York.

KEEGSTRA, K., K. W. TALMIDGE, W. D. BAUER, and P. ALBERSHEIM. 1973. The structure of plant cell walls. *Plant Physiol*. 51:188–196.

KOLATTUKUDY, P. E. 1980. Biopolyester membranes in plants: cutin and suberin. *Science* 208:990–1000.

LOTT, J. N. A. 1976. *A Scanning Electron Microscope Study of Green Plants*. The C. V. Mosby Company, St. Louis, Missouri.

MATILE, P. 1975. *The Lytic Component of Plant Cells*. Springer-Verlag, New York.

MEANS, A. R., and J. R. DEDMAN. 1980. Calmodulin—an intracellular calcium receptor. *Nature* 285:73–77.

MORRÉ, D. J., and H. H. MOLLENHAUER. 1974. The endomembrane concept: a functional integration of endoplasmic reticulum and Golgi apparatus. Pages 84–137 *in* A. W. Robards, ed. *Dynamic Aspects of Plant Ultrastructure*. McGraw-Hill Book Company, London.

SINGER, S. J., and G. L. NICOLSON. 1972. The fluid mosaic model of the structure of cell membranes. *Science* 175:720–731.

SWANSON, C. P., and P. L. WEBSTER. 1977. *The Cell*, 4th ed. Prentice-Hall, Englewood Cliffs, N.J.

TOLBERT, N. E., ed. 1980. *The Plant Cell*, vol. 1 *in* P. K. Stumpf and E. E. Conn, eds. *The Biochemistry of Plants*. Academic Press, New York.

3

Cellular Constituents
and Their Biosynthesis

One of the major attributes of a living plant is its capacity to carry out metabolic activities. Although the term metabolism is used to denote a variety of cellular reactions, essentially metabolism refers to two types of activities: Those reactions wherein storage products are hydrolyzed and broken down into metabolically active small molecules and those reactions leading to biosynthesis of new cellular material. The first type is often referred to as a catabolic reaction whereas the latter is known as an anabolic reaction. All these metabolic reactions are catalyzed by specific proteins, known as enzymes, and it is quite understandable that much of modern biology today is concerned with the problems of protein structure, function, and synthesis. Another major aspect of metabolism is the study of the mechanisms by which an organism can control its cellular activities. Metabolic control may be achieved in a number of different ways, including interaction of individual members of metabolic pathways, by enzyme synthesis and inactivation and degradation, by hormones and growth regulators, and by inorganic ions. Our understanding of metabolic control mechanisms in plants is at an elementary level and further studies in this field will be profitable.

A meristematic plant cell is capable of carrying out a large number of enzyme-catalyzed reactions. The exact number of enzymes is not important, but clearly many metabolic reactions will be proceeding simultaneously within a cell and a high degree of organization must exist if all reactions are to be carried out effectively and efficiently. Chapter 2 described how the cell is, in fact, highly structured and organized, containing such organelles as the nucleus, mitochondria, plastids, spherosomes, ribosomes, Golgi bodies, glyoxysomes, and the endoplasmic reticulum. All these structures are characterized by membranes and a striking feature of the architecture of the cell is the large surface area provided for carrying out the metabolic activities.

Besides the studies on cellular organization, chemical studies have revealed that many cell organelles have a unique chemical composition. The nucleus, for instance, contains sizable amounts of deoxyribonucleic acid in addition to the other organic molecules characteristic of living cells—polysaccharides, proteins, lipids, and so forth. Similarly, chloroplasts contain relatively large amounts of chlorophyll and carotene in addition to the other organic molecules.

An important aspect of metabolism is the description of the organic molecules found in organs, tissues, and cells. The chemical study of living matter has a long history and until the middle of the nineteenth century organic chemistry was primarily a study of natural products. It was recognized, however, that despite the bewildering array of organic molecules present in living matter a relatively few major kinds of compounds are common to all plants, animals, and microoganisms—namely, amino acids, proteins, lipids, carbohydrates, nucleotides, and porphyrins. Many other kinds of molecules are present in living matter, of course, but it is possible to understand the basic concepts of metabolism in terms of a few classes of organic molecules.

In this chapter a survey is made of the major types of compounds found in plants. In addition, the biosynthesis of these compounds is discussed and the processes involved in the control of metabolism are described. The major purpose of this chapter is to collect many of the points elucidated in courses in organic chemistry and biochemistry but applied specifically to plants.

ORGANIC MOLECULES IN PLANTS

Living matter consists primarily of proteins, amino acids, nucleotides, lipids, carbohydrates, and porphyrins. Table 3-1 shows that several of these compounds are interrelated. Thus proteins (polymers) are composed of many molecules of amino acids (monomers) whereas nucleotides may form nucleic acids and carbohydrates may exist as monomers (monosaccharides) or polymers (polysaccharides). The monomeric units are readily metabolized and transported within the plant to regions in which metabolic reactions convert them into structural (e.g., cellulose, protein) or storage (e.g., starch) polymers or into enzymes (proteins) or are organized into cells and eventually to tissues, organ systems, and ultimately the complete multicellular plant.

Lipids, porphyrins, lignin, cutin, and suberin do not bear the monomer–polymer relationship shown by the proteins, nucleic acids, and carbohydrates. Table 3-1 shows, however, that these molecules are composed of a relatively small number of building blocks. With both kinds of molecules—the large polymers and the lipids, porphyrins, lignin, cutin, and suberin—metabolic reactions transform the small building blocks into larger molecules.

All amino acids contain nitrogen and a few contain sulfur. Nucleotides have nitrogen and phosphorus in addition to carbon, hydrogen, and oxygen. Some lipids contain only carbon, hydrogen, and oxygen, but others also have nitrogen and phosphorus. The porphyrins consist of carbon, hydrogen, oxygen, and nitrogen and of metallic elements, magnesium in chlorophyll and iron in cytochrome. The synthesis of the polymeric molecules involves the formation of chemical bonds between carbon

TABLE 3-1 SOME IMPORTANT BIOLOGICAL MOLECULES IN PLANTS

Conventional name	Monomers (major elements)	Polymers[a]
Proteins	Amino acids (C, H, O, N, several contain S)	Proteins
Nucleic acids	Nucleotides (C, H, O, N, P)	Deoxyribonucleic acid (DNA) Ribonucleic acid (RNA)
Carbohydrates	Monosaccharides; e.g., glucose, fructose, xylose (C, H, O)	Polysaccharides such as sucrose, starch, cellulose, hemicellulose, pectins

Plants also contain other biopolymers with a more complex arrangement of monomers than noted above.

Conventional name	Building blocks, monomers (major elements)
Lipids, (neutral fat, phospholipids, etc.)	Glycerol, fatty acids, nitrogen bases, phosphoric acid (C, H, O, N, P)
Lignin	Coniferyl alcohol and related phenolics (C, H, O)
Porphyrins (chlorophyll, cytochromes, phytochrome)	Pyrroles (C, H, O, N) Mg in chlorophyll, Fe in cytochromes
Cutin	C_{16} and C_{18} fatty acids, phenolics (C, H, O)
Suberin	C_{16} and C_{18} hydroxylated fatty acids, phenolics, C_{18} alcohols, C_{10} and C_{18} dicarboxylic acids (C, H, O)

[a] Polymers are large macromolecules composed of repeating units of a small molecule or monomer.

atoms or between carbon, oxygen, and nitrogen atoms. Such molecules are relatively stable and the chemical bonds are broken by enzymic activity or by harsh chemical treatment (high temperature, pressure, acid or alkaline conditions). Other important biological molecules are stabilized through the action of weak chemical bonds (e.g., hydrogen bonds). These weak bonds may be broken by mild environmental conditions, such as changes in hydrogen ion concentration or temperature. The double-stranded DNA molecule is stabilized by hydrogen bonding, as is the hemicellulose fraction of primary cell walls where certain polysaccharides are associated with cellulose molecules. Another example of weak chemical bonding is the maintenance of some intrinsic protein molecules within lipid bilayers of membranes.

Proteins

Proteins are large macromolecules ranging in molecular weight from several thousand to several million. A protein with a molecular weight of 5000 contains about 50 amino acid monomers whereas one with a molecular weight of 100,000 might contain 1000

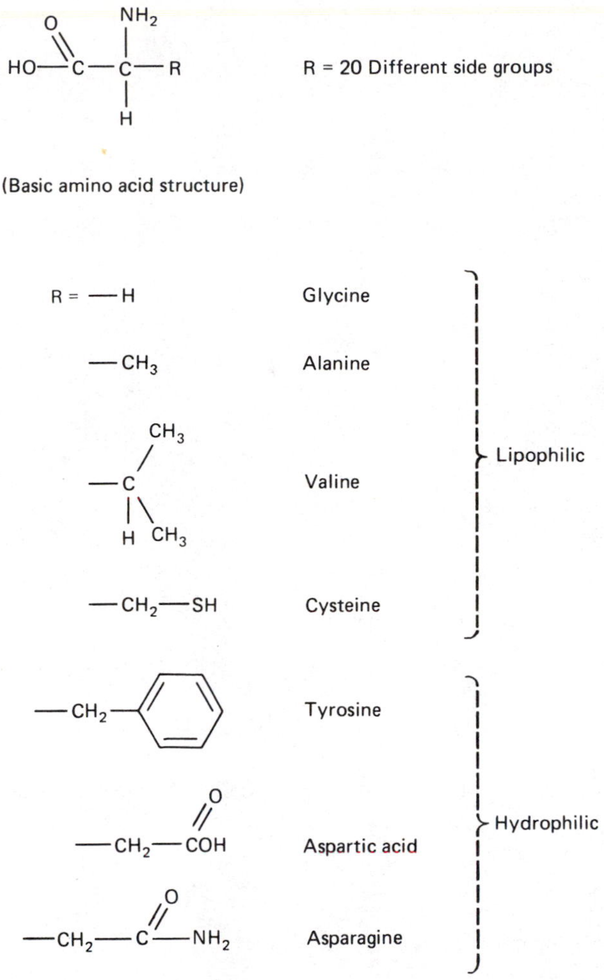

(Basic amino acid structure)

R = 20 Different side groups

Figure 3-1 Some of the amino acids incorporated into protein.

amino acid monomers. In proteins 20 different amino acids commonly occur. And 20 to 40 additional amino acids are found in the free state in plant extracts but are rarely incorporated into protein. The basic amino acid structure is shown in Figure 3-1, where it is seen that the molecule contains an amino group (—NH₂) adjacent to a carboxylic acid group (—COOH). Attached to this common structure, known as an α-amino carboxylic acid, are different side groups (R). Also shown in Figure 3-1 are several of the characteristic side groups present in amino acids. Because of their chemical configuration, these side groups are differentially soluble in lipid-rich (lipophilic) or water-rich materials (hydrophilic). Some side groups are basic in nature; others are acidic. The differences in solubility and basicity or acidity have an important role in the structural and enzymic functions of proteins.

Proteins are formed from amino acid monomers through the splitting out of a

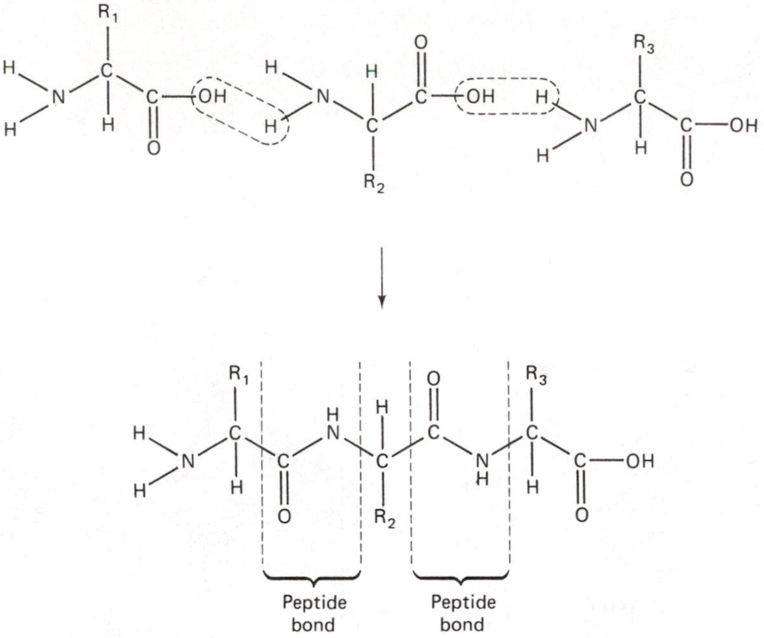

Figure 3-2   The interaction of three amino acids to form a tripeptide molecule. Two peptide bonds are formed by the splitting out of two molecules of water.

molecule of water from two adjacent amino acids (Figure 3-2). The bond is known as a peptide bond. The repeating unit

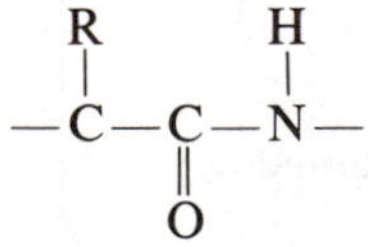

$$
\begin{array}{c}
\quad R \qquad\quad H \\
\quad | \qquad\quad | \\
-C-C-N- \\
\quad || \\
\quad O
\end{array}
$$

serves as the backbone of a long chain of amino acid monomers in the protein molecule. The chemical and physical properties of the protein depend on the nature of the R groups attached to the backbone. Because 20 different amino acids may be incorporated into a protein and these amino acids may be linked in different sequences, it is not difficult to see why such a larger number of different protein molecules can be produced. Consider a dipeptide containing two amino acids. Because there are 20 different amino acids, they may be combined in 400 (20×20) different ways. Most proteins contain hundreds of amino acids monomers and the number of different protein molecules that can be derived from such combinations is astronomical. Millions of different proteins are possible and biological specificity depends on this fact.

The amino acid sequence in a protein determines its structure and function. For this reason, one of the major problems in modern biology is the determination of the amino acid sequence in biologically active proteins. The problem is laborious and time consuming because of the size and complexity of most proteins. Insulin, a mammalian protein (hormone) required for the maintenance of critical blood sugar levels, for instance, is a relatively small protein with a molecular weight of 5733. When hydrolyzed with acid, insulin is broken down into its constituent amino acid monomers and found to contain a total of 51—of 17 different kinds. Such information, however, tells nothing of the sequence of amino acids in the insulin molecule nor how the protein is organized into a biologically active molecule. In 1954, after 10 years of research in collaboration with a number of associates, the British biochemist Sanger described the amino acid sequence of insulin. With improvements in analytical techniques, the amino acid sequences in many additional proteins have been ascertained. The protein moiety of cytochrome *c* has been thoroughly invest-igated, for example. Cytochrome *c* is a respiratory factor found in animals, microorganisms, and plants. A study of its amino acid sequences in different organisms has provided information on the evolutionary development and degree of relatedness of the organisms. Similar studies on plastocyanin, a small protein with 99 amino acid monomers, isolated from chloroplasts have been used to determine systematic relationships of different plant families. The technique of amino acid sequencing of biologically important proteins will play an important role in future biosystematic studies.

The amino acid sequence in ribonuclease, an enzyme catalyzing the breakdown of RNA, is shown in Figure 3-3. A few of the side chain groups of the 124 amino acid monomers are shown. The primary structure of ribonuclease is determined by the sequence of the 124 amino acid monomers. The primary structure is responsible for the higher levels of organization of the protein through the interactions of the side groups attached to the backbone of peptide linkages. The ribonuclease is folded and held in position at four points by —S—S— linkages between cysteine molecules.

The ribonuclease molecule attains a coiled configuration, or secondary structure, by further interactions between the amino acid monomers. The polypeptide chain (primary structure) is coiled in a helix to give the secondary structure. This helix, in turn, is folded so as to give the tertiary structure.

Roles of proteins in plants   The central position of proteins in biological processes is well known. Some proteins are found in large quantities in storage organs, such as the endosperm or cotyledons. Simple proteins, such as albumins (e.g., β-amylase), globulins (e.g., α-amylase), glutelins (glutelin from wheat and oryzenin from rice), and prolamins (gliadin from wheat and zein from maize), are examples of storage proteins. During germination the storage proteins are hydrolyzed by enzymes to small peptides and amino acids that nourish the embryonic axis in early growth and development.

Some proteins appear to function primarily as structural units or protective structures. The best-known examples are those found in animals (hair, feathers, wool), but structural proteins are also found in plant cuticles and some fibers.

Other proteins, such as glycoproteins (protein plus carbohydrates) and glyco-

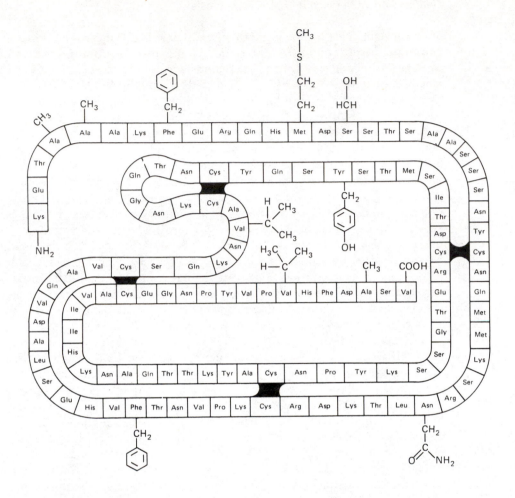

Amino acid symbols
Ala — alanine
Arg — arginine
Asp — aspartic acid
Asn — asparagine
Cys — cysteine

Gln — glutamine
Glu — glutamic acid
Gly — glycine
His — histidine
Ile — isoleucine

Leu — leucine
Lys — lysine
Met — methionine
Phe — phenylalanine
Pro — proline

Ser — serine
Thr — threonine
Trp — tryptophan *
Tyr — tyrosine
Val — valine

*Tryptophan not present in ribonuclease

Figure 3-3 The primary structure of ribonuclease is represented by the sequence of 124 amino acid groups (19 different amino acids). A few R groups are shown to indicate the complex nature of the molecule. The chain is folded and held in place at four sites (dark blocks) by disulfide links (—S—S—) between cysteine (Cys) molecules.

lipids (protein plus lipids), are found at membrane surfaces, where they take part in membrane organization and function. Glycoproteins, for example, have been found to play important roles in cell recognition systems and cell defense mechanisms against fungi and microorganisms.

Nucleoproteins (nucleic acid plus protein) are present in the nucleus, where they figure importantly in the transmission of information during nuclear division and in protein synthesis. The best-known roles of proteins are as enzymes in catabolic and anabolic processes. This aspect of protein structure and function is discussed in a later section.

Distribution of proteins in plants   The technique of electrophoresis has been used to study the distribution of proteins in different parts of the plant (see Figure 3-4). The technique is based on the fact that proteins differ in their electrical charge (because of the nature of the amino acid side groups, some being basic and others acidic) and therefore migrate at different velocities in an electrical field. Protein size also influences the mobility of the protein. As shown in Figure 3-4, soluble proteins were extracted from different parts of 3-day-old pea seedlings by a buffer solution of pH 8.3. These proteins were then separated by gel electrophoresis. Clearly the various tissue segments have a characteristic protein profile. Some proteins seem similar in almost all parts of the plant (e.g., at R_f 0.28 and 0.42) whereas others occur only in particular tissues. Each protein band probably represents a population of proteins with similar electrophoretic properties.

Additional information on the enzymic properties of proteins can be obtained by allowing the proteins separated by electrophoresis to react with various substrates. Some of the protein bands shown in Figure 3-4 may catalyze starch or protein breakdown or other specific reactions. Biochemical techniques are now available for making rather detailed studies of the protein composition of plant tissues and organs.

Enzymes   All enzymes are proteins and are subject to the numerous reactions that proteins undergo. Their activity may be altered by changes in temperature, hydrogen ion concentration, heavy metals and so forth. Enzymes, however, have the ability to catalyze biochemical reactions and it is this property that makes enzymic proteins unique.

As pointed out earlier, a single cell contains many different enzymes, each capable of catalyzing a specific chemical reaction. From the preceding description of the amino acid composition of proteins it can be seen that it is possible to have a large number of different protein molecules, each with its specific and unique sequence of amino acids. The primary structure of a protein, however, does not explain the catalytic activity of enzymes. The answer to this question must reside in the secondary, tertiary, or quaternary levels of protein organization.

A great deal of evidence supports the idea that only a small portion of an enzyme molecule is directly involved in catalyzing a biochemical reaction. Enzymologists speak of the active site of a protein as that particular part of the protein that binds to

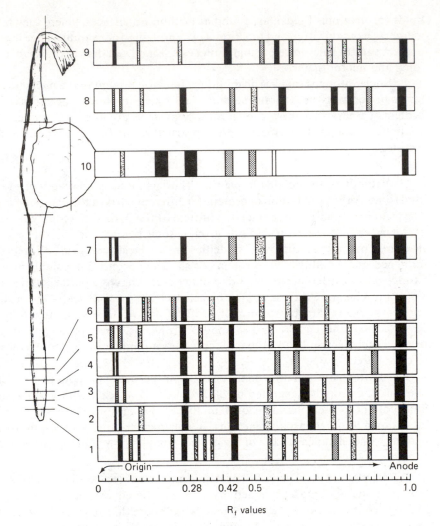

Figure 3-4 Diagrammatic interpretation of the electrophoretic separation on acryl-amide gel of the soluble proteins of 3-day-old pea seedlings extracted by tris-glycine buffer (0.1 M, pH 8.3): successive mm sections from the terminal part of the root (1–6), hypocotyl (7), epicotyl (8), plumule (9), and cotyledons (10). (After Steward et al., 1965; (Figure 3.)

the substrate to form an enzyme-substrate complex. Some enzymes apparently possess only one active site per protein molecule whereas others may have several such sites. The active site might be imagined as consisting of a particular sequence of amino acids in the protein chain, which, in turn, is folded so as to react specifically with a substrate. The active site and the substrate have a three-dimensional relationship that enables them to fit together in the enzyme-substrate complex.

Consider, for example, the enzyme ribonuclease. Through the amino acid sequence, ribonuclease is folded into a definite configuration that is stabilized by —S—S— bonds between eight cysteine molecules. Ribonuclease catalyzes the splitting of ribonucleic acid, specifically between a phosphate group and a pyrimidine nucleotide group. As shown in Figure 3-5, it might be visualized that a specific part of the ribonuclease molecule (active site), because of its shape or the nature of the R groups of the amino acids in this region, is able to form bonds (shown by dotted lines) between the phosphate and pyrimidine groups of the substrate. This process is followed by a change in the enzyme configuration, thereby breaking the chemical bond between the phosphate group and the ribose sugar and forming free nucleosides and inorganic phosphate.

The entire ribonuclease molecule is not essential for enzymic activity. The end of the protein terminating with lysine (N terminal) is apparently necessary, but the end

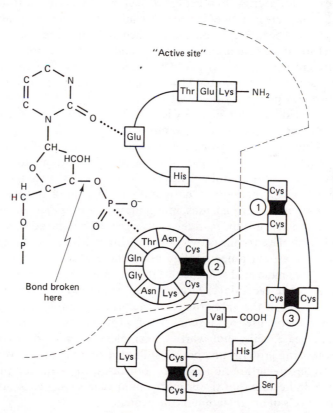

Figure 3-5   Model to show the active site (- - -) of ribonuclease and the binding (. . . .) of a ribonucleotide to the active site. The bond is broken as indicated, releasing inorganic phosphate and a ribonucleoside. From this model it is seen that only a small portion of the protein is involved in the enzymic reaction. See text for additional discussion.

terminating in valine (C terminal) is not. If all four cysteine-cysteine ($-S-S-$) bonds are reduced, thus allowing the protein to unfold, enzymic activity is lost. If the cysteine-cysteine bonds at positions 1 and 2 (Figure 3-5) remain intact while the other two are reduced, enzyme activity is retained. Other portions of the molecule can be modified by chemical treatment, thus changing its enzymic activity. If the histidine molecules (Figure 3-5) are oxidized, for instance, ribonuclease activity is destroyed. Although the precise identity of amino acids in the active site, or their shape, is not known at this time, the active site depicted in Figure 3-5 shows how enzyme-substrate specificity might be achieved in this protein molecule.

Some proteins lack catalytic activity unless a nonprotein moiety is associated with the protein. In some instances, the nonprotein moiety is firmly bound to the protein and cannot be removed without destroying the primary structure of the protein. In such cases, the nonprotein portion of the enzyme is referred to as a prosthetic group. Frequently vitamins like riboflavin, thiamin, and pyridoxin function as prosthetic groups. In other cases, such metals as zinc or iron are prosthetic groups.

There are instances in which the nonprotein moiety is not an integral part of the enzyme molecule but is loosely attached and can be removed without destroying the structure of the protein. If the protein and nonprotein portions are separated by mild treatment, enzyme activity is lost, but if the two components are brought together, enzyme activity is restored. The term coenzyme is frequently used to designate the nonprotein moiety in such instances, especially if it is an organic molecule. If the loosely held nonprotein moiety is a metal ion, such as K^+, Ca^{2+}, Mg^{2+}, the term metal cofactor or metal activator is commonly used.

The prosthetic groups, coenzymes, or metal cofactors can be visualized as helping to facilitate the binding of the protein molecule to the substrate to form the enzyme-substrate complex. The protein molecule is not a rigid structure but is flexible and prone to change its configuration, depending on changes in the immediate environment. Hydrogen ions, inorganic ions, or organic molecules may interact with the amino acid groups in order to change the configuration of the active site, thereby modifying the enzymic activity of the protein. Similarly, such factors may induce the protein to assume a new configuration and enable it to acquire enzymic activity.

The amino acid sequence establishes not only the primary structure of a protein but also the secondary, tertiary, and quaternary levels of organization through interactions of the side groups of the primary amino acid sequence. As noted, the primary structure of a number of proteins has been ascertained, insulin and ribonuclease being noteworthy examples. Techniques have been developed for synthesizing proteins and both insulin and ribonuclease have been synthesized from their constituent amino acids. In both cases, the synthetic proteins are biologically active, indicating that the primary amino acid sequence is sufficient to establish the other levels of organization of the protein.

Proteins have also been synthesized through the use of enzyme systems and other subcellular components in cell-free extracts. Before discussing this technique, a vital component of the protein-synthesizing mechanism—the nucleic acids—is described.

TABLE 3-2   COMPOSITION OF DNA AND RNA

Unit molecules	DNA	RNA
Purines	Adenine (A)	Adenine (A)
	Guanine (G)	Guanine (G)
Pyrimidines	Cytosine (C)	Cytosine (C)
	Thymine (T)	Uracil (U)
Pentose sugar	Deoxyribose	Ribose
Phosphate	Phosphate	Phosphate

Nucleic Acids

Two kinds of nucleic acid are found in cells, deoxyribonucleic acid (DNA) and ribonucleic acid (RNA). Both are large macromolecules; DNA may have a molecular weight up to 2 billion whereas RNA exists in several forms varying in size from 25,000 to 2 million. Nucleic acid is composed of a large number of nucleotide monomers, which, in turn, consist of a nitrogen base, sugar, and phosphoric acid. A comparison of the composition of DNA and RNA nucleotides is shown in Table 3-2. Two kinds of nitrogen bases are present, purines and pyrimidines. Both DNA and RNA contain the two purines, adenine and guanine, and one pyrimidine, cytosine. The pyrimidine thymine is present in DNA, but RNA contains a different pyrimidine, uracil. The two nucleic acids also differ with respect to the pentose sugar component; DNA contains deoxyribose while RNA contains ribose. Both DNA and RNA contain phosphoric acid. The nitrogen bases, sugar, and phosphoric acid are arranged in a linear chain as shown in Figure 3-6.

Deoxyribonucleic acid (DNA)   As shown in Table 3-2, DNA is composed of the purines adenine and guanine, the pyrimidines cytosine and thymine, deoxyribose, and phosphoric acid. Structures of the purines and pyrimidines appear in Figures 3-7 and 3-11. In 1963 Watson and Crick proposed that the DNA molecule is a double-stranded structure (see Figure 3-8). They suggested that because of the shape and size of the purines and pyrimidines, there is a pairing between them in the double strand of DNA so that thymine (T) pairs with adenine (A) and cytosine (C) with guanine (G). On the basis of previous work on the coiled structure of protein (as a helical structure) and similar studies by Wilkins on purified DNA, Watson and Crick further proposed that double-stranded DNA is twisted so that each strand of DNA assumes the helical structure shown in Figure 3-9. The two strands are held together by hydrogen bonds between the pairs of nitrogen bases.

From the time that nucleic acid was first isolated from cell nuclei in 1868 the fact that is has some important role in cellular activities has been recognized. The early nucleic acid samples were found to contain considerable quantities of protein. Cytological studies made it clear that this nucleoprotein was concentrated in the

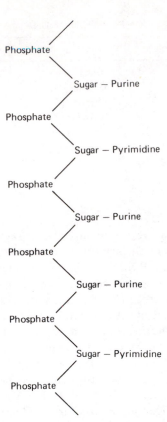

Figure 3-6   Basic chemical structure of nucleic acids.

chromosomes of the cell and that it must be somehow involved in chromosome behavior. Genetic evidence showed that chromosomes are the site of genetic information within the cell and that, when the cell divided, this information was passed intact to the daughter cells. Furthermore, evidence showed that the chromosomes replicated themselves and then separated, with each daughter cell being provided with an exact copy of the chromosome set present in the mother cell.

The Watson–Crick DNA molecule provides a model for the replication process, as shown in Figure 3-10. At the time of nuclear division the double-stranded DNA molecule separates and each strand then serves as a template for the formation of a new strand, which is synthesized from a pool of nucleotides within the cell. The synthesis of the nucleotides and the formation of the new DNA strand require the action of a number of enzymes.

Ribonucleic acid (RNA)   The RNA molecule is also built up of nucleotides to form a long single-stranded polynucleotide chain. The structures of two such nucleotides are shown in Figure 3-11. There is some evidence that portions of the

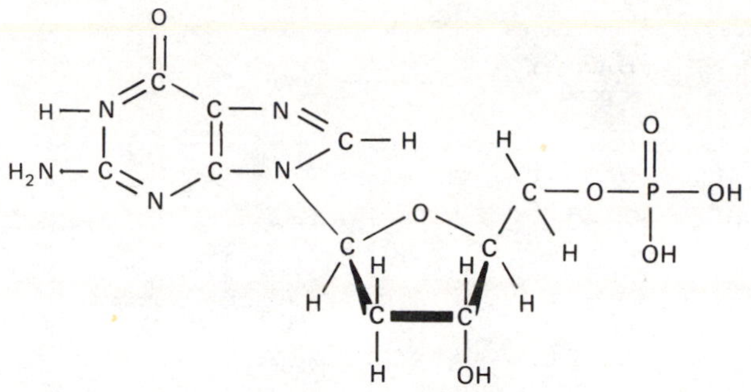

Deoxyguanosine monophosphate (dGMP)

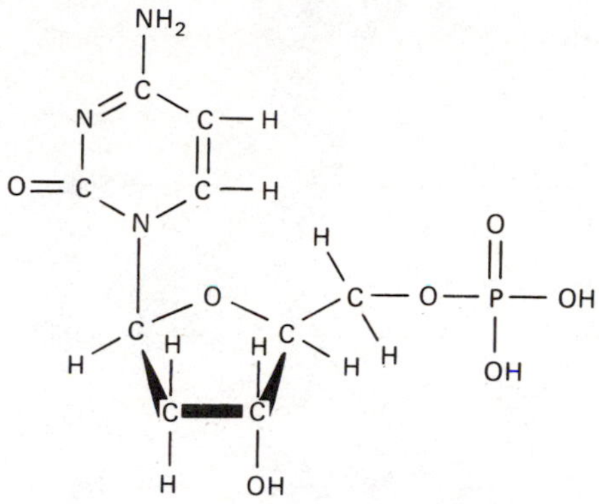

Deoxycytidine monophosphate (dCMP)

Figure 3-7 Structures of a purine deoxynucleotide (*d*GMP) and a pyrimidine deoxynucleotide (*d*CMP).

polynucleotide may be folded in such a way that double-stranded segments with complementary base pairing exist, as noted in the DNA molecule. Other segments of the RNA molecule are apparently present only as single strands. As already shown in Table 3-2, the nitrogen base composition of RNA differs from that of DNA, with uracil replacing thymine; moreover, the pentose sugar ribose is present rather than deoxyribose. The pattern of distribution of RNA within the cell also differs from that of DNA. Small amounts of RNA are found in the nucleus, but most of it is found in the cytoplasm. Three forms of cytoplasmic RNA are recognized; messenger RNA (*m* RNA), transfer or soluble RNA (*t* RNA), and ribosomal RNA (*r* RNA). Messenger RNA is a large molecule composed of numerous nucleotides whereas *t* RNA is a small molecule consisting of between 70 and 80 nucleotides. The ribosomal RNA is associated with protein in a subcellular organelle, the ribosome. All forms of RNA are involved in protein synthesis.

DNA and RNA distribution in cells Both DNA and RNA are involved in protein synthesis, one of the major activities of growing and dividing cells. While the nucleus

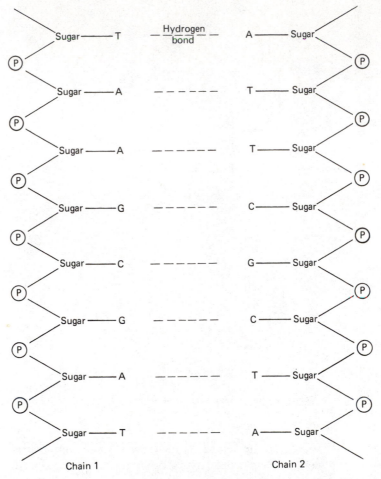

Figure 3-8 Base pairing between complementary bases in the two strands of DNA. Chain 1 has a base sequence of TAAGCGAT, which specifies the base sequence of chain 2 as ATTCGCTA.

serves as the major source of genetic information for protein synthesis, both mitochondria and chloroplasts also contain DNA and RNA and carry out protein synthesis. Mitochondria and chloroplasts, however, cannot synthesize all their characteristic proteins and must rely on genetic information in the nucleus and on proteins synthesized in the cytoplasm. The chloroplast enzyme ribulose bisphosphate carboxylase (Chapter 6), the major enzyme of photosynthetic carbon dioxide assimilation, is composed of light and heavy subunits, for example. The heavy subunit is coded by chloroplast DNA and is synthesized in the organelle whereas the light subunit is coded by nuclear DNA and is synthesized in the cytoplasm and then moved into the chloroplast, where the active enzyme is organized.

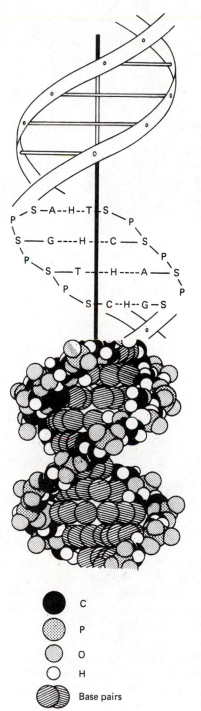

Figure 3-9 Watson and Crick model of DNA molecule. (After Swanson, 1964; (Figure 5.9, p. 27.)

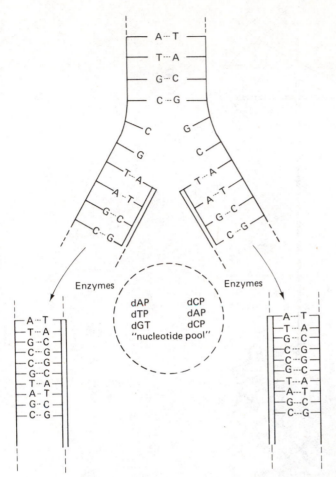

Figure 3-10 Model of replication of DNA molecule during nuclear division. Abbreviations: A, adenine; T, thymine; C, cytosine; G, guanosine; *d*, deoxyribose.

Other important nucleotides   In addition to the nucleotides in DNA and RNA, several other nucleotides are found in cells. The term nucleotide is a broad one and includes nitrogenous bases other than the purines and pyrimidines. Most of these nucleotides occur in rather small amounts in cells and generally function in metabolism as prosthetic groups of enzymes or as enzyme cofactors (Table 4-2). The structure of a few of these nucleotides is given here; their specific functions are discussed in more detail in other chapters.

Adenosine triphosphate (ATP)   Adenosine triphosphate is a nucleotide composed of adenine, ribose, and phosphoric acid, as was shown in Figure 3-11. Moreover, two additional phosphate groups are attached to the phosphate group of adenosine monophosphate.

Adenosine monophosphate (AMP)

Uridine monophosphate (UMP)

Figure 3-11 Structure of a purine ribonucleotide (AMP) and a pyrimidine ribonucleotide (UMP).

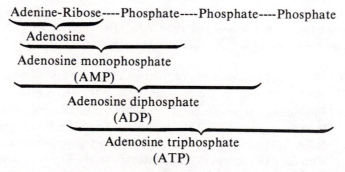

Adenine-Ribose----Phosphate----Phosphate----Phosphate

Adenosine

Adenosine monophosphate (AMP)

Adenosine diphosphate (ADP)

Adenosine triphosphate (ATP)

ATP is an important cellular constituent and participates in both catabolic and anabolic reactions. The phosphate groups (either the terminal group or the two terminal groups, —P or —P—P) interact in metabolic processes. For example,

$$\text{Glucose} + \text{ATP} \xrightarrow{\text{enz}} \text{glucose-P} + \text{ADP}$$

$$\text{Amino acid} + \text{ATP} \xrightarrow{\text{enz}} \text{amino acid-P—P} + \text{AMP}$$

Neither free glucose nor free amino acids are metabolically active, but through interactions with ATP they become phosphorylated and capable of participating in catabolic or anabolic reations (see Chapter 4).

Other nucleotide triphosphates   In addition to adenine, the other purines and pyrimidines (Table 3-2) can also form diphosphates and triphosphates similar to ADP and ATP. Thus cells contain guanosine triphosphate (GTP), uridine diphosphate (UDP), cytosine triphosphate (CTP), and so on. These nucleotides serve as a source of phosphate groups just as ATP does. UTP is specifically involved in sucrose and starch biosynthesis, for instance; GTP is necessary in cellulose biosynthesis; and CTP is specifically required for the biosynthesis of fats.

Nucleotides that contain vitamins   The discussion on proteins pointed out that many enzymes have associated with them prosthetic groups that are essential for their biological activity. Many organic prosthetic groups are nucleotides that contain one of the B vitamins—for instance, nicotinamide, riboflavin, thiamine, folic acid, pantothenic acid, and pyridoxine. (See Table 4-2 for the structures of several of these nucleotides.)

METABOLIC INTERMEDIATES AND BIOSYNTHESIS

Protein Synthesis

Because of the involvement of proteins in so many crucial processes of growth and differentiation, biologists have intensively investigated the mechanism of protein synthesis. It is not enough to consider how free amino acids are assembled into a large protein molecule; the problem of arranging the amino acids into a unique sequence to form a specific protein must also be studied. It is believed that a pattern or *template* is essential for this process and that both DNA and RNA participate. Here are the current ideas on how this process is accomplished.

1. The information necessary for specifying the amino acid sequence of proteins resides in the sequence of nucleotides in the DNA molecule.
2. The nucleotide sequence of DNA is read off or transcribed by a form of RNA known as messenger RNA (*m*RNA). The reaction is catalyzed by an enzyme, RNA polymerase. The *m*RNA, following transcription of the DNA message, has a nucleotide sequence complementary to the DNA nucleotide sequence.
3. *m*RNA moves from the nucleus into the cytoplasm, where it becomes associated with the ribosomes. The ribosomes are composed of protein and another form of RNA, ribosomal RNA (*r*RNA).
4. Amino acids, formed by metabolic processes in the cytoplasm, are activated by ATP and then interact with another form of RNA, transfer or soluble RNA (*t*RNA). The formation of the activated amino acid-*t*RNA complex is catalyzed

by an enzyme specific for each amino acid. There is at least one *t* RNA for each amino acid.

5. *t* RNA with its specific amino acid becomes oriented to the *m* RNA on ribosomes according to their complementary nucleotide sequence.

6. Amino acids are polymerized to form a polypeptide chain under the influence of a peptide-forming enzyme.

7. The secondary, tertiary, and quaternary structure of the protein develops as a result of the amino acid sequence and the environmental conditions.

A schematic representation of the foregoing features of protein synthesis is shown in Figure 3-12. Evidence in support of this scheme has come from a number of sources. Protein synthesis in vitro has been accomplished by using plant extracts containing ribosomes, *m* RNA, *t* RNA, activating enzymes, ATP and GTP, and a mixture of 20 amino acids.

Figure 3-12 shows that nucleic acid synthesis and protein synthesis may be controlled by the availability of nucleotides, ATP, or amino acids. If other metabolic

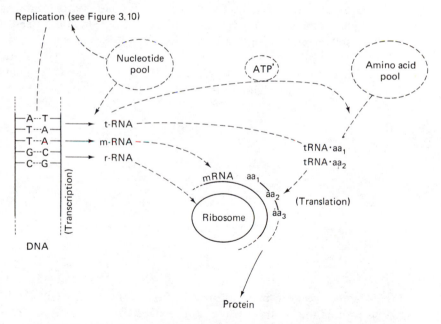

Figure 3-12 Some interrelationships between DNA, RNA, and protein. The genetic information in DNA in the form of nucleotide triplets is transcribed as several RNAs (*t* RNA, *m* RNA, *r* RNA). Transfer RNA (*t* RNA) couples with a specific amino acid and the *t* RNA · *aa* complex is then aligned (translated) with the messenger (*m* RNA) on the ribosome. The DNA is in the nucleus and the transcription of appropriate RNAs also takes place in the nucleus. The RNAs move into the cytoplasm, where protein synthesis (translation) occurs.

processes prevent the ready access of these compounds to the sites of nucleic acid and protein synthesis, severe restrictions are quickly established. Consequently, a number of different ways of regulating protein synthesis exist. Because proteins play a crucial role in cellular processes, the regulation of development is believed to occur through mechanisms involving protein synthesis and protein activity. This problem is discussed more fully at the end of the chapter.

Carbohydrates

Carbohydrates differ from the bulk of the other plant constituents in that they primarily contain only carbon, hydrogen, and oxygen (Table 3-1). Many, however, are present as phosphorylated derivatives and a few contain nitrogen. Carbohydrates that participate in metabolic reactions generally are present as monosaccharides, such as glucose, fructose, and arabinose, whereas the storage and structural carbohydrates are found as polysaccharides.

One feature that distinguish plants from animals is the presence in plants of a rigid cell wall composed mainly of cellulose. The cellulose polymer contains several thousand glucose monomers arranged so as to form long microfibrils that impart structural rigidity to the cell wall. Plants also contain starch, another polymer of glucose. The starch molecule contains several thousand glucose molecules and it is not a fibrillar structure like cellulose.

Monosaccharides   The carbohydrate monomer is known as a monosaccharide (Table 3-1). The simplest monosaccharide is glyceraldehyde, a three-carbon (triose) sugar. Glyceraldehyde exists in two forms.

D-Glyceraldehyde   L-Glyceraldehyde

The difference between the two forms lies in the arrangement of the —OH and —H groups around the central carbon atom marked with an asterisk. When the —OH group is to the right of this carbon atom, the triose is referred to as D-glyceraldehyde; when the —OH group lies to the left, the triose is L-glyceraldehyde. Because monosaccharides can be considered derivatives of D- or L-glyceraldehyde by the addition of more carbon atoms to lengthen the carbon chain, the various monosaccharides belong to either the D or L series. Almost all the biologically active carbohydrates have the D configuration.

As the number of carbon atoms in a monosaccharide increases, the number of sugars possible within each series increases because of the orientation of the —H and —OH groups around the carbon atoms, as shown for the hexoses (six carbon atoms) in Figure 3-13. D-Glucose, D-galactose, D-mannose, and D-fructose are isomers because they have the same chemical composition ($C_6H_{12}O_6$) but differ in their physical properties. Actually, there are 16 hexose isomers, 8 in the D series and 8 in the L series. Only the 4 shown in Figure 3-13 commonly occur in plants, however, it should be noted that the —OH group on the C-5 position in all four monosaccharides lies to the right as in D-glyceraldehyde; so all belong to the D series of sugars. In living matter the straight-chain forms of sugar are almost nonexistent; instead they are present in the ring forms shown in Figure 3-13.

Two monosaccharides of considerable importance in cells are ribose and deoxyribose, five-carbon sugars (pentoses) that are constituents of the nucleic acids. Their configurations are illustrated in Figures 3-7 and 3-11.

Sugar alcohols and cyclitols   Many plants store sugar alcohols rather than pentose or hexose sugars. These molecules, also referred to as polyols, have a strong affinity for water and have been found to accumulate in plants under moisture and low-temperature stress. Polyols appear to serve as protective agents against damage from environmental stress. Figure 3-14 shows the structure of several polyols (sorbitol, glycerol).

Inositol is another polyhydroxylated molecule commonly found in plants. The term *cyclitol* is used to refer to this class of molecules. Inositol exists in several isomeric forms; *myo*-inositol is the form commonly present in plants. The structural configuration of *myo*-inositol is the same as D-glucose and it has been shown that *myo*-inositol is synthesized from D-glucose-6-phosphate in plants.

Inositol is found as a constituent of membranes as phosphatidylinositol. It also has been found esterified to indole acetic acid and is believed to play a role in indole acetic acid storage and transport. Probably the most widely distributed form of inositol is the hexaphosphate, phytic acid (Figure 3-14). In plant tissues phytic acid generally occurs as a potassium, calcium, and magnesium salt and is known as phytin. Phytin is found in the endosperm and aleurone cells of many seeds. During germination phytin is hydrolyzed by the enzyme phytase to give inositol, which participates in numerous biosynthetic pathways essential to the developing embryonic axis. Also produced from phytic acid hydrolysis are phosphate ions—essential for anabolic reactions—and potassium, calcium, and magnesium ions. Phosphate, calcium, potassium, and magnesium are required in rather substantial quantities during early embryonic axis growth until a functioning root system is formed. Phytic acid is discussed further in Chapter 16.

Disaccharides   A disaccharide is formed when two monosaccharides become linked together in a glycosidic bond through the removal of a molecule of water from the hydroxyl groups of the monosaccharides. Sucrose, composed of glucose and fructose (see Figure 3-13 for structural formula), is the most widely distributed

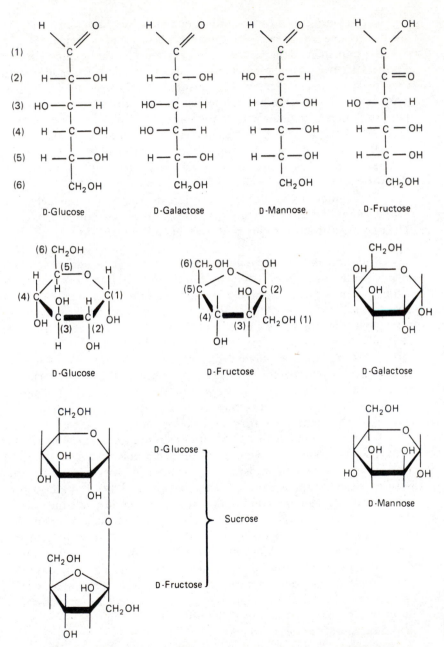

Figure 3-13 Structural configuration of some carbohydrates. The ring configurations should be interpreted as though the ring lies in the plane of the page with the heavy dark lines pointing toward the reader. The groups pointing up lie above the plane of the page whereas the groups pointing down lie below the plane of the page.

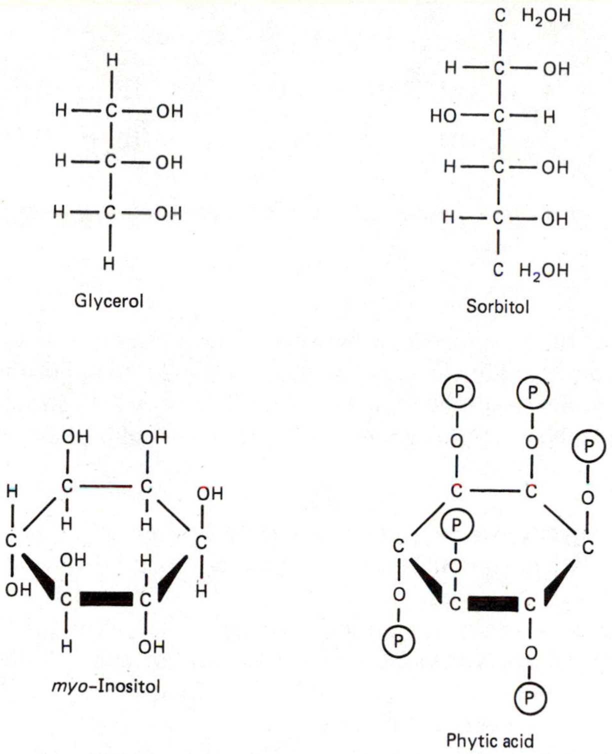

Glycerol

Sorbitol

myo-Inositol

Phytic acid

Figure 3-14 Configurations of two sugar alcohols; glycerol and sorbitol, and *myo* inositol. Phytic acid is the hexaphosphate of *myo* inositol.

disaccharide in the plant world and one of the major products of photosynthesis. In such plants as sugarcane and sugar beet sucrose may reach concentrations as high as 15 to 20%. In many other plants starch is the major storage form of carbohydrate. Sucrose is also the major translocation form of carbohydrate in plants. In some plants, however, sucrose is accompanied by a trisaccharide, raffinose (glucose-fructose-galactose), in the phloem cells where carbohydrate translocation occurs. All these sugars apparently serve equally well in translocation.

Phosphorylated sugars Several of the monosaccharides, as well as sucrose, are found in storage organs and in vascular transport elements as free sugars. Before entering into metabolic reactions, however, these sugars must be phosphorylated to form sugar phosphates. As shown below, the phosphorylated sugars can be visualized as being formed through the interaction of a molecule of phosphoric acid with a sugar molecule in a manner analogous to that described for the formation of a disaccharide.

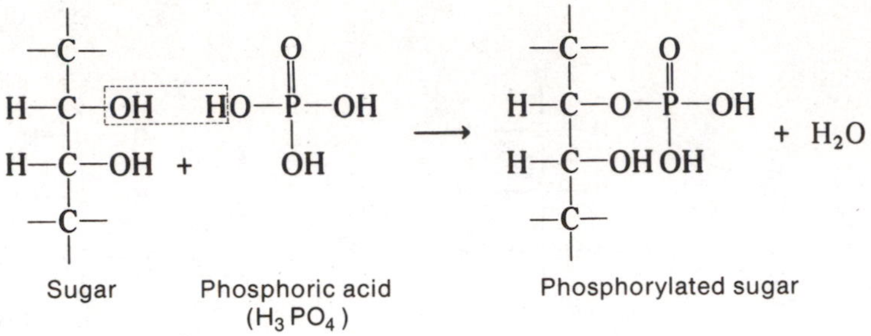

Because the reaction occurs between an acid (phosphoric acid) and an alcohol (sugar), the product is known as an ester. Accordingly, the phosphorylated sugars are often referred to as phosphate esters. Figure 3-15 shows the structural configuration of two important sugar phosphate esters, glucose-6-phosphate and fructose-1,6-bisphosphate.

Polysaccharides (macromolecules) The addition of monosaccharides, together with the elimination of water molecules, produces long chains of carbohydrates known as polysaccharides. Polysaccharides may consist of a single monosaccharide or they may contain several different monosaccharides. Two of the most widely distributed polysaccharides in plants—starch and cellulose—are composed of D-glucose molecules. Two forms of glucose are involved, however, and this fact results in two quite different macromolecules.

    D-Glucose may be represented by the structural formula shown in Figure 3-16(a). In nature, however, glucose is present in a ring structure as shown in Figure 3-16(b) and (c). As a result of ring formation, the —OH and —H groups on the first carbon atom

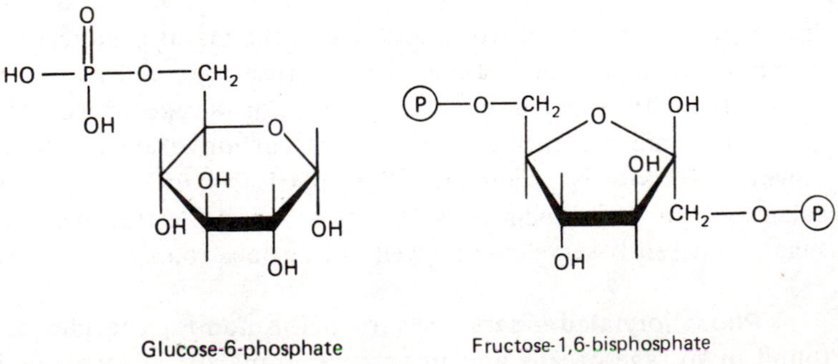

Glucose-6-phosphate   Fructose-1,6-bisphosphate

Figure 3-15 Structural configurations of glucose-6-phosphate and fructose-1,6-bisphosphate. The symbol Ⓟ on fructose-1,6-bisphosphate denotes a phosphate group $\left(-P\begin{smallmatrix} \diagup OH \\ \| \diagdown OH \\ O \end{smallmatrix} \right)$

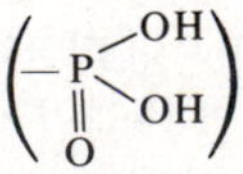

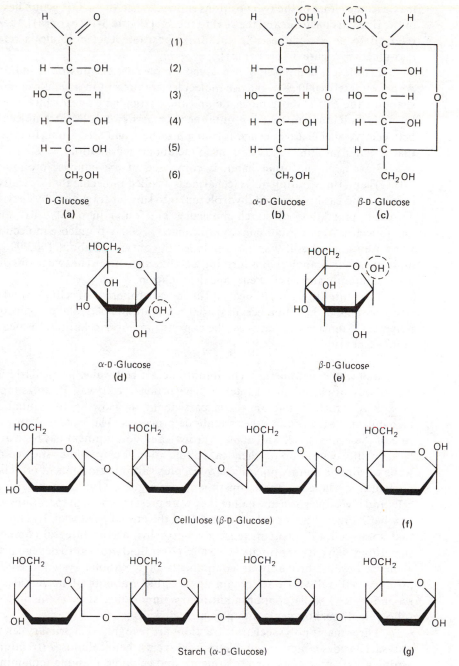

Figure 3-16 Structural configurations of cellulose (f) and starch (g) are based on the difference in shape of β-D-glucose (e) and α-D-glucose (d). The relative position of the —OH group around the first carbon atom of glucose [(b) and (c) or (d) and (e)] determines the shape of the glucose polymer.

may assume either of the two positions shown. When the —OH group lies to the right of the first carbon atom and the —H to the left [Figure 3-16(b) or (d)], the molecule is referred to as α-D-glucose. In the other position, the molecule is referred to as β-D-glucose [Figure 3-16(c) or (e)].

Cellulose is composed of β-D-glucose molecules, as illustrated in Figure 3-16(f). Approximately 6000 β-D-glucose molecules are present in a cellulose molecule. The shape of the β-D-glucose molecule and the particular way in which the polymer is formed result in a long, twisting, unbranched macromolecule. Intramolecular bonding between parallel chains occurs, forming a rather rigid structural, fiberlike molecule that is water insoluble and resistant to metabolic degradation.

Starch, on the other hand, is composed of α-D-glucose monomers [Figure 3-16(g)] and the resulting polysaccharide is a coiled molecule that is relatively soluble in water and easily degraded (hydrolyzed) into smaller fragments by enzymic activity. Two different kinds of starch molecules are found in plant cells—amylose and amylopectin. Amylose contains approximately 2000 α-D-glucose molecules arranged as an unbranched coil whereas amylopectin consists of about 100,000 α-D-glucose molecules with branch chains forming a netlike structure. The two forms differ in their water solubility and ease of enzymic degradation.

The quite different physical and chemical properties of cellulose and starch arise solely because of the differences in spatial configuration of α- and β-D-glucose and their polymers. Similar differences can be expected in polysaccharides formed from other monosaccharides.

Cell wall constituents   The importance of cell walls, particularly primary cell walls, was discussed in Chapter 2. The primary cell wall is not simply a static, protective structure but an active participant in growth, differentiation, cellular recognition systems, and other metabolic processes. The increased awareness of the role of the primary cell wall in plant growth and development has been accompanied by new information on its chemical composition. Early studies showed primary cell walls to contain mainly polysaccharides, plus variable amounts of protein, lipid, and nucleotides. These last three constituents are regularly found in metabolically active cells and it was questioned whether they were present in wall preparations as impurities or whether they were actual components of the primary cell wall. It is now recognized that some cell wall proteins (e.g., glycoproteins) are an integral component of the structured elements of primary cell walls. Glycolipids appear to be regularly present in some cell walls, probably as components of metabolic systems. Highly purified primary cell walls do not contain appreciable amounts of nucleotides. Of course secondary cell walls contain additional components, such as lignin. Most of the following discussion deals with primary cell walls.

Three major polysaccharide fractions are recognized in primary cell walls: pectic polysaccharides (fraction soluble in weak acid), hemicelluloses (fraction soluble in strong alkali after weak acid treatment), and cellulose (residue remaining after acid and alkali extraction). A major feature of pectic polysaccharides is the presence of galacturonic acid (Figure 3-17) residues. Also present are galactose, arabinose, and

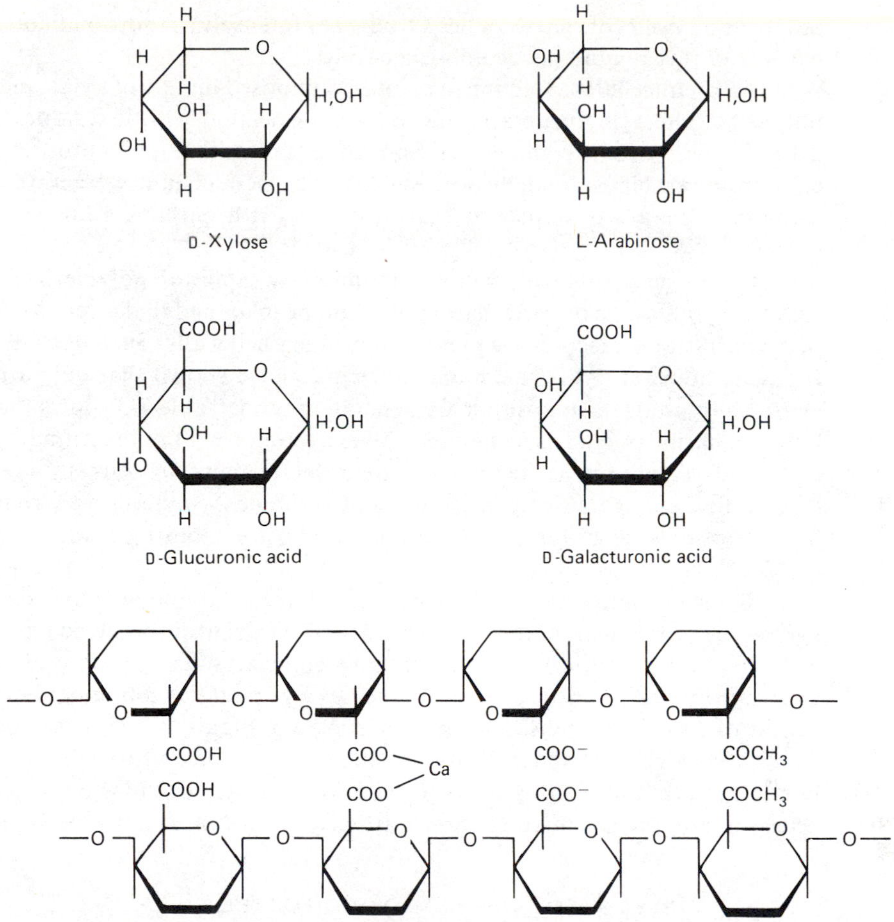

D-Xylose

L-Arabinose

D-Glucuronic acid

D-Galacturonic acid

Figure 3-17 *Upper*: Structures of several important building blocks of cell wall polysaccharides. D-xylose and L-arabinose are pentose sugars; D-glucuronic acid and D-galacturonic acid are uronic acid derivatives of glucose and galactose. *Lower*: Two chains of galacturonic acid showing cross linking between the chains by bonding with Ca^{2+}. If the carboxy groups are methylated (as at far right), cross linking cannot occur.

rhamnose (deoxy derivative of mannose) molecules. In many cell wall preparations a rhamnogalacturonan polymer forms the backbone of the pectic polysaccharide fraction with polymers of arabinose, galactose, and arabinose-galactose attached by covalent bonds to form a network of structural material.

The preceding description of pectic polysaccharides applies to the primary cell walls of dicotyledonous plants. The primary cell walls of monocotyledonous plants contain a much smaller quantity of pectic polysaccharides and what is present seems mostly galacturonans with perhaps small amounts of arabinogalactans. The primary cell walls of monocots have not been studied as intensely as those of dicots. Additional work may reveal more pectic polysaccharides.

The hemicellulose fraction of dicots is composed mainly of xylan and xyloglucan polymers. The xyloglucan appears to have associated with it residues of fucose (a deoxyhexose related to galactose). Some plants also have a glucuronic acid polymer containing arabinose and xylose residues. The hemicellulose fraction of monocot primary cell walls is similar to that of dicots, with perhaps more xylan polymers present.

From the number of monomeric units available for polymerization to hemicelluloses and pectic polysaccharides it might be imagined that a very large number of polysaccharide molecules are present in primary cell walls. Such does not seem to be the case, however. Albersheim and colleagues have shown that only a few polysaccharides of unique composition are generally found. Table 3-3 shows that only eight polysaccharide (including cellulose) polymers are present in the primary cell walls of cultured sycamore cells. Only seven monomeric units are present—galactose, galacturonic acid, glucose, glucuronic acid, arabinose, xylose, and rhamnose. The formation of such polymers of definite composition must be under some kind of metabolic control.

Table 3-3 shows that a glycoprotein rich in the amino acid hydroxyproline is a prominent component of the primary cell wall. The role of this glycoprotein in the cell wall is not understood. Albersheim and associates have proposed that the hydroxyproline-rich glycoprotein is an integral and essential component of the cell wall of cultured sycamore cells. Their model, shown in Figure 3-18, is based on the data of Table 3-3. Cellulose microfibrils are shown to be encapsulated in xyloglucan, a hemicellulose. The binding between cellulose and xyloglucan is by hydrogen bonds. These bonds are sensitive to temperature, changes in pH, and pressure and may

TABLE 3-3 POLYMER COMPOSITION OF THE WALLS OF SUSPENSION-CULTURED SYCAMORE (*Acer pseudoplatanus*) CELLS

Wall component	Wt.	% of cell wall
A. Pectin polysaccharides	34	–
Rhamnogalacturonan I	–	7
Homogalacturonan	–	6
Arabinan	–	9
Galactan and possible arabinogalactan	–	9
Rhamnogalacturonan II	–	3
B. Hemicelluloses	24	–
Xyloglucan	–	19
Glucuronoarabinoxylan	–	5
C. Cellulose	23	–
D. Hydroxyproline-rich glycoprotein	19	–

From The primary cell walls of flowering plants by A. Darvill, M. McNeil, P. Albersheim, and D. P. Delmar, 1980. *In* N. E. Tolbert, editor, *The Plant Cell.* Academic Press, New York.

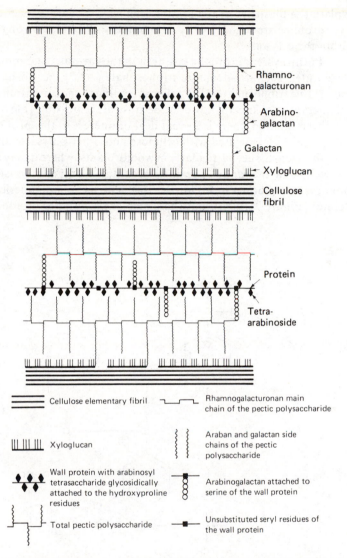

Figure 3-18 Tentative structure of primary cell walls of sycamore. The model is based on data in Table 3-3, which presents the polymer composition of cell walls. The various structural units are depicted in their approximate proportions. Protein is an integral component and two amino acid residues, hydroxyproline and serine, are thought to serve as bridging units between the protein and polysaccharides. (After Keegstra et al., 1973; Figure 8.)

explain the plasticity of the primary cell wall. The other pectic polysaccharides and glycoproteins are bound by covalent bonds to provide a strong backbone or network within the cell wall.

Pathways leading to the synthesis of the various sugar monomers involve several nucleotides, such as adenosine triphosphate (ATP), guanosine triphosphate (GTP), and uridine triphosphate (UTP). Those leading to the formation of the specific cell wall polymers are not known, however. Because of the possible role of cell wall formation in cell differentiation, there is a great deal of interest in this problem at present.

Lignin is associated with cellulose and other polysaccharides of secondary cell walls of xylem tissue, particularly in woody plants, where it may account for 25% of the dry weight of the wood. Lignin is a complex polymer consisting of a number of monomers, as shown in Figure 3-19(a). The basic monomeric molecule is coniferyl alcohol [Figure 3-19(b)], but there also are a number of additional closely related

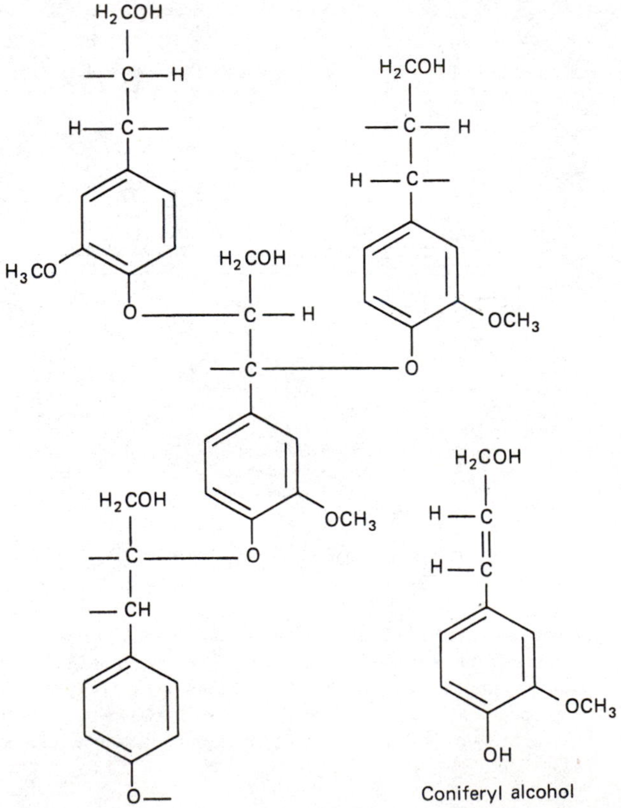

Figure 3-19 A portion of the lignin macromolecule is shown in (a). The basic repeating unit may be visualized as coniferyl alcohol, as shown in (b), or a structurally related derivative.

molecules. The six-membered ring in coniferyl alcohol is quite different from that noted in the hexose sugars. Recent studies have shown that this ring structure, known as a benzene ring, is synthesized in plants from a tetrose sugar and phosphoenolpyruvic acid, which is closely related to the carbohydrates. So although lignin cannot properly be called a carbohydrate, it is found in the plant cell in close association with polysaccharides. During the chemical processing of wood pulp for the manufacture of paper and similar products great quantities of lignin are accumulated. Today it is generally considered a waste product and its disposal presents quite a problem. Developing methods of converting lignin to usable compounds would eliminate a serious waste-disposal situation.

Synthesis of carbohydrates   During photosynthesis carbon dioxide is assimilated and converted into triose phosphate, a mixture a glyceraldehyde-3-phosphate and dihydroxyacetone phosphate. As shown in Chapter 6, part of the triose phosphate is used to regenerate ribulose-1,5-bisphosphate, a carbon dioxide acceptor in photosynthesis, whereas the remainder of the triose is converted into fructose-1-phosphate. Free glucose and fructose may be formed from the phosphorylated hexose sugars through the action of specific phosphatase enzymes. The reactions are summarized in Figure 3-20.

The formation of sucrose, starch, cellulose, and some of the intermediates in the synthesis of pectins and hemicellulose involves the nucleotides uridine triphosphate (UTP), guanosine triphosphate (GTP), and ATP. Glucose-1-phosphate reacts with the appropriate nucleotide to form uridine diphosphoglucose (UDPG), guanosine diphosphoglucose (GDPG), or adenosine diphosphoglucose (ADPG).

$$\text{Glucose-1-P} + \begin{Bmatrix} \text{UTP} & \text{UDPG} \\ \text{ATP} & \longrightarrow & \text{ADPG} \\ \text{GTP} & \text{GDPG} \end{Bmatrix} + \text{PP (pyrophosphate)}$$

Sometimes UDPG, GDPG, and ADPG are referred to as *active glucose*. UDPG then participates in the synthesis of sucrose by one of the following two pathways:

$$\text{UDPG} + \text{fructose} \longrightarrow \text{sucrose} + \text{UDP} \tag{3-1}$$

$$\text{UDPG} + \text{fructose-6-phosphate} \longrightarrow \text{UDP} + \text{sucrose phosphate} \tag{3-2}$$

The sucrose phosphate formed by reaction (3-2) may next be dephosphorylated to yield sucrose. Enzymes catalyzing both reactions have been isolated from a number of different plants and plant tissues, but nothing is known of the conditions that control the particular pathway followed. Free sucrose is the major translocation form of sugar in the phloem cells. Because sucrose is synthesized in the cytosol and must be transported through a number of membranes before it enters the transport system, it is possible that the synthesis of sucrose phosphate [reaction (3-2)] provides for a readily transportable carbohydrate. If free sucrose is synthesized [reaction (3-1)], the sucrose must by phosphorylated before it can be transported through a membrane. Reaction (3-2) may be a more economical system from the standpoint of metabolic reactions required to transport sucrose away from the site of synthesis.

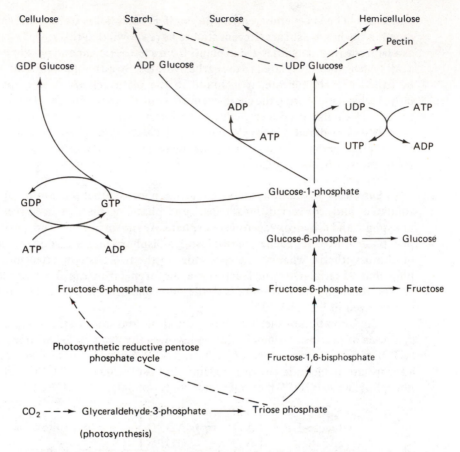

Figure 3-20 Pathways involved in the synthesis of carbohydrates from triose phosphate. Pathways marked --- involve a number of intermediates. Abbreviations: ATP, adenosine triphosphate; UTP, uridine triphosphate; GTP, guanosine triphosphate; UDP, uridine diphosphate; GDP, guanosine diphosphate.

Starch synthesis may also occur by way of UDPG, but the bulk of the starch is formed by a synthetic pathway involving adenosine diphosphoglucose (ADPG).

$$\text{ADPG} + \text{starch "primer"} \longrightarrow \text{starch} \cdot \text{glucose} + \text{ADP}$$

The "primer" is a small preformed unit of starch to which the enzyme adds glucose molecules, thereby lengthening the chain of the starch molecule.

Cellulose is synthesized in much the same manner as starch except that guanosine diphosphoglucose (GDPG) functions as the glucose donor.

$$\text{GDPG} + \text{cellulose "primer"} \longrightarrow \text{cellulose} \cdot \text{glucose} + \text{GDP}$$

Both starch and cellulose are composed of glucose molecules, but starch contains α-D-glucose whereas cellulose contains β-D-glucose. It is not clear whether the different nucleotide requirements for starch (ADPG) and cellulose (GDPG) synthesis are related in some way to differences in configuration of the α- and βD-glucose molecules.

The biosynthetic pathways involved in the synthesis of pectins and hemi-celluloses are less well known than those of sucrose, starch, and cellulose. The cell wall components consist of polymers of galacturonic and glucuronic acid, galactose, glucose, mannose, rhamnose, xylose, and arabinose. Recent studies have shown that the monosaccharides are converted into "active" derivatives by reacting with UTP and then into glucans, galactans, mannans, and so forth in much the same way as starch and cellulose are synthesized. The discovery that the polysaccharides of primary cell walls contain definite and characteristic proportions of monosaccharides raises a question concerning the source of the information within the cell for specifying these precise configurations. Until now only proteins and nucleic acids were considered *informational* molecules. It will be of interest to follow future discoveries in this field to see if polysaccharides are also found to be sources of cellular information.

All the pathways from glucose-1-phosphate to starch, cellulose, sucrose, and the other polysaccharides occur by way of phosphorylated intermediates. Although the immediate phosphate source is UTP, GTP, and possibly other nucleotides, the ultimate phosphorylated intermediate is ATP as shown.

$$\begin{Bmatrix} UDP \\ GDP \\ CDP \end{Bmatrix} + ATP \longrightarrow \begin{Bmatrix} UTP \\ GTP \\ CTP \end{Bmatrix} + ADP$$

During photosynthesis ATP is synthesized from ADP in the chloroplast by photo-phosphorylation (Chapter 6). Carbohydrate synthesis can also occur in nonphoto-synthetic cells; here the ATP is synthesized from ADP in respiratory processes through oxidative phosphorylation in the mitochondria (Chapter 4).

Lipids

Lipids are a heterogeneous group of molecules lacking the well-defined monomeric units noted for protein, nucleotides, and carbohydrates (Table 3-1). All contain atoms of carbon, oxygen, and hydrogen. Many also contain nitrogen, phosphorus, and sulfur.

Fats and fatlike molecules are generally classified as "lipids" on the basis of their solubility properties, being insoluble in water but soluble in such materials as chloroform, benzene, petroleum ether, and similar solvents. The fats, sometimes referred to as neutral fats, serve as reserve food storage materials, primarily in seeds, whereas the fatlike materials, mainly phospholipids and glycolipids, are constituents of all membranes. The cuticular waxes are also lipids but are quite different in their composition from the fats.

The lipids are macromolecules but not in the same sense as the proteins, nucleic acids, and polysaccharides. The latter molecules are formed from unit molecules that interact to form chemical bonds between the monomers. They can also be isolated and their molecular weights determined. The lipids, on the other hand, do not form chemical bonds between individual unit molecules to give a large macromolecule that can be isolated and its molecular weight determined. The bonds that hold the lipid macromolecules together are not as strong as those that occur in the other biological polymers. Nevertheless, insofar as their function is concerned, the membrane-bound lipids may be regarded as macromolecules. Other lipids are found as fat droplets in spherosomes or as constituents of other cellular components.

The unit molecules of a fat are quite diverse but commonly consist of glycerol (a trihydric alcohol) and long-chain fatty acids. The fatty acids have the general formula $CH_3(CH_2)_xCOOH$, where x is usually an even number; for example, palmitic acid, a saturated fatty acid, is $CH_3(CH_2)_{14}COOH$. As shown in Figure 3-21(a), glycerol and the fatty acids interact to form a triglceride accompanied by the elimination of three water molecules. It is not necessary that all three of the hydroxyl groups of glycerol be esterified by fatty acids. Thus many fats are mono- and diglycerides rather than triglycerides. The fatty acids in a fat may be identical, but more frequently they are different. Usually the naturally occurring fats are complex mixtures of saturated and unsaturated fatty acids of varying chain length.

Phospholipids are another important class of lipids. The structure of phosphatidic acid [Figure 3-21(b)] illustrates their basic chemical composition. The phosphate group of phosphatidic acid may be further substituted to give other phospholipids, as shown in Figure 3-21(c). Here a choline molecule is attached to the phosphate group to give a phospholipid known as a lecithin. The phospholipids are important constituents of all membranes, where they are present in relatively large amounts. The phosphate group (in phosphatidic acid) and the choline phosphate are water soluble (hydrophilic) whereas the fatty acids are insoluble in water (hydrophobic). At an oil–water interface the phospholipid molecule is oriented so that the hydrophilic portion of the molecule is embedded in the water phase and the hydrophobic portion is embedded in the oil phase. The resulting orientation of the phospholipid molecule aids in forming a membrane separating the oil and water phases.

Another type of plant lipid—a glycolipid—is illustrated in Figure 3-21(d). In such lipids one of the hydroxyl groups of glycerol forms a bond with a sugar molecule. Chloroplast membranes are particularly rich in glycolipids, with the sugar generally being galactose or deoxyglucose.

The plant waxes are long-chain fatty acids, alcohols, and ketones of considerable complexity. The waxes are found covering stems, leaves, fruits, flower petals, and underground organs, where they function in conserving moisture, controlling the flow of materials into and out of the plant, and protecting delicate cells. Embedded in the waxy coverings are other lipid polymers—cutin and suberin. Both biopolymers are composed of fatty acids and long-chain alcohols joined through ester linkages. A few phenolic molecules are esterified to the cutin biopolymer whereas numerous phenolic molecules are attached to suberin.

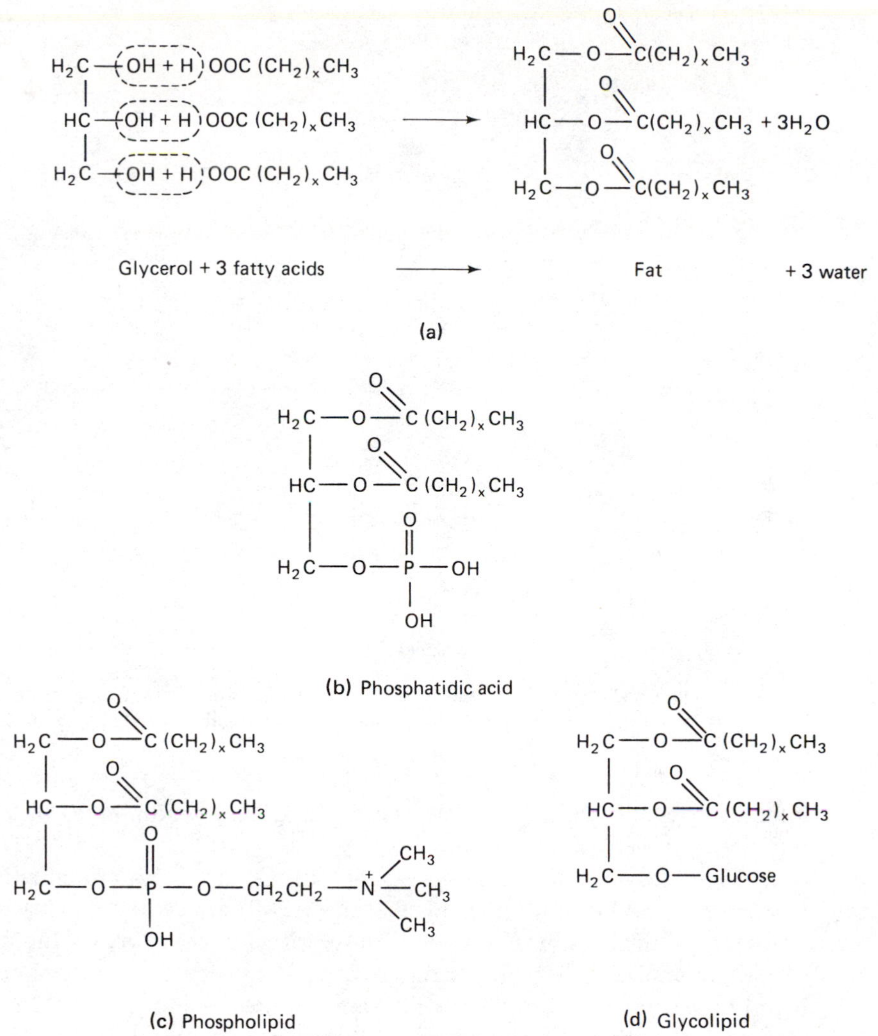

$$H_2C \text{—} (OH + H) OOC (CH_2)_x CH_3$$
$$HC \text{—} (OH + H) OOC (CH_2)_x CH_3 \longrightarrow$$
$$H_2C \text{—} (OH + H) OOC (CH_2)_x CH_3$$

$$H_2C \text{—} O \text{—} \overset{O}{\overset{\|}{C}}(CH_2)_x CH_3$$
$$HC \text{—} O \text{—} \overset{O}{\overset{\|}{C}}(CH_2)_x CH_3 + 3H_2O$$
$$H_2C \text{—} O \text{—} \overset{O}{\overset{\|}{C}}(CH_2)_x CH_3$$

Glycerol + 3 fatty acids $\longrightarrow$ Fat + 3 water

(a)

(b) Phosphatidic acid

(c) Phospholipid

(d) Glycolipid

Figure 3-21   The formation of a fat from glycerol and fatty acids (a). The structures of several other lipids are also shown: phosphatidic acid (b), phospholipid (c), and glycolipid (d).

Biosynthesis of lipids   The basic lipid unit molecule is phosphatidic acid, which is synthesized from glycerol and fatty acids (Figure 3-22). Glycerol phosphate originates from triose phosphate, formed during either photosynthesis or the catabolic breakdown of glucose. The fatty acids are synthesized from acetyl-CoA by a process that requires both ATP and reduced NADP. Fatty acids are also the precursors of waxes. Phosphatidic acid may form triglycerides (neutral fats), or phospholipids by

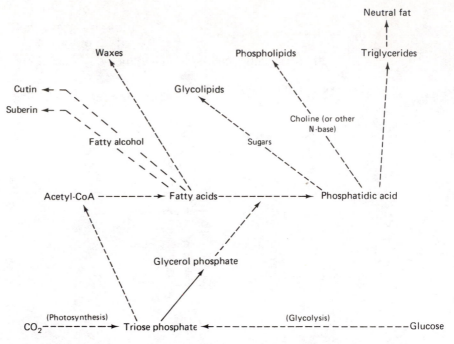

Figure 3-22   Some pathways leading to the synthesis of neutral fats, phospholipids, waxes, cutin, and suberin. Pathways indicated --- contain several intermediates.

interacting with choline or some other nitrogenous compound, or glycolipids by interacting with sugars.

The biosynthetic pathways shown in Figure 3-22 are greatly simplified. Coenzyme A, ATP, and other nucleotides (CTP) are required at various stages of synthesis. Many membranes, such as those found in chloroplasts and mitochondria, contain complex and highly structured lipid molecules. It is unlikely that such membranes are formed by wholly random processes. Chloroplasts and mitochondria contain DNA and RNA and it is possible that these informational molecules function in organizing the complex lipid structures.

Porphyrins

Porphyrins are composed of pyrrole rings. Chlorophyll and heme contain four pyrrole rings linked together to form a tetrapyrrole, as shown in Figure 3-23. Chlorophyll contains an atom of magnesium that is bound (chelated) with the nitrogen atom of the pyrrole rings. A long-chain alcohol, phytyl alcohol, is esterified to one of the four pyrrole rings. Heme is also a tetrapyrrole but contains an atom of iron. The heme molecule, in association with specific proteins, functions in several different kinds of metabolic reactions. The heme protein, cytochrome, functions in electron transport

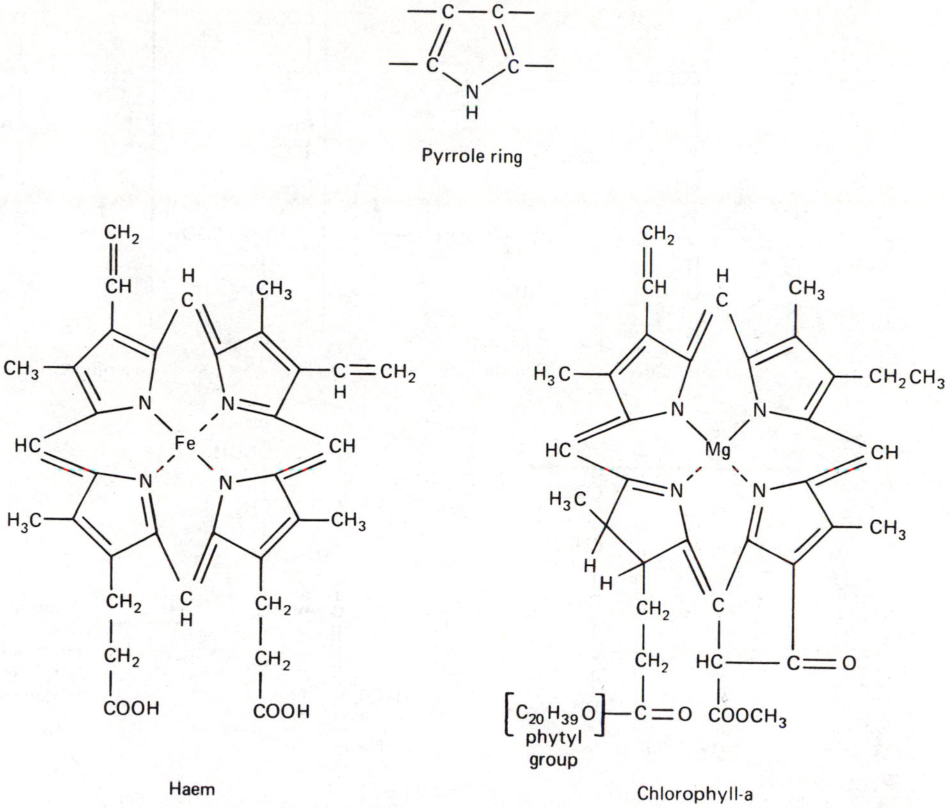

Pyrrole ring

Haem

Chlorophyll-a

Figure 3-23 Structural formulas of a pyrrole ring (top) and two important tetrapyrroles, heme and chlorophyll *a*.

reactions, both in the chloroplast and mitochondrion. Catalase, a heme protein, is an enzyme that breaks down hydrogen peroxide to oxygen and water.

Another porphyrin-containing protein, phytochrome, is belived to consist of four pyrrole rings linked together to form a straight-chain tetrapyrrole rather than a ring structure. Protein is linked to the tetrapyrrole to form the active phytochrome molecule. Phytochrome is involved in seed germination, photoperiodism, and many other photomorphogenetic responses. Its role in plant development is discussed in later chapters.

Porphyrin biosynthesis   The biosynthesis of porphyrins illustrates a general feature of the biosynthetic mechanisms involved in synthesizing complex molecules. The starting materials are simple molecules, usually arising from intermediates in respiration or photosynthesis. By a series of steps involving the addition or removal of water, carbon dioxide, oxygen atoms, hydrogen atoms, phosphate groups, and other

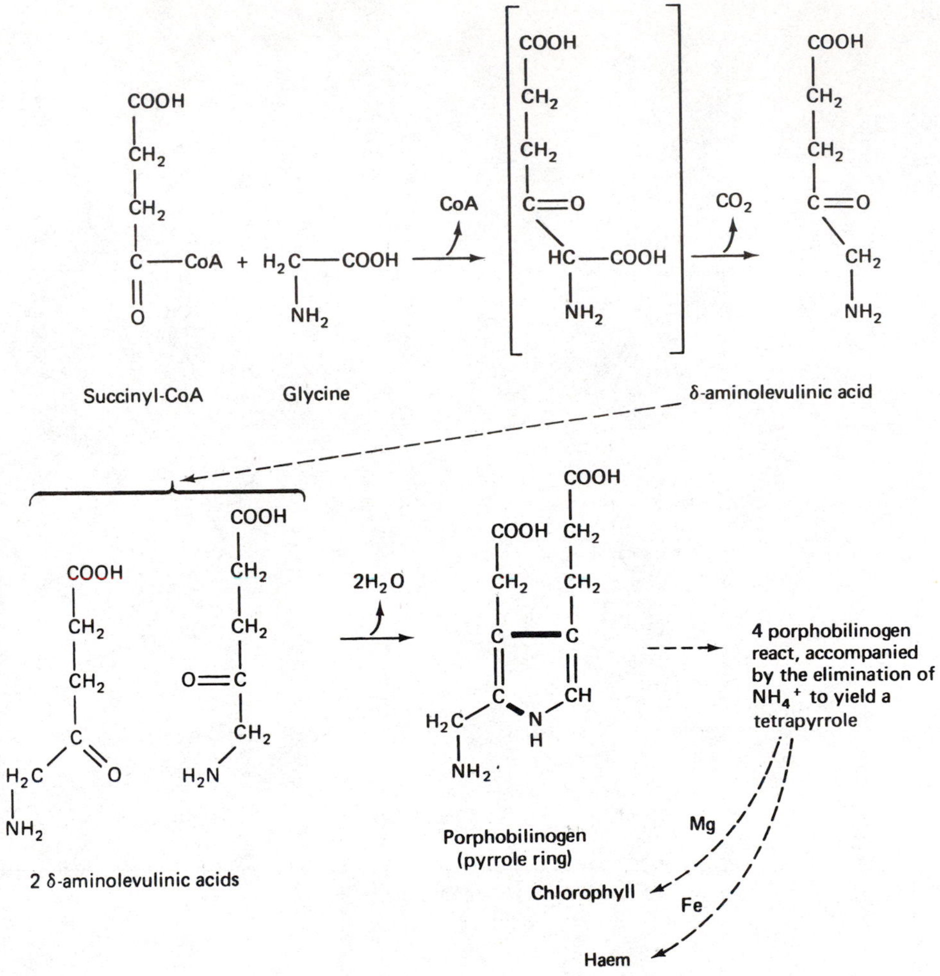

Figure 3-24 Some early steps in the biosynthesis of porphyrins.

simple groups, complex molecules are assembled and organized. This process is shown in Figure 3-24 for the porphyrins. Succinic acid, a member of the tricarboxylic acid (TCA) cycle, is activated through the addition of CoA and then reacts with the simple amino acid, glycine, to form δ-aminolevulinic acid. Two of these molecules unite with the elimination of two molecules of water to form a molecule containing a pyrrole ring. Through further reactions, the pyrrole rings are completed and joined to form a tetrapyrrole. Each reaction is catalyzed by a specific enzyme.

When the role of chlorophyll in photosynthesis was recognized about a hundred years ago, organic chemists undertook to synthesize the molecule. It proved a difficult task and it was not until 1960 that Robert Woodward reported the successful chemical synthesis of chlorophyll a. The chemical synthesis involves the use of high temperatures and pressures, plus the use of acids and other drastic conditions. The plant carries on the synthesis of chlorophyll under ordinary conditions of temperature and pressure and the physiological conditions of the cell.

Organic Acids

Plants contain another class of compounds, the organic acids, which are important in many different kinds of metabolic processes. Figure 3-25 shows the structural formulas of a few of the plant acids. Organic acids, such as malic, succinic, citric, and *a*-ketoglutaric, are present in catalytic amounts in all living cells, where they participate in the TCA cycle (see Chapter 4). Under certain conditions such acids as malic and citric accumulate in large quantities in the vacuoles of plants, where they impart a sour taste to the plant tissue. Other organic acids contribute to the flavor of fruits. The metabolic significance of the organic acids is discussed at various places throughout the book.

Interrelations between Catabolic and Anabolic Processes

Some interrelations between catabolic processes (degradative) and anabolic processes (synthetic) are shown in Figure 3-26. One of the major pathways of catabolism involves

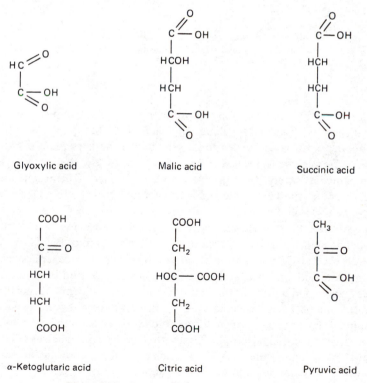

Figure 3-25   Some organic acids found in plants.

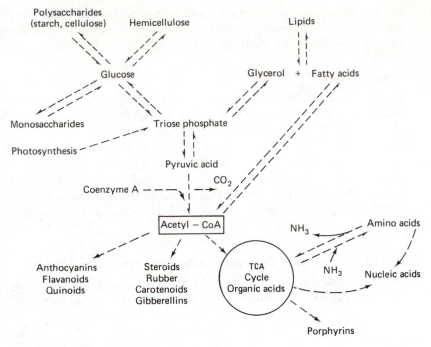

Figure 3-26 Some metabolic intermediates and their participation in biosynthetic pathways. The --- signifies that the pathways involve numerous intermediates that are not shown.

the breakdown of glucose by way of triose phosphate to pyruvic acid (glycolysis). This pathway is discussed in detail in Chapter 4. Many of the intermediates serve as substrates for reactions leading to the biosynthesis of plant constituents.

If pyruvic acid is further metabolized by the TCA cycle, the important intermediate acetyl-CoA (active acetate) is formed. During aerobic respiration acetyl-CoA enters the TCA cycle and is further metabolized to CO_2 and H_2O, accompanied by the production of ATP via electron transport (see Chapter 4). Organic acids in the TCA cycle react with ammonia to form amino acids and eventually proteins and nucleic acids. Porphyrins also originate from some of the TCA cycle intermediates.

Acetyl-CoA is also a precursor of other complex plant constituents, such as phytosteroids, rubber, anthocyanins, gibberellins, and carotenoids. It is seen to be a key intermediate in both anabolic and catabolic metabolism. The anabolic and catabolic pathways are catalyzed by different enzymes and usually occur in different subcellular organelles or in different locations within the cell. Both types of reactions involve the participation of ATP or other nucleotides, such as GTP, UTP, or CTP. The nucleotides, particularly ATP, provide common metabolic intermediates and serve to couple the catabolic and anabolic reactions.

CONTROL OF METABOLIC PROCESSES

From the preceding material it is clear that cells contain many metabolic intermediates and numerous enzymes. The activities of the cell, however, do not occur haphazardly but are highly organized so that the processes of catabolism and anabolism are carried out in an orderly and efficient manner. Studies on the mechanisms responsible for regulating growth and development give some insight into the way in which these processes are controlled.

Control Through Structure and Organization

The degree of complexity of an organism profoundly influences the nature of the metabolic processes carried out. Thus a unicellular green alga contends with different biochemical and physiological problems than a maize plant or an oak tree. The unicellular organism has no structural or conducting elements and the formation of the chemical precursors of phloem and xylem cells is not necessary. Also, the unicell lives in a watery environment and has no special problems of conserving water or obtaining solutes or getting rid of waste materials. Multicellular organisms must have a method for communication between various tissues and organs. A circulatory system and chemical messengers (growth substances or hormones) function at this level of organization. The various tissues and organs of multicellular organisms compete for water and essential metabolites, thereby establishing preferential pathways of metabolism. A developing fruit diverts water, carbohydrates, and other metabolites from leaves, stems, and roots, for instance. The young fruit may be likened to a "sink" into which flow nutrients from the rest of the plant. Actively growing leaves and meristems also act as sinks.

Other regulatory mechanisms will undoubtedly be found to operate in multicellular organisms. Why is a leaf palisade parenchyma cell different from a cambial cell or a root parenchyma cell? All the cells presumably contain the same genetic information. Obviously the genetic information involved in chloroplast formation is "turned off" in the cambial or root cells. Research on these and similar problems constitutes much of modern plant biology today.

Control Through Competition for Metabolites

At the cellular level, metabolism may be controlled by mechanisms that involve the availability of metabolites to the metabolic system. Consider two cells located within a meristem, for example. If one cell is preferentially supplied with a metabolite, such as sucrose, because it lies nearer to a vascular bundle than the other or because of differential membrane permeability, it is clear that the cell with the enhanced sucrose supply might have an advantage over the other. A similar situation may exist between two cells with respect to inorganic ions, growth regulators, amino acids, nucleotides, or water.

Within a single cell the various metabolites are not randomly distributed. Subcellular membranes control the movement of metabolites in the cell. In addition, the cell contains organelles, such as mitochondria, ribosomes, chloroplasts, Golgi bodies, and microbodies bounded by membranes that also regulate the flow of metabolites. The movement of metabolites within the cell or between cells may represent a very efficient regulatory mechanism for controlling metabolism.

Control by Enzymes

The basic level of metabolic control operates at the enzyme level. As noted, the cell is compartmentalized into mitochondria, plastids, and other organelles and each compartment contains its particular complement of enzymes. Furthermore, most enzymes seem organized into multienzyme systems so that the product of one enzyme serves as the substrate for another enzyme, the substrate being slightly modified at each step and immediately utilized by the next enzyme in the pathway until the end product is formed. The enzymes in these multienzyme systems are organized on membrane surfaces.

Two basic methods of control appear to operate at the enzyme level. The activity of an enzyme may be regulated or the synthesis of an enzyme may be regulated.

Enzyme activity   The rates of enzyme-catalyzed reactions vary and differences in rates may be an effective control mechanism. Imagine that the rate of activity of one enzyme in a metabolic pathway is very slow compared to the rates of other enzymes. The rate of formation of the final product thus will be regulated by the speed of the slowest enzyme in the pathway. A specific enzyme in a pathway may be particularly sensitive to temperature and a small change in temperature may alter the entire metabolic pathway.

Theoretically enzyme reactions are reversible; actually, they are essentially undirectional because of the organization of the metabolic system. Many enzymes are organized at membrane surfaces. If the substrate and product are differentially transported in the membrane, the product may not be able to react with the enzyme because it cannot move across the membrane. An enzyme reaction may also be irreversible because the product participates in an irreversible process, say the formation of a storage product, such as fat or starch. The storage product may be converted again into the initial product but by way of a new series of enzymes.

Enzymic activity is also modified by metabolites and other small molecules. The active site of an enzyme (Figure 3-5) binds not only substrate but also other molecules. If the substrate is unable to attach to the active site, it will not be changed. Some of the molecules that bind to the active site and inhibit enzyme activity are similar in structure and size to the substrate molecule. Such molecules are known as competitive inhibitors and their effect can be negated through the addition of more substrate molecules. Other enzyme inhibitors actually denature or otherwise inactivate the active site (e.g., heavy metals, such as mercury and lead). Such molecules are known as noncompetitive inhibitors and their effects cannot be negated by adding more substrate molecules.

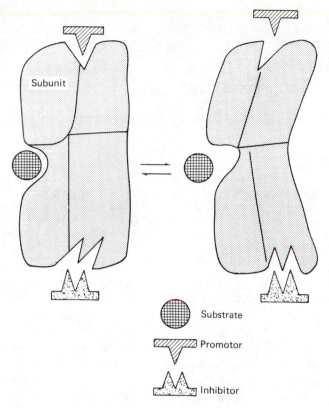

Subunit

Substrate

Promotor

Inhibitor

Figure 3-27 Schematic representation of the action of an allosteric protein (enzyme). When the promotor molecule is bound to the enzyme (left), the conformation of the enzyme allows the substrate to form an enzyme-substrate complex. When the inhibitor molecule is bound to the enzyme (right), the conformation (shape) of the enzyme is altered, the substrate cannot form an enzyme-substrate complex and enzyme activity is blocked.

A new class of enzyme regulators has been recognized. These are molecules that act on the enzyme at a site other than the active site and are referred to as allosteric compounds (the term *allo* meaning other). The nature of the allosteric effect is illustrated in Figure 3-27. Enzymes regulated by an allosteric mechanism appear to be composed of several subunits and the arrangement of the subunits to form an active enzyme depends on the binding of allosteric molecules to the enzyme subunits. In Figure 3-27 the binding of a promotor or inhibitor (allosteric molecules) to the enzyme determines whether the enzyme can react with the appropriate substrate and thereby participate in the formation of a specific product.

The action of allosteric enzymes in metabolic regulation is illustrated in Figure 3-28, a multienzyme metabolic pathway leading from molecule A to product L. Operation of the pathway depends on the coordinated activity of all the enzymes. If the

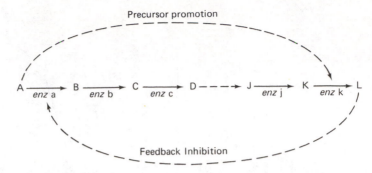

Figure 3-28 Regulation of enzyme activity by allosteric effects on specific enzymes in a metabolic pathway. Molecule A, the first molecule in a multienzyme pathway leading to the formation of L, acts as an allosteric effector on **enz** k. The product of the multienzyme pathway, L, acts as an allosteric inhibitor of **enz** a and thereby "shuts down" the entire pathway. Note that neither A nor L is a substrate for the enzyme which it influences.

activity of a single enzyme is slowed down or stopped, the formation of product L is regulated. It has been observed in many such pathways that product L inhibits the activity of the first enzyme (*enz a*) in the pathway. Note that L is not a substrate for *enz a* and that its influence on *enz a* is by way of an allosteric inhibition. Such an inhibition is referred to as feedback inhibition. The pathway between A and L might be regulated by feedback inhibition at any point—for example, at *enz c* or *enz j*—but such a regulatory mechanism is not so efficient in terms of conserving intermediates as control applied to the first enzyme (*enz a*) in the pathway.

Allosteric regulation of the metabolic pathway shown in Figure 3-28 has been found to operate in another manner. The initial substrate, A, acts as an allosteric effector on *enz k*, thus increasing the conversion of K to L. Such an action changes the concentration levels of the other molecules in the pathway (B, C, D, and J) and tends to increase the rate of conversion of A to L.

In either of the cases shown, it should be noted that the molecule changing the enzyme activity is not the substrate of the enzyme. Molecules other than those directly participating as substrate or product in a metabolic pathway can also interact with allosteric enzymes. Inorganic ions, growth regulators, adenylates (ATP, ADP, AMP), and coenzymes (NAD, NADP) have been found to have allosteric effects. The range of interactions between allosteric proteins and effector or inhibitor molecules seems broad enough to encompass the regulation of a large number of enzymes.

It might be noted that allosteric proteins located in cellular membranes have been suggested as participating in the transport of materials into or out of the cell. Further discussion of this aspect of allosteric protein activity is found in Chapter 11.

Enzyme synthesis Metabolic control is also obtained by regulation of the synthesis of enzymes (Figure 3-29). This method of metabolic regulation differs from the preceding one (Figure 3-28) in that the actual absence or presence of specific

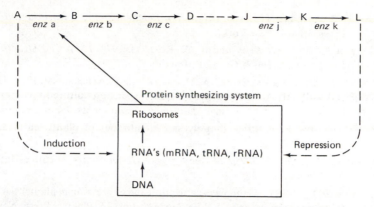

Figure 3-29   Regulation of enzyme synthesis by molecules in a metabolic pathway. A, the first molecule in the pathway leading to the formation of L, interacts with the protein-synthesizing system and induces the synthesis of enz *a*. The final product of the pathway, L, interacts with the protein-synthesizing system and represses the synthesis of enz *a*.

enzymes is involved rather than the activity of an already-present enzyme. The protein synthesizing system (described earlier, Figure 3-12) is induced (turned on) or repressed (turned off) by molecules A and L, respectively. It is not necessary that A and L only have an effect on enz *a*; the other enzymes (enz *b*, enz *c*, etc.) in the pathway might be similarly turned on or off and the molecules in the pathway (B, C, D, J, or K) might act as inducers or repressors. Molecules other than those directly involved in a metabolic pathway might interact as corepressors or coinducers of protein synthesis. One of the best-known examples in plants is the role of gibberellic acid in inducing the synthesis of α-amylase and several other enzymes during seed germination. This topic is discussed further in Chapter 16.

It should be clear from what has been said of metabolic control mechanisms that various regulatory systems are available. The precise location and mode of action of many regulatory systems in plants are not known at present. It is certain, however, that many of the systems described function in plants and the study of metabolic control is a vigorous segment of modern plant physiological research.

REFERENCES

BONNER, J., and J. E. VARNER, eds. 1976. *Plant Biochemistry*, 3rd ed. Academic Press, New York and London.

DARVILL, A., M. MCNEIL, P. ALBERSHEIM, and D. A. DELMAR. 1980. The primary cell walls of flowering plants. *In* N. E. Tolbert, ed. *The Plant Cell*, vol. 1 *in* P. K. Stumpf and E. E. Conn, eds. *The Biochemistry of Plants, a Comprehensive Treatise*. Academic Press, New York.

GOODWIN, T. W., and E. I. MERCER. 1972. *Introduction to Plant Biochemistry*. Pergamon Press, Oxford and Elmsford, N.Y.

GURR, M. I., and A. T. JAMES. 1971. *Lipid Biochemistry: An Introduction.* Cornell University Press, Ithaca, New York.

HALL, J. L., T. J. FLOWERS, and R. W. ROBERTS. 1974. *Plant Cell Structure and Metabolism.* Longmans, Green & Co. Ltd., London.

KEEGSTRA, K., K. W. TALMIDGE, W. D. BAUER, and P. ALBERSHEIM. 1973. The structure of plant cell walls. III A model of the walls of suspension-cultured sycamore cells based on the interconnections of the macromolecular components. *Plant Physiol.* 51:188–196.

KOLATTUKUDY, P. E. 1980. Biopolyester membranes of plants: cutin and suberin. *Science* 208:990–1000.

KROGMANN, D. W. 1973. *The Biochemistry of Green Plants.* Prentice-Hall, Englewood Cliffs, New Jersey.

LEAVER, C. J., ed. 1980. Genome organization and expression in plants. NATO Advanced Study Institutes Series: Series A, Life Sciences, vol. 29. Plenum Press, New York.

LEHNINGER, A. L. 1975. *Biochemistry,* 2nd ed. Worth Publishers, New York.

STEWARD, F. C., R. L. LYNDON, and J. T. BARBER. 1965. Acrylamide gel electrophoresis of soluble proteins: a study on pea seedlings in relation to development. *Am. J. Bot.* 52:155–164.

STUMPF, P. K., and E. E. CONN, editors-in-chief. 1980. *The Biochemistry of Plants, a Comprehensive Treatise.* 8 vol. Academic Press, New York.

SWANSON, C. P., and P. L. WEBSTER. 1977. *The Cell,* 4th ed. Prentice-Hall, Englewood Cliffs, New Jersey.

4

Respiration

The concept of respiration had its beginning many millennia ago when humans first became aware that the act of breathing was essential to animal life. The word *respiration* is derived from the latin *respirare* (literally, to breathe). To the early Renaissance medical scientists and anatomists, respiration meant the inhalation and exhalation of air into and out of lungs of animals. Later, in the seventeenth century, it was realized that absorption of air by blood in lung capillaries resulted in the conversion of purplish blue (venous) to scarlet red (arterial) blood.

It was not until late in the eighteenth century that the chemistry of respiration began to be understood. First, there was the discovery that air consists of several different gases. [It is known today that nitrogen (N_2) (sometimes referred to as *dinitrogen* to emphasize that there are two atoms in each molecule) constitutes approximately 79% of air by volume, oxygen (O_2) (sometimes referred to as *dioxygen* to emphasize that there are two atoms in each molecule) about 21%, and carbon dioxide (CO_2) about 0.033%; only trace amounts of a few other gases of no physiological importance (e.g., argon, neon, etc.) are present in air.] The fact that plants have the ability to "purify" air that was "vitiated" by the burning of candles or exhalations of animals was discovered by the Englishman Priestley in 1772; a few years later Priestley's "purifying" gas was called oxygen. Also, at that time it was realized that oxygen is the gas that revivifies blood and that the breathing process involves the uptake of oxygen and release of carbon dioxide.

In the 1780s the French chemist Lavoisier—often called the "father of chemistry" because he was the first to establish the foundations on which modern chemistry reposes—reasoned that respiration is a form of combustion, which he correctly interpreted as the chemical combination of a burning substance with oxygen. Also,

Lavoisier showed that air expired in breathing contained water vapor. Furthermore, he deduced that hydrogen combines with oxygen in respiration in living tissues. Thus he was the first to point out that not only was carbon dioxide produced in respiration but water, too.

Lavoisier's experimental findings led his contemporaries, especially the Dutch plant physiologist Ingen-Housz, to establish at the end of the eighteenth century that living plants, not animals alone, exchange oxygen and carbon dioxide with the external atmosphere. In fact, the nature of gas exchange in both plant respiration and photosynthesis came to be understood in modern terms at the end of the eighteenth century. And by the middle of the nineteenth century the concept that each and every growing cell of a higher plant respires at all time, in the light as well as the dark—absorbing O_2, oxidizing carbonaceous substances, releasing CO_2, and producing water—was firmly established.

In this chapter we are concerned not only with gas exchanges that occur in respiration but also with the metabolic events associated with respiration. Two important features of respiratory metabolism should be noted now. First, several intermediate metabolites of respiration can serve as starting materials for the biosynthesis of essential cell constituents not directly related to respiration. Second, respiratory metabolism accounts for the generation of adenosine triphosphate (ATP) from adenosine diphosphate (ADP); as we will see, ATP has a critical role in many aspects of cellular metabolism apart from respiration. Thus a knowledge of respiratory metabolism is basic to a study of all metabolic events in plant cells; for this reason, respiration is considered in this book even before photosynthesis (Chapter 6).

It should be emphasized at the outset that the discussion in this chapter is concerned with "dark respiration"—that is, the various respiratory processes that occur in all living cells, in the dark as well as the light. The special kind of respiratory metabolism that occurs only in illuminated green leaves—referred to as *photorespiration*—is considered separately in Chapter 7. We will see (Figure 7-2 and related discussion) that the rate of photorespiration increases with increasing concentration of O_2 in the external atmosphere from very low levels (about 2%) up to and higher than 21% (the natural concentration of O_2 in the atmosphere). On the other hand, those respiratory processes considered in this chapter [e.g., see reaction (4-1)] are saturated at about 2%; that is, they proceed at the same rate whether the concentration of O_2 in the external atmosphere is 2 or 21%.

SUMMARY REACTION OF RESPIRATION

Glucose can be considered the substance most commonly respired in plant cells. It is consumed in cellular respiration according to an overall reaction usually written

$$C_6H_{12}O_6 + 6\ O_2 \longrightarrow 6\ CO_2 + 6\ H_2O \qquad (4\text{-}1)$$

Reaction (4-1), however, obscures the fact that oxygen gas does not react directly with glucose in respiration. Instead water molecules are added to intermediate products of glucose degradation (one molecule of water for each carbon atom in the glucose molecule) and hydrogen atoms in the intermediate products are transferred to O_2, which is reduced to water. The details of metabolic reactions in respiration are considered later. An overall summary reaction more in harmony with these biochemical considerations is

$$C_6H_{12}O_6 + 6\ H_2O + 6\ O_2 \longrightarrow 6\ CO_2 + 12\ H_2O \tag{4-2}$$

The addition of six molecules of water to each side of the equation is algebraically meaningless but important from the biochemical point of view.

Reaction (4-2)—formally the reverse of the summary reaction of photosynthesis—proceeds in individual plant cells in a series of subreactions, each of which results in the breakdown of more complex to simpler molecules and each of which is catalyzed by a specific enzyme:

$$C_6H_{12}O_6 \longrightarrow A \longrightarrow B \longrightarrow C \longrightarrow \text{etc.} \xrightarrow{+H_2O} \xrightarrow{+O_2} CO_2 + H_2O$$

$$\tag{4-3}$$

Many compounds formed in intermediate steps of respiration [indicated by letters A, B, and C in reation (4-3)] may serve as precursors for the biosynthesis of a variety of cellular constituents. This aspect of respiration is considered in further detail later in this chapter.

RESPIRATORY GAS EXCHANGE

Transfer of the respiratory gases (O_2 and CO_2) between a plant and its external environment involves movement through intercellular spaces, plus the cell wall, cytoplasm, and various membranes of individual plant cells. Although the plasma membrane of plant cells is generally considered highly permeable to both gases (Chapter 11), the transfer of respiratory gases through the protoplasts of plant cells probably includes metabolically controlled transport processes as well as the physical process of diffusion.

O_2 has a rather low solubility in water. Therefore a deficiency of O_2 is likely to occur in waterlogged soils. Under these conditions the growth of many plants may be impaired. In fact, many plants will die if severe oxygen deficiency around the roots persists for extended periods of time. Certain plant species that grow naturally with their roots submerged in water, however, can withstand conditions of waterlogging. In water lilies that grow in flooded soils, for example, O_2 moves mostly by mass flow (see pp. 293–294) through a network of internal gas spaces down the petioles of newly emerged leaves to the roots, forcing a simultaneous flow of respired gases (mainly CO_2) from the roots and rhizomes up the petioles of the older emergent leaves to the atmosphere (Dacey, 1980). Similarly, in rice continuous gas spaces extend longitudinally through the plant body and constitute 5 to 30% of the root tissue (Barber et al.,

1962). O_2 enters this important hydrophytic crop plant in the aerial shoot portion and moves through the internal gaseous passageways to the cells of the root.

Measurement of respiratory gas exchange provides the most convenient means of determining respiration rates. Generally a sample (e.g., an individual plant, or a plant part, or tissue slices, or cell suspensions, or perhaps a tissue homogenate, or a preparation of cellular organelles, such as mitochondria) is placed in a sealed chamber and the rate of uptake of O_2 or the rate of release of CO_2 in the chamber is monitored.

A number of methods exist for measuring respiratory CO_2 release and O_2 uptake by plant tissues. Release as CO_2 can be measured by passing an airstream from a sealed chamber containing plant tissue into an alkaline solution, where changes in pH or electrical conductivity are recorded. From these recorded changes the quantity of CO_2 released by the tissues can be calculated.

Manometric methods are also used to measure rates of O_2 uptake or CO_2 release by plant tissues (for details of techniques, consult Umbreit et al., 1971). In these methods sliced or minced samples of plant tissues are suspended in water in a flask. Changes in the amount of respiratory gases in the gas phase within the flask are reflected in changes in its pressure; these pressure changes are measured with a manometer (i.e., an instrument that measures changes in gas pressure).

Other methods utilize more sophisticated instruments. The infrared CO_2 analyzer (or the paramagnetic O_2 analyzer) is capable of measuring changes in CO_2 (or O_2) in the atmosphere in chambers in which whole plants or bulky tissues, such as large fruits, are sealed. In addition, a membrane-bound polarographic electrode sensitive to O_2 can measure changes in the concentration of O_2 in a solution (e.g., in which mitochondria are suspended); this instrument has been adapted recently to measurements of oxygen evolution in the gas phase above leaf discs (Delieu and Walker, 1981). It should also be mentioned that mass spectrometric methods are often used, especially in recent years, in studies of gas exchanges between plant tissues and the external atmosphere.

Respiratory rates of tissues of higher plants are usually much lower than those of many animal tissues (e.g., heart, liver). A well-known exception is the extremely high rate of respiration of the inflorescence of certain species of the arum lily family (see p. 123). Considerable variation exists in the respiratory rates of plant tissues, depending on the species of plant; the kind of tissue, whether meristematic, dormant, mature, or senescent; and whether root, fruit, flower, and so on. The respiration data shown in Table 4-1 are selected mainly for the purpose of illustrating the range of "average" values under "average" conditions.

Comparisons of respiration rates in Table 4-1 are based on a unit of wet weight of tissue, but other units can be used as bases of comparison. These include dry weight of tissue, protein-nitrogen content of tissue, the whole organism, and the individual cell. Although the magnitude of the rate of respiration depends on the base of comparison selected, no general agreement exists among plant physiologists about which one is best. Selection depends on the type of investigation, the kinds of data to be gathered, and the resources available to the investigator.

Respiratory rates vary with changes in certain environmental factors. Tempera-

TABLE 4-1 RESPIRATION RATES OF A FEW PLANT TISSUES (AT APPROXIMATELY ROOM TEMPERATURE)[a]

	Respiration rate (microliters per 100 mg fresh weight per hour)	
	Q_{O_2}	Q_{CO_2}
Alfalfa seed (resting)	36	–
Alfalfa seed (germinating)	106	–
Sweet potato root (segment)	96	–
Red maple stem (cambium)	22	–
Red maple stem (phloem)	17	–
Potato tuber, intact	–	0.3–0.6
Maize leaf, intact	–	54–68
Cucumber pistil	–	29–48
Yucca petal	–	44–67
Apple fruit, intact	–	2–5

[a] The symbol Q (at the head of the columns) is often used in respiration studies to refer to the rate of O_2 uptake (or CO_2 release) by a given quantity of tissue during a unit of time.

Data selected from Altman and Dittmer, 1964, pp. 228–232.

ture has a marked effect. Respiration rates of actively growing tissues increase sharply with increases in temperature in the biological range up to about 40 to 45° C. Light affects the rate of respiration of green leaves (see Chapter 7). In addition, respiration rates are affected by a variety of other factors, including wounding of tissues, application of herbicides to tissues, and the intentional addition of respiratory inhibitor chemicals (e.g., cyanide ion, carbon monoxide) to plant tissues in experimental work. The rate of respiration of a plant tissue is often a sensitive indicator of its health or vigor. When fungal infestation occurs, for example, the respiration rate of the tissue rises above the normal even before visible signs of disease are apparent.

Sliced tissues exhibit certain peculiarities in respiration rates. When thin disks, about 1 mm thick and perhaps 1 cm in diameter, are cut from bulky storage organs (e.g., potato tuber, carrot root, beet root, sweet potato root), they develop a marked increase in the rate of respiration when incubated for several hours in aerated water at room temperature. This respiratory increment during *aging* of the sliced tissues is called *wound* or *wound-induced* respiration and may be three- to fivefold greater (or even more in some tissues) in 24 h than the initial respiration of the sliced disks. Moreover, other physiological changes are known to take place during the aging of sliced disks, including increases in the rates of synthesis of RNA, DNA, and protein; in the ability to absorb ions from an external solution; and in increasing resistance to inhibition by respiratory poisons, such as cyanide. It appears that slicing and aeration of quiescent storage tissues induce a rapid metabolic activation and a development of the membrane systems in the wounded tissues (Kahl, 1974).

An especially interesting examples of a change in respiratory rate during development occurs in certain fruits. When an apple fruit, for instance, reaches its maximum size but is still green, the respiratory rate declines to a low value. Then there is a sharp increase in the rate of respiration during a relatively short period before ripening (see Figure 18-11). This rise in respiration rate accompanying the ripening of certain fruits is known as the *climacteric*. The climacteric rise terminates at the climacteric peak, at which time the apple fruit is soft and edible. Storage of apple fruits in atmospheres low in O_2—about 3% O_2, 5% CO_2, and 92% N_2—has been found to postpone the climacteric and thereby prolong their lifetimes.

RESPIRATORY SUBSTRATES

A *respiratory substrate* is any organic plant constituent oxidized partially (to more oxidized compounds) or completely (to CO_2 and water) in respiratory metabolism (see following examples). Generally a respiratory substrate designates a substance that accumulates in relatively large amounts in plant cells rather than one that is an intermediate breakdown product [see reaction (4-3)]. Breakdown products [indicated by letters A, B, and C in reaction (4-3)] usually are referred to as "intermediate metabolites of respiration" or simply as "respiratory metabolites."

Carbohydrates are the principal respiratory substrates in cells of higher plants. The most important respiratory substrates among carbohydrates are sucrose and starch. Sucrose (a disaccharide consisting of glucose and fructose—that is, two hexoses that are building blocks from which complex carbohydrates are synthesized) and starch (a polymer of glucose) are important forms of carbohydrate storage in plant cells. Sucrose as well as fructose and glucose are the principal soluble sugars in plant cells. Moreover, sucrose is the principal form in which organic materials are translocated within the plant body (see Chapter 11). Starch occurs frequently in plant cells and often serves as a carbohydrate storage reserve. Inulin (a polymer of fructose) also functions as a carbohydrate storage reserve that can be utilized by plant cells as a respiratory substrate, but its distribution among plant species is much less than that of starch.

The pathways involved in the *synthesis* of complex carbohydrates, such as starch and sucrose, from simpler ones and the role of phosphorylated derivatives in these biosyntheses were outlined in Chapter 3. But here is should be emphasized that plant cells are also able readily to degrade sucrose and starch into their hexose components, glucose and fructose. Interconversion of glucose, fructose, sucrose, and starch is achieved in plant cells through enzyme-catalyzed transformations of their phosphorylated derivatives. Some metabolic pathways known to be involved in the interconversion of glucose, fructose, sucrose, and starch are outlined in Figure 4-1. Of these four substances, glucose is generally regarded as the "starting point" for the respiratory metabolism of carbohydrates; we will follow this convention in the next sections.

In addition to carbohydrates, other substances sometimes serve as respiratory substrates in some plant tissues. Certain seeds—the best example is castor bean—are very rich in fat reserves that are contained in the endosperm tissue surrounding the embryo, for instance. During the first several days of germination these fats are

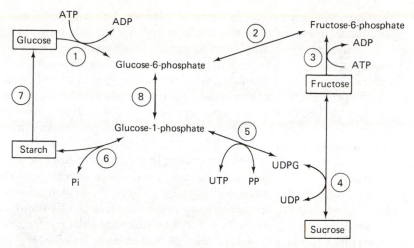

Figure 4-1 Diagrammatic representation of some metabolic reactions (both reversible and irreversible, as indicated by the arrows) in which glucose, fructose, sucrose, and starch are known to participate, illustrating the interconvertibility of these four substances in plant cells. Abbreviations: Pi, phosphoric acid (H_3PO_4); PP, pyrophosphoric acid ($H_4P_2O_7$); UDP, uridine diphosphate; UTP, uridine triphosphate; UDPG, uridine diphosphoglucose; ADP, adenosine diphosphate; ATP, adenosine triphosphate. The enzymes are 1, glucokinase; 2, phosphohexose isomerase; 3, fructokinase; 4, UDPG-fructose-transglycosylase; 5, UTP-glucose-1-phosphate-uridyltransferase; 6, starch phosphorylase; 7, α-amylase, β-amylase, and maltase; 8, phosphoglucomutase. [*Note:* The reactions shown here are not the only ones in which starch, sucrose, glucose, and fructose participate. Not shown, for instance, is the *main* mechanism of sucrose synthesis in plant cells—the reaction of fructose-6-phosphate with UDPG to form sucrose-6-phosphate, which breaks down (irreversibly) to sucrose. Also not shown is the degradation of sucrose to glucose and fructose through the agency of the enzyme invertase. Finally, fructose-6-phosphate and glucose-6-phosphate can be degraded to fructose and glucose, respectively, in the presence of appropriate *phosphatase* enzymes.]

converted mostly to sucrose, which is subsequently absorbed and respired by the growing embryo. Thus the respiratory metabolism in the endosperm of these germinating fatty seeds is predominantly that of conversion of fats to sucrose whereas the growing embryo respires sucrose to CO_2 and water. (The conversion of fats to sucrose in the endosperm tissue of germinating fatty seeds is considered in Chapter 16.)

Moreover, organic acids may be utilized as respiratory substrates in some tissues under certain circumstances. A four-carbon organic acid known as malic acid ($COOHCHOHCH_2COOH$), for example, accumulates in the leaves of succulent plants belonging to the Crassulaceae and other families during the night hours (pp. 175–177. Also, when succulent plants are kept *continuously* in the dark for relatively long periods of time (e.g., 2 to 3 days), a process of *dark deacidification* sets in. Malic acid is respired to CO_2 and water via a mechanism considered later (see Figure 4-6).

Glycolic acid ($CH_2OHCOOH$), a two-carbon organic acid, is also utilized as a respiratory substrate. We will see in Chapter 7 that glycolic acid is produced in illuminated leaves of most higher plants and respired, as shown in Figure 7-3.

Proteins are seldom respired except in special cases. In detached leaves, for example, protein degradation parallels the breakdown of carbohydrates. Proteins also serve as respiratory substrates during the early stages of germination of seeds high in reserve proteins. When proteins serve as respiratory substrates, they are degraded first to amino acids; then amino acids are converted to various intermediates of carbo-hydrate respiration. In effect, amino acids are respired by pathways utilized in the respiration of glucose.

Respiratory substrates may change in a matter of a few hours. It has been found (Jacobson, et al., 1970) that a marked transition in respiratory substrate occurs in sliced potato tissues that exhibit the phenomenon of wound respiration (p. 89). Lipid substances are the chief respiratory substrate in freshly sliced disks from potato tuber. But during aging for 24 h starch becomes the chief respiratory substrate in the sliced disks, just as in the intact tuber. No final answer is available at present to explain this change in respiratory substrate. It seems reasonable to speculate, however, that lipid material in cellular membranes, especially in the endoplasmic reticulum, undergoes temporary disruption and disintegration as a result of slicing. Then this damaged lipid material serves as the respiratory substrate for several hours during aging until the membranes are metabolically repaired.

One technique used to deduce the nature of the respiratory substrate in a plant tissue is to calculate the *respiratory quotient* for the tissue. The *respiratory quotient* (RQ) is the ratio of the moles of CO_2 released by a tissue over a period of time (e.g., an hour) to the moles of O_2 taken up. (Determination of RQ for green tissues requires that measurement of respiratory gas exchange be made when the tissues are kept in the dark; this procedure eliminates photosynthetic gas exchange.)

The magniture of the RQ of a tissue depends on the oxidation state of the substrate utilized in respiration. This situation can be illustrated by the following three examples.

1. If glucose were the sole respiratory substrate, the respiratory quotient would be 1.0.

$$C_6H_{12}O_6 + 6\ O_2 \longrightarrow 6\ CO_2 + 6\ H_2O$$
Glucose

(4-4)

$$RQ = \frac{6 \text{ moles of carbon dioxide}}{6 \text{ moles of oxygen}} = 1.0$$

2. If malic acid (an organic acid that accumulates in relatively large amounts in leaves of succulent plants kept for 2 to 3 days in the dark (see p. 91) were the sole respiratory substrate, the respiratory quotient would be 1.33.

$$C_4H_6O_5 + 3\ O_2 \longrightarrow 4\ CO_2 + 3\ H_2O$$
Malic acid

(4-5)

$$RQ = \frac{4 \text{ moles of carbon dioxide}}{3 \text{ moles of oxygen}} = 1.33$$

3. If palmitic acid (a fatty acid converted to sucrose in the endosperm tissue during early stages of germination of fatty seeds; see Chapter 16) were the sole respiratory substrate, the respiratory quotient would be 0.36.

$$C_{16}H_{32}O_2 + 11\ O_2 \longrightarrow C_{12}H_{22}O_{11} + 4\ CO_2 + 5\ H_2O$$

Palmitic acid   Sucrose

(4-6)

$$RQ = \frac{4 \text{ moles of carbon dioxide}}{11 \text{ moles of oxygen}} = 0.36$$

NATURE OF BIOLOGICAL OXIDATION

Because respiration deals with the oxidation of respiratory substrates and intermediate metabolites, it is necessary to consider the nature of biological oxidation.

General Considerations

First, let us consider the *state of oxidation* of carbonaceous compounds. The state of oxidation is determined by the relative numbers of oxygen and hydrogen atoms in the compound. Methane (CH_4), for example, is the most reduced of all one-carbon compounds whereas CO_2 is the most oxidized. Glucose ($C_6H_{12}O_6$), with one oxygen atom for every carbon atom, has an oxidation state midway between CH_4 and CO_2.

The classical example of oxidation is the direct addition of oxygen to the substrate oxidized, as in the combustion of coal.

$$C + O_2 \longrightarrow CO_2 \qquad\qquad (4-7)$$

But this type of oxidation occurs to only a limited extent in biological systems [i.e., in oxygen fixation reactions (pp. 283–284)]. Two other types of oxidation are more common. One involves the removal of electrons from the substance oxidized. In the oxidation of ferrous iron to ferric iron, for example, ferrous iron loses an electron.

$$Fe^{2+} \longrightarrow Fe^{3+} + e^- \qquad\qquad (4-8)$$

The third type of oxidation involves the removal of hydrogen from the substance oxidized. In the oxidation of succinic acid to fumaric acid, for instance, succinic acid undergoes dehydrogenation as follows:

$$HOOCCH_2CH_2COOH \longrightarrow HOOCCH{=}CHCOOH + 2\ H \qquad (4-9)$$

Succinic acid   Fumaric acid

Also, the oxidation of a substance with an aldehyde group (—CHO) can be regarded as the removal of hydrogen from the substance. The first step is the addition of water to the aldehyde group. (*Note*: The addition of water to, or abstraction from, an organic molecule does not change its state or oxidation.) The second step is the dehydrogenation of the intermediate and formation of a carboxyl group (—COOH). The oxidation of acetaldehyde to acetic acid is an example.

$$CH_3CHO \xrightarrow{+H_2O} CH_3CH(OH)_2 \longrightarrow CH_3COOH + 2\ H \qquad (4\text{-}10)$$

$$\underset{\text{Acetaldehyde}}{} \qquad \underset{\substack{\text{Hydrate} \\ \text{intermediate}}}{} \qquad \underset{\substack{\text{Acetic} \\ \text{acid}}}{}$$

Neither free electrons nor free hydrogen atoms are produced in reactions (4-8), (4-9), and (4-10). Instead a substance called an *electron acceptor* or a *hydrogen acceptor* combines with the electrons or hydrogens. Such a substance is itself reduced when it accepts either hydrogens or electrons.

Although the three types of oxidation mentioned—the addition of oxygen and the removal of electrons or hydrogens—appear quite different, actually all have one feature in common. In each, electrons are *removed* from the substance oxidized. This fact is obvious in reaction (4-8), where electrons are removed as electrons. But in the dehydrogenation reactions (4-9) and (4-10), the removal of hydrogens can be regarded as the removal of electrons because a hydrogen atom consists of a proton *and* an electron (i.e., $H = H^+ + e^-$).

Even the direct addition of oxygen to carbon in reaction (4-7) can be regarded as the removal of electrons from the carbon atom and their transfer to the carbon-oxygen bonds in the CO_2 molecule. The oxygen atom, having a greater affinity for electrons than the carbon atom, attracts them more powerfully and so possesses them to a greater extent in the CO_2 molecule than does the carbon atom. Thus in reaction (4-7) carbon is oxidized because it loses electrons.

Whenever a substance is oxidized, another is reduced. Reduction—the reverse of oxidation—consists of the *addition* of electrons to the substance that is reduced. In the formation of CO_2 from carbon and O_2 in reaction (4-7), O_2 is reduced because it gains electrons. In the reduction of O_2 to water in respiration [see reactions (4-1) and (4-2)], O_2 accepts electrons and protons as follows:

$$O_2 + 4\ H^+ + 4\ e^- \longrightarrow 2\ H_2O \qquad (4\text{-}11)$$

The reductive reaction [reaction (4-11)] and also the oxidative reactions (4-8) to (4-10) are *partial* reactions. Neither the electron acceptors [in the oxidative reactions (4-8) to (4-10)] nor the electron donor [in the reductive reaction (4-11)] are included in the reactions as written. As we will see, two partial reactions occur in conjunction with one another and must be added together to give a *complete* oxidation-reduction reaction [for example, see reactions (4-16) and (4-18)].

Electron and Hydrogen Carriers

Certain plant constituents participate in oxidation-reduction reactions in plant cells by accepting electrons or hydrogen atoms from one metabolite and donating them to another. These transfer agents are known as *electron* or *hydrogen carriers*. Four of the best-known electron carriers in plants cells are

1. nicotinamide-adenine dinucleotide (the oxidized and reduced forms are written NAD^+ and NADH, respectively)

2. nicotinamide-adenine dinucleotide phosphate (the oxidized and reduced forms are written $NADP^+$ and NADPH, respectively)

3. flavin-adenine dinucleotide (the oxidized and reduced forms are written FAD and $FADH_2$)

4. flavin mononucleotide (the oxidized and reduced forms are written FMN and $FMNH_2$)

All these compounds are chemically related to one another (i.e., all are nucleotides; their formulas (except that of FMN) were given in Chapter 3. (FMN has the same reactive group as FAD, but the adenine and ribose groups, and one of the phosphate groups, are lacking.) Each of these four compounds is a nonprotein organic molecule required for the activity of specific enzymes. [But protein molecules (e.g., cytochrome c, see p. 96, and ferredoxin (see p. 261) may also serve as electron carriers in plant cells.] Thus the four compounds considered here are *enzyme cofactors* as well as electron or hydrogen carriers. Each undergoes specific changes in its reactive groups during oxidation and reduction, as shown in Table 4-2. (Because the reactive group in NAD^+/NADH and $NADP^+$/NADPH is exactly the same, only one is shown in Table 4-2; similarly, only one of the two flavin carriers is shown in Table 4-2.) The partial oxidation-reduction reactions are

$$NADH = NAD^+ + H^+ + 2e^- \qquad (4\text{-}12)^1$$

$$NADPH = NADP^+ + H^+ + 2e^- \qquad (4\text{-}13)^1$$

$$FADH_2 = FAD + 2H^+ + 2e^- \qquad (4\text{-}14)$$

$$FMNH_2 = FMN + 2H^+ + 2e^- \qquad (4\text{-}15)$$

Also shown in Table 4-2 are the oxidized and reduced forms of ubiquinone (Ub), a carrier that functions in the respiratory chain (p. 113); ubiquinone is also called coenzyme Q. The reactive group of ubiquinone is attached to 10 isoprene units, as indicated in Table 4-2. Ubiquinone and also FAD and FMN may be called "hydrogen carriers" as well as electron carriers because they add or lose hydrogen atoms when reduced or oxidized. But, of course, as indicated earlier (p. 94), a hydrogen atom consists of a proton and an electron (i.e., $H = H^+ + e^-$).

Many oxidation-reduction enzymes are known to require one or another of the cofactors in Table 4-2: These cofactors are also known as prosthetic groups (see Chapter 3) and some will be mentioned later. But it will be useful here to give a specific example. The enzyme that catalyzes the oxidation of succinic acid to fumaric acid in

---

[1] The reduced form of the nicotinamide nucleotide coenzymes is best written as NADH or NADPH because *one* equivalent of acid (i.e., one proton) is released on oxidation [see reactions (4-12) and (4-13)]. On the other hand, the oxidized form is best written as NAD^+ and $NADP^+$ because the nitrogen atom in the nicotinamide portion of the molecule has a formal charge of +1 (see Table 4-2).

TABLE 4-2 OXIDIZED AND REDUCED FORMS OF THREE IMPORTANT HYDROGEN AND ELECTRON CARRIERS

Oxidized Form	Reduced Form

NAD⁺ / NADH structures:
Adenine — Ribose—Phosphate—Phosphate—Ribose — NAD^+

Adenine — Ribose—Phosphate—Phosphate—Ribose — NADH

Adenine — Ribose—Phosphate—Phosphate—Ribitol — FAD

Adenine — Ribose—Phosphate—Phosphate—Ribitol — $FADH_2$

Ub (ox)   Ub (red)

Note that the reduced forms of FAD and ubiquinone (Ub) have two more hydrogen atoms (indicated by asterisks) than the oxidized forms.

reaction (4-9) is succinic acid dehydrogenase; the cofactor that functions with this enzyme is flavinadenine dinucleotide. The *complete* reaction catalyzed by succinic dehydrogenase is the sum of the partial reactions (4-9) and (4-14).

$$\text{Succinic acid} + \text{FAD} \xrightarrow[\text{dehydrogenase}]{\text{succinic}} \text{fumaric acid} + \text{FADH}_2 \qquad (4\text{-}16)$$

In addition to the preceding cofactors, the respiratory proteins known as the cytochromes also function as electron carriers in oxidation-reduction reactions in plant cells. But in contrast to reactions (4-12), (4-13), and (4-14), which occur by the removal or addition of *two* electrons simultaneously, oxidation and reduction of the cytochromes occur by removal or addition of *one* electron at a time from an iron atom in the cytochrome molecule [see reaction (4-8)]. The role of the cytochromes in respiration is considered later in this chapter, but it will be useful now to note that a respiratory protein known as cytochrome *c* accepts electrons, one at a time, from certain respiratory metabolites (e.g., $NADH^+$) and donates them to O_2. This electron transfer is catalyzed by the enzyme cytochrome oxidase. The partial reaction catalyzed by this enzyme was already indicated [reaction (4-11)] and is repeated here.

$$O_2 + 4 H^+ + 4 e^- \longrightarrow 2 H_2O$$

The other partial reaction, multiplied by 4 because each molecule of O_2 accepts four electrons, is as follows:

$$4 \text{ Cyt } c(Fe^{2+}) \longrightarrow 4 \text{ Cyt } c(Fe^{3+}) + 4 e^- \qquad (4\text{-}17)$$

The *complete* reaction catalyzed by cytochrome oxidase is the sum of the partial reactions (4-11) and (4-17).

$$4 \text{ Cyt } c(Fe^{2+}) + O_2 + 4 H^+ \xrightarrow[\text{oxidase}]{\text{cytochrome}} 2 H_2O + 4 \text{ Cyt } c(Fe^{3+}) \qquad (4\text{-}18)$$

The oxidizing and reducing potentials in millivolts of each of the several partial reactions of respiratory metabolism [e.g., reactions (4-9) and (4-11) to (4-15) and also reaction (4-17)] have been measured. When partial reactions are tabulated according to their oxidizing and reducing potentials (i.e., their *redox potentials*), strong reducing agents with relatively high potentials (e.g., NADH) are listed at or near the top of the table and strong oxidizing agents with low potentials (e.g., O_2) at the bottom. So, for instance, the partial reaction $NADH/NAD^+$ [reaction (4-12)] has a high potential and the partial reaction H_2O/O_2 [reaction (4-11)] has a low potential. We will see (pp. 110–113) that NADH is oxidized to NAD^+ by O_2 in the *respiratory chain*. The complete reaction is the sum of partial reactions (4-11) and (4-12).

$$2 \text{ NADH} + O_2 + 2 H^+ \longrightarrow 2 \text{ NAD}^+ + 2 H_2O \qquad (4\text{-}19)$$

Also, succinic acid is oxidized to fumaric acid by O_2 in the respiratory chain (see pp. 110–113). The complete reaction is the sum of the partial reactions (4-9) and (4-11).

$$2 \text{ Succinic acid} + O_2 \longrightarrow 2 \text{ fumaric acid} + 2 H_2O \qquad (4\text{-}20)$$

OUTLINE OF RESPIRATORY METABOLISM

As already indicated in reaction (4-3), the breakdown of a respiratory substrate in respiration proceeds through a series of reactions, each catalyzed by a specific enzyme. The sequence of reactions is generally referred to as *respiratory metabolism*. Different kinds of reactions are involved, including phosphorylations, hydrations, oxidations, decarboxylations, isomerizations, group transfers, and cleavage reactions. But in spite of this seeming complexity, the basic pattern of respiratory metabolism can be described in relatively simple terms and is outlined in Figure 4-2.

Starting with glucose, which we have seen (pp. 86–87) can be considered to be the starting point for respiratory metabolism in plant cells, pyruvic acid $(CH_3COCOOH)$— a key respiratory metabolite—is formed in a series of reactions referred to collectively as *glycolysis*. (Later we will see that the *oxidative pentose phosphate pathway* is another sequence for the initial phase of glucose breakdown.)

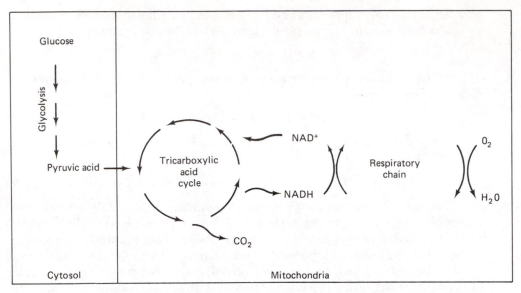

Figure 4-2 Abbreviated outline of the major pathway of respiratory metabolism in plant cells.

Pyruvic acid produced by glycolysis (or via the oxidative pentose phosphate pathway and glycolysis) is oxidized in a sequence of reactions referred to as the *tricarboxylic acid cycle.* Several intermediates of the tricarboxylic acid cycle undergo decarboxylations so that CO_2 is produced. Other intermediates are oxidized through the agency of the coenzyme NAD^+ and hence NADH is produced. NADH is oxidized via the respiratory chain, so that NAD^+ is regenerated. O_2 is the final electron acceptor in the respiratory chain and is reduced to water [see reaction (4-19)]. The tricarboxylic acid cycle and the respiratory chain are located in mitochondria, whereas enzymes of glycolysis are in the cytosol (see Figure 4-2).

The several sequences shown in Figure 4-2, beginning with glycolysis and/or the oxidative pentose phosphate pathway, progressing through the tricarboxylic acid cycle, and ending with the respiratory chain, are discussed in the following sections.

GLYCOLYTIC PATHWAY OF GLUCOSE DEGRADATION TO PYRUVIC ACID

In the initial phase of respiratory metabolism glucose is degraded to pyruvic acid. (Recall from the preceding section that glucose is conveniently designated as the starting point in respiratory metabolism.) The component reactions and enzymes are the same in all living cells of higher plants, animals, and microorganisms. The sequence of reactions in the conversion of glucose to pyruvic acid was first elucidated in the early part of the twentieth century of preparations from yeast and muscle cells. In mammalian skeletal muscle the pathway actually starts with glycogen, a polymer of glucose. Glycogen is the principal reserve carbohydrate in muscle cells and has a close structural relation to starch, the reserve carbohydrate in most plant cells. Because of these early investigations, it became customary to refer to this pathway as *glycolysis* (literally, the degradation of glycogen). Also, the pathway is called the EMP pathway in honor of three of the biochemists (Embden, Meyerhof, and Parnas) who contributed most to its elucidation. Glycolysis is the major pathway in plant cells for the initial phase of respiratory metabolism of hexoses. The enzymes of glycolysis are present not only in the soluble portion of the cytoplasm (i.e., the cytosol) but also in chloroplasts and (in developing seeds) proplastids (ap Rees, 1980). [Enzymes numbered 8, 9, and 10 in Figure 4-3 have been reported not to be present in chloroplasts (Kow et al., 1982).]

Glycolysis proceeds by way of phosphate derivatives. In the first step of the pathway glucose is phosphorylated. The phosphorylating agent in ATP, the formula of which was given in Chapter 3. ATP engages in reactions in which its terminal phosphate group is transferred to a receptor molecule; as a result, ADP is formed. In the reaction in which glucose is phosphorylated by ATP, the enzyme is hexokinase and the products are glucose-6-phosphate and ADP.

$$\text{Glucose + ATP} \xrightarrow{\text{hexokinase}} \text{glucose-6-phosphate + ADP} \tag{4-21}$$

The individual reactions in the glycolytic pathway are shown in Figure 4-3. Following the phosphorylation of glucose by ATP, glucose-6-phosphate undergoes isomerization to fructose-6-phosphate. Fructose-6-phosphate is phosphorylated, by ATP, to fructose-1,6-bisphosphate. The latter substance is cleaved into two triose phosphates, which are interconvertible. The next reaction is catalyzed by an oxidizing enzyme—glyceraldehyde-3-phosphate dehydrogenase (also known as triose phosphate dehydrogenase). In this reaction the aldehyde group (—CHO) in glyceraldehyde-3-phosphate is oxidized to a carboxyl group (—COOH). The oxidizing agent is NAD^+. Also, this reaction is accompanied by the addition of phosphoric acid (H_3PO_4); phosphoric acid is abbreviated Pi. The complete reaction catalyzed by triose phosphate dehydrogenase[2] is as follows:

$$\text{Glyceraldehyde-3-phosphate + Pi + NAD}^+ \longrightarrow$$
$$\text{1,3-bisphosphoglyceric acid + NADH + H}^+ \tag{4-22}$$

The product of reaction (4-22), 1,3-bisphosphoglyceric acid, reacts with ADP; ATP and 3-phosphoglyceric acid are formed. Two further reactions result in the formation of phosphoenol pyruvic acid, which reacts with ADP to form ATP and pyruvic acid.

In summary, for each molecule of glucose consumed in glycolysis, 2 molecules of pyruvic and 2 molecules of NADH are formed. Also, 2 molecules of ATP are consumed (in the first and third reactions in Figure 4-3) and 4 molecules of ATP are regenerated [in the seventh and tenth reactions in Figure 4-3]. Thus there is a net gain of 2 molecules of ATP in the conversion of 1 molecule of glucose to 2 molecules of pyruvic acid.

---

[2]The NAD-linked triose phosphate dehydrogenase that functions in glycolysis is not the same enzyme as the NADP-linked triose phosphate dehydrogenase in chloroplasts; the latter enzyme participates in the photosynthetic carbon reduction cycle (reductive pentose phosphate cycle) (pp. 168–169).

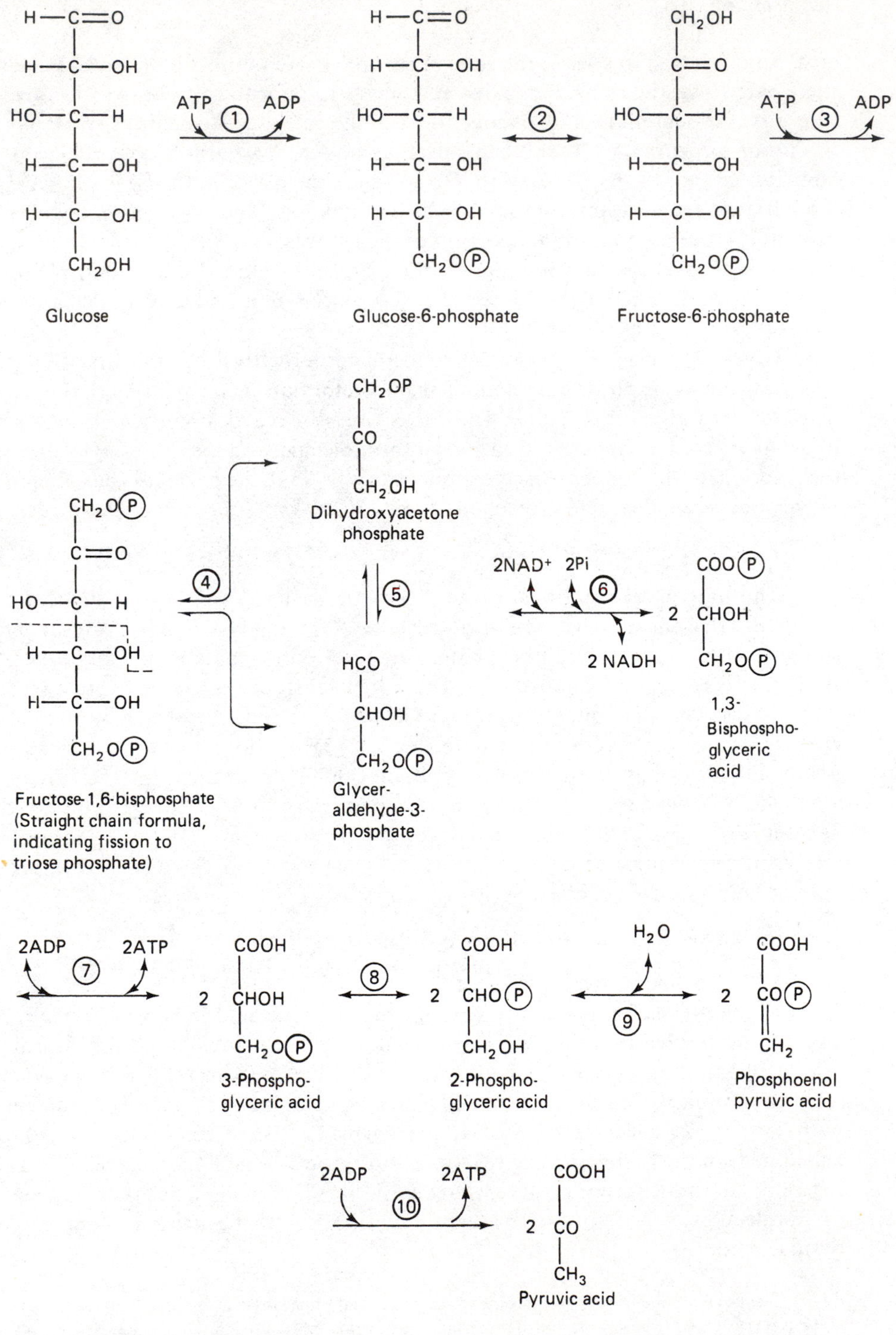

$$H-C=O$$
$$H-C-OH$$
$$HO-C-H$$
$$H-C-OH$$
$$H-C-OH$$
$$CH_2OH$$

ATP ⟶ ① ⟶ ADP

$$H-C=O$$
$$H-C-OH$$
$$HO-C-H$$
$$H-C-OH$$
$$H-C-OH$$
$$CH_2O\,\text{Ⓟ}$$

② ⟷

$$CH_2OH$$
$$C=O$$
$$HO-C-H$$
$$H-C-OH$$
$$H-C-OH$$
$$CH_2O\,\text{Ⓟ}$$

ATP ⟶ ③ ⟶ ADP

Glucose   Glucose-6-phosphate   Fructose-6-phosphate

$$CH_2OP$$
$$CO$$
$$CH_2OH$$

Dihydroxyacetone
phosphate

$$CH_2O\,\text{Ⓟ}$$
$$C=O$$
$$HO-C-H$$
$$H-C-OH$$
$$H-C-OH$$
$$CH_2O\,\text{Ⓟ}$$

④

⑤

$$HCO$$
$$CHOH$$
$$CH_2O\,\text{Ⓟ}$$

2NAD⁺ 2Pi

⑥

2 NADH

$$COO\,\text{Ⓟ}$$
$$2\ CHOH$$
$$CH_2O\,\text{Ⓟ}$$

1,3-
Bisphospho-
glyceric
acid

Fructose-1,6-bisphosphate
(Straight chain formula,
indicating fission to
triose phosphate)

Glycer-
aldehyde-3-
phosphate

2ADP ⟶ ⑦ ⟶ 2ATP

$$COOH$$
$$2\ CHOH$$
$$CH_2O\,\text{Ⓟ}$$

3-Phospho-
glyceric acid

⑧ ⟷

$$COOH$$
$$2\ CHO\,\text{Ⓟ}$$
$$CH_2OH$$

2-Phospho-
glyceric acid

$$H_2O$$

⑨

$$COOH$$
$$2\ CO\,\text{Ⓟ}$$
$$CH_2$$

Phosphoenol
pyruvic acid

2ADP ⟶ ⑩ ⟶ 2ATP

$$COOH$$
$$2\ CO$$
$$CH_3$$

Pyruvic acid

100

One of the important features of the glycolytic pathway is the production of intermediates, which may be used as starting materials for the biosynthesis of complex plant constituents. Glyceraldehyde-3-phosphate, for example, may be diverted away from the glycolytic pathway by conversion to glycerol or serine, in reactions unrelated to glycolysis. Glycerol formed in this manner may be used as a starting material for the synthesis of lipids; serine is an amino acid that may be used in the synthesis of proteins.

Because NAD^+ is consumed in glycolysis, the continued metabolism of glucose via this pathway requires the regeneration of NAD^+ from NADH. Regeneration of NAD^+ from NADH in cells of higher plants takes place in the respiratory chain (see pp. 110–113).

All reactions of the glycolytic pathway [except for the first, third, and last (see Figure 4-3)] are reversible. Thus a plant cell can synthesize fructose-1,6-bisphosphate from phosphoenolpyruvic acid by a reversal of glycolysis. Furthermore, fructose-1,6-bisphosphate, once formed, may be dephosphorylated through the activity of one of the dephosphorylating enzymes (phosphatases) present in all plant cells. The product of dephosphorylation, fructose-6-phosphate, may be converted, in turn, to glucose, sucrose, or starch (see Figure 4-1). So glycolysis represents not only a pathway of carbohydrate degradation to pyruvic acid but also one of synthesis of glucose, sucrose, and starch from intermediates of the glycolytic pathway.

OXIDATIVE PENTOSE PHOSPHATE PATHWAY

In cells of higher plants two pathways are known to catalyze the initial phase of oxidation of hexoses produced by the breakdown of sucrose and starch: (a) the glycolytic pathway (considered earlier) and (b) the oxidative pentose phosphate pathway, also sometimes called the hexose monophosphate pathway or the phosphogluconic acid pathway. (The *oxidative* pentose phosphate pathway discussed here should not be confused with the *reductive* pentose phosphate pathway; the latter accounts for the photosynthetic assimilation of CO_2 in chloroplasts and is described in Chapter 6.) Regarding the relative contributions of the oxidative pentose phosphate pathway and glycolysis to hexose oxidation, it has been estimated that glycolysis accounts for perhaps two-thirds of the total hexose oxidized and the oxidative pentose phosphate pathway for no more than a third (Stitt and ap Rees, 1978).

The enzymes of the oxidative pentose phosphate pathway are located in the soluble portion of the cytoplasm (i.e., the cytosol) of plant cells, in chloroplasts, and (in developing seeds) in proplastids, just as are the enzymes of glycolysis (ap Rees, 1980). The oxidative pentose phosphate pathway is shown in Figure 4-4. Assuming again, as

Figure 4-3 Reactions of the glycolytic pathway. The symbol Ⓟ is used to designate the phosphate group ($-H_2PO_3$) whereas Pi denotes phosphoric acid (H_3PO_4). All sugars have the D configuration. The enzymes are 1, hexokinase; 2, phosphohexokinase; 3, phosphofructokinase; 4, aldolase; 5, triose phosphate isomerase; 6, triose phosphate dehydrogenase; 7, phosphoglyceryl kinase; 8, phosphoglyceromutase; 9, enolase; 10, pyruvate kinase. (*Note:* The formulas of the several acids are shown in the un-ionized form, but actually exist in plant cells as ions, e.g., pyruvate, 3-phosphoglycerate, etc.)

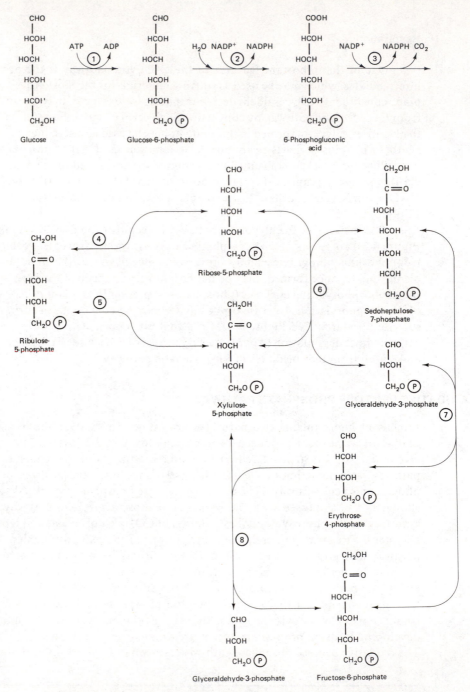

Figure 4-4 Oxidative pentose phosphate pathway; reactions start with glucose. The symbol (P) designates the phosphate group (—H_2PO_3). The enzymes are 1, hexokinase; 2, glucose-6-phosphate dehydrogenase; 3, 6-phosphogluconic acid dehydrogenase; 4, phosphoriboisomerase; 5, phosphoketopentose isomerase; 6, 7, and 8, transketolases.

in the glycolytic pathway, that glucose is the starting material for respiratory metabolism, the pathway begins with the phosphorylation of glucose by ATP, just as in glycolysis; the product is glucose-6-phosphate. Then the aldehyde group (—CHO) in glucose-6-phosphate is converted to a carboxyl group (—COOH) by addition of water to the aldehyde group and dehydrogenation of the intermediate addition product, in a manner analogous to that indicated earlier [see reaction (4-10)]. The coenzyme is $NADP^+$. The reaction is

$$\text{Glucose-6-phosphate} + H_2O + NADP^+ \longrightarrow$$
$$\text{6-phosphogluconic acid} + NADPH + H^+ \qquad (4\text{-}23)$$

The product of reaction (4-23), 6-phosphogluconic acid, undergoes *oxidative decarboxylation* (i.e., oxidation accompanied by the release of CO_2 from the carboxyl group). Again the oxidizing agent is $NADP^+$. The reaction is

$$\text{6-Phosphogluconic acid} + NADP^+ \longrightarrow$$
$$\text{ribulose-5-phosphate} + CO_2 + NADPH + H^+ \qquad (4\text{-}24)$$

The product of reaction (4-24) is a phosphorylated derivative of ribulose, a pentose sugar.

Following the formation of ribulose-5-phosphate, subsequent reactions are nonoxidative and include several isomerizations and group transfers. Moreover, the oxidative reactions [reactions (4-23) and (4-24)] leading to the formation of ribulose-5-phosphate are irreversible whereas the nonoxidative reactions starting with ribulose-5-phosphate are reversible. Starting with ribulose-5-phosphate, the sequence of reactions is as follows. Ribulose-5-phosphate undergoes transformation to two other phosphorylated pentoses—xylulose-5-phosphate and ribose-5-phosphate (see Figure 4-4). Then these two phosphorylated pentoses react with each other to form equimolar amounts of glyceraldehyde-3-phosphate (a triose-phosphate) and sedo-heptulose-7-phosphate (a septose-phosphate). These latter two compounds react together to form equimolar amounts of erythrose-4-phosphate (a tetrose-phosphate) and fructose-6-phosphate (a hexose-phosphate). Finally, erythrose-4-phosphate reacts with xylulose-5-phosphate to form equimolar amounts of fructose-6-phosphate and glyceraldehyde-3-phosphate.

Although glyceraldehyde-3-phosphate and fructose-6-phosphate are the end products of the oxidative pentose phosphate pathway (see Figure 4-4), both of these compounds also are members of the glycolytic pathway (see Figure 4-3). Therefore, once produced by the oxidative pentose phosphate pathway, they may be further metabolized by the glycolytic pathway. Still another sequence suggested recently to occur in chloroplasts (Kow et al., 1982) (see Table 4-3) is cyclic—the starting material, glucose-6-phosphate, is regenerated by means of reactions b, c, d, and e in Table 4-3—and represents a means of oxidizing glucose-6-phosphate [and substances which may be converted to glucose-6-phosphate, e.g., starch (see Figure 4-1)] to CO_2 and reduced coenzyme. But of course, when a coenzyme is reduced, the reduced coenzyme must be oxidized before a process such as shown in Table 4-3 can continue.

TABLE 4-3 OXIDATION OF GLUCOSE-6-PHOSPHATE TO CO_2 AND REDUCED COENZYME BY THE OXIDATIVE PENTOSE PHOSPHATE PATHWAY. THE REACTION SEQUENCE BEGINS WITH 6 MOLECULES OF GLUCOSE-6-PHOSPHATE. [The symbol P designates the phosphate group ($-H_2PO_3$) and Pi designates phosphoric acid (H_3PO_4).]

	Reaction numbers in Figure 4-4
(a) 6 glucose-6-P + 12 NADP$^+$ + 6 H_2O $\longrightarrow$ 6 CO_2 + 12 NADPH $\qquad$ + 6 ribulose-5-P + 12 H$^+$	2 and 3
(b) 4 ribulose-5-P $\longrightarrow$ 2 fructose-6-P + 2 erythrose-4-P	4, 5, 6, 7
(c) 2 ribulose-5-P + 2 erythrose-4-P $\longrightarrow$ 2 fructose-6-P $\qquad$ + 2 glyceraldehyde-3-P	5 and 8
(d) 4 fructose-6-P $\longrightarrow$ 4 glucose-6-P	glycolysis
(e) 2 glyceraldehyde-3-P + H_2O $\longrightarrow$ glucose-6-P + Pi	glycolysis

Sum: glucose-6-P + 7 H_2O + 12 NADP$^+$ $\longrightarrow$ 6 CO_2 + 12 NADPH + 12 H$^+$ + Pi

We will see on page 262 that one of the ways to oxidize reduced coenzyme (i.e., NADPH) generated by the sequence of reactions shown in Table 4-3 is by the reduction of nitrite to ammonia. The coupling of glucose-6-phosphate oxidation (see summary reaction in Table 4-3) to nitrite reduction (through NADP$^+$/NADPH interconversion) was proposed recently by Kow et al., 1982.

Whereas the oxidation-reduction coenzyme in glycolysis is nicotinamide-adenine dinucleotide (NAD$^+$/NADH), that in the oxidative pentose phosphate pathway is nicotinamide-adenine dinucleotide phosphate (NADP$^+$/NADPH). NADH formed in glycolysis is oxidized in the respiratory chain (pp. 110–113), but NADPH formed in the oxidative pentose phosphate pathway is not. Instead the function of NADPH is to supply reducing power required specifically by many enzymes that carry out reductive biosynthetic reactions [e.g., biosynthesis of lignin precursors (ap Rees, 1980)]. Utilization of NADPH in reductive biosynthetic reactions serves to regenerate NADP$^+$, which is required for the continued operation of the oxidative pentose phosphate pathway.

The formation of NADPH (in contrast to NADH in glycolysis) is an important function of the oxidative pentose phosphate pathway. A second function is the production of intermediates that may serve as starting materials for the biosynthesis of more complex cell constituents. Erythrose-4-phosphate, for example, may be used as the starting material for the biosynthesis of complex phenolic compounds; pentose sugars (e.g., ribose-5-phosphate and xylulose-5-phosphate) are used for the biosynthesis of nucleotides and nucleic acids. The oxidation of glucose-6-phosphate and substances convertible into glucose-6-phosphate (e.g., starch) is still another function of the oxidative pentose phosphate pathway (see Table 4-3 and related text discussion). Finally, it should be noted that the oxidative pentose phosphate pathway does not generate ATP, as does the glycolytic pathway.

ANAEROBIC RESPIRATION

Pyruvic acid produced via the glycolytic pathway (or indirectly via the oxidative pentose phosphate pathway) may undergo different metabolic fates in cells of higher plants, depending on whether O_2 is present. Under aerobic conditions (i.e., when O_2 is present) mitochondrial enzyme systems catalyze the oxidation of pyruvic acid to CO_2 and water. This is the usual fate of pyruvic acid in tissues of higher plants. On the other hand, plant tissues in atmospheres devoid of O_2 convert pyruvic acid to CO_2 and ethyl alcohol.

Tissues of higher plants deprived of O_2 generally are capable of surviving at least for a few hours. In fact, *anaerobiosis* (i.e., life in the absence of O_2) may occur in certain tissues even under normal growing conditions. Certain seeds, for example, are subject to anaerobiosis during the first several hours of germination when immersed in water. The intact testa of pea seeds has been found to limit the uptake of O_2 by the developing embryo; only when the testa is split by the emerging radicle is O_2 able to diffuse to the embryo (Spragg and Yemm, 1959).

In plant cells maintained under anaerobic conditions pyruvic acid produced from glucose via the glycolytic pathway is converted to CO_2 and ethyl alcohol. Ethyl alcohol accumulates within these cells because it is metabolized rather sluggishly. The anaerobic breakdown of glucose to CO_2 and ethyl alcohol is a form of respiration referred to as *fermentation*. Fermentation (literally, a chemical change accompanied by effervescence) is the type of respiration carried on normally by yeast cells and accounts for the production of alcohol in alcoholic beverages and the production of CO_2 needed for the rising of bread and bread products.

The conversion of pyruvic acid to CO_2 and ethyl alcohol in higher plant cells under anaerobic conditions is catalyzed by soluble cytosolic enzymes. The first step is the decarboxylation of pyruvic acid (i.e., a release of CO_2 from pyruvic acid). The enzyme that catalyzes this decarboxylation reaction is pyruvic acid decarboxylase.

$$\begin{matrix} \text{COOH} \\ | \\ \text{C} = 0 \\ | \\ \text{CH}_3 \end{matrix} \longrightarrow \text{CO}_2 + \underset{\text{Acetaldehyde}}{\text{CH}_3\text{CHO}} \qquad (4\text{-}25)$$

Pyruvic
acid

In the second step, the product of reaction (4-25) is reduced to ethyl alcohol. The reducing agent is NADH and the enzyme catalyzing the reaction is alcohol dehydrogenase.

$$\underset{\text{Acetaldehyde}}{\text{CH}_3\text{CHO}} + \text{NADH} + \text{H}^+ \longrightarrow \underset{\text{Ethyl alcohol}}{\text{CH}_3\text{CH}_2\text{OH}} + \text{NAD}^+ \qquad (4\text{-}26)$$

In the production of CO_2 and ethyl alcohol from glucose by fermentation, NADH, which is produced in glycolysis—in the dehydrogenation of glyceraldehyde-3-phosphate (see Figure 4-3)—is consumed in equimolar amounts in the reduction of acetaldehyde to ethyl alcohol [reaction (4-26)]. Thus there is a closed system with respect to this coenzyme, as indicated in Figure 4-5.

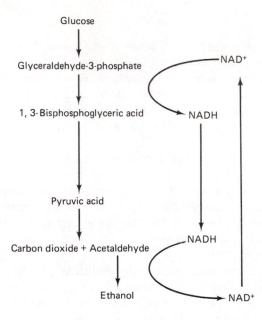

Figure 4-5 Abbreviated outline of the pathway of alcohol fermentation. Because nicotinamide-adenine dinucleotide is oxidized and reduced over and over again, only a trace amount is needed in the conversion of large quantities of glucose to CO_2 and ethyl alcohol.

The summary reaction for fermentation is as follows:

$$C_6H_{12}O_6 \longrightarrow 2\,C_2H_5OH + 2\,CO_2 \tag{4-27}$$

Glucose   Ethyl alcohol

There are several differences between fermentation and aerobic respiration. Fermentation accounts for only a partial breakdown of glucose (i.e., to ethyl alcohol and CO_2). In aerobic respiration, on the other hand, pyruvic acid is oxidized completely to CO_2 and water. Another difference between fermentation and the aerobic respiration of glucose is the yield of ATP. In fermentation there is a net gain of only two molecules of ATP for each molecule of glucose degraded to pyruvic acid via the glycolytic pathway; this fact was noted earlier. But we will see that many more molecules of ATP are generated when glucose is oxidized completely to CO_2 and water under aerobic conditions. Finally, NADH is oxidized to NAD^+ rather sluggishly in fermentation; vigorous oxidation of NADH occurs only in mitochondria in aerobic respiration.

TRICARBOXYLIC ACID CYCLE

Earlier we saw that the respiratory metabolism of glucose yields pyruvic acid. Now attention must be given to the metabolism of pyruvic acid under aerobic conditions.

In normal circumstances aerobic conditions prevail in plant cells (i.e., O_2 is present). Under these conditions pyruvic acid is oxidized to CO_2 and water in cellular organelles known as mitochondria. Mitochondria have all the enzymes necessary for the complete oxidation of pyruvic acid to CO_2 and water.

The metabolic conversion of pyruvic acid to CO_2 and water proceeds through a pathway that consists of a series of dicarboxylic and tricarboxylic acids. These intermediate acids are consumed and regenerated continually in a sequence of reactions that takes place in a cyclic manner. This cyclic sequence of reactions was described in broad outline in the 1930s and is known as the *tricarboxylic acid cycle* [or the *citric acid cycle* because citric acid (a tricarboxylic acid) is one of the intermediates]. The tricarboxylic acid (TCA) cycle is sometimes referred to as the *Krebs cycle* in honor of the English biochemist who first recognized its functional role. The reactions of the TCA cycle are shown in Figure 4-6. All the enzymes of TCA cycle are in the mitochondrial *matrix*, which is bounded by the inner mitochondrial membrane.

Before it enters the TCA cycle, pyruvic acid undergoes oxidative decarboxylation in a coordinated series of reactions catalyzed by a multicomponent complex of several enzymes that form one supramolecular unit. This enzyme complex is called the pyruvic acid dehydrogenase complex or simply pyruvic dehydrogenase. The reactions catalyzed by pyruvic dehydrogenase require the participation of several coenzymes, including NAD^+ and coenzyme A (CoA). (For the formula of coenzyme A, see Chapter 3.) The overall reaction catalyzed by pyruvic dehydrogenase is

$$\text{Pyruvic acid} + \text{CoA} + \text{NAD}^+ \longrightarrow \text{acetyl CoA} + CO_2 + \text{NADH} + H^+ \qquad (4\text{-}28)$$

The TCA cycle starts with the condensation of the two-carbon acetyl group (CH_3CO-) in acetyl coenzyme A—the product of reaction (4-28)—with oxaloacetic acid (a four-carbon acid) and water to yield citric acid (a six-carbon acid). As shown in Figure 4-6, citric acid undergoes isomerization to isocitric acid; this process is accomplished by a dehydration step followed by a hydration step, the result being an interchange of an H and OH. Each of these two steps is catalyzed by one enzyme (aconitase); the intermediate is *cis*-aconitic acid. Then isocitric acid undergoes oxidative decarboxylation to α-ketoglutaric acid; NAD^+ is the oxidizing agent.[3]

The next step in the TCA cycle (see Figure 4-6) is the conversion of α-ketoglutaric acid to succinyl coenzyme A. This conversion is similar to the conversion of pyruvic acid to acetyl coenzyme A [see reaction (4-28)] and consists of a coordinated series of subreactions catalyzed by a multicomponent complex of several enzymes. The overall reaction is

$$\alpha\text{-Ketoglutaric acid} + \text{CoA} + \text{NAD}^+ \longrightarrow$$
$$\text{succinyl CoA} + CO_2 + \text{NADH} + H^+ \qquad (4\text{-}29)$$

The product of reaction (4-29), succinyl coenzyme A, is cleaved to succinic acid and coenzyme A in a reaction that requires the participation of guanosine diphosphate (GDP), phosphoric acid (Pi), and guanosine triphosphate (GTP).

$$\text{Succinyl-CoA} + \text{GDP} + \text{Pi} \longrightarrow \text{succinic acid} + \text{GTP} + \text{CoA} \qquad (4\text{-}30)$$

The formulas of GDP and GTP were considered in Chapter 3.

---

[3]There are two isocitric dehydrogenases, one specific to NAD and the other specific to NADP. The NAD-specific isocitric dehydrogenase is an intramitrochondrial enzyme functional in the TCA cycle. The NADP-linked enzyme is present in extramitochondrial cytoplasm and has a different metabolic role.

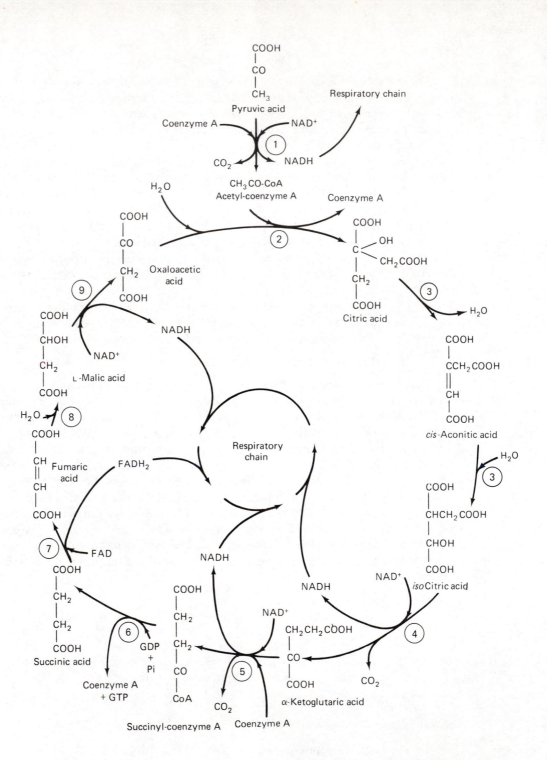

COOH
|
CO
|
CH$_3$
Pyruvic acid

Respiratory chain

Coenzyme A — NAD$^+$

①

CO$_2$ — NADH

H$_2$O

CH$_3$CO-CoA
Acetyl-coenzyme A

Coenzyme A

COOH
|
CO
|
CH$_2$ Oxaloacetic
| acid
COOH

②

COOH
|
C—OH
| \
| CH$_2$COOH
CH$_2$
|
COOH
Citric acid

③

H$_2$O

⑨

COOH
|
CHOH
|
CH$_2$
|
COOH
L-Malic acid

NADH

NAD$^+$

COOH
|
CCH$_2$COOH
||
CH
|
COOH
cis-Aconitic acid

③

H$_2$O

H$_2$O — ⑧

COOH
|
CH   Fumaric
|| acid
CH
|
COOH

FADH$_2$

Respiratory
chain

NADH

NAD$^+$

COOH
|
CHCH$_2$COOH
|
CHOH
|
COOH
isoCitric acid

⑦

FAD

COOH
|
CH$_2$
|
CH$_2$
|
COOH
Succinic acid

⑥

GDP
+
Pi

Coenzyme A
+ GTP

COOH
|
CH$_2$
|
CH$_2$
|
CO
|
CoA
Succinyl-coenzyme A

⑤

NADH

NAD$^+$

CO$_2$

Coenzyme A

COOH
|
CH$_2$
|
CH$_2$CH$_2$COOH
|
CO
|
COOH
α-Ketoglutaric acid

NADH

NAD$^+$

④

CO$_2$

In a reaction not part of the TCA cycle, GTP formed in reaction (4-30) donates its terminal phosphate group to ADP to form ATP; the enzyme that catalyzes this reaction is nucleoside diphosphokinase.

$$GTP + ADP \longrightarrow GDP + ATP \qquad (4\text{-}31)$$

Succinic acid, the product of reaction (4-30), is oxidized to fumaric acid; the enzyme catalyzing this reaction (i.e., succinic dehydrogenase) is a nonheme iron protein that has as its reactive group the coenzyme flavin-adenine dinucleotide [see reactions (4-14) and (4-16)]. Next, fumaric acid is converted to malic acid by the addition of water. Finally, malic acid is oxidized to oxaloacetic acid. All reactions of the TCA cycle are reversible except the first (i.e., the condensation of acetyl-CoA, water, and oxaloacetic acid to yield citric acid); thus the TCA cycle proceeds in the direction of regeneration of oxaloacetic acid (i.e., clockwise in Figure 4-6).

In summary, there are four oxidative reactions in the TCA cycle (see Figure 4-6). In each, two electrons are removed from the substance that undergoes oxidation. The electrons are transferred via NADH and $FADH_2$ to O_2 through the respiratory chain (discussed below). Also, CO_2 is produced in two decarboxylation reactions in the cycle For one turn of the cycle, one molecule of acetic acid (in the form of acetyl coenzyme A) is oxidized according to the following overall reaction.

$$CH_3COOH + 2\,O_2 \longrightarrow 2\,CO_2 + 2\,H_2O \qquad (4\text{-}32)$$

The O_2 shown in reaction (4-32) is taken up in the respiratory chain; no O_2 is taken up in the TCA cycle itself.

In addition to its role in the degradation of acetic acid to CO_2 and water, the TCA cycle has another function. Several of the intermediates of the cycle are starting materials for the biosynthesis of some of the constituents of plant cells. Oxaloacetic acid, for instance, is the starting material for the biosynthesis of aspartic acid, an amino acid; a few other amino acids are synthesized from aspartic acid. Another intermediate of the cycle, α-ketoglutaric acid, is the precursor for the biosynthesis of glutamic acid and several other amino acids. Still another intermediate of the cycle, succinyl coenzyme A, is a precursor of porphyrins.

The continued oxidation of acetic acid (as acetyl coenzyme A) via the TCA cycle requires the continued replenishment of oxaloacetic acid, the first member of the cycle. One way that this step may be accomplished is as follows. When intermediates of the TCA cycle are diverted away from the cycle—to be used as starting materials for biosynthesis of cellular constituents in plant cells—then oxaloacetic acid is replenished

---

Figure 4-6   Reactions of the tricarboxylic acid cycle. Abbreviations: GDP, guanosine diphosphate; GTP, guanosine triphosphate; Pi, phosphoric acid (H_3PO_4); CoA, coenzyme A. The enzymes are 1, pyruvic dehydrogenase (a multicomponent "complex" of several enzymes); 2, citric acid synthetase; 3, aconitase; 4, isocitric acid dehydrogenase; 5, α-ketoglutaric acid dehydrogenase (a multicomponent "complex" of several enzymes); 6, succinyl-coenzyme A synthetase; 7, succinic acid dehydrogenase; 8, fumarase; 9, malic acid dehydrogenase. (*Note*: The formulas of the several acids are shown in the un-ionized form but actually exist in plant cells as ions—for instance, succinate, malate, isocitrate.)

by an *anaplerotic* (Latin, *ana*, again; *pleroticus*, filling up: literally, a "filling in") reaction catalyzed by phosphoenol pyruvic acid carboxylase, a cytosolic enzyme.

$$
\begin{array}{ccc}
\text{COOH} & & \text{COOH} \\
| & & | \\
\text{C}-\text{O}\textcircled{P} + \text{HCO}_3^- \longrightarrow & \text{C}=\text{O} & + \text{H}_2\text{PO}_4^- \\
\| & & | \\
\text{CH}_2 & & \text{CH}_2 \\
& & | \\
& & \text{COOH}
\end{array}
\qquad (4\text{-}33)
$$

| Phosphoenol pyruvic acid | Bicarbonate ion | Oxaloacetic acid | Phosphate ion |

In view of the fact that phosphoenol pyruvic acid is a member of the glycolytic pathway (see Figure 4-3), it is clear that three of the four carbon atoms in oxaloacetic acid in reaction (4-33) are obtained from members of the glycolytic pathway (and/or the oxidative phosphate pentose pathway). The fourth carbon in oxaloacetic acid in this reaction is supplied by either respiratory CO_2 or CO_2 in the external atmosphere; CO_2 becomes bicarbonate ion. Also note that reaction (4-33) accounts for the initial fixation of atmospheric CO_2 in C_4 photosynthesis (see reaction 6-8).

In addition to the need to replenish oxaloacetic acid, the continued operation of the TCA cycle requires NAD^+ be regenerated from NADH and FAD be regenerated from $FADH_2$. The regeneration of these two cofactors takes place in the respiratory chain.

RESPIRATORY CHAIN

We have already seen that NAD^+ accepts electrons from specific organic acids in the TCA cycle (Figure 4-6) and also from glyceraldehyde-3-phosphate in the glycolytic pathway (Figure 4-1). As a result, NAD^+ is reduced to NADH. But NADH is only a temporary storage place for electrons. The reoxidation of NADH to NAD^+ under aerobic conditions in plant cells occurs by transfer of electrons from NADH to O_2 via a system of mitochondrial enzymes and electron carriers referred to as the *respiratory chain*.

The respiratory chain also transfers electrons from succinic acid (via the coenzyme flavin-adenine dinucleotide) to O_2. Thus succinic acid is oxidized to fumaric acid. (Both succinic and fumaric acids, it will be recalled, are members of the TCA cycle.)

The two electron transfer reactions catalyzed by the respiratory chain were considered earlier [see reactions (4-19) and (4-20)] and each is repeated here except that each is divided by two simply to indicate the fate of one molecule of substrate (NADH or succinic acid).

$$
\text{NADH} + \tfrac{1}{2}\text{O}_2 + \text{H}^+ \longrightarrow \text{H}_2\text{O} + \text{NAD}^+ \qquad (4\text{-}34)
$$

$$
\underset{\text{Succinic acid}}{\text{COOHCH}_2\text{CH}_2\text{COOH}} + \tfrac{1}{2}\text{O}_2 \longrightarrow \text{H}_2\text{O} + \underset{\text{Fumaric acid}}{\text{COOHCH}=\text{CHCOOH}} \qquad (4\text{-}35)
$$

The respiratory chain consists of a number of individual components, each capable of rapid and reversible oxidation and reduction and thus acting as an electron carrier. At one end of the chain NADH (or succinic acid) donates electrons to the oxidized form of the first carrier. By accepting electrons, the first carrier is reduced. Reoxidation of the first carrier is accomplished by reaction with a second carrier. The oxidized form of the second carrier accepts electrons from the first carrier and so is reduced. This process of sequential oxidation and reduction of individual carriers proceeds along the chain until electrons from the terminal carrier are passed to O_2, which acts as the final electron acceptor and is reduced to water. In effect, the respiratory chain is an *electron transport system* that consists of several separate steps involving coupled oxidation-reduction pairs. Only the last component interacts with O_2. [*Note*: Hydrogen ions required for the reduction of O_2 to water (see reaction 4-11) are supplied by oxidation of NADH (see reaction 4-12) and/or succinic acid (see reaction 4-9); in the latter reaction recall that a hydrogen atom consists of a hydrogen ion and an electron (pp. 94–95).]

The sequence of reactions in the respiratory chain can be depicted with curved arrows.

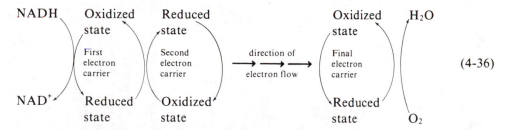

$$(4\text{-}36)$$

The curved arrows emphasize that the individual components of the electron transport system act in a cyclical manner; each is alternately oxidized and reduced over and over again. When NADH (or succinic acid) is produced by mitochondria in a plant cell and O_2 is not limiting, the transfer of electrons proceeds along the respiratory chain.

Many other electron transport systems, in addition to the respiratory chain, are known to have important roles in cell metabolism. Some are mentioned later in this chapter and elsewhere (e.g., see Chapter 9). Curved arrows [as in reaction (4-36)] will be used to indicate the sequence of oxidation-reduction reactions.

The enzymes and electron (and hydrogen) carriers of the respiratory chain are located in the inner membrane of the mitochondrion, and include flavins, several nonheme iron compounds including soluble iron-sulfur proteins, quinones, and hemes. Some of these compounds are prosthetic groups of enzymes.

Complete unanimity of opinion among plant physiologists regarding the exact sequence of events involved in electron transport in the respiratory chain does not yet exist. But the sequence shown in Figure 4-7 is regarded as probable. The first reaction is the oxidation of NADH (by NADH dehydrogenase) [and/or succinic acid (by succinic acid dehydrogenase)]. Both enzymes are flavoproteins and also contain nonheme iron prosthetic groups, the first has FMN as its prosthetic group, the second

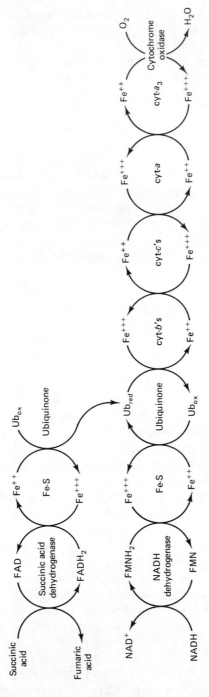

Figure 4-7 A schematic representation of the respiratory chain in mitochondria of higher plants. Succinic acid is oxidized to fumaric acid and/or NADH is oxidized to NAD$^+$. The final electron acceptor in this pathway is O$_2$, which is reduced to water. There are several cytochromes b (indicated as cyt b's), and more than one cytochrome c (indicated as cyt c's). [*Note*: For evidence that an additional flavoprotein component (not shown here) is present between ubiquinone and cytochromes b, consult Palmer, 1976, p. 144.]

FAD. Two electrons are transferred from NADH (or succinic acid) to flavin groups [e.g., see reaction (4-16)]. Then these electron pairs are passed sequentially to iron-sulfur proteins (which also contain nonheme iron groups) and then to ubiquinone, the formula of which was given in Table 4-2.

The electron carriers between ubiquinone and O_2 are *cytochromes* (see Figure 4-7). The cytochromes are proteins in which iron is present in a porphyrin ring (pp. 74–75). Because the iron-porphyrin complex is called heme, the cytochromes are called hemoproteins. The heme group is a one-electron carrier [i.e., Fe^{2+} (ferrous) $\rightleftharpoons$ Fe^{3+} (ferric form) $+ e^-$], in contrast to the flavins, Fe—S proteins, and ubiquinone, all of which are two-electron carriers. It is at ubiquinone that the respiratory chain changes from a two-electron to a one-electron transport system (see Table 4-2 and related text discussion).

Cytochromes are classified in several groups (e.g., cytochromes a, cytochromes b, cytochromes c), according to the structure and properties of the heme group. Different cytochromes within each group are identified on the basis of differences in light absorption spectra. There are two cytochromes a and several cytochromes b and cytochromes c in the respiratory chain of plant mitochondria. In addition to the mitochondrial cytochromes, other cytochromes are located in chloroplasts and function in electron transport in photosynthesis (see Chapter 6). Also, certain other cytochromes are present in the endoplasmic reticulum.

Only the terminal cytochrome, cytochrome a_3, reacts with O_2 (see Figure 4-7). Attempts to separate cytochrome a_3 from cytochrome a have not been successful; cytochrome a and cytochrome a_3 together constitute the enzyme known as cytochrome c oxidase. (Cytochrome c oxidase is often called cytochrome oxidase; we will adopt this convention.) Cytochrome oxidase catalyzes the transfer of electrons from the last cytochrome c in the respiratory chain to O_2. This reaction was written earlier [reaction (4-18)] and is repeated here.

$$4 \text{ Cyt } c(Fe^{2+}) + O_2 + 4 \text{ H}^+ \xrightarrow[\text{oxidase}]{\text{cytochrome}} 2 \text{ H}_2O + 4 \text{ Cyt } c(Fe^{3+})$$

Water produced in this and other metabolic reactions associated with respiration [e.g., see reactions (4-19) and (4-20)] is often called *respiratory water*. Respiratory water represents only a small fraction of the total water in plant cells.

The enzyme cytochrome oxidase, unlike other cytochrome pigments, contains two atoms of copper in addition to two molecules of heme iron. Both heme iron and copper undergo reversible changes in their oxidation state ($Cu^+ \rightleftharpoons Cu^{2+} + e^-$, $Fe^{2+} \rightleftharpoons Fe^{3+} + e^-$) during electron transport by cytochrome oxidase. Cytochrome oxidase has a high affinity for O_2; in fact, this enzyme is saturated at O_2 levels of 2% or less. [It will be recalled (pp. 85–86) that the natural level of O_2 in the atmosphere is 21%.]

Even though cytochromes are present in plant cells in extremely minute amounts, it is possible with suitable instrumentation to study them in vivo. Young root tips or very thin slices of young storage tissues provide suitable material; such tissues are fairly transparent and do not contain such pigments as chlorophylls, which mask the cytochromes. When these tissues are examined in a low-dispersion spectroscope (an instrument that disperses white light into a spectrum), cytochromes in the reduced form can be recognized by a display of three narrow, well-defined absorption bands in the visible region of the spectrum. The reduced forms of the cytochromes arise when O_2 is withheld from the tissues (e.g., by flushing with N_2). Under these conditions the relatively large amounts of reduced substrates normally present in cells serve to keep all respiratory chain components, including the cytochromes, in the reduced state. On the other hand, when the tissue is exposed to O_2, the cytochromes are oxidized and the absorption bands become faint and indistinct instead of sharp and well defined.

The oxidized and reduced cytochromes can also be examined in living tissues with a sensitive spectrophotometer. Spectrophotometers differ from spectroscopes in that radiation of any desired narrow range of wavelengths can be isolated and focused on the material to be examined. The transmitted radiation is detected by a sensitive phototube, where it produces a photocurrent that is amplified electronically and recorded. Use of a spectrophotometer instead of a spectroscope enables the investigator to obtain far more precise and accurate quantitative data.

Evidence for the order in which the individual components function in the respiratory chain (see Figure 4-7) has been obtained in several ways. One guideline is the measurement of redox potentials (p. 97). Because each individual component is capable of alternate oxidation and reduction, it has been possible to measure the redox potentials of those components that can be isolated. This information has been useful in assigning individual components to definite positions in the respiratory chain.

Another important tool for investigating the sequence of reactions in the respiratory chain is the use of specific poisoning agents (i.e., inhibitors). Several substances (e.g., cyanide ion, carbon monoxide, certain steroids, the insecticide rotenone, the antibiotic antimycin A, and the barbituate drug sodium amytal) are capable of blocking electron flow because they react with one or another of the individual components in the respiratory chain. Each inhibitor has the effect of dividing the respiratory chain into segments on the two sides of the specific site of inhibition. Thus these inhibitors permit a study of the components involved in *isolated segments* of the respiratory chain.

Sophisticated photoelectronic techniques have been applied to the design of instrumentation for recording spectra of individual electron carriers as they function in intact mitochondria. These techniques make it possible to determine, with millisecond time resolution, which of the components of the respiratory chain are in successively more oxidized forms in an aerobic suspension of mitochondria when NADH (or succinic acid) is added. Alternatively, when O_2 is admitted to an anaerobic suspension of mitochondria in the presence of excess NADH (or succinic acid), the rate and sequence of reoxidation of the individual components can be measured. This method of measuring the time sequence of reduction of the individual components of a fully oxidized chain (or oxidation of components of a fully reduced chain) has been useful in determining the reaction sequences in the respiratory chain.

A fourth method of determining the sequence of reactions in the respiratory chain is by comminution of mitochondria isolated from plant tissues, the object being to prepare functional fractions free of one another and of structural proteins. Isolated mitochondria can be comminuted by a variety of techniques, including sonication (exposure to high-frequency sound waves), extraction with organic solvents, treatment with detergents and dispersing agents, brief exposure to solutions of alkali or other corrosive chemicals, mechanical shaking and grinding, and so forth. Homogeneous preparations obtained after these treatments can be studied separately with respect to their properties. Also, it is possible to obtain much valuable information concerning the respiratory chain by adding together the separated fractions and studying the properties of the "reconstituted" systems.

SIGNIFICANCE OF ATP

We saw in preceding sections that the pathway of respiratory metabolism starts with the conversion of glucose to pyruvic acid by glycolytic enzymes (and/or enzymes of the oxidative pentose phosphate pathway) located in the soluble portion of the cytoplasm of plant cells. Pyruvic acid is degraded to acetic acid and the latter to CO_2 and water by TCA enzymes located within the matrix of mitochondria. The respiratory chain in the inner membrane of mitochondria accounts for the oxidation of NADH and succinic acid and the reduction of O_2 to water; moreover, during these latter processes ATP is generated from ADP. The significance of ATP will be considered first. The generation of ATP is the subject of the next section. [The letter P in ATP designates the phosphate group: $-H_2PO_3$. The formulas of ATP (adenosine triphosphate) and ADP (adenosine diphosphate) were discussed in Chapter 3.]

ATP can be regarded as the energy "currency" of living cells. The energy *liberated* in the hydrolysis of the terminal phosphate group of ATP to yield ADP is utilized in living cells to drive reactions in which energy is *consumed*. How this process takes place can be illustrated by the formation of glucose-6-phosphate from glucose in living cells [see reactions (4-37), (4-38), and (4-39) below].

First, consider the direct phosphorylation of glucose to glucose-6-phosphate by phosphoric acid (Pi).

$$\text{Glucose} + \text{Pi} \longrightarrow \text{glucose-6-phosphate} + H_2O \qquad (4\text{-}37)$$

Reaction (4-37) will not proceed of its own accord as written (i.e., from left to right); no enzyme will catalyze this reaction. The energy requirement for reaction (4-37) has been measured and is +3100 calories per mole of reactants at 25° C; this value is called the free energy change for the reaction.

Reactions for which the free energy change is *positive* [e.g., reaction (4-37)] will not proceed of their own accord. But when the free energy change of a reaction is *negative*, the reaction will proceed. (Reactions for which the free energy change is near zero are reversible and will proceed either from left to right or right to left.)

Now consider the hydrolysis of the terminal phosphate group of ATP. This reaction

$$\text{ATP} + H_2O \longrightarrow \text{ADP} + \text{Pi}, \qquad (4\text{-}38)$$

catalyzed by the enzyme ATPase, proceeds of its own accord with liberation of energy; the free energy change is -7800 calories per mole at $25°$ C.

Reactions (4-37) and (4-38) can be added together; their free energy values are also additive.

$$\text{Glucose} + \text{ATP} \longrightarrow \text{glucose-6-phosphate} + \text{ADP} \tag{4-39}$$

The free energy change in reaction (4-39) [which was written earlier; see reaction (4-21)] is -4700 calories per mole at $25°$ C $(+3100 - 7800 = -4700)$.

Reaction (4-39) has a negative free energy change. Therefore it proceeds of its own accord from left to right in the presence of the appropriate enzyme. In effect, the free energy liberated in the hydrolysis of ATP [reaction (4-38)] is utilized to drive the synthesis of glucose-6-phosphate [reaction (4-37)].

It was the American biochemist F. Lipmann who first realized that the free energy of hydrolysis of ATP [reaction (4-38)] was large and negative and this property permitted reaction (4-38) to be coupled to and drive energy-consuming reactions [e.g., reaction (4-37)]. Of course, coupling of this type must be catalyzed by an appropriate enzyme—hexokinase is the enzyme that catalyzes reaction (4-39)—but it is the large negative free energy change in reaction (4-39) that is fundamentally important. Unless the free energy change in a coupled reaction [e.g., reaction (4-39)] is negative, the reaction will not proceed no matter how much enzyme is present.

The term "high-energy bond" or "energy-rich bond" is sometimes used to designate the terminal phosphate bond of ATP. This usage is incorrect, however. The free energy change in reaction (4-38) refers not to chemical bond energy but to the free energy of hydrolysis of ATP.

Reactions in which ATP is utilized in living cells can be generalized as follows:

$$\text{ATP} + \text{X} \longrightarrow \text{X—}\textcircled{P} + \text{ADP} \tag{4-40}$$

where X is the substance that is phosphorylated and X—$\textcircled{P}$ is the phosphorylated derivative of X. (The symbol $\textcircled{P}$ refers to the phosphate group: $-H_2PO_3$). When ATP donates its terminal phosphate group to an acceptor [e.g., X in reaction (4-40)], ADP is formed.

The specific role of ATP in living cells is essentially that of a phosphorylating agent. By phosphorylating metabolites with which ATP is able to react (i.e., by donating its terminal phosphate group to the metabolite), the chemical reactivity of the metabolite is enormously increased—that is, the phosphorylated metabolite is able to enter into reactions in which it could not otherwise participate.

In addition to its role in phosphorylating glucose to glucose-6-phosphate [see reaction (4-39)], ATP is also required in numerous biosynthetic reaction sequences leading to the formation of essential cellular constituents. Several specific examples will be mentioned in Chapter 9—for instance, in the assimilation of ammonia (pp. 263–264), in nitrogen fixation (pp. 276–277), and in sulfate assimilation (pp. 278–279). Moreover, ATP is needed in active transport processes (pp. 324–325) and in cytoplasmic streaming (p. 294). Also, it is worthwhile to note that ATP is utilized in bioluminescence phenomena (in certain bacteria and in such insects as fireflies) and muscle contraction (in muscle cells in animals).

The steady-state concentrations of ATP and ADP in living cells are in the order of 10^{-3} to 10^{-4} M. These two substances can be regarded as *coupling agents*—that is, substances utilized in one metabolic reaction and regenerated in another. By being utilized and regenerated cyclically, a coupling agent permits one metabolic reaction to proceed at the expense of the other even though the concentration of the coupling agent may be low. Several coupling agents were encountered earlier, including $NAD^+/NADH$, $FAD/FADH_2$, and $NADP^+/NADPH$.

GENERATION OF ATP

The living cell generates ATP from ADP in two different ways: transphosphorylation reaction, and oxidative and photosynthetic phosphorylation. In the former a phosphate group is transferred from another already phosphorylated compound to ADP. Two examples of transphosphorylation reactions were encountered in the glycolytic pathway. The compound known as 1,3-bisphosphoglyceric acid (Figure 4-3) is able to transfer one of its phosphate groups to ADP as follows:

$$
\begin{array}{c}
\text{CH}_2\text{O}\textcircled{P} \\
| \\
\text{HCOH} \quad + \text{ADP} \\
| \\
\text{COO}\textcircled{P}
\end{array}
\longrightarrow
\begin{array}{c}
\text{CH}_2\text{O}\textcircled{P} \\
| \\
\text{HCOH} \quad + \text{ATP} + \text{H}_2\text{O} \\
| \\
\text{COOH}
\end{array}
\qquad (4\text{-}41)
$$

1,3-Bisphosphoglyceric acid   3-Phosphoglyceric acid

In the second example of a transphosphorylation reaction, phosphoenolpyruvic acid donates its phosphate group to ADP (see Figure 4-3).

$$
\begin{array}{c}
\text{CH}_2 \\
\| \\
\text{CO}\textcircled{P} \quad + \text{ADP} \\
| \\
\text{COOH}
\end{array}
\longrightarrow
\begin{array}{c}
\text{CH}_3 \\
| \\
\text{C}=\text{O} \quad + \text{ATP} + \text{H}_2\text{O} \\
| \\
\text{COOH}
\end{array}
\qquad (4\text{-}42)
$$

Phosphoenol   Pyruvic
pyruvic acid   acid

In each reaction [(4-41) and (4-42)] the energy of hydrolysis of the phosphate group ($-\text{COO}\textcircled{P}$ in 1,3-bisphosphoglyceric acid and $-\text{CO}\textcircled{P}$ in phosphoenol pyruvic acid) is even more negative than the energy of hydrolysis of ATP [reaction (4-38)]; so the free energy change in each of these two transphosphorylation reactions [(4-41) and (4-42)] is negative, as would be expected from the preceding discussion.[4]

Transphosphorylation reactions, such as those shown in reactions (4-41) and (4-42), account for only a small fraction of the ATP generated in living cells of higher plants. Generation of ATP from ADP in respiratory metabolism occurs mainly in oxidative phosphorylation. *Oxidative phosphorylation* refers to the generation of

---

[4] Reaction (4-31) is still another transphosphorylation reaction encountered earlier. GTP in reaction (4-31) is formed from GDP, Pi, and succinyl-CoA in reaction (4-30).

ATP from ADP in conjunction with electron transfer in the mitochondrial respiratory chain. A similar process occurs in electron transport in the light reactions in chloroplasts and is called *photosynthetic phosphorylation* (see Chapter 6).

The synthesis of ATP from ADP and Pi in root cells takes place mainly by oxidative phosphorylation in mitochondria. Also in green leaves in the dark, a similar situation exists. But when green leaves are illuminated, synthesis of ATP by photosynthetic phosphorylation in chloroplasts considerably exceeds that by oxidative phosphorylation in mitochondria (see Hampp et al., 1982, and references cited there).

Oxidative phosphorylation accompanies the oxidation of either NADH or succinic acid in the respiratory chain. In the case of NADH, three molecules of ATP are generated for each molecule of NADH oxidized to NAD^+ in the respiratory chain. Thus the overall reaction for oxidative phosphorylation (with NADH as respiratory substrate) is

$$NADH + H^+ + 3\,ADP + 3\,Pi + \tfrac{1}{2}O_2 \longrightarrow NAD^+ + 4\,H_2O + 3\,ATP \quad (4\text{-}43)$$

[*Note*: Pi denotes phosphoric acid (H_3PO_4).]

Reaction (4-43) is the sum of reactions written earlier—that is, reaction (4-34) added to the *reverse* of reaction (4-38), the latter multiplied by three.

Because succinic acid enters the respiratory chain at a different point than NADH (see Figure 4-7), only two molecules of ATP are generated for each molecule of succinic acid oxidized to fumaric acid in the respiratory chain. The overall reaction for oxidative phosphorylation (with succinic acid as respiratory substrate) is

$$Succinic\ acid + 2\,ADP + 2\,Pi + \tfrac{1}{2}O_2 \longrightarrow fumaric\ acid + 2\,ATP + 3\,H_2O \quad (4\text{-}44)$$

Reaction (4-44) is the sum of reactions already written earlier—that is, reaction (4-35) added to the *reverse* of reaction (4-38), the latter multiplied by two. [Of course, it is possible to write FAD in place of succinic acid and $FADH_2$ in place of fumaric acid because $FAD/FADH_2$ is the cofactor of succinic acid dehydrogenase (see Figures 4-6 and 4-7).]

Reactions (4-43) and (4-44) are simply formal statements of reactants and products and indicate nothing regarding the reaction mechanism. The free energy change in each of these two reactions is negative, as would be expected from the preceding discussion. But the source of the energy required to *reverse* reaction (4-38) is not indicated in reactions (4-43) and (4-44); this matter is considered in the next section.

The processes of transphosphorylation and oxidative phosphorylation together account for the generation of more than 30 molecules of ATP from ADP during the aerobic oxidation of one molecule of glucose. It will be a useful exercise for the reader to calculate the exact number. It can be done by assuming that each glucose molecule is respired completely to CO_2 and water first through the glycolytic pathway (see Figure 4-3), then via the conversion of pyruvic acid to acetyl-CoA (see Figure 4-6), and, finally, through the TCA cycle (see Figure 4-6). Beginning with the oxidation of pyruvic acid, 8 molecules of NADH are generated from NAD^+ (i.e., 4 molecules of NADH for each of the 2 molecules of pyruvic acid that is oxidized first to acetyl-CoA and then to CO_2 and water). The oxidation of each of these 8 molecules of NADH to NAD^+ in the respiratory chain leads to the generation of 3 molecules of ATP from ADP in oxidative phosphorylation (i.e., a total of 24 molecules of ATP). To these 24 molecules of ATP can be added 4 molecules of ATP generated in the oxidation of succinic acid to fumaric acid via the cofactor flavin-adenine dinucleotide and 2 molecules of ATP generated from GTP [see reaction (4-31)]. To these 30 molecules of ATP can be added 2 molecules of ATP generated from ADP by transphosphorylation reactions in the glycolytic pathway. Finally, 2 molecules of NADH are generated from NAD^+ in the glycolytic pathway. [Oxidation of cytosolic NADH occurs in mitochondria, either via the respiratory chain (in some plants) or by a NADH dehydrogenase unrelated to ATP generation (Moore and Rich, 1980).]

Oxidative phosphorylation and electron transport in the respiratory chain are interrelated processes, each depending on the other. The rate of electron transport (and the rate of oxidation of NADH or succinic acid or the rate of reduction of O_2) is controlled by the availability of ADP. Conversely, the rate at which ATP is generated from ADP and Pi is controlled by the rate of oxidation of NADH (or succinic acid) by O_2.

The interdependence of electron transport and ATP generation from ADP and Pi in oxidative phosphorylation can be demonstrated in vitro if sufficient precautions are taken to preserve the structural integrity of mitochondria during their isolation from plant cells. Damage to mitochondria may result in *uncoupling* of electron transport from phosphorylation. That is, damaged isolated mitochondria may be able to oxidize NADH or succinic acid but may not be able to convert ADP to ATP. Uncoupling of electron transport from phosphorylation can also occur by applying one or another of several chemical agents to isolated mitochondria. At appropriate concentrations, these uncoupling agents—two of the best known are 2,4-dinitrophenol (DNP) and carbonyl cyanide *p*-trifluoromethoxyphenylhydrazone (FCCP)—permit electron transport to continue at a normal rate, or sometimes even at an enhanced rate, but abolish the ability of the mitochondria to generate ATP from ADP. The way in which these uncoupling agents operate is discussed in the next section.

THE CHEMIOSMOTIC THEORY

The chemiosmotic theory proposed by P. Mitchell, the English biochemist, in the early 1960s is almost universally accepted as the mechanism that accounts for the coupling between redox reactions in the respiratory chain and oxidative phosphorylation. In this theory, as it relates to mitochondrial ATP formation from ADP and Pi, the flow of electrons from NADH (and/or succinic acid) to O_2 through the respiratory chain in the inner mitochondrial membrane is accompanied by extrusion of protons from the mitochondrial matrix, thereby generating a proton concentration gradient across the membrane. So a proton electrochemical gradient (i.e., a proton motive force[5]) is built up. In effect, the energy of electron flow in the respiratory chain [i.e., the drop in redox potential between NADH (and/or succinic acid) and O_2 (see discussion of redox potential on p. 97, and also reactions (4-19) and (4-20)] is partially conserved as a proton motive force across the inner mitochondrial membrane. The return flow of protons down their electrochemical gradient, through an ATPase back into the mitochondrial matrix, drives the formation of ATP from ADP and Pi. [The enzyme known as ATPase catalyzes not only the hydrolysis of ATP, as shown in reaction (4-38), but also the reverse reaction. Reaction (4-38) results in a release of energy; the reverse reaction requires an input of energy.]

The transfer of protons across the inner mitochondrial membrane, from the inner to the outer surface, is widely assumed to occur through the alternate reduction and oxidation of ubiquinone, one of the components of the respiratory chain. Ubiquinone picks up and releases a hydrogen atom during alternate reduction and oxidation (see Table 4-2 and related discussion in the text); moreover, it is visualized to move from one side of the protoplasmic membrane to the other as it picks up and releases a hydrogen atom. In this way, it not only transfers electrons through the respiratory chain (see Figure 4-7), but also accounts for the transfer of protons across the inner mitochondrial membrane.

Proton motive force has been likened to an electrical circuit, with electrons replaced by protons. The transmembrane electrochemical gradient is energetically equivalent to about 230 mv. Even though protons flow back through ATPase (and thereby bring about ATP formation from ADP and Pi), proton motive force is decreased only slightly, partly because phosphorylation results in an increase in the respiration rate and hence in proton extrusion. The force and flow of protons across the membrane through ATPase are called *proticity*, in analogy to the flow of an electrical current (i.e., electricity). In the chemiosmotic theory proticity (i.e., the difference in electrochemical potential of protons across the membrane) is used directly to drive the energetically unfavorable formation of ATP from ADP and Pi in the ATPase enzyme system. The word *chemiosmosis* was originally based on the fact that there is a conversion of chemical energy [e.g., in the oxidation of NADH by O_2, as shown in reaction (4-19)] to osmotic energy (i.e., the difference in the concentration of protons on the two sides of the inner mitochondrial membrane).

The inner mitochondrial membrane illustrated in Figure 4-8 consists of lipid material through which most solutes in general and protons in particular cannot pass except at points where the respiratory chain and ATPase systems are located (and also at specific transport sites not shown in the diagram). [*Note*: Only O_2, CO_2, and water among physiologically active substances are thought to have the ability to diffuse

---

[5]Proton motive force (pmf) consists of two components, an electric potential difference and a pH difference: pmf $= \Delta\Psi - 2.3\, RT/F \Delta pH$, where the uppercase Greek letter psi (Ψ) is the electric potential, R the ideal gas constant (8.3×10^7 ergs mol^{-1} deg^{-1}), T the absolute temperature (in °K), and F the Faraday constant (96500×10^7 ergs mol^{-1} $volt^{-1}$). When electric potential is expressed in millivolts, $2.3\, RT/F$ has a value of about 60 at 25°C.

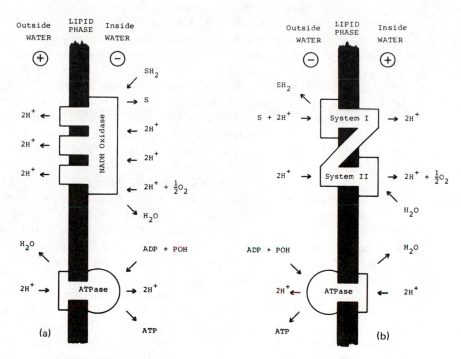

Figure 4-8 Diagrammatic representation of chemisomotic mechanisms in oxidative and photosynthetic phosphorylation systems. Diagram (a) shows oxidative phosphorylation in the inner membrane of the mitochondrion. The term *NADH Oxidase* indicates the respiratory chain system that oxidizes NADH. SH_2 represents reduced substrate (e.g., NADH); S, oxidized substrate (e.g., NAD^+); POH, phosphoric acid (i.e., Pi). Diagram (b) shows photosynthetic phosphorylation in the thylakoid membrane of the chloroplast. System I and system II designate photosystems I and II in the photosynthetic electron transfer chain (see Figure 6-10). Abbreviations: S, oxidized substrate (e.g., $NADP^+$); SH_2, reduced substrate (e.g., NADPH); POH, phosphoric acid (i.e., Pi). For further discussion, consult the text. [After P. Mitchell, *Trends in Biochemical Sciences* 3 (March 1978), p. N60.]

freely across the inner mitochondrial membrane. The outer mitochondrial membrane (not shown in Figure 4-8) permits almost all solutes to diffuse across it.] Also shown in Figure 4-8 are diagrammatic representations of the respiratory chain (labeled NADH Oxidase) and the ATPase enzyme complex; both systems are "plugged through" the membrane. Such substrates as NADH and succinic acid donate not only electrons, as noted, but also protons [see reactions (4-9) and (4-12) and related discussion]. Protons are pumped across the membrane and thereby generate the electrochemical proton gradient required for the formation of ATP by ATPase. For each molecule of NADH oxidized by O_2 in the respiratory chain, six protons are shown in Figure 4-8 to be extruded. The exact number of protons (whether six or more or less), however, and, in fact, the detailed mechanism by which these protons are transferred across the membrane during the operation of the respiratory chain—and the detailed mechanism by which the inward flow of protons through ATPase leads to ATP formation from ADP and Pi—are not yet completely understood and so will not be considered here.[6]

The fact that oxidative phosphorylation comes to a halt when uncoupling agents (see p. 119) are applied to isolated mitochondria is widely regarded as confirming the existence of the protonic chemiosmotic mechanism. Uncouplers are unnatural complex organic molecules which dissolve in protoplasmic membranes (i.e., are lipid soluble) and, being weak acids, act as artificial transmembrane proton carriers; that is to say, they permit protons to move rapidly down the electrochemical gradient, thereby bringing about the collapse of the proton gradient across the membrane. In effect, the proton gradient is "short circuited" by the uncoupler. [A proton uncoupler is often called a *protonophore* (Greek, *phore*, bearer).]

The chemiosmotic theory also explains the formation of ATP from ADP and Pi in the thylakoid membranes of chloroplasts. Figure 4-8 shows photosystems I and II, and also ATPase, "plugged through" a thylakoid membrane. As in oxidative phosphorylation, protons are transported across the membrane through the flow of electrons in the photosynthetic electron transfer chain and thereby create a proton motive force across the membrane.[7] The energy required for the flow of electrons from water to $NADP^+$ in the photosynthetic electron transfer chain—and for the transport of protons across the membrane—is supplied by absorption of light. The transport of protons—driven by electron flow—in the thylakoid membrane is *inward* so that the *inside* of the thylakoid becomes acidic on illumination (see Figure 4-8). Four protons, not six as in oxidative phosphorylation, are thought to be transported across the thylakoid membrane for each molecule of $NADP^+$ reduced to NADPH (see Figure 4-8). But as with oxidative phosphorylation, the detailed mechanism of proton transfer across the thylakoid membrane is not completely understood and so will not be considered here (see footnote 6). Finally, as in oxidative phosphorylation, the return flow of protons down the proton electrochemical gradient through the ATPase complex provides the energy needed for the formation of ATP from ADP and Pi.

OXIDASES OTHER THAN CYTOCHROME OXIDASE

An oxidase is an enzyme that catalyzes the transfer of electrons from substrate to O_2. The best-known example is cytochrome oxidase. Cytochrome oxidase, it will be recalled, is a mitochondrial enzyme that terminates the respiratory chain (Figure 4-7).

[6] It is not possible here to develop all the details of the chemiosmotic theory. Indeed, some are still controversial and have not yet been proved experimentally. The interested reader is referred to Hinkle and McCarty, 1978, for a popular explanation; for a more sophisticated account, see Mitchell, 1979. Also a recent book (Nicholls, 1982) has useful information concerning the chemiosmotic theory.

[7] The actual transfer of protons across the thylakoid membrane occurs through the alternate reduction and oxidation of plastoquinone, one of the components of the photosynthetic electron transport chain (Figure 6-10 and related discussion). Plastoquinone has a role similar to ubiquinone in oxidative phosphorylation in mitochondria (p. 113).

Through the agency of cytochrome oxidase, electrons are transferred from certain respiratory substrates (e.g., NADH) to O_2, which is reduced to water.

The important role that cytochrome oxidase has in O_2 uptake by living plant cells is widely recognized. In fact, it is generally acknowledged that cytochrome oxidase accounts for the bulk of the O_2 uptake by most living cells of higher plants throughout most of their lifetimes. Yet it is well established that oxidases other than cytochrome oxidase also participate in O_2 uptake by plant cells. In the following pages we consider several of these oxidases.

The alternate oxidase   An oxidase presently known as the *alternate oxidase* has been demonstrated in the arum lily family (Araceae), a family of plants to which belong such well-known horticultural specimens as the calla lily and jack-in-the-pulpit. For a few hours during the 2- to 4-day period of onset of flowering, the terminal portion of the fleshy inflorescence (spadix) of certain members of this family undergoes a remarkable increase in its rate of respiration, in some cases reaching values of 1000 to 5000 μL of O_2 100 mg fresh tissue $^{-1}$ h^{-1}. (Compare this value to those in Table 4-1.) Considerable heat is produced during the rapid oxidation of starch reserves; the tissue may reach temperatures as high as 40° C when the temperature of the surrounding air is only 20° C. Furthermore, the high rate of O_2 uptake by this tissue is completely insensitive to known inhibitors of cytochrome oxidase (e.g., cyanide ion at a concentration of about 10^{-3} M). The alternate pathway of electron flow to O_2, terminated by the alternate oxidase, becomes operative in the inflorescence of the arum lily during the period of rapid increase in respiratory rate. The nature of the control mechanism for the "switch" from O_2 uptake mediated by cytochrome oxidase to that mediated by the alternate oxidase is unknown.

In addition to the operation of the alternate pathway in the spadix of arum lilies, other examples have been reported. A few are mentioned here.

1. Immediately after preparation of sliced disks of many storage organs, the respiration is sensitive to cyanide but becomes insensitive after several hours of aging (see p. 89), indicating the induction of the alternate pathway during aging (Kahl, 1974).

2. After the first 12 hours of germination of chick pea seeds, there is a shift from cyanide-sensitive respiration to the alternate pathway; the maximum development of the alternate pathway is reached at about 80 hours of germination (Burguillo and Nicolás, 1977). Whether this shift to the alternate pathway is common in germination of other seeds is not now known.

3. By storing whole potato tubers for several days in an atmosphere of pure O_2 to which is added small amounts of ethylene (a plant growth substance considered in Chapter 14), the alternative pathway was induced in sliced disks prepared from the tubers (Day et al., 1978).

4. The increase in respiratory rate during the climacteric of certain fruits (see Figure 18-11) is accompanied by the implementation of the alternate pathway (Solomos, 1977).

It appears probable that mitochondria in tissues exhibiting the alternate pathway have two *parallel* electron transport chains from substrate (e.g., NADH) to O_2. One is the cyanide-sensitive respiratory chain terminated by cytochrome oxidase. The second (alternate) pathway is terminated by the alternate oxidase. Some evidence leads to the conclusion that the alternate pathway of electron transport to O_2 *branches* away from the cytochrome respiratory chain at ubiquinone (see Figure 4-7), but none of the electron transport components has been identified yet (Kelly, 1982).

The alternate pathway produces only a little ATP from ADP and Pi; the energy of oxidation of substrate (e.g., NADH) by O_2 appears to be dissipated as heat. It has been suggested that this situation is significant in arum lilies, where the heat may serve to volatilize insect attractants and thereby influence flowering and pollination (Moore and Rich, 1980). But the physiological significance of the alternate pathway in the other tissues mentioned remains obscure. Also the presumed widespread distribution of the alternate oxidase among plant species (see items 1 to 4 above) has been questioned recently (Kelly, 1982).

Glycolic acid oxidase The peroxisomes of green leaves contain a flavin-containing oxidase known as glycolic acid oxidase. Unlike cytochrome oxidase (and also the alternate oxidase), which catalyzes the reduction of O_2 to water [e.g., see reaction (4-18)], glycolic acid oxidase reduces O_2 to hydrogen peroxide (H_2O_2). The reaction is

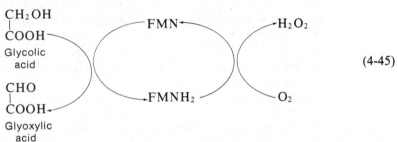

$$(4\text{-}45)$$

The coenzyme for glycolic acid oxidase is flavin-adenine mononucleotide [see reaction (4-15)]. Glycolic acid oxidase has an important role in photorespiration (see Figure 7-4 and related discussion).

H_2O_2 produces in reaction (4-45) is decomposed by a reaction catalyzed by the enzyme catalase, which is present in all plant cells.

$$2 \ H_2O_2 \ \longrightarrow \ 2 \ H_2O + O_2 \qquad\qquad (4\text{-}46)$$

Catalase is not classified as an oxidase. It is a soluble heme protein; unlike the cytochromes, catalase is not bound to membranes.

Flavin oxidase Another flavin-containing oxidase, referred to simply as flavin oxidase, has a role in the oxidative degradation of fatty acids in tissues of certain seedlings. This enzyme is present in organelles known as glyoxysomes. (The role of glyoxysomes in the utilization of fatty acids as respiratory substrates will be considered in Chapter 16.) The coenzyme for flavin oxidase is flavin-adenine dinucleotide [see reaction (4-14) and Table 4-2]. The reaction catalyzed by flavin oxidase may be depicted as follows:

$$\text{Reduced fatty acid} \diagdown\diagup \text{FAD} \diagdown\diagup \text{H}_2\text{O}_2$$
$$\text{Oxidized fatty acid} \diagup\diagdown \text{FADH}_2 \diagup\diagdown \text{O}_2 \tag{4-47}$$

H_2O_2 produced in this reaction is decomposed by catalase [see reaction 4-46)].

Phenolase (diphenol oxidase) The brownish or blackish discoloration that develops in certain tissues shortly after injury by cutting, mincing, or bruising is the result of the activity of a copper-containing oxidase known as diphenol oxidase. The term phenolase is also used. Tissues in which browning reactions occur are familiar to everyone. Examples include the peel of certain fruits (e.g., banana) and the fleshy, edible portions of fruits and vegetables, including peach, white potato, banana, mushroom, pear, eggplant, and most varieties of apple and avocado.

The substrates of phenolase are o-dihydroxyphenols. They are complex derivatives of phenol (a hydroxyl derivative of benzene) and include various phenolics, flavonoids, and tannins. One of the better known phenolic substrates of phenolase is chlorogenic acid, commonly found in many leaves and fruits. Chlorogenic acid has the following formula—the leftmost hydroxyl groups are ortho (o) to one another.

Oxidation of an o-dihydroxyphenol (e.g., chlorogenic acid) by O_2 (i.e., phenolase activity) results in the production of the corresponding o-quinone. These quinones are very reactive and condense with amino acids and proteins present in plant tissues, thus resulting in the production of complex brown polymers. (Polymers are compounds formed by the condensation of smaller units.) The production of quinones and polymers is the basis of the browning reaction mentioned earlier. The reaction sequence is:

$$\tag{4-48}$$

Browning reactions occur as the result of the unregulated breakdown of cell structure. In undamaged cells phenolase substrates are present in vacuoles whereas phenolase itself is outside the vacuole. On the slightest damage to cell structure, phenolase comes in contact with its substrate and the browning reaction occurs. In the processing of tea leaves, the darkening reactions are hastened by passing the leaves between rollers. This procedure does not break the cell walls but results in disorganizing the protoplasm and thus permits phenols in the vacuole to penetrate to phenolase outside the vacuole.

Ascorbic acid oxidase   Ascorbic acid oxidase is a copper-containing enzyme that has been extracted from several plant tissues. Its substrate, ascorbic acid (i.e., vitamin C), is probably present in at least trace amounts in all living plant cells; some tissues (e.g., citrus fruits) contain appreciable quantities. The reaction catalyzed by ascorbic acid oxidase is

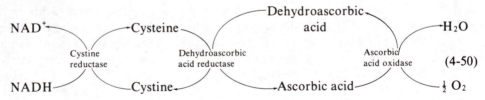

$$
\text{Ascorbic acid} + \tfrac{1}{2} O_2 \rightarrow \text{Dehydroascorbic acid} + H_2O
\tag{4-49}
$$

Several years ago it was suggested that ascorbic acid oxidase might function as a terminal oxidase, in competition with cytochrome oxidase. This suggestion was based on the finding that a mixture containing ascorbic acid, cystine, and cysteine (two sulfur-containing amino acids readily interconvertible into one another by oxidation-reduction), and enzyme preparations of ascorbic acid oxidase, dehydroascorbic acid reductase, and cystine reductase, is capable of oxidizing NADH, as shown in the following sequence.

$$
\begin{array}{ccccccc}
NAD^+ & & Cysteine & & Dehydroascorbic\ acid & & H_2O \\
 & \text{Cystine reductase} & & \text{Dehydroascorbic acid reductase} & & \text{Ascorbic acid oxidase} & \\
NADH & & Cystine & & Ascorbic\ acid & & \tfrac{1}{2}O_2
\end{array}
\tag{4-50}
$$

However, the demonstration of an electron transport system in vitro has little bearing on whether the system operates in vivo. Thus the possibility that the electron transport sequence shown in reaction (4-50) functions in vivo has little experimental support at present. In fact, little is known today about the function of ascorbic acid oxidase in plant cells.

IAA oxidase   It has long been known that plant tissues contain enzymes capable of oxidizing IAA. (The symbol IAA designates indole-3-acetic acid, a plant growth substance considered in Chapter 14.) Oxidation products of IAA do not act as plant growth substances. So enzyme systems that oxidize IAA are thought to represent regulatory mechanisms for the control of the concentration of IAA in vivo.

Although the term *IAA oxidase* is used widely, enzymes that oxidize IAA are not true oxidases (see the definition of an oxidase at the beginning of this section). Instead IAA oxidases belong to a class of enzymes known as peroxidases. Thus IAA oxidase is properly called *IAA oxidase peroxidase*. A peroxidase catalyzes the oxidation of various substrates, including IAA, by hydrogen peroxide. The overall reaction catalyzed by a peroxidase may be written

$$AH_2 + H_2O_2 \longrightarrow A + 2 H_2O \tag{4-51}$$

where AH_2 is the substrate (e.g., IAA) and A is the oxidized product. Note that catalase [see reaction 4-46)] is actually a peroxidase, where $AH_2 = H_2O_2$.

REFERENCES

ALTMAN, P. L., and D. S. DITTMER. 1964. Biology data book. *Fed. Am. Soc. Expt. Biol.*, Washington, D.C.

AP REES, T. 1974. Pathways of carbohydrate breakdown in higher plants. *In* D. H. Northcote, ed. *Internl. Rev. Biochem.* 11:89–127.

AP REES, T. 1980. Integration of pathways of synthesis and degradation of hexose phosphates. Pages 1–42 *in* P. K. Stumpf and E. E. Conn, eds. *The Biochemistry of Plants, a Comprehensive Treatise*, vol. 3. Academic Press, New York.

BANKS, B. E. C., and C. A. VERNON. 1970. Reassessment of the role of ATP in vivo. *J. Theor. Biol.* 29:301–326.

BARBER, D. A., M. EVERT, and N. T. W. EVANS. 1962. The movement of ^{15}O through barley and rice plants. *J. Expt. Bot.* 13:397–403.

BEEVERS, H. 1961. *Respiratory Metabolism in Plants*. Row, Peterson and Company, Evanston, Ill.

BURGUILLO, P. F., and G. NICOLAS. 1977. Appearance of an alternate pathway cyanide-resistant during germination of seeds of *Cicer arietinum*. *Plant Physiol.* 60:524–527.

DACEY, J. W. H. 1980. Internal winds in water lilies: an adaptation for life in anaerobic sediments. *Science* 210:1017–1019.

DAVIES, D. D., ed. Metabolism and respiration. *The Biochemistry of Plants, a Comprehensive Treatise*, vol. 2. Academic Press, New York.

DAY, D. A., G. F. ARRON, R. E. CHRISTOFFERSEN, and G. C. LATIES. 1978. Effect of ethylene and carbon dixoide on potato metabolism. Stimulation of tuber and mitochondrial respiration and inducement of the alternative path. *Plant Physiol.* 62:820–825.

DELIEU, T., and D. A. WALKER. 1981. Polarographic measurement of photosynthetic oxygen evolution by leaf discs. New Phytol. 89:165–178.

GOODWIN, T. W., and E. I. MERCER. 1972. *Introduction to Plant Biochemistry*. Pergamon Press, Elmsford, N.Y.

HAMPP, R., M. GOLLER, and H. ZIEGLER. 1982. Adenylate levels, energy charge, and phosphorylation potential during dark-light and light–dark transition in chloroplasts, mitochondria, and cytosol of mesophyll protoplasts from *Avena sativa* L. Plant Physiol. 69:448–455.

HINKLE, P. C., and R. E. MCCARTY. 1978. How cells make ATP. *Sci. Am.* 238 (3):104–123.

JACOBSON, B. S., B. N. SMITH, S. EPSTEIN, and G. G. LATIES. 1970. The prevalence of carbon-13 in respiratory carbon dioxide as an indicator of the type of endogenous substrate. *J. Gen. Physiol.* 55:1–17.

KAHL, G. 1974. Metabolism in plant storage tissue slices. *Bot. Rev.* 40:263–314.

KELLY, G. J. 1982. How widespread are cyanide-resistant mitochondria in plants? *Trends Biochem. Sciences* 7:233.

KOW, Y. W., D. L. ERBES, and M. GIBBS. 1982. Chloroplast respiration. A means of supplying oxidized pyridine nucleotide for dark chloroplastic metabolism. Plant Physiol. 69: 442–447.

MITCHELL, P. 1979. Keilin's respiratory chain concept and its chemiosmotic consequences. *Science* 206:1148–1159.

MOORE, A. L., and R. R. RICH. 1980. The bioenergetics of plant mitochondria. *Trends Biochem. Sciences* 5:284–288.

NICHOLLS, D. G. 1982. *Bioenergetics. An Introduction to the Chemiosmotic Theory.* Academic Press, New York.

PALMER, J. M. 1976. The organization and regulation of electron transport in plant mitochondria *Ann. Rev. Plant Physiol.* 27:133–157.

SOLOMOS, T. 1977. Cyanide-resistant respiration in higher plants. *Ann. Rev. Plant Physiol.* 28:279–297.

SPRAGG, S. P., and W. W. YEMM. 1959. Respiratory mechanism and the changes in glutathione and ascorbic acid in germinating peas. *J. Expt. Bot.* 10:409–425.

STITT, M., and T. AP REES. 1978. Pathways of carbohydrate oxidation in leaves of *Pisum sativum* and *Triticum aestivum*. *Phytochemistry* 17:1251–1256.

UMBREIT, W. W., P. H. BURRIS, and J. F. STAUFFER. 1971. *Manometric Techniques*, 5th ed. Burgess Publishing Company, Minneapolis.

WALKER, J. R. L. 1975. *The Biology of Plant Phenolics*. Edward Arnold, London.

WISKICH, J. T. 1977. Mitochondrial metabolite transport. *Ann. Rev. Plant Physiol.* 28:45–69.

5

Photophysiology

A number of physiological processes in plants and animals are directly affected or controlled by radiant energy. Photosynthesis in plants and vision in animals are the best-known examples. Other plant processes also affected by radiation include protoplasmic streaming, flower induction, seed germination, chlorophyll biogenesis, bending of organs as a result of irradiation (phototropism), and numerous growth reactions. The term *radiation biology* has been given to the study of the action of radiant energy on biological systems. Some effects of radiation, such as those of x rays and gamma rays, may be lethal to living organisms. This chapter, however, is concerned primarily with the effects of only a narrow range of radiant energy on physiological processes of plants. The study of these reactions has been termed *photophysiology*.

RADIANT ENERGY

The energy of the sun is derived from a nuclear reaction involving the conversion of hydrogen atoms into helium atoms. This reaction, which takes place at high temperatures and pressures, produces a broad spectrum of radiant energy. Each second the sun sends into space energy equivalent to about 1 million times that of the earth's coal, natural gas, and petroleum supplies before they were tapped by humans. The earth, however, intercepts only a small fraction of the energy poured into space by the sun.

The energy of the sun passes through space as electromagnetic radiation with waves of varying lengths as shown in Figure 5-1. Various portions of the electro-

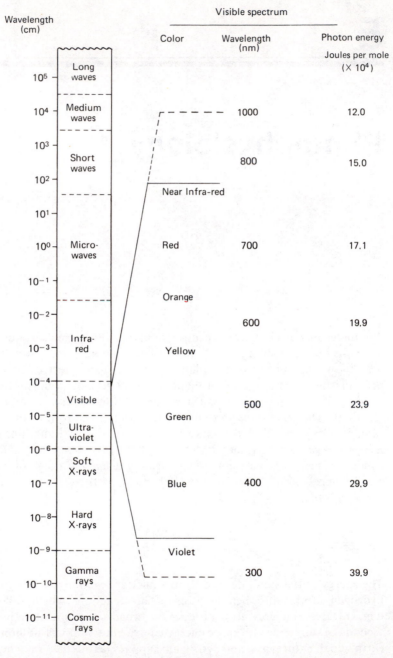

Figure 5-1 The electromagnetic spectrum. Region of photophysiology lies between 280 and 1000 nm. Visible portion of spectrum lies between 380 and 780 nm. The joule is the unit of energy in the SI system (see appendix).

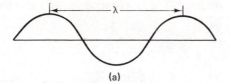 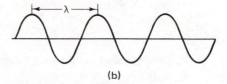

(a)   (b)

Figure 5-2   The wave nature of electromagnetic radiation. The length of the wave (λ)
in (a) is twice that of the wave in (b). The frequency of the waves in (b) is twice that in
(a).

magnetic spectrum are indicated, such as gamma rays, ultraviolet, visible, and
infrared. Each of these types of radiation is recognized by the specific nature of its
interaction with matter.

Electromagnetic radiation can be characterized by either wavelength or frequency. The relationship between the two is shown in Figure 5-2. The wavelength is the
distance between successive points of the same wave; the wavelength of wave (a) in
Figure 5-2 is twice that of wave (b). The frequency is the number of waves that pass a
given point in unit time, usually 1 s; the frequency of wave (a) is one-half that of wave
(b). Wavelength and frequency are related by

$$c = \lambda \nu \qquad \qquad (5\text{-}1)$$

where c = the velocity of electromagnetic radiation in a vacuum (this is a universal
constant equal to 3×10^8 m s^{-1})
λ = wavelength (meters)
ν = frequency (waves per second)

Electromagnetic radiation also behaves as though it consisted of small packets of
energy known as photons or quanta. The energy of a photon is directly proportional to
the frequency of the wave and inversely proportional to the wavelength:

$$E = h\nu \qquad \qquad (5\text{-}2)$$

{because $\nu = c/\lambda$ [Eq. (5-1)] by substitution}

$$E = \frac{hc}{\lambda}$$

where E = the energy of a photon of radiant energy
h = Planck's constant (this is a universal constant equal to 6.62×10^{-34} Js)

The energy of a photon of particular wavelength can be measured in ergs, joules,
electron volts, or gram-calories. Biologists most frequently refer to wavelengths in
terms of nanometers (nm, 10^{-9} m) and energy in terms of joules per mole of photons (J
mol photons). The energies of photons of several wavelengths in the photophysiology
range are shown in Figure 5-1. The energy of a photon of short wavelength is much
greater than the energy of a photon of longer wavelength. For each mole of matter that

reacts with radiation in a primary photochemical event, 1 mole[1] of photons is absorbed.

Terminology of Radiant Energy

Two systems of nomenclature and units are used in describing and characterizing radiant energy. One, the photometric system, is based on the sensitivity and brightness response of the human eye. The system is used by illumination engineers and photographers in evaluating space lighting and determining film exposure in cameras. Instruments like illumination meters or photographic light meters are available for such purposes. The instruments are calibrated in lux, which is an illuminance of one lumen per square meter. In the United States foot-candles are frequently used; this is an illuminance of one lumen per square foot (1 foot-candle is equivalent to 10.76 lux). A lumen is a unit of luminous flux emitted by a standard source known as a candela (a base unit of SI as noted in the Appendix).

In the photometric system the eye or a detecting device constructed to respond like the eye is used in measuring the incoming radiant energy. The eye responds to radiant energy between wavelengths of 380 and 700 nm with a maximum sensitivity at 550 nm (seen as green). Plants do not have a pigment similar to that of the eye and so it is not appropriate to use the photometric system in evaluating or measuring photoresponses in plants. Nevertheless, because plant photoprocesses occur in the wavelengths between 380 and 700 nm and because photometric light meters are readily available and comparatively inexpensive, plant physiologists often report radiant energy in terms of lux or foot-candles. In such usage it is essential to describe the light source—for instance, the sun, 100 W tungsten lamp, 40 W cool white fluorescent lamp—the distance between the source and the plant, and the type of light meter used. From such instructions other investigators are able to duplicate the illuminance conditions at the plant surface. It should be recognized, however, that photometric units are not recommended for evaluating photoprocesses in plants. For such purposes, the second system, the radiometric system, should be used.

In the radiometric system the unit of radiant energy is the joule (J). The amount of energy (energy density) falling on a unit surface is measured as $J\,m^{-2}$. The time rate of flow of energy is measured as a joule per second ($J\,s^{-1}$) and the amount of radiant flux intercepted per unit area is measured as a joule per square meter per second ($J\,m^{-2}\,s^{-1}$). Because $J\,s^{-1}$ equals a watt (W), the preceding term is simplified to a watt per square meter ($W\,m^{-2}$). The amount of radiant flux intercepted by a unit surface area is termed irradiance.

Instruments known as radiometers are available for measuring irradiance over a

---

[1] In the International System of Units (SI) the mole is a basic unit (see appendix) of substance. It is defined as follows. The mole is the amount of substance of a system that contains as many elementary entities as there are atoms in 0.012 kilogram of carbon-12. When the mole (abbr. mol) is used, the elementary entities must be specified and may be atoms, molecules, ions, photons, electrons, other particles, or specified groups of such particles. It can be noted that because 0.012 kg of carbon-12 contains the Avogadro number of atoms (6.02252×10^{23}), this is the number of elementary units contained in one mole.

range of wavelengths. By the selection of appropriate filters, specific wavelengths can be isolated and measured—that is, 300 to 1000 nm, 400 to 700 nm, 700 to 800 nm, and so on. Such measurements are useful because, as will be shown, plants contain a number of different pigments that absorb radiant energy at different wavelengths.

As described previously, radiant energy can be measured as $W\,m^{-2}$ (300 to 1000 nm), $W\,m^{-2}$ (400 to 700 nm), and so forth. As noted, however, radiant energy interacts with matter as discrete quanta or photons. Because one photon interacts with one molecule of matter, it is useful to have a measure of the number of photons (photon density) in the radiant energy intercepted by a unit surface or other object—for example, leaf, plant. This value, $mol\,m^{-2}$, can be measured by widely available instruments known as quantum sensors. Ideally it would be helpful to know the number of photons at each wavelength. (Recall from Figure 5-1 that the energy and number of photons vary with wavelength.) Yet it is technically difficult to determine this factor and what is usually measured is the number of photons within certain wavelengths. In the case of photosynthetically active radiation (PAR), the photon flux density is the number of photons in the 400 to 700 nm waveband incident per unit surface area per unit time—$\mu mol\,m^{-2}\,s^{-1}$ (PAR).

Table 5-1 presents a summary of the commonly used measurements of radiant energy at plant surfaces (or in growth chambers, etc.). Illuminance is the flux of lumens per square meter of surface area (lux). This unit is in the photometric system and is of limited value in studying plant photoprocesses. The other two measurements are in the radiometric system. The first measures the flux density of radiant power in $W\,m^{-2}$ whereas the second measures photon flux as $\mu mol\,m^{-2}\,s^{-1}$. These three measurements are of quite different characteristics of radiant energy and they are not easily related to each other. Hansen and Biggs (1979) have calculated some conversion factors that are

TABLE 5-1 SOME COMMONLY USED MEASUREMENTS OF RADIANT ENERGY IN PLANT PHYSIOLOGICAL STUDIES

Entity measured	Units	Remarks
Illuminance[a] (photometric)	lux (lumen m^{-2})	Density of lumens. Specify wavebands.
Irradiance[a] (radiometric)	$W\,m^{-2}$	Density of radiant power. Specify wavebands.
Irradiance[a] (radiometric)	$\mu mol\,m^{-2}\,s^{-1}$	Flux density[b] of photons. Specify wavebands.

[a] Illuminance (photometric) and irradiance (radiometric) are equivalent terms in the two systems and refer to density of lumens or power intercepted per unit area.

[b] Flux is the term used to denote the rate of flow of a substance and is expressed in units of quantity through a unit surface area per unit time.

NOTE: The flux density of photons frequently is reported as $\mu E\,m^{-2}\,s^{-1}$ rather than $\mu mol\,m^{-2}\,s^{-1}$. An einstein (abbreviated E) is defined as a mole of photons. Either term may be used—that is, micromole of photons (μmol photons) or microeinstein (μE).]

useful: W m^{-2} is approximately equivalent to 4.6 μ mol m^{-2} s^{-1} and 1000 lux (k lux) is approximately equivalent to 19.53 μ mol m^{-2} s^{-1}. Both conversion factors apply over the 400 to 700 nm range (PAR). The publication of Hansen and Biggs, listed in the References, should be referred to for the assumptions made and procedures used in calculating these conversion factors.

Measurement of Radiant Energy

Instruments for measuring radiant energy usually contain a detector that converts the incident radiant flux into an electrical signal that is then displayed on a suitable meter. The two commonly used instruments are the thermodetector and the photoelectric detector. With a thermodetector, the incoming radiant energy is converted to heat, the heat produced being proportional to the energy absorbed. The heat produced is converted to an electrical signal for measurement purposes.

Photoelectric detectors use semiconductors, such as selenium or silicon, sandwiched between two metal electrodes. Incident radiant energy generates an electromotive force proportional to the energy absorbed. Selenium generally is used in photographic light meters whereas silicon is used in radiometers. In both cases, it is necessary to use appropriate filters to fit the sensor to its intended use. In the case of the photographic light meter, the sensitivity of the human eye is matched. With the radiometer, radiant energy in the wavebands between 400 and 700 nm (PAR) may be isolated and measured or some other appropriate wavebands may be used. By the appropriate selection of sensors, the following data can be determined: solar radiation (W m^{-2}), PAR in radiant flux density (W m^{-2}), PAR in photon density (μ mol m^{-2} s^{-1}), near-infrared (70 nm bandwidth centered at 780 nm) radiant flux density (W m^{-2}), and illuminance in lux.

Plant physiologists commonly use growth chambers in experimental work. Various radiant energy sources are used and it is essential that the investigator thoroughly describe the environmental conditions, particularly the radiant energy regime, so that other workers can grow plants under comparable conditions. The following information should be provided concerning the radiant energy source: a description of the lamps—manufacturer, type, input wattage, number and average age of bulbs, rated voltage, line voltage, and temperature of the light bank—and a description of the measuring equipment—manufacturer, model, type of sensor used, waveband observed, and location of the sensor in the growing room.

Radiation Biology

Radiations from wavebands between 280 and 1000 nm mediate a number of different physiological processes, considered later in the chapter. It will be useful at this point, however, to give an overall view of the major physiological responses of plants before discussing the basic principles of interactions between radiant energy and matter.

1. Radiation between 700 and 1000 nm (far-red). Many plants irradiated solely at these wavelengths exhibit an increase in height through internode elongation.

The far-red form of phytochrome (P_{fr}) absorbs in this region with a peak at 730 nm. Bacteriochlorophyll absorbs in this region, thereby driving bacterial photosynthesis.

2. Radiation between 610 and 700 nm (red). Photosynthetic activity is at a peak through the absorption of radiant energy by chlorophyll. A number of enzymes linked to the biosynthesis of plant constituents are also regulated through chlorophyll excitation. The red form of phytochrome (P_r) absorbs here with a peak at 660 nm, thereby initiating many morphogenetic reactions.

3. Radiation between 510 and 610 nm (green-yellow). Light at these wavelengths has minimal effect on plant growth and development. Photosynthesis and photomorphogenetic responses are at a minimum. The maximum sensitivity of the human eye is at 550 nm.

4. Radiation between 400 and 510 nm (blue). Chlorophyll absorbs radiant energy of these wavelengths. Also, other photosynthetic pigments, such as phycocyanin, phycoerythrin, and carotene, have peaks in absorption in these wavebands. Phototropic movements of plants are promoted by the absorption of radiant energy of these wavebands. Phytochrome, carotenoids, and flavoproteins also absorb in this region of the spectrum. The so-called high irradiance reactions (HIR) (pp. 155–156) are initiated in this region.

5. Radiation between 280 and 400 nm (ultraviolet, UV). Plants grown in light of this spectral composition usually have thickened leaves and a compact, shortened growth habit (rosette form). Many inorganic and organic molecules absorb radiant energy at these wavelengths. Radiant energy at wavelengths of 280 to 315 nm is responsible for "sunburn" in humans.

6. Radiation shorter than 280 nm (short UV). These radiations are lethal to plants and many other forms of life. Microorganisms are particularly susceptible to these wavelengths and may be killed on exposure to germicidal lamps that emit energy at 254 nm. Such lamps are often used in transfer chambers and growth rooms to reduce bacterial and fungal contamination. Precautions should be taken to shield this short UV radiant energy from the eyes. Nucleic acids and proteins strongly absorb these radiations.

INTERACTIONS BETWEEN RADIANT ENERGY AND MATTER

When a photon impinges on matter several things may occur, depending on the energy of the photon and the nature of the matter. Several possible effects are illustrated in Figure 5-3. If the photons are energetic enough, such as those from cosmic rays, gamma rays, or x rays, electrons may be ejected from the target molecules. Frequently chemical bonds are broken, leaving fragmented molecules [Figure 5-3, reaction (a)]. These fragmented molecules, known as *free radicals*, are short lived and very reactive. Even if only a few free radicals are produced by irradiation, they may react with neighboring molecules to produce additional free radicals. The mutagenic and often lethal effects of irradiation from nuclear reactors, x-ray machines, and radioisotopes on biological materials are due to the production of free radicals (primarily from water molecules) and their subsequent interactions with nucleic acids, proteins, and other important biological molecules.

At the other end of the electromagnetic spectrum are the less energetic long-wave radiations, such as infrared, radar, and radio. Photons with these energies cause shifts in the vibrational and/or rotational configuration of atoms and molecules [Figure 5-3, reaction (b)] in such a way as to produce thermal energy, or heat, in the irradiated material.

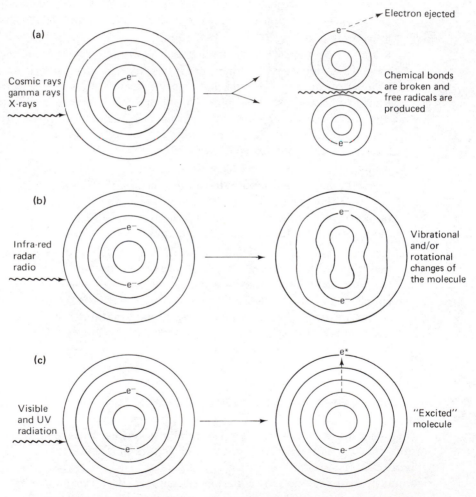

Figure 5-3 Representation of the possible effects of radiant energy when it interacts with matter. The target material at the left contains molecules that may undergo reactions (a), (b), or (c), depending on the energy of the incident photons (cosmic ray, radar, visible, etc.).

Between these two extremes of the electromagnetic spectrum lies the narrow spectral region labeled "photophysiology" in Figure 5-1. Included in this region are those wavelengths referred to as visible. The visible spectrum includes electromagnetic radiation with wavelengths ranging from 380 to 780 nm, commonly known as "light." This portion of the spectrum is visible because it is perceived by a pigment in the human eye. On absorbing a photon of radiant energy, the pigment becomes *excited* and vision results from further reactions initiated by the excited pigment molecules. Plant photophysiological processes occur over a somewhat broader range of wavelengths— 280 to 1000 nm—than do those of the eye. For this reason, it is not correct to speak of the effect of "light" on plants because it is too restrictive. The term radiant energy is preferred when speaking of photophysiological reactions. If it is understood that "light" for plants includes wavelengths from 280 to 1000 nm, however, the term light is convenient. Both terms, radiant energy and light, will be used here.

The energies of the photons in the photophysiology region of the spectrum are sufficient to displace electrons within the outer valence shells of the receptor molecules, thereby producing excited molecules [Figure 5-3, reaction (c)]. These excited molecules, in turn, interact with other molecules to produce changes. Specific details of the role of excited molecules in plant processes is discussed later.

It is noteworthy that the physiological effects of radiant energy on biological systems occur over such a narrow range of wavelengths between 280 and 1000 nm. It is not so strange, however, when we realize that it is predominantly these wavelengths that reach the earth's surface from the sun. Wavelengths shorter than 280 nm and longer than 1000 nm are largely filtered out by atmospheric absorption, the short wavelengths primarily by ozone and oxygen and the long wavelengths by ozone, water vapor, and carbon dioxide. The wavelength band between 280 and 1000 nm is sometimes referred to as a "window" through which radiant energy reaches the earth.

Photochemistry

A photophysiological (or photochemical) reaction begins with the absorption of a photon by an atom or molecule of the target material. The law of photochemical equivalence states that in a primary photochemical event one atom or molecule is activated for each photon absorbed. The number of excited atoms or molecules thus equals the number of photons absorbed. Furthermore, the total energy of a photon is used in exciting the target molecule; a one-half or one-third photon is not used. The simple one-to-one relationship of one molecule activated per one photon absorbed is seldom observed in biological reactions because a single excited molecule may trigger a whole chain of secondary reactions. It is the complex secondary events that present difficult problems of interpretation.

In biological systems the molecule absorbing radiant energy in the visible range is referred to as a *pigment*. To the human eye these pigments are colored: chlorophyll is green, carotenoids are red or yellow, and phytochrome is blue. Pigments are complex organic molecules composed of many atoms of hydrogen, oxygen, carbon, nitrogen, and so on, but in the discussion that follows the pigment will be considered a single molecule for simplicity. The structure of such a simplified pigment molecule is depicted in Figure 5-4. The electrons have different energies, depending on their distance from the nucleus. The nearer the electrons to the nucleus, the greater is the pull or attraction of the nucleus on the electron. Electrons with similar energies revolve about the nucleus at specific distances as though enclosed in a shell. If a photon of appropriate energy strikes the pigment, an electron in an inner shell may be raised to an outer shell and the pigment is said to be in an *excited* state. Energy is emitted if the electron then drops from the excited state to an inner shell.

Each electron possesses a certain amount of energy related to its distance from the nucleus. In addition, some of the energy of each electron arises from the fact that the electron is spinning on its own axis. There are two possible directions of spin: a right-handed spin, designated ⌒↘ , and a left-handed spin designated ↙⌒↘ . Except for free radicals, even numbers of electrons are present in each shell of the atoms of stable molecules, and electrons are paired in such a way that the spins cancel each other, that is, ⌒↘↙ ⌒↘ .

With this simplified picture of atomic structure, it is now possible to visualize what happens when radiant energy is absorbed by a pigment molecule. The excited pigment may react in a number of different ways, depending on the electronic structure of the pigment and the specific energy of the radiation. Some of these reactions are shown schematically in Figure 5-5 in which the fates of two electrons in the pigment

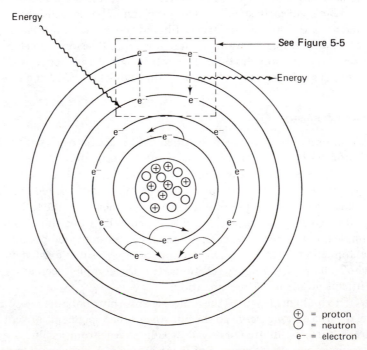

Figure 5-4 An atomic model showing valence electrons in various energy levels (shells) around the nucleus composed of protons and neutrons (see text for details).

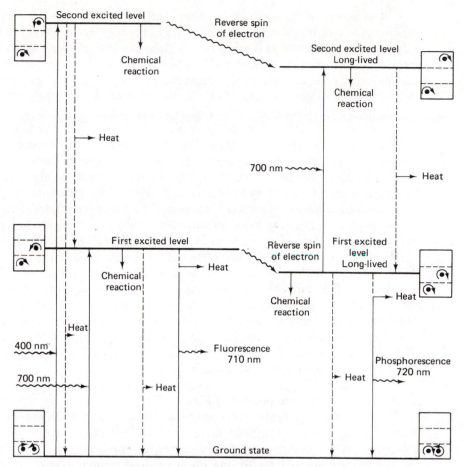

Figure 5-5 Energy levels of a hypothetical pigment molecule (see text for details). (Effects of two wavelengths, 400 and 700 nm.)

molecule are considered. As shown at the lower left of the figure, the two electrons are in the same energy level and they are spinning in opposite directions—a condition of maximum stability (also referred to as ground state). If a sufficiently energetic photon (e.g., 700 nm) strikes one of the electrons, the electron acquires the energy of the photon and moves to a new energy level (first excited level). The two electrons are now in different energy levels but are still spinning in opposite directions—the pigment is said to be in an excited state. If the absorbed photon carries enough energy, the electron may be raised to the second excited level. In Figure 5-5 red light at a wavelength of 700 nm (equivalent to 17.1 J per mole—see Figure 5-1) has raised the electron to the first excited level, whereas more energetic blue light of shorter wavelength (400 nm, equivalent to 29.9 J per mole) has raised an electron to the second excited level.

The energy of the electron in either the first or second excited level may be dissipated in several different ways, as shown in Figure 5-5. The energy of the excited electron may be emitted as thermal energy (heat) and the pigment molecule thereby returns to ground state. Frequently, however, a portion of the energy is emitted as thermal energy and the balance as radiant energy. If this situation occurs, the emitted radiant energy is less energetic than the absorbed radiant energy and of a longer wavelength. If this radiant energy emission occurs within 10^{-8} to 10^{-9} s, the process is called *fluorescence*. As shown in Figure 5-5, electrons in both the first and second excited levels have returned to ground state by emission of thermal energy and by fluorescence at 710 nm (an arbitrary wavelength for purposes of illustration). There is no fluorescence from the second excited level but rather a thermal relaxation to the first excited level, followed by fluorescence in going to the ground state.

Each pigment molecule has a characteristic fluorescence emission spectrum. Chlorophyll *a*, for example, has a unique fluorescence (depending on characteristics of the solvent used) that is the same regardless of the wavelength (350 to 700 nm) of radiant energy used to excite the molecule. If the energy of the excited molecule is emitted as thermal energy or fluorescence, this energy is unavailable for transformation into chemical work. Because the lifetime of the excited molecule is short (approximately 10^{-8} to 10^{-9} s), the likelihood that a chemical reaction will occur depends on the concentrations of the pigment molecules and the reacting molecules and on the structural organization of the photoreceptor. The pigment molecules should be in close proximity (within distances of 10 to 50 nm) to the other reacting molecules so as to allow a transfer of energy between them.

That the pigment molecule may react in a different manner than that described is shown on the right-hand portion of Figure 5-5. An electron in the first excited level may dissipate some energy in reversing spin and drop into a slightly lower energy level. The two electrons are now in different energy levels but with spins that are in the same direction. Such an electron is said to be in the first long-lived level. It may be raised to the second long-lived level by the absorption of radiant energy of appropriate wavelength—for example, 700 nm, as shown in Figure 5-5. Whether the electron is in the first or second long-lived state, energy loss occurs by thermal emission of radiant energy in processes similar to those previously described. Yet there is one major difference—electrons in the long-lived levels do not emit energy so quickly because the electron must first reverse spin before dropping to the ground state. In some cases, the excited molecule persists for as long as 10^{-3}s before energy emission occurs. This long-lived radiant emission process, *phosphorescence*, is also of a longer wavelength— for example, 720nm—than that absorbed and, in this respect, is similar to fluorescence. Thus the major difference between the two processes is that in phosphorescence there is an appreciable delay in the reemission of radiant energy. A pigment with an excited electron in the long-lived state has a greater probability of reacting with another chemical in its immediate neighborhood and thus initiating a photophysiological reaction than a pigment with an electron in the short-lived electronic configuration.

If all the energy of an excited pigment molecule is being channeled into a photochemical or photophysiological reaction, no energy will be emitted as fluorescence or phosphorescence. Pigments in a leaf that is actively carying out photo-synthesis, for example, may show almost no fluorescence, but if the chlorophyll *a* of the leaf is extracted and then irradiated, the solution exhibits fluorescence because the chlorophyll is uncoupled from the photophysiological reaction. Radiant energy emission by fluorescence or phosphorescence can be used to study some photo-physiological reactions in intact plant material. Again using the example of photosynthesis, if a particular chemical applied to a leaf increases fluorescence following irradiation, the conclusion is usually that the chemical is uncoupling the photophysiological reaction in some manner and is thereby allowing excited chlorophyll to lose energy by fluorescent emission. Such studies have been extremely useful in studying many photophysiological reactions in plants.

Action Spectra

It should be apparent that so far the discussion on photophysiology has been in general terms and that simple systems have been described. In a plant or any living organism, however, a great number of pigments have the ability to absorb radiant energy. In fact, practically all large organic molecules (polymers) absorb radiation in wavelengths not far removed from the visible range (proteins absorb at 280 nm, nucleic acids absorb at 260 nm, etc.). Although many pigments respond differently to radiant energy, it is difficult to decide which specific pigment is actually coupled with the chemical events that follow light absorption and constitute a photophysiological reaction. A common

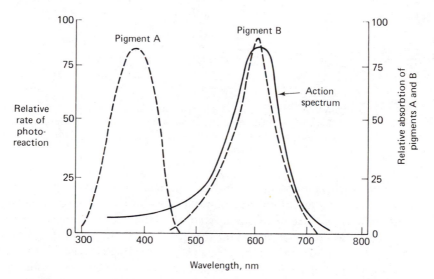

Figure 5-6 Action spectrum of a hypothetical photoreaction and absorption spectra of two pigments isolated from the plant. The close correspondence of the action spectrum (physiological activity) with the absorption spectrum of pigment B supports the view that pigment B is the pigment that "drives" the photoreaction.

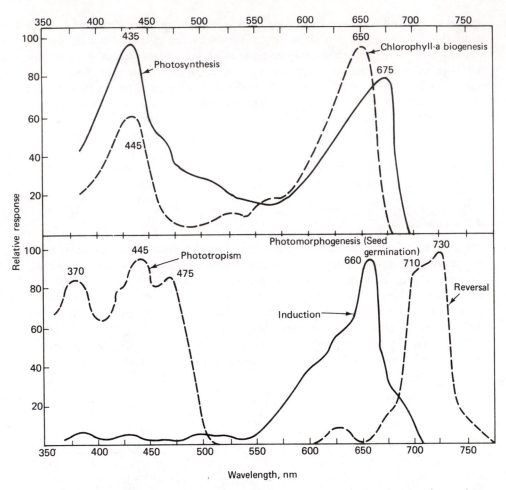

Figure 5-7   Action spectra for some plant photophysiological reactions—photosynthesis, chlorophyll *a* biogenesis, phototropism, red induction, and far-red reversal of photomorphogenesis (seed germination). Adjusted to an arbitrary value of 100 units of response at the peak activity. (After Withrow, 1959; Figure 5.)

procedure in attempting to identify the pigment involved in a particular photoreaction is to determine an action spectrum. This step is accomplished by measuring the rate of the photoreaction, or of some narrowly delimited photophysiological response, at different wavelengths of radiant energy.

Once an action spectrum for a photophysiological reaction has been determined, the next step is to compare this action spectrum with the absorption spectrum of pigments isolated from the plant. A hypothetical situation is illustrated in Figure 5-6. The photoreaction is seen to have a peak in activity at about 600 nm. Two pigments, A and B, were isolated from the plant and their absorption spectra determined. Pigment

A has a peak in absorption at 395 nm and pigment B at 600 nm. The close correspondence between the absorption spectrum of pigment B and the action spectrum strongly supports the notion that pigment B is responsible for absorbing the radiant energy used to drive the photoreaction.

The action spectra of a number of different photophysiological reactions in plants are shown in Figure 5-7. These reactions will be described in the following sections. The pigments responsible for capturing radiant energy used in photosynthesis and chlorophyll *a* biogenesis are reasonably well known. In the case of photo-morphogenetic processes, such as the promotion or inhibition of seed germination, phytochrome has been identified as the responsible pigment. The pigment or pigments involved in phototropism are not known with certainty. In addition, there are probably a number of physiological reactions in plants not now recognized as involving radiant energy. Recent improvements in techniques of isolating and separating plant pigments, as well as identifying the physiological reactions to which they may be coupled, will aid in identifying new photoprocesses. It should be noted that although action spectra and pigment absorption spectra are useful tools in physiological research, it is difficult to duplicate the in vitro pigment environment in which they may be bound to proteins or membrane components. Because action and absorption spectra generally are measured on isolated systems, the actual localization and internal environment may be quite different.

SOME PHOTOPHYSIOLOGICAL PROCESSES IN PLANTS

Chlorophyll *a* Synthesis

One of the final steps in the biosynthesis of chlorophyll *a* involves the addition of two hydrogen atoms to protochlorophyll in a photophysiological reaction (Figure 5-8). With a few notable exceptions, dark-grown plants lack chlorophyll *a*. Such plants are usually yellowish in color and possess an elongated, spindly growth habit. Leaf development is also usually strongly repressed. Plants displaying these characteristics are said to be etiolated (Figure 17–18). A brief exposure to radiant energy of appropriate wavelengths is sufficient to induce the formation of chlorophyll *a*. A large number of photomorphogenetic events, including changes in the major chlorophyll absorption maxima, follow protochlorophyll transformation, leading eventually to a green plant. The close correspondence of the action spectrum of chlorophyll *a* formation and the absorption spectrum of protochlorophyll indicates that proto-chlorophyll is the pigment responsible for absorbing the radiant energy. Thus energy of the excited protochlorophyll is coupled with a hydrogenation reaction, resulting in the formation of chlorophyll *a*. The hydrogenation is probably an enzymic reaction in which protochlorophyll is bound to a protein. This pigment-protein complex then would be transformed to a chlorophyll *a* protein complex. Nothing is known at present about the nature of the hydrogen donor in the hydrogenation reaction.

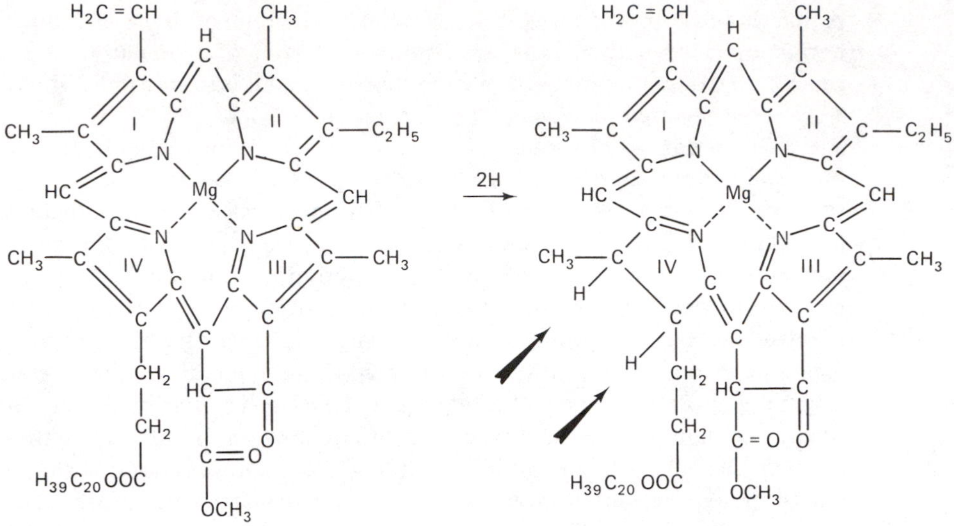

Protochlorophyll   Chlorophyll-a ($C_{55}H_{72}O_5N_4Mg$)

Figure 5-8 Structural formulas of protochlorophyll and chlorophyll *a*. Two hydrogen atoms are attached at the points shown to convert protochlorophyll to chlorophyll *a* in a photochemical reaction.

Photosynthesis

Photosynthesis is one of the most thoroughly studied photophysiological reactions in all of biology. It has been recognized for well over a hundred years that photosynthesis involves a complex series of reactions that result in the conversion of radiant energy into chemical energy. The details of our present understanding of photosynthesis will be discussed in Chapter 6, but some of the photophysiological aspects are considered briefly here.

Early plant physiologists understood that only leaves and stems—the green portions of plants—were involved in photosynthesis. It was later discovered that the green color was due to an organic molecule, chlorophyll, and the general consensus was that chlorophyll absorbed radiant energy. The first action spectrum of photosynthesis, determined in the 1880s by Engelmann, showed that maximum photosynthetic activity occurred in the blue and red regions of the spectrum, coinciding with the maximum absorption peaks of isolated chlorophyll. It was also recognized that many additional pigments were present in green plants and it was difficult to decide whether all these pigments were active in photosynthesis or whether some might be inactive.

Further information regarding the role of the various leaf pigments in photosynthesis came from studies of their fluorescence spectra. Recall that an excited pigment molecule may return to ground state by the emission of radiant energy of a longer wavelength than that which it absorbed. Thus when chlorophyll *a* is irradiated, it will emit energy as fluorescence with the characteristic spectrum shown in Figure 5-9. The maximum fluorescence occurs at 668 nm, which is slightly greater than the absorption maximum at 662 nm; there is also a second smaller fluorescence shoulder at 723 nm. Suppose that we have an instrument that can detect radiant energy at a wavelength of 723 nm (the second peak of the fluorescence spectrum) and that we irradiate a leaf with radiant energy at 620 nm; then the instrument tuned in at 723 nm will detect radiant energy emission. Next, assume that we irradiate the leaf at 450 nm, where chlorophyll *a* also strongly absorbs radiant energy; again, the instrument tuned in at 723 nm will detect fluorescence emission. Then suppose that the leaf is irradiated at 500 nm, where chlorophyll *a* absorbs poorly but β-carotene absorbs radiant energy very well. Once again the instrument tuned in at 723 nm detects the fluorescence emission spectrum characteristic of chlorophyll *a*. In short, the fluorescence spectrum characteristic of chlorophyll *a* is emitted regardless of the wavelength of radiant energy used to irradiate the leaf, as long as it falls between 400 and 680 nm.

Studies of this type led to the conclusion that several leaf pigments in addition to the chlorophylls participate in the absorption of radiant energy used in photosynthesis. Chlorophyll *a* participates directly in the reactions leading to the conversion of radiant energy into chemical energy whereas the other pigments (the so-called accessory pigments) transfer their excitation energy to chlorophyll *a*. The transfer of excitation energy between the pigments is thought to occur by a process known as *inductive resonance*.

This process may be visualized as follows. Consider two pigments, A and B, which are converted to excited molecules (A* and B*) through the absorption of radiant energy. Pigment A has a peak in absorption at a shorter wavelength than B.

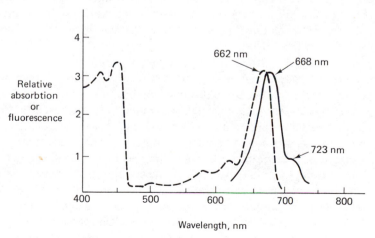

Figure 5-9   The absorption (- - -) spectrum and fluorescence (——) spectrum of chlorophyll *a*.

Furthermore, both pigments are arranged so that they are in close proximity, as might occur in a structural unit, such as a chloroplast. If pigment A is irradiated at an appropriate wavelength, it is converted to an excited state A*. As shown in Figure 5-6, an excited pigment can lose excitation energy several different ways—by fluorescence, thermal degradation, or a chemical reaction. Another method is to transfer the excitation energy to another pigment, which in turn, is excited—that is, by inductive resonance.

$$A + radiant\ energy \longrightarrow A*$$

$$A* + B \longrightarrow A + B*$$

Several restrictions are placed on this process of energy transfer. The absorption maximum of pigment B will be at a longer wavelength than that of pigment A. Inductive resonance does not occur from B* to A. Upon transfer of energy to pigment B from pigment A, pigment B* can then participate in any reaction characteristic of the excited state.

Some progress has been made in our understanding of the localization of the photosynthetic pigments within the chloroplast and their organization with other chloroplast components. The very early events of photosynthesis involved in the transfer of energy from an excited chlorophyll *a* molecule to other chloroplast molecules, however, have not been clearly delineated at present. Photosynthesis is discussed further in Chapter 6.

Phototropism

It has been known for many years that plants bend toward light. The response results from differential growth of the irradiated plant organ. This growth response, known as *phototropism*, was studied in great detail in the nineteenth century by Charles Darwin. He noted that the coleoptiles of grass seedlings were especially responsive to light and that, when the coleoptile tip was unilaterally irradiated, the coleoptile curved toward the light source.

Red light has little effect on phototropism, but as shown in Figure 5-10, blue light exerts a marked effect on coleoptile curvature. The pigment responsible for absorbing radiant energy active in phototropism has not been positively identified. Two pigments, carotenoids and riboflavin, have been suggested as the photoreceptors. Their absorption spectra are shown in Figure 5-10. The carotenoids occur widely in plants and may exist in a number of isomeric forms. A comparison of the absorption spectrum of β-carotene with the action spectrum of coleoptile curvature in the oat plant (*Avena*) shows a reasonably good correspondence between 400 and 500 nm but a wide divergence in the short wavelengths between 300 and 400 nm. It has been argued that in the plant the carotenoid is "bound" to form a pigment complex, the absorption spectrum of which may approximate the action spectrum more closely than does the pure carotenoid in solution. Little or no evidence exists on this point, however, so that it has not yet been proved.

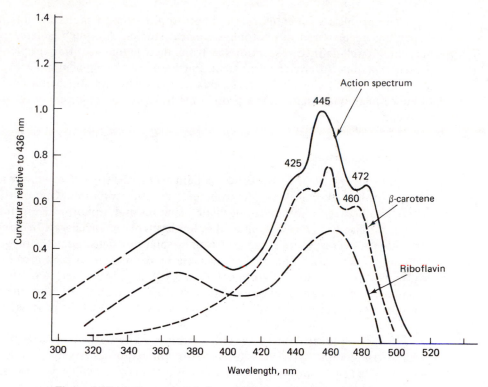

Figure 5-10 Action spectrum for positive curvature of *Avena* coleoptiles and absorption spectra of β-carotene and riboflavin. (After Thimann and Curry, 1961; Figures 4 and 5.)

Riboflavin also is widely distributed in plant material, and, like the carotenoids, it is a yellow pigment that absorbs light between 300 and 500 nm. It has an absorption spectrum with a peak at about 360 nm, which corresponds closely to the 360-nm peak in the phototropism action spectrum. The peak around 460 nm is flat for riboflavin, however, lacking the fine structure shown for the action spectrum. Riboflavin is one of a number of flavin compounds present in biological material, usually in a bound form with protein. Studies have shown that a flavoprotein located in isolated membrane fractions is reduced when irradiated in the blue region of the spectrum. The reduced flavoprotein then reduces a b-type cytochrome also present in the membrane fraction. These results suggest that the blue photoreceptor involved in phototropism is a pigment complex composed of a flavoprotein and a heme protein (cytochrome). As a result of light absorption by the pigment complex, an electrochemical gradient is established across the membrane, leading to a change in distribution of indole-3-acetic acid (see Chapter 14) within the tissue. The unequal distribution of indole-3-acetic acid is reflected in the amount of growth substance transported to the growing regions and produces a curvature of the coleoptile. Additional details of the actual growth phenomena are given in Chapter 17.

The problem of identifying the phototropic receptor molecule illustrates the difficulties encountered in photophysiological studies. Although isolated pigments may show a well-defined absorption spectrum, these pigments within the plant may be present in a quite different environment. Furthermore, the pigment molecules may be localized in specific subcellular fractions, such as membranes. Biochemical techniques of subcellular fractionation and protein analysis will be required to resolve these problems.

Photomorphogenesis

Photomorphogenetic processes utilize radiant energy to "trigger" or initiate reactions that control or alter growth, development, or differentiation. One of the major differences between photomorphogenetic reactions and a photoprocess like photosynthesis is that the former are initiated by low levels of radiant energy. In many plants, for example, photosynthesis may be driven by solar flux densities in the range 1000 to 1200 W m^{-2} whereas a photomorphogenetic reaction, such as seed germination, is triggered at flux densities of around 0.01 to 0.1 W m^{-2}. Moreover, in photosynthesis radiant energy must be supplied continuously whereas in photomorphogenesis a brief exposure to an appropriate radiation source may suffice to set the process in motion.

Responses to red light   For many years it was known that plants respond to low levels of irradiation by several different growth responses, such as etiolation and bending. That such photoresponses have a wavelength dependency is a more recent discovery based originally on studies of seed germination. It was observed that the seeds of some plant species did not germinate if maintained in the dark in a fully imbibed condition. If such seeds were given a brief exposure to light, however, germination proceeded in a normal manner. Seeds of many weed species, for instance, buried in the soil where they had access to moisture and other suitable environmental conditions, germinated only if they were turned up to the soil surface (cultivation, plowing, erosion) where they were exposed to light. Lacking a light exposure, such seeds might remain buried for years, still viable and able to germinate once given a light treatment.

In the 1930s two scientists in the Washington, D.C., area, L. H. Flint of the U.S. Department of Agriculture and E. D. McAlister of the Smithsonian Institution, were studying the germination of light-sensitive seeds under light of different wavelengths obtained from a series of filters. They reported in 1935 that the seeds of lettuce (*Lactuca sativa* cv. Grand Rapids) were promoted to germinate by light in the spectral region of 525 to 700 nm. The optimal promotive effect was noted around 660 nm, light commonly referred to as red light. They also found that light in the spectral region of 700 to 820 nm, referred to as far-red light, has no promotive influence on the germination of Grand Rapids lettuce seed.

These observations were confirmed and expanded in the 1950s by a group of plant scientists under the leadership of H. A. Borthwick and S. B. Hendricks at the U.S. Department of Agriculture Plant Industry Station in Beltsville, Maryland. This group constructed a large spectrograph that enabled them to expose seeds or other plant parts to rather specific wavelengths under conditions of relatively high flux densities. They were thus able to measure plant responses in spectral regions of comparable incident energies. It will be recalled that the photons from short wavelengths are more energetic than those from long wavelengths. The Beltsville group determined the action spectrum of germination of light-sensitive seeds.

Dry seeds of Grand Rapids lettuce do not germinate; nor do dry seeds respond to light. Seeds were allowed to imbibe water from water-soaked blotters in Petri dishes for 16 hours in complete darkness. If left in darkness for a further 32 hours, a few seeds were observed to germinate. But if at the end of 16 hours of dark imbibition the seeds were exposed to a brief flash of white light or sunlight and then returned to darkness for an additional 32 hours, all seeds germinated. If the dark period was interrupted at the end of 16 hours with equal incident energies of light of varying wavelengths rather than white light or sunlight, seed germination was promoted by red light (maximum effectiveness at 660 nm). Far-red light did not promote germination. If seeds promoted to germinate by red light were given an immediate exposure to far-red light (maximum effectiveness at 730 nm), however, seed germination was inhibited.

The promotion of seed germination in red light (660 to 680 nm) and the inhibition of seed germination in far-red light (700 to 740 nm) are reversible, as shown in Table 5-2. It is seen that germination depends on the kind of light last presented to the seed. To account for this pattern of response to red and far-red light, Borthwick, Hendricks, and associates proposed that the seeds contain a pigment that can exist in two forms. When the pigment receives red light, it is converted to the far-red absorbing form. The far-red absorbing form is converted to the red absorbing form when irradiated with far-red light. In 1960 Borthwick and Hendricks named the pigment phytochrome.

Phytochrome   The Beltsville group studied the participation of phytochrome in other photomorphogenetic responses, such as flower initiation and stem elongation.

TABLE 5-2   LETTUCE SEED GERMINATION AFTER EXPOSURE TO RED (R) AND FAR-RED (FR) RADIATION IN SEQUENCE

Irradiation schedule	Final irradiation	Germination (%)
Control (dark)		8.5
R (660 nm)	R	98
R + FR (730 nm)	FR	54
R + FR + R	R	100
R + FR + R + FR	FR	43
R + FR + R + FR + R	R	99
R + FR + R + FR + R + FR	FR	54
R + FR + R + FR + R + FR + R	R	98

After Borthwick et al., 1952.

Other investigators have added still more phytochrome-mediated physiological responses, as shown in Table 5-3. The involvement of phytochrome in all these photoprocesses is indicated because they are reversibly promoted or inhibited by red and far-red light in a manner similar to that shown by lettuce seed germination in Table 5-2. Furthermore, these photomorphogenetic responses all have similar action spectra. A generalized action spectrum for red and far-red light effects on photomorphogenetic responses is shown in Figure 5-11. The figure represents a composite of numerous action spectra determined for some of the photoresponses listed in Table 5-3.

Following the recognition that phytochrome is the pigment responsible for mediating the red and far-red light effects, attempts were made to isolate the pigment. A spectrophotometer was developed that irradiated plant material at 660 or 730 nm, and from the differences in light absorption at those two wavelengths it was possible to obtain an idea of the amount of phytochrome present in a sample of plant material. This instrument enabled the Beltsville group to screen numerous plants or plant parts for phytochrome. Dark-grown (etiolated) tissue of oats, wheat, rye, and maize proved to be good sources of phytochrome, and from these tissues suitable procedures were developed for isolating and purifying phytochrome. The absorption spectrum of the red-absorbing (P_r) and far-red-absorbing (P_{fr}) forms of phytochrome isolated from etiolated maize tissue is shown in Figure 5-12. Note that the absorption spectrum of phytochrome is similar to the generalized action spectrum for phytochrome activity shown in Figure 5-11.

With the development of a suitable measuring instrument, it became feasible to isolate and purify sufficient phytochrome for studies of its chemical composition. It was found that phytochrome consists of two components: a protein and a chromophore that gives phytochrome its light-absorbing properties. The protein appears to be composed of subunits. Moreover, the proteins of phytochromes isolated from different plant species are not identical in amino acid composition but seem to be more or less similar in size and configuration.

Information on the chemical nature of the phytochrome chromophore has come from studies of algal chromoproteins. These algal chromoproteins have absorption spectra very similar to phytochrome, as shown in Figure 5-12. Here the absorption spectrum of allophycocyanin, an algal chromoprotein isolated from *Porphyra smithii*, is shown with the absorption spectrum of the red-absorbing and far-red-absorbing forms of phytochrome. The shapes of the two absorption spectra are similar, with allophycocyanin having an absorption maximum at a wavelength slightly less than 660 nm, the absorption maximum for phytochrome. Considerable information is available on the chemical nature of the algal chromophores and it has been possible to come up with a suggestion concerning the chemical identity of the phytochrome chromophore. It is believed to be a tetrapyrrole closely related in structure to the bile pigments in animals. A possible structure is shown in Figure 5-13. Four pyrrole rings are arranged in a linear manner. With the absorption of red light, the distribution of double bonds in the chromophore is thought to shift so that the molecule is changed to the far-red-absorbing form. With the absorption of far-red light, there is an internal redistribution of double bonds to give rise to the red-absorbing form. The chromophore is firmly

TABLE 5-3 SOME PHYTOCHROME-MEDIATED PHOTORESPONSES

1. Elongation (leaf, petiole, stem)	13. Formation of tracheary elements
2. Hypocotyl hook unfolding	14. Changes in rate of cell respiration
3. Sex expression	15. Synthesis of anthocyanin
4. Bud dormancy	16. Increase in protein synthesis
5. Root development	17. Increase in RNA synthesis
6. Rhizome formation	18. Auxin catabolism
7. Bulb formation	19. Permeability of cell membranes
8. Leaf abscission	20. Photoperiodism
9. Succulency	21. Seed respiration
10. Enlargement of cotyledons	22. Changes in membrane conformation
11. Seed germination	23. Changes in cell turgor
12. Flower induction	

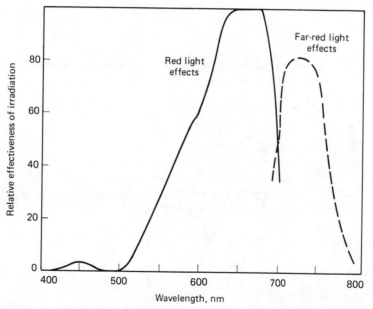

Figure 5-11 Action spectrum for the promotion (left) and inhibition (right) of germination of Grand Rapids lettuce seed. For the promotion experiment, seeds were soaked in water for 16 hours in darkness and then exposed to radiation of different wavelengths. After irradiation, the seeds were returned to darkness for 32 hours and the germination percentage determined. In the inhibition experiments seeds were soaked in water for 16 hours in darkness and then exposed to radiation of 580 to 660 nm for several minutes, after which they were irradiated for several minutes in different wavelengths, returned to darkness for 32 hours and then examined for germination. (After Borthwick et al., 1952.)

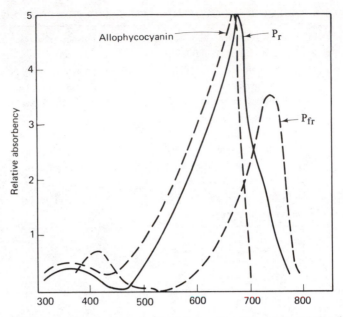

Figure 5-12   Absorption spectrum for the red-absorbing (P_r) and far-red absorbing (P_{fr}) forms of phytochrome isolated from etiolated maize seedlings. Also, the absorption spectrum of allophycocyanin, a pigment isolated from the alga *Porphyra smithii*. (From Hendricks and Borthwick, 1963; Figure 7.)

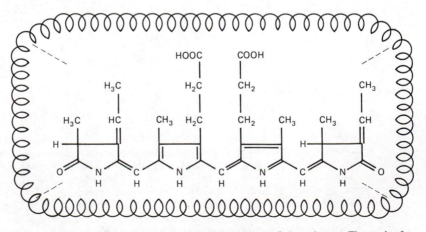

Figure 5-13   A suggested structure of the chromophore of phytochrome. The mode of attachment of the chromophore to the protein component is not known.

attached to the phytochrome protein and it is possible that the conformation of the protein is changed as the chromophore goes from the red to the far-red form.

Phytochrome in the plant kingdom has been demonstrated in green and red algae, desmids, bryophytes, angiosperms, and gymnosperms. In most instances, the presence of phytochrome is inferred from the physiological response of the organism to red and far-red light. Similarly, phytochrome has been detected in roots, coleoptiles, stems, hypocotyls, cotyledons, petioles, leaf blades, vegetative buds, floral tissues, seeds, and developing fruits of higher plants. In a few cases, phytochrome has been isolated from plant tissues. Within the cell, phytochrome may be located at membrane surfaces, but it is not known if phytochrome is associated with specific organelles, such as mitochondria, chloroplasts, the endoplasmic reticulum. Phytochrome is soluble (and thus extracted) at alkaline pHs where organelles are not stable; thus localization is a difficult problem because phytochrome is precipitated at pHs that maintain organelle integrity during isolation.

Photochemical transformations of phytochrome   From their physiological studies on seed germination and flowering Borthwick and Hendricks suggested that the two forms of phytochrome (P) are interconvertible as follows:

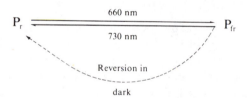

Phytochrome in the P_{fr} form slowly reverts to P_r in darkness. Thus P_{fr} is rapidly converted to P_r by far-red light (730 nm) and slowly converted to P_r in darkness. Some P_{fr} might also be lost by degradative processes. Inasmuch as P_{fr} is thought to be the biologically active form of phytochrome, the level of P_{fr} in a tissue or organ depends on both light and darkness.

From the absorption spectra of P_r and P_{fr} (Figure 5-12) it is seen that both forms absorb radiation over much the same spectral range. Thus both P_r and P_{fr} absorb light at 660 or 730 nm, and following monochromatic irradiation at 660 or 730 nm, or some intermediate wavelengths, a certain ratio of P_{fr} to P_r is established. The ratio of P_{fr} to P_r following irradiation at 730 nm is about 0.03, for example, indicating that a small but measurable amount of P_{fr} is present. It is possible, then, to establish variable levels of P_{fr} to P_r in plant tissue by irradiation at different wavelengths and different physiological responses are obtained, depending on the ratio of P_{fr} to P_r present.

The interconversions of P_r and P_{fr} are more complex than just described. Furthermore, biologically active P_{fr} is stabilized by reacting with a substance [X] whose identity is unknown. [X] may be a substance like ATP, NAD, or cytochrome, or a molecule whose concentration is dependent on a light-mediated reaction. Or P_{fr} may bind to some protein in a membrane. Whatever [X] might be, the amount of $[P_{fr} \cdot X]$

and the length of time it is present in the tissues are important factors in phytochrome-mediated responses.

P_r appears to be synthesized continuously in light-grown plants. The degradation of phytochrome proceeds from the P_{fr} form. A scheme summarizing the ideas presented is shown in Figure 5-14.

The sun is the source of the natural radiant energy environment in which plants grow. Plants are therefore exposed to all wavelengths of radiant energy and, insofar as the phytochrome system is concerned, both red and far-red light are present in sunlight. Because of differences in energy levels of red and far-red light and because of the differential sensitivity of phytochrome in the P_r and P_{fr} forms, following an exposure to sunlight most of the phytochrome is in the P_{fr} form. With darkness the P_{fr} form may be destroyed or may decay to P_r; or if it has interacted with X to form X · P_{fr}, the biologically active form X · P_{fr} participates in some photomorphogenetic process.

When sunlight passes through clouds or through leaves or soil, some of the wavelengths will be partially absorbed or reflected so that the emerging light will have quite a different wavelength composition than unfiltered sunlight. Also, the position of the sun in the sky will have an influence on the spectral composition of the light reaching the surface of the earth. During winter months, when the sun is low, more far-red light strikes the earth, whereas during the summer, with the sun high in the sky, more red light strikes the earth.

Mode of action of phytochrome Some of the processes listed in Table 5-3 occur rapidly after appropriate levels of P_r and P_{fr} are achieved (changes in membrane permeability, changes in cell turgor, enzyme synthesis) whereas others may not be apparent for hours, days, or weeks (flower induction, bulb formation). How is phytochrome involved in such a variety of different biochemical and physiological responses? The initial event appears to be a change in membrane permeability, but it is not known if this change occurs at the plasma membrane or at an organelle membrane. The phytochrome model illustrated in Figure 5-13 shows two components—chromophore and protein. If the protein is an integral part of membrane protein, a

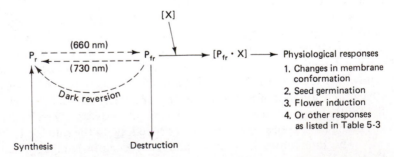

Figure 5-14 Some interactions between P_r and P_{fr} leading to physiological responses. The dashed lines between P_r and P_{fr} represent intermediate states of phytochrome. [X] represents an unknown substance that stabilizes P_{fr} to a biologically active form [P_{fr} · X].

change in the chromophore through the establishment of different levels of P_r and P_{fr} may alter the conformation of phytochrome protein. Such a change in membrane conformation will alter the permeability of the membrane to water and solutes (H^+, K^+ Ca^{2+}). Thus rapid changes in turgor pressure and solute content occur.

The initial change in membrane conformation following phytochrome interconversion will set in motion other biochemical processes. A closely associated membrane protein, for example, may have enzymic properties that are activated by phytochrome changes. Or changes in solute concentrations of such ions as K^+, H^+, and Ca^{2+} may modify enzyme activity and set in motion a chain of biochemical reactions that eventually are observed later as organ initiation—for instance, flower induction, bulb formation. Another possibility is that a particular ratio of P_r to P_{fr} activates specific genes and the eventual synthesis of new enzymes. The initial event again may be a change in membrane conformation following phytochrome interconversion.

High irradiance responses Phytochrome reactions and phototropism are characterized by their sensitivity to low irradiances. Some photomorphogenetic processes, however, proceed only under high irradiances achieved over long periods of time. These reactions are referred to as high irradiance responses (HIR). An example of an HIR is cotyledon enlargement in *Sinapis alba*. The action spectrum (Figure 5-15) shows a peak in activity around 730 nm similar to P_{fr} but no peak at 660 nm. The peak in activity in the blue region around 450 nm is characteristic of HIRs. Not shown in

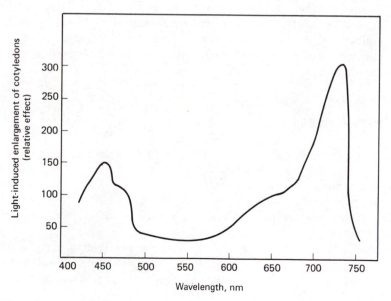

Figure 5-15 Action spectrum for high intensity reaction (HIR) regulating the light-induced enlargement of the cotyledons of mustard seedlings (*Sinapis alba*). (After Mohr, 1972)

Figure 5-15 is the fact that many HIRs also have a peak in activity in the near UV region around 370 nm.

A major distinction between HIR and phytochrome responses is that phytochrome-mediated reactions are red and far-red irradiation reversible (as illustrated in Table 5-2 for lettuce seed germination) whereas HIRs are not. It is not clear if the HIRs can be attributed to some special property of phytochrome when irradiated for long periods of time at high irradiancies or if some other pigment is responsible. The fact that plants in nature are exposed to sunlight with high levels of radiant energy acting over long periods of time suggests that HIRs are important in plant growth and development.

Photoregulation of Enzyme Activity

In addition to the enzymes mediated by phytochrome described earlier, another class of photoregulated enzymes is found in green plants. The enzymes are localized in the stromal fraction (pp. 32–33) of chloroplasts, where they catalyze some of the reactions of CO_2 assimilation and starch synthesis. The enzymes are activated by light and deactivated by darkness through the effect of radiant energy on chlorophyll in the granal fraction of chloroplasts. When chloroplasts are irradiated, chlorophyll undergoes changes (pp. 179–182) that lead to alterations in the chemical environment of the stroma. The pH shifts from 7 to 8, concentrations of ATP and NADPH increase, concentration of Mg^{2+} increases, and certain cofactors are reduced. In darkness these changes are reversed. Because of the changes in the chemical environment of the stroma, enzymes in the stroma are either activated or deactivated. Moreover, light may have a direct influence on enzyme conformation (activity) by modifying some of the amino acid components of the enzyme. The specific roles of the photoregulated chloroplast enzymes in CO_2 assimilation and starch synthesis are described in the next chapter.

REFERENCES

ANDERSON, L. E. 1979. Interaction between photochemistry and activity of enzymes. Pages 271–281 *in* M. Gibbs and E. Latzko, eds. *Encyclopedia of Plant Physiology* II, vol. 6. Springer-Verlag, New York.

BORTHWICK, H. A., S. B. HENDRICKS, M. W. PARKER, E. H. TOOLE, and V. K. TOOLE. 1952. A reversible photoreaction controlling seed germination. *Proc. Natl. Acad. Sci. U.S.* 38:662–666.

BRIGGS, W. R. 1976. The nature of the blue light photoreceptor in higher plants and fungi. *In* H. Smith, ed. *Light and Plant Development.* Butterworth, London–Boston.

BUCHANAN, B. B. 1980. Role of light in the regulation of chloroplast enzymes. *Ann. Rev. Plant Physiol.* 34:341–374.

BUCHANAN, B. B., R. A. WOLOSIUK, and P. SCHÜRMANN. 1979. Thioredoxin and enzyme regulation. *Trends Biochem. Sciences* 4:93–95.

CLAYTON, R. R. 1970. Light and living matter. Vol. 1: *The Physical Part*. Vol. 2: *The Biological Part*. McGraw-Hill Book Company, New York.

COLEMAN, R. A., and L. H. PRATT. 1974. Phytochrome: immunological assay of synthesis and destruction in plants. *Planta* 119:221–231.

CORRELL, D. J., J. L. EDWARDS, and W. SHROPSHIRE, Jr. 1977. Phytochrome bibliography. Smithsonian Institution, Washington, D.C.

FLINT, L. H., and E. D. McALISTER. 1935. Wavelengths of radiation in the visible spectrum inhibiting the germination of light-sensitive seed. Smithsonian Miscellaneous Publication No. 94. Smithsonian Institution, Washington, D.C.

FONDEVILLE, J. C., M. J. SCHNEIDER, H. A. BORTHWICK, and S. B. HENDRICKS. 1967. Photocontrol of *Mimosa pudica* L. leaf movements. *Planta* 75:228–238.

HALE II, C. C., and S. ROUX. 1980. Photoreversible calcium fluxes induced by phytochrome in oat coleoptile cells. *Plant Physiol.* 65:658–662.

HANSEN, M. C., and W. H. BIGGS. 1979. Units and terminology used in radiation measurements in conjunction with the plant sciences. Li-Cor, Inc., Lincoln, Nebraska.

HENDRICKS, S. B., and H. A. BORTHWICK. 1965. The physiological functions of phytochrome. Pages 405–436 *in* T. W. Goodwin, ed. *Chemistry and Biochemistry of Plant Pigments*. Academic Press, New York.

MITRAKOS, K., and W. SHROPSHIRE, Jr., eds. 1972. *Phytochrome*. Academic Press, New York.

MOHR. H. 1972. *Lectures on Photomorphogenesis*. Springer-Verlag, New York.

SMITH, H. 1974. The biochemistry of photomorphogenesis. Pages 159–197 *in Plant Biochemistry*, vol. 11. MTP International Review of Science, Biochemistry Series One. Butterworth & Co. Ltd. and University Park Press, Baltimore.

SMITH, H. 1975. *Phytochrome and Photomorphogenesis*. McGraw-Hill Book Company, New York–London.

SMITH, H., ed. 1976. *Light and Plant Development*. Butterworths, London-Boston.

SMITH, K. C., ed. 1977. *The Science of Photobiology*. Plenum Press, New York.

THIMANN, K. W., and G. M. CURRY. 1961. Phototropism. Pages 646–672 *in* W. D. McELROY and B. GLASS, eds. *Light and Life*. The Johns Hopkins Press, Baltimore.

VINCE-PRUE, D. 1975. *Photoperiodism in Plants*. McGraw-Hill Book Company, London–New York.

WITHROW, R. B., ed. 1959. Photoperiodism and related phenomena in plants and animals. Publication No. 55. American Association for the Advancement of Science, Washington, D.C.

6

Photosynthesis

Green plants capture solar energy and convert it to chemical energy by the process known as photosynthesis. During photosynthesis carbon dioxide and water are transformed into simple carbohydrates and oxygen gas is liberated into the atmosphere. The simple carbohydrates produced in photosynthesis are converted, by additional metabolic processes, into lipids, nucleic acids, proteins, and other organic molecules. These organic molecules, in turn, are elaborated into leaves, stems, roots, tubers, fruits, seeds, and other tissue and organ systems.

Plants and plant products are the major food sources for almost all other organisms of the earth. The total mass of living organisms (plants, insects, mammals, etc.), referred to as the biosphere, is small compared with the nonliving portions of the earth. For example, the earth's crust (lithosphere) weighs 1.5×10^{22} kg, the ocean (hydrosphere) 1.4×10^{22} kg, and the atmosphere 5.1×10^{18} kg, whereas the biosphere amounts to only 1.2×10^{15} kg (dry weight). Despite the disparity in weights, the activities of the organisms within the biosphere contribute significantly to the maintenance and activity of the lithosphere, hydrosphere, and atmosphere.

Estimates of the amounts of carbon fixed annually by green plants are given in Table 6-1. Two kinds of estimates are shown. The values in column 3 are rate measurements—that is, the amounts of carbon fixed per unit area of land surface per year. Such values are commonly referred to as productivity estimates. The values in column 4 measure the total amount of carbon fixed annually by a particular plant community. These values are referred to as plant production and are determined by multiplying plant productivity (column 3) by the area of the community (column 2). The productivity values given in Table 6-1 are for net productivity—the amount of carbon assimilated into plant material after the loss of carbon by respiratory processes.

TABLE 6-1   AMOUNTS OF CARBON FIXED ANNUALLY BY GREEN PLANTS

Plant community (1)	Area (km^2) (2)	Carbon fixed[a] kg m^{-2} yr^{-1} (3)	kg yr^{-1} (4)
Forests	44×10^6	1.4	0.62 $\times 10^{14}$
Arable land	23×10^6	0.91	0.21 $\times 10^{14}$
Grasslands	27×10^6	0.91	0.25 $\times 10^{14}$
Desert	33×10^6	0.045	0.015 $\times 10^{14}$
Tundra and ice	22×10^6	0.02	0.005 $\times 10^{14}$
Total land	149×10^6		1.10 $\times 10^{14}$
Freshwater lakes, rivers	0.2×10^6	0.1	0.0002 $\times 10^{14}$
Ocean	361×10^6	0.08	0.293 $\times 10^{14}$
Total for earth	510×10^6	0.27 (mean)	1.39 $\times 10^{14}$

[a]The values for carbon fixed represent net production—that is, the amount of carbon fixed in excess of that utilized in respiratory processes.

Data from a number of sources. See Bowen, 1966.

Approximately 30% of the carbon fixed by green plants is lost through respiration.

Table 6-1 shows that productivity varies rather widely in different plant communities. Some of the variability is explained by actual differences in the photosynthetic process itself among plants, but such differences are usually small. Most of the differences in productivity are due to differences in environmental factors, such as temperature, water supply, and nutrients, or to morphological characteristics. Consider the productivity of trees in two forest communities, one consisting of coniferous (evergreen) trees and the other of deciduous trees. The evergreen trees hold their leaves throughout the year and fix carbon at a rate of 1.4 kg of carbon per square meter per year. Deciduous trees, on the other hand, are devoid of leaves for about 6 months out of the year and fix carbon at a rate of 0.54 kg of carbon per square meter per year. Or compare two evergreen forest communities, one in the tropics and the other in a temperate climate. Trees in the tropical forest fix carbon at the rate of 2.3 kg of carbon per square meter per year compared to 1.4 kg of carbon per square meter per year by the evergreen trees growing in a temperate climate.

Thus the productivity values of trees range between 0.54 and 2.3 kg of carbon fixed annually per square meter of land surface. The mean value of forest productivity is about 1.4 kg of carbon per square meter per year, a rate of carbon fixation resulting in the production of 0.62×10^{14} of carbon annually on 44×10^6 km^2 of land surface. Although occupying only about 30% of the total land surface, forests account for more than half the carbon fixed annually by terrestrial plants. The major reason for the high production by forest communities as a whole is the high rate of carbon fixation in tropical trees.

Under comparable environmental conditions plants on arable land (agricultural crops) and native grasses have similar rates of carbon fixation—0.91 kg of carbon per square meter per year (Table 6-1). The productivity of desert plants is limited by high temperatures and lack of moisture whereas the productivity of plants growing in the tundra and ice is limited by low temperature. Available soil moisture also tends to be limited in cold climates.

About 70% of the surface of the earth is covered with water. Freshwater lakes, streams, and rivers account for a small fraction of the aquatic habitats and the total amount of carbon fixed annually is small compared to that fixed by plants growing in the oceans of the earth. About 90% of the oceans are deeper than 200 m and productivity is low—about 0.07 kg of carbon fixed per square meter per year. Plants growing in the shallow seas and continental shelf fix carbon at rates comparable to land plants. Plants of deep seas are mostly unicellular algae and are quite numerous. This fact, together with the area of the deep seas, accounts for most of the total production of aquatic habitats.

The total amount of carbon fixed annually by terrestrial and aquatic plants is 1.39×10^{14} kg of carbon (Table 6-1). To obtain an idea of what such a figure means, it is useful to convert it to more familiar numbers. A metric ton (10^6 g) is equivalent to 2200 lb. If the production values in Table 6-1 are converted to metric tons, land plants fix 110×10^9 tons of carbon annually and aquatic plants fix 29×10^9 tons of carbon annually for a total of 139×10^9 tons of carbon per year. Compared to the 3×10^9 tons of carbon used annually by humans in the form of coal, petroleum, and natural gas, clearly plant photosynthesis is a "big business."

EFFICIENCY OF PLANTS IN CONVERTING RADIANT ENERGY INTO CHEMICAL ENERGY

The production of organic matter by photosynthesis depends on the availability of inorganic nutrients, adequate supplies of water and carbon dioxide, favorable temperature, radiant energy, and the absence of toxic substances from the immediate environment. These factors are part of the environment and may vary widely, giving rise to different levels of plant productivity. Internal factors, such as kinds of pigments, enzyme levels, and the degree of organization of the photosynthetic apparatus, also influence productivity. Taken together, the external and internal factors can be evaluated in terms of the efficiency of the plant in converting solar energy into chemical energy. The question is: How much of the radiant energy from the sun that falls on a plant is converted into plant organic matter?

Each hectare[1] of the surface of the earth receives approximately 168×10^9 J of energy daily. This energy covers a broad spectrum ranging from short wavelengths of ultraviolet radiation to long-wavelength infrared radiation. Plants, however, can only utilize wavelengths lying between 400 and 700 nm in photosynthesis. So only a

---

[1]One hectare = 10,000 m^2 = 2.47 acres.

relatively small portion of the radiant energy reaching the earth's surface is being used by plants.

To determine the efficiency of the plant in converting solar energy into chemical energy, it is necessary to measure how much plant material is produced in unit time (year, month, week) on a unit of land (acre, square meter). The energy content of the plant material is then determined by combustion and compared to the amount of solar energy that fell on the plants. The efficiency of conversion is calculated as follows:

$$\text{Efficiency of energy conversion} = \frac{\text{energy content of plant material}}{\text{solar energy available}}$$

Determinations of this kind have been made for a number of plants and plant communities in different parts of the world and the values found to be between 2 and 2.5% for crops being grown under intensive agricultural conditions, such as wheat in the Netherlands and rice in Japan. Under ordinary agricultural practices the efficiency of energy conversion varies between 0.1 and 1.0%. Values as high as 6 to 10% are found in some crop plants over brief periods of time. Under laboratory conditions efficiencies as high as 20 to 25% have been observed.

These figures show that the efficiency of energy conversion by plants varies widely from values as low as 0.1% to as high as 25%. Several questions arise. Is there an upper limit to the efficiency of energy conversion? Why are there such low values? These questions are examined more closely in the balance of the chapter, but several points can be mentioned here. On theoretical grounds, the maximum efficiency of energy conversion is believed to be between 25 and 30%. There is an upper limit and it has been achieved in certain plants for brief periods of time under optimal conditions.

The low levels of energy conversion (0.1 to 1.0%) attained by native vegetation or under ordinary agricultural conditions are low because of limiting environmental factors—lack of water, low or high temperatures, deficiencies of inorganic nutrients, pests (weeds, fungi, bacteria, rats), faulty seeds, poor cultural practices. By improving farming practices, especially irrigation and fertilization, energy conversion (and crop yield) can be improved ten- to twentyfold.

Under intensive agricultural conditions energy conversion efficiencies do not exceed 2 to 2.5%. The major reason for such low values is the fact that most crop plants display their maximal photosynthetic efficiency for relatively brief periods of time during the life cycle of the plant. Very young plants do not have sufficient leaf area to carry out high rates of photosynthesis whereas older plants are subject to senescence. Still another factor limiting plant productivity is the low level of carbon dioxide in the atmosphere. These and other factors relating to plant productivity are examined in more detail later after some aspects of the nature of the photosynthetic process are considered.

EARLY PHOTOSYNTHESIS RESEARCH

Although many early speculations about the nature of photosynthesis existed, it was not until the latter half of the eighteenth century that any progress was made. In a brief burst of activity that spanned about 30 years, the nature of the overall process of photosynthesis was elucidated. By 1804 it was possible to write the following equations as representing the chemical events of photosynthesis:

$$\text{Carbon dioxide} + \text{water} + \text{light} \xrightarrow[\text{plant}]{\text{green}} \text{organic matter} + \text{oxygen} \qquad (6\text{-}1)$$

$$\text{or} \qquad CO_2 + H_2O \xrightarrow[\text{green plant}]{\text{radiant energy}} \text{carbohydrate } (CH_2O) + O_2$$

During this same period in history chemists recognized that the atmosphere is composed of several different gases—carbon dioxide, oxygen, nitrogen—a fact that aided in the development of the photosynthetic equation.

It remained for Robert Mayer, a German scientist, to point out in 1845 that photosynthesis is of major significance in the biological world because it provides the means whereby solar energy in the form of electromagnetic radiation is converted into a useful form of chemical energy as plant organic matter.

Progress since 1845 has been uneven. There have been long periods of comparative quiet, but since about 1930 the pace has quickened and during the last 20 years, in particular, progress has been rapid. Many of the advances have depended on discoveries in fields outside of plant physiology, in the areas of biochemistry, microbiology, organic chemistry, and physics. Scientists from many different disciplines have worked on photosynthesis and, although a great deal is known about certain aspects of the photosynthetic process, other aspects are virtually unknown. New techniques, ideas, and skills are being applied to these unresolved problems.

Early in the nineteenth century it was recognized that the synthesis of carbohydrates, lipids, proteins, and other organic molecules proceeds through a number of chemical intermediates and attempts were made to isolate and identify these compounds. Particular attention was given to the identity of the so-called first product of photosynthesis. Moreover, considerable attention was paid to the chemical nature of the pigments of green plants. Two major green pigments, chlorophyll *a* and chlorophyll *b*, were isolated and their structures eventually elucidated.

Another class of pigments, the carotenoids, was found to be closely associated with the chlorophylls in the chloroplasts. An additional class of photosynthetic pigments is present in algae. These pigments, known as phycobilins, are open-chain tetrapyrroles (see Chapter 2) and are of two kinds, phycocyanins (blue in color) and phycoerythrins (red in color). All photosynthetic pigments are lipid soluble and water insoluble in contrast to several classes of water-soluble pigments, such as anthocyanins and flavones, found in vacuoles. Only the chlorophylls, carotenoids, and phycobilins participate in photosynthesis by absorbing radiant energy, which is then coupled to the reduction of carbon dioxide, evolution of oxygen, and the formation of organic matter as described in the following sections.

Effect of Environmental Factors on Photosynthesis

Despite the early difficulties encountered in studying the chemical events of photosynthesis, considerable information was gained by examining the effects of environmental factors, such as temperature, irradiance, and CO_2 concentration, on photosynthesis. Some typical results are illustrated in Figure 6-1.

Photosynthesis does not occur in the absence of light, but as the irradiance increases, the rate of photosynthesis increases. At CO_2 concentrations of 330 ppm, the level of CO_2 in natural air, the rate of photosynthesis does not increase indefinitely with increasing irradiance but levels off at irradiancies of 200 to 300 W m^{-2} (lower curve, Figure 6-1). The photosynthetic system is said to be *light saturated*. Radiant energy of this intensity is roughly equivalent to 20% that of full sunlight (1000 W m^{-2}). The maximum rates of photosynthesis occur at irradiancies considerably below full sunlight when the level of CO_2 approximates that present in natural air. Also note that temperature, either 20 or 30°C, has little influence on the rate of photosynthesis at CO_2 concentrations of 330 ppm.

When the CO_2 concentration is increased to 660 ppm (middle curve, Figure 6-1),

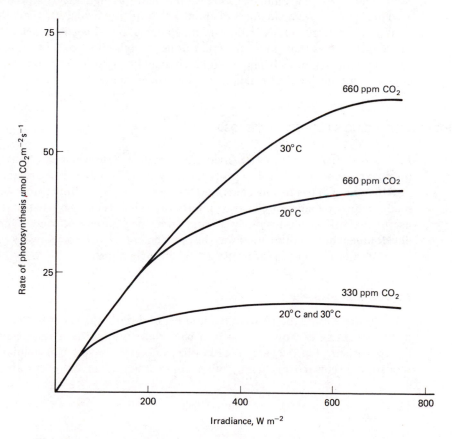

Figure 6-1 Rate of photosynthesis in an excised leaf as influenced by irradiance, CO_2 concentration, and temperature. [*Note*: An irradiance of 800 W m^{-2} is comparable to an illuminance of 85 klux. Full summer sunlight approximates 1000 W m^{-2} (110 klux).]

the rate of photosynthesis increases as the irradiance increases up to levels of about 500 W m^{-2}. The rate then levels off and additional radiant energy is without effect (light-saturation). If, however, the temperature is increased from 20 to 30° with CO_2 concentration remaining at 660 ppm, the rate of photosynthesis is markedly increased (upper curve, Figure 6-1) as irradiancies are increased. This situation contrasts sharply with the influence of temperature on the rate of photosynthesis at 330 ppm CO_2 (lower curve), where a 10° increase in temperature was without influence on the rate of photosynthesis.

The different effect of temperature on CO_2 assimilation at low and high levels of CO_2 is explained as follows. When the CO_2 concentration is low (330 ppm in Figure 6-1), the rate of photosynthesis increases as irradiance increases up to about 300 W m^{-2}. Because temperature does not increase the rate of photosynthesis from 0 to 300 W m^{-2}, it can be concluded that the factor limiting photosynthesis is some reaction that depends on radiant energy. Light reactions are physical reactions with a temperature coefficient[2] (Q_{10}) of around one. On the other hand, at higher CO_2 concentrations (660 ppm in Figure 6-1) at irradiances of about 500 W m^{-2}, an increase in temperature of 10° doubles the rate of CO_2 assimilation. Biochemical reactions have a Q_{10} of 2 or more. The conclusion is that under the conditions of high CO_2 concentration the process limiting photosynthesis is related to biochemical reactions. Additional details of the physical and biochemical reactions are described later.

PHOTOSYNTHESIS STUDIES SINCE 1930

Early work on photosynthesis provided a great deal of information about the nature of photosynthesis but it failed to give any answers as to how plants are able to convert radiant energy into chemical energy. During the 1930s a number of discoveries were made, many in fields unrelated to photosynthesis, which suggested that it might be useful to look at photosynthesis from some new points of view. One important development during this period was the rapid emergence of biochemistry as well as new methods and techniques for investigating biological systems.

Studies with Bacteria

It was found during the 1930s that a number of different kinds of bacteria are capable of transforming carbon dioxide into bacterial protoplasm. These bacteria are not photosynthetic, but they are able to utilize CO_2 and convert it to cellular material. It was soon discovered that many other kinds of organisms have this ability and that

---

[2]Temperature coefficient (Q_{10}) is defined as follows:

$$Q_{10} = \frac{\text{rate of process at X}° + 10°}{\text{rate of process at X}°}$$

In the above example X° = 20° C.

higher green plants are not unique in their capacity to assimilate CO_2. Enzyme systems capable of assimilating CO_2 in vitro were isolated from a number of these organisms. Plant biochemists subsequently found similar enzyme systems in plants, and, on the basis of the comparative biochemistry of the bacterial and plant systems, considerable insight was gained into the nature of the biochemical phases of CO_2 assimilation.

Although valuable because of their contribution to the biochemistry of the carbon dioxide assimilation processes in plants, these studies did not furnish any clue regarding the nature of the photochemical reaction in photosynthesis. Valuable information on this point, however, came from studies with another group of bacteria, the photosynthetic bacteria. These organisms are characterized by the ability to utilize radiant energy for the conversion of CO_2 into bacterial protoplasm, although they are unable to evolve oxygen. They have a pigment similar in structure to chlorophyll a, bacteriochlorophyll, which functions in bacterial photosynthesis as does chlorophyll a in the higher plants. One group of photosynthetic bacteria, the green sulfur bacteria, was found to assimilate carbon dioxide by using hydrogen sulfide (H_2S) as a hydrogen source according to the equation

$$CO_2 + 2\ H_2S \xrightarrow[\substack{\text{green sulfur bacteria}\\ \text{(bacteriochlorophyll)}}]{\text{radiant energy}} CH_2O + H_2O + 2\ S \qquad (6\text{-}2)$$

C. B. van Niel, a microbiologist who was largely responsible for showing the importance of photosynthetic bacteria in photosynthesis research, pointed out the similarity of Eq. (6-2) to that representing green plant photosynthesis in Eq. (6-1). To make the equations comparable, it is necessary to add a molecule of water to each side of Eq. (6-1) as follows:

$$CO_2 + 2\ H_2O \xrightarrow[\substack{\text{green plants}\\ \text{(chlorophyll } a)}]{\text{radiant energy}} CH_2O + H_2O + O_2 \qquad (6\text{-}3)$$

By adding n molecules of water to each side of Eqs. (6-2) and (6-3), a generalized equation for photosynthesis in higher plants and bacteria can be written,

$$CO_2 + 2\ H_2A + nH_2O \xrightarrow[\text{photosynthetic organism}]{\text{radiant energy}} CH_2O + 2\ A + (n + 1)\ H_2O \quad (6\text{-}4)$$

Here H_2A represents a hydrogen donor; in green plants this is water; in green sulfur bacteria it is hydrogen sulfide. In some groups of photosynthetic bacteria the hydrogen donor is hydrogen gas whereas in others it may be an organic acid. The action of light might be thought to produce a reducing entity [H] and an oxidizing entity [O] from water.

Another feature to be noted from van Niel's work is the source of oxygen gas liberated during photosynthesis. From Eq. (6-1) it can be seen that the oxygen gas might be thought to arise from either a splitting of a CO_2 molecule or by a splitting of two water molecules ($2\ H_2O \longrightarrow O_2 + 4\ [H]$). A comparison of Eqs. (6-2) and (6-3) reveals that the hydrogen donor—water in green plant photosynthesis—is the source of oxygen liberated during photosynthesis in green plants. Experiments with the oxygen isotope, ^{18}O, wherein ^{18}O-labeled water or ^{18}O-labeled CO_2 was fed to plants have confirmed the idea that water is the source of the oxygen produced in photosynthesis.

Oxygen Production by Isolated Chloroplasts (Hill Reaction)

During the 1930s biochemists had some success in studying the mechanism of respiration in yeast and mammals by using cell homogenates. It was found that the homogenates retained the ability to catalyze the oxidation of a number of substrates and several of the respiratory intermediates were identified. When similar techniques were applied to leaf homogenates with the expectation that some of the photosynthetic intermediates might be identified, little success ensued. Carbon dioxide assimilation by isolated chloroplasts or leaf homogenates could not be demonstrated. In 1939, however, Robert Hill showed that leaf fragments or chloroplasts are capable of evolving oxygen when illuminated if the extracts are supplied with suitable hydrogen acceptors. Hill found salts of ferric iron (ferrioxalate, ferricyanide) to be satisfactory hydrogen acceptors. With ferricyanide, the reaction may be written

$$2\ H_2O + 4\ [Fe(CN)_6]^{3-} \longrightarrow 4\ [Fe(CN)_6]^{4-} + 4\ H^+ + O_2 \qquad (6\text{-}5)$$

This reaction, now referred to as the *Hill reaction*, shows clearly that water is the source of the oxygen evolved during photosynthesis. In green plant photosynthesis, carbon dioxide—not ferricyanide—serves as the hydrogen acceptor. But Hill and coworkers were unable to demonstrate that their chloroplast preparations reduced CO_2. But as discussed later, subsequent research has shown that isolated chloroplasts can carry out such a reaction. The techniques available to Hill and other early workers were not sensitive enough to demonstrate such processes.

Partial Reactions of Photosynthesis

The studies described in Figure 6-1 showed that photosynthesis involves two quite different types of reactions—a physical reaction closely tied to the absorption of radiant energy and a chemical reaction (or reactions) concerned with the assimilation of CO_2 and its conversion to carbohydrates and other cellular constituents. The work of van Niel and Hill provided some additional insight into the reactions in green plant photosynthesis, which are shown schematically as follows:

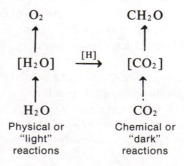

The bracketed molecules denote a complex of reactions that result ultimately in the evolution of oxygen and the formation of reduced carbon dioxide. This scheme is consistent with the idea that the reduction of carbon dioxide in green plants is not directly associated with radiant energy but with a product of the light reaction. Some details of these "dark" and "light" reactions are now discussed.

THE PATH OF CARBON IN PHOTOSYNTHESIS

The Calvin Cycle or Reductive Pentose Phosphate Cycle (C_3 Plants)

Inasmuch as plants contain numerous simple molecules that might arise directly from carbon dioxide during photosynthesis, a procedure is needed for specifically tagging the carbon dioxide before it enters the plant. The identity of the compounds formed from the tagged carbon dioxide can then be determined. Such a tag, or label, for carbon dioxide is provided by isotopic carbon. The most widely used radioactive isotope of carbon, ^{14}C, was discovered in the early 1940s and in 1945 a group of plant scientists at the University of California at Berkeley under the leadership of Melvin Calvin and Andrew A. Benson initiated a research program to follow the path of carbon in photosynthesis.

The general experimental approach is as follows. Plants are grown under suitable conditions of nutrition, temperature, and light so that a steady rate of photosynthesis is established in the presence of air whose carbon dioxide is almost entirely $^{12}CO_2$. Carbon dioxide labeled with $^{14}C(^{14}CO_2)$ is then added and the plant is allowed to carry out photosynthesis under the appropriate conditions of temperature and illumination. At various times after the introduction of the $^{14}CO_2$, plant samples are removed and killed by immersing them in boiling ethanol. This treatment stops enzymic reactions and extracts the soluble compounds from the plant. The alcohol-soluble extract is then analyzed to ascertain which compounds contain ^{14}C. In early experiments the Berkeley group used analytical procedures appropriate for organic molecules—extraction, precipitation, crystallization, etc.—but such methods were difficult and not especially suitable for isolating very small quantities of materials. In many experiments milligram amounts and less of organic molecules were available for analysis.

About the time that $^{14}CO_2$ experiments were started a new analytical procedure was introduced to biologists—paper chromatography. The separation of plant pigments by column chromatography had been achieved in the late 1800s, but the use of filter paper as a supporting medium for chromatography was not introduced until the middle 1940s. In this procedure the material to be analyzed is added near the edge of a piece of filter paper, which is then placed in an appropriate organic solvent. The solvent moves over the sample by capillary action and displaces the different constitutents according to their relative solubility in the solvent. Different solvent mixtures may be used, thereby achieving varying degrees of separation of the constituents in the original extract.

The advantages of the technique of paper chromatography for analyzing the ^{14}C-labeled plant extracts are many. Milligram and microgram quantities of material are sufficient for separation and identification. Of special significance is the fact that the ^{14}C-labeled compounds can be differentiated from the unlabeled compounds. As noted, plants contain many small organic molecules. Suppose, however, that only a few of these molecules incorporate ^{14}C during the early stages of photosynthesis whereas the bulk of the carbon compounds are not labeled. Paper chromatography separates all the organic molecules, labeled and unlabeled. If the chromatogram is then examined with an electronic device—for instance, a Geiger-Müller tube—the compounds containing radioactive ^{14}C are detected while the others are not. Or the chromatogram may be exposed to an x-ray film, and the ^{14}C-containing compounds will produce an image on the film, but the unlabeled compounds will not.

The preceding techniques were used by Calvin, Benson, and associates to separate, isolate, and identify the organic molecules involved in the fixation of carbon dioxide. As plant material, they used cultures of the unicellular algae *Chlorella* and *Scenedesmus*. Experiments with detached leaves provided additional support for their conclusions. Several dozen ^{14}C-labeled molecules were found following a few minutes photosynthesis in the presence of $^{14}CO_2$. As the times of exposure of $^{14}CO_2$ were shortened, the number of labeled compounds decreased until at 2 s the bulk of the radioactivity was found in a single compound, glyceric acid-3-phosphate (phosphoglyceric acid, PGA). Furthermore, the radioactivity from $^{14}CO_2$ was found in the carboxyl group of PGA, suggesting that the CO_2 was added to some acceptor molecule as follows:

$$\text{Acceptor} + C*O_2{}^3 \longrightarrow \text{Ⓟ} OCH_2CHOHC*OOH \qquad (6\text{-}6)$$
$$\text{PGA}$$

The identity of the acceptor molecule was ascertained from several kinds of experiments. Using paper chromatography, the amounts and radioactivity of the various compounds were followed under different photosynthetic conditions (high-low CO_2, light-darkness, etc.). It was noted that one compound decreased in concentration when CO_2 was present, accompanied by an increase in the concentrations of PGA. The same compound was found in high concentration in photosynthetic cells in the absence of light. From such experiments it was deduced that the acceptor molecule was ribulose-1,5-bisphosphate (RuBP), a phosphorylated pentose sugar.

Ribulose-1,5-bisphosphate reacts with CO_2 to form two molecules of glyceric acid-3-phosphate (PGA) in an enzymic reaction catalyzed by ribulose bisphosphate carboxylase (RuBPCase). See Equation (6-7). Because the initial carbon dioxide fixation product is a three-carbon compound (glyceric acid-3-phosphate), plants with this fixation pattern are often referred to as C_3 plants. It will be seen later that another pattern of carbon dioxide assimilation features a four-carbon compound as the initial fixation compound. These latter plants are know as C_4 plants.

Some of the PGA formed is converted to starch and/or sugar, but most is recycled to regenerate RuBP, the CO_2 acceptor. The identities of the intermediates in

---

[3]The asterisk (*) indicates that the carbon atom is isotopically labeled with ^{14}C.

$$
\begin{array}{c}
\text{CH}_2\text{O}\,\textcircled{P} \\
| \\
\text{C}{=}\text{O} \\
| \\
\text{H}{-}\text{C}{-}\text{OH} \\
| \\
\text{H}{-}\text{C}{-}\text{OH} \\
| \\
\text{CH}_2\text{O}\,\textcircled{P}
\end{array}
\;+\; \text{C*O}_2
\;\xrightarrow{\text{RuBPCase}}\;
\left[\begin{array}{c}\text{Labile}\\ \text{inter-}\\ \text{mediate}\end{array}\right]
\;\xrightarrow{+\text{H}_2\text{O}}\;
\begin{array}{c}
\text{CH}_2\text{O}\,\textcircled{P} \\
| \\
\text{H}{-}\text{C}{-}\text{OH} \\
| \\
\text{C*OOH} \\
+ \\
\text{COOH} \\
| \\
\text{H}{-}\text{C}{-}\text{OH} \\
| \\
\text{CH}_2\text{O}\,\textcircled{P}
\end{array}
\qquad (6\text{-}7)
$$

Ribulose-1-5-   Glyceric acid-3-
bisphosphate   phosphate

this cyclic process were elucidated by Calvin, Benson, Bassham, and colleagues at the University of California, Berkeley. The reaction sequence, shown in Figure 6-2, is referred to as the Calvin cycle or the reductive pentose phosphate cycle.[4] In the cycle as outlined in Figure 6-2, three CO_2 molecules (reaction 1) are converted to six molecules of GAP (glyceraldehyde phosphate). Glyceraldehyde phosphate is in equilibrium with dihydroxyacetone phosphate (DHAP). Three molecules of GAP and two molecules of DHAP are recycled to form the CO_2 acceptor RuBP, leaving one molecule of a three-carbon intermediate (DHAP) to be further metabolized to starch or sugar. Thus there is a net gain of organic carbon (DHAP) in the operation of the reductive pentose phosphate cycle.

The formation of dihydroxyacetone phosphate and the resynthesis of RuBP involve the interconversion of a number of different phosphorylated carbohydrate intermediates. Figure 6-2 shows that two molecules of NADPH and three molecules of ATP are needed for every CO_2 molecule assimilated (6 NADPH and 9 ATP per 3 CO_2). All the reactions are mediated by specific enzymes that have been isolated and identified. It has been possible to reconstitute a cell-free preparation containing enzymes of the reductive pentose phosphate cycle, RuBP, ATP, NADPH, and CO_2, and show the synthesis of hexose phosphate (fructose-6-P). Light is not necessary to accomplish the synthesis of hexose phosphate. It will be shown in a later section that ATP and NADPH are generated from ADP and inorganic phosphate and from oxidized NADP, respectively, in the light.

The suggestions of van Niel that the role of light in photosynthesis is to produce a "reducing" entity that is used in the reduction of carbon dioxide was corroborated by further studies by Calvin and associates. They grew plants in the light in the presence of unlabeled carbon dioxide, after which the plants were darkened and immediately supplied with $^{14}CO_2$. It was found that the ^{14}C was incorporated into the same products (glyceric acid-3-phosphate, hexose phosphates, pentose phosphates, etc.) as were formed in the light. These results are consistent with the notion that during the illumination period a stable reducing entity is formed that is later able to reduce carbon dioxide in the dark.

[4] The term "photosynthetic carbon reduction cycle" also is used.

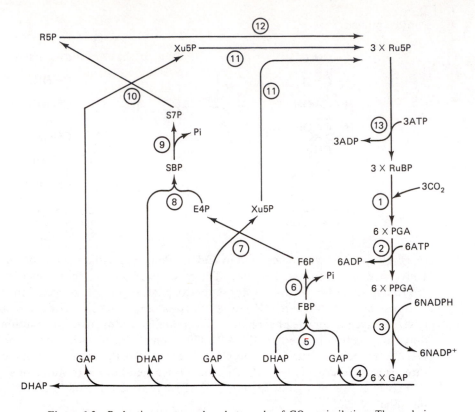

Figure 6-2 Reductive pentose phosphate cycle of CO_2 assimilation. The cycle is catalyzed by enzymes identified by number $\bigcirc$. Carbon dioxide adds to ribulose bisphosphate (RuBP) in the presence of ribulose bisphosphate carboxylase $\textcircled{1}$ to form 2 molecules of glyceric acid-3-phosphate (PGA). In the formulation 3 molecules of CO_2 are shown to react with 3 molecules of RuBP to form 6 molecules of PGA. This is done to illustrate the synthesis of a three-carbon product—dihydroxyacetone phosphate (DHAP), which can be metabolized further to sugar. The other molecules regenerate the CO_2 acceptor, RuBP. The enzyme phosphoglyceric acid kinase $\textcircled{2}$ in the presence of ATP forms diphosphoglyceric acid (PPGA), which is then reduced by NADPH to glyceraldehyde phosphate (GAP) in the presence of triosephosphate dehydrogenase $\textcircled{3}$. The enzyme triosephosphate isomerase $\textcircled{4}$ catalyzes the inter-conversion of glyceraldehyde phosphate (GAP) and dihydroxyacetone phosphate (DHAP). In the presence of aldolase $\textcircled{5}$, GAP and DHAP are converted to fructose bisphosphate (FBP). A phosphate group is removed from FBP through the action of phosphatase $\textcircled{6}$ to form fructose-6-phosphate (FGP). Transketolase $\textcircled{7}$ removes the top two carbon atoms from FGP and transfers them to GAP to yield xylulose-5-phosphate (Xu5P). The remaining four-carbon fragment from FGP becomes erythrose-4-phosphate (E4P). E4P and DHAP interact in the presence of aldolase $\textcircled{8}$ to form sedoheptulose bisphosphate (SBP). Phosphatase $\textcircled{9}$ removes a phosphate group from SBP, forming sedoheptulose-7-phosphate (S7B). In a reaction similar to $\textcircled{7}$, transketolase $\textcircled{10}$ catalyzes the transfer of two carbon atoms from S7P to GAP to form Xu5P. The five-carbon fragment from S7P becomes ribose-5-phosphate (R5P). An enzyme, epimerase $\textcircled{11}$, converts Xu5P to ribulose-5-phosphate (Ru5P). R5P is converted to Ru5P through the action of pentose phosphate isomerase $\textcircled{12}$. Finally, Ru5P is transformed to RuBP, the CO_2 acceptor molecule, through the action of phosphoribulokinase $\textcircled{13}$ and ATP. Enzymes $\textcircled{1}$, $\textcircled{3}$, $\textcircled{6}$, $\textcircled{9}$, and $\textcircled{13}$ are activated by light as discussed in a later section. (After R. McC. Lilley and D. A. Walker, 1979.)

Although much of the research leading to the formulation of the reductive pentose phosphate cycle of photosynthesis was carried out on unicellular algae, studies with higher green plants established that the cycle occurs in many monocots and dicots, including such plants as wheat, rice, sugar beets, spinach, soybeans, and tobacco. Two techniques are used to determine the mode of carbon dioxide fixation in plants. The first depends on the rapid appearance of ^{14}C-labeled glyceric acid-3-phosphate following a brief exposure of the plant to ^{14}C-labeled carbon dioxide. The chromatographic procedures described earlier for establishing the carbon reduction cycle are used to isolate and identify the products of the reaction.

The second technique of ascertaining the pathway of carbon dioxide fixation is based on the identification of one or more of the enzymes catalyzing reactions in the cycle (Figure 6-2). Of these, ribulose-1,5-bisphosphate carboxylase (RuBPCase) is of special importance because it catalyzes the addition of carbon dioxide to ribulose-1,5-bisphosphate with the formation of glyceric acid-3-phosphate. RuBPCase has been shown to be present in the soluble protein fraction of leaves of plants and it is inferred that these plants fix carbon dioxide by the reductive pentose phosphate cycle. The soluble protein fraction containing RuBPCase can also be used to study the rate of CO_2 fixation and thereby provide information concerning the relative amounts of CO_2 fixed by different species of plants.

Ribulose bisphosphate carboxylase. RuBPCase has an interesting history. Years before its role in CO_2 assimilation was recognized, Wildman and Bonner found that leaves contained a soluble protein, which they called Fraction I Protein. Growing and expanding leaves of many species of plants were found to contain this protein, which subsequently was found to be identical to RuBP carboxylase. The enzyme is present in all plants containing chlorophyll a as well as many photosynthetic algae and bacteria. Not only is RuBPCase responsible for the fixation of CO_2 during photosynthesis, but is also acts as an oxygenase in the fixation of O_2. This role in oxygen fixation is discussed in the next chapter on photorespiration. Because of its widespread occurrence in plants and because it is found in such relatively high concentrations in the soluble protein fraction of leaves, RuBPCase has been called the most abundant protein in the world.

RuBPCase has been isolated and purified from the leaves of many plants and in 1971 the protein was crystallized from tobacco leaves in Wildman's laboratory. Yields of crystalline RuBPCase in the range of 3 to 4 mg dry protein per kilogram of fresh tobacco leaves have been obtained. The protein occurs in the stromal fraction of chloroplasts at concentrations as high as 300 mg/ml. Wildman has suggested that Fraction I Protein be considered for human consumption because of its excellent balance of amino acids and the relative ease of extraction from green leaves.

Crystalline RuBPCase has a molecular weight of 557,000 and is composed of subunits—eight large subunits (MW 55,800 each) and eight small subunits (MW 12,000 each). The large subunits are synthesized within the chloroplast on genetic information coded on chloroplast DNA. The small subunits are synthesized outside the chloroplast in the cytosol on genetic information coded on nuclear DNA. The small subunits then penetrate the chloroplast envelope and the complete protein (RuBPCase) is assembled in the stroma. A great deal of interesting research has been done on RuBPCase with respect to the amino acid composition of the protein subunits, the DNA and RNA segments of the chloroplast and nuclear genomes, and the evolutionary development of enzyme activity.

C₄-Dicarboxylic Acid Pathway (C₄ Cycle)

In some plants ^{14}C-labeled glyceric acid-3-phosphate is not one of the early labeled compounds observed following brief exposures to $^{14}CO_2$. In 1965 Kortschak and colleagues at the Sugarcane Experiment Station in Hawaii found, for example, that when they supplied $^{14}CO_2$ to sugar cane leaves, the labeled carbon appeared in malic acid and aspartic acid. Further work on sugar cane and later on other plants by M. D. Hatch and C. R. Slack in Australia has shown that there is an additional pathway of carbon dioxide assimilation in certain groups of plants.

The initial reaction in this pathway, referred to as the C₄-dicarboxylic acid pathway (C₄ cycle), occurs as follows:

$$
\begin{array}{c}
\text{COOH} \\
|\\
\text{CO}\textcircled{P} \\
||\\
\text{CH}_2
\end{array}
\;+\; C^*O_2 + H_2O
\;\xrightarrow[\substack{\text{carboxylase}\\(\text{PEPCase})}]{\substack{\text{phosphoenol}\\\text{pyruvic acid}}}\;
\begin{array}{c}
\text{COOH} \\
|\\
\text{C}=\text{O} \\
|\\
\text{CH}_2 \\
|\\
\text{C}^*\text{OOH}
\end{array}
\;+\; H_3PO_4 \qquad (6\text{-}8)
$$

Phosphoenol-
pyruvic acid   Oxaloacetic
                                                                acid

Carbon dioxide reacts with phosphoenolpyruvic acid in the presence of the enzyme phosphoenolpyruvic acid carboxylase (PEPCase) to form oxaloacetic acid, a four-carbon dicarboxylic acid. Note that in this reaction PEPCase plays the key role of CO_2 assimilation, as does RuBPCase in the reductive pentose phosphate pathway in C_3 plants. As shown in reaction (6-8), if labeled CO_2 is used, labeled carbon appears in one of the carboxyl groups of oxaloacetic acid.

Further reactions of the C₄ cycle are as follows:

$$\text{Oxaloacetic acid} + \text{NADP}_{red} \xrightarrow[\text{dehydrogenase}]{\text{malic}} \text{malic acid} + \text{NADP}_{ox} \qquad (6\text{-}9)$$

$$\text{Malic acid} + \text{NADP}_{ox} \xrightarrow[\text{enzyme}]{\text{malic}} \text{pyruvic acid} + CO_2 + \text{NADP}_{red} \qquad (6\text{-}10)$$

There are several variations among C₄ plants for the metabolism of oxaloacetic acid, but reactions (6-9) and (6-10) illustrate the major points, the formation of pyruvic acid and CO_2. Pyruvic acid is phosphorylated to yield phosphoenolpyruvic acid, the CO_2 acceptor in reaction (6-8) whereas CO_2 is further metabolized in the reductive pentose phosphate cycle by way of RuBP carboxylase. The entire reaction sequence is shown in Figure 6-3. Note that the C₄ cycle regenerates the CO_2 acceptor, phosphoenolpyruvic

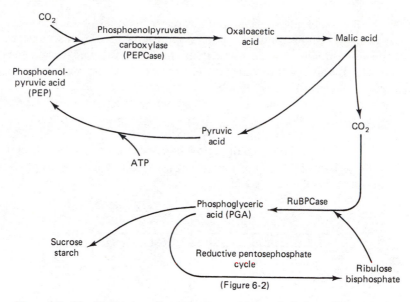

Figure 6-3 The C_4-dicarboxylic acid pathway of CO_2 assimilation. Carbon dioxide reacts with phosphoenolpyruvic acid (PEP) in the presence of the enzyme PEPCase to form oxaloacetic acid. The oxaloacetic acid is transformed to malic acid (in some C_4 plants to aspartic acid), which is decarboxylated to yield pyruvic acid and CO_2. Pyruvic acid is phosphorylated (ATP) to regenerate phosphoenolpyruvic acid and the CO_2 is further metabolized via RuBPCase in the reductive pentose phosphate pathway (Figure 6-2) to sucrose, starch, and other products.

acid, but does not result in a net increase in carbon, as is accomplished in the reductive pentose phosphate pathway (Figure 6-2).

Plants displaying C_4 metabolism possess two distinct arrays of enzymes for converting CO_2 to starch, sugar, and other assimilation products—the C_4 pathway featuring PEP carboxylase and the C_3 pathway with RuBP carboxylase. Furthermore, the two CO_2-assimilating systems are located within different leaf cells.

The C_3 and C_4 plants possess a characteristic leaf anatomy (see Figure 7-2, Chapter 7). In C_4 plants, such as maize and crabgrass, the conducting tissues, or vascular bundles, are surrounded by a layer of parenchyma cells known as bundle sheath cells. This layer lacks intercellular spaces between the individual cells of the sheath and between the sheath cells and the enclosed vascular tissue. Also, the bundle sheath extends over the ends of the vascular bundles within the leaf tissue. Thus all substances moving between leaf mesophyll cells and leaf vascular tissue must necessarily pass through the bundle sheath cells surrounding the vascular bundles. The bundle sheath cells in the leaves of C_4 plants contain chloroplasts and other organelles, such as mitochondria and peroxisomes. On the other hand, the bundle sheath cells in leaves of C_3 plants are either completely absent or, when present, have fewer organelles (chloroplasts, mitochondria) than the surrounding cells or none at all.

By using appropriate separation techniques, bundle sheath cells and leaf mesophyll cells of C_4 plants, as well as chloroplasts from these cells, have been isolated and individually studied as to their ultrastructure and biochemical properties. If isolated bundle sheath cells or their chloroplasts are supplied with $^{14}CO_2$, labeled products characteristic of the reductive pentose phosphate cycle are observed— glyceric acid-3-phosphate, sugar phosphates, sucrose, starch. The enzyme RuBPCase is present in bundle sheath chloroplasts. If leaf mesophyll cells or their chloroplasts, on the other hand, are supplied $^{14}CO_2$, the first products of CO_2 assimilation are found to be oxaloacetic acid and malic acid (or aspartic acid in some C_4 plants), compounds characteristic of the C_4 pathway. PEPCase is found in leaf mesophyll cells of C_4 plants but not in bundle sheath cells.

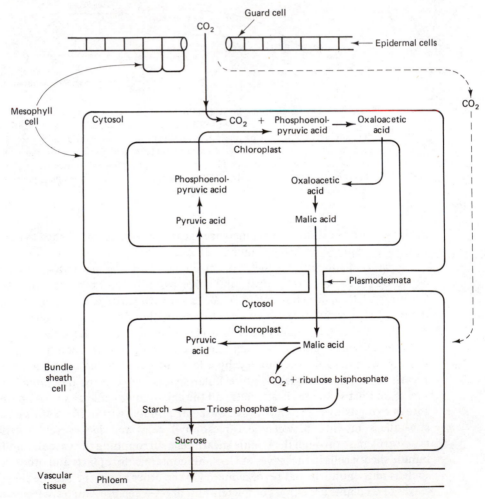

Figure 6-4 Movement of metabolites between mesophyll and bundle sheath cells in the leaves of C_4 plants. (See text for details.)

The mesophyll and bundle sheath cells are in close contact with one another (Figure 2-9) and are connected by cytoplasmic connections (plasmodesmata). Biochemical studies have revealed that leaf mesophyll chloroplasts and bundle sheath chloroplasts contain specific unique arrays of enzymes, indicating that the chloroplasts from these two kinds of cells perform different metabolic functions. It has been suggested that there is an exchange of metabolites between the two kinds of cells similar to that shown in Figure 6-4. As illustrated, CO_2 fixed in mesophyll cells by PEPCase is converted to malic acid, which is then exported to the bundle sheath cells. Here malic acid is decarboxylated to produce pyruvic acid and CO_2. The pyruvic acid is exported back to the mesophyll cells, where it is converted to phosphoenolpyruvic acid, the CO_2 acceptor in the C_4 pathway. The CO_2 from the decarboxylation of malic acid is further metabolized by way of RuBPCase and the reductive pentose phosphate pathway to sucrose and starch. Starch may accumulate in the bundle sheath chloroplasts or sucrose, synthesized in the cytosol, may be exported into the vascular tissue, where it is transported to other parts of the plant.

Figure 6-4 shows that there is a cyclic movement of carbon by way of malic acid and pyruvic acid between leaf mesophyll cells and leaf bundle sheath cells. The net effect of the cycle is that CO_2 is delivered to the RuBPCase and associated enzymes for further metabolism to sucrose and starch.

It is possible that some CO_2 entering the leaves of C_4 plants may diffuse directly to the bundle sheath cells, where it can enter the reductive pentose phosphate pathway. The amount of CO_2 fixation, however, seems to be relatively minor compared to that accomplished by PEPCase and the C_4 pathway. The flow of metabolites between the leaf mesophyll cells and the bundle sheath cells may be viewed as an efficient method of supplying CO_2 to the RuBPCase-reductive pentose phosphate cycle in bundle sheath chloroplasts. More will be said of this topic in the next chapter.

Small amounts of organic acids are formed in the dark in nonphotosynthetic tissues (e.g., roots) through the action of PEPCase. The primary product, oxaloacetic acid, is a key metabolite in the TCA cycle. This cycle provides for the formation and interconversion of organic acids used in a variety of metabolic processes (see Chapters 3 and 4 and reaction 4-33).

Crassulacean Acid Metabolism (CAM)

It has long been recognized that the leaves of plants belonging to the Crassulacean family (e.g., *Kalanchoe*, *Sedum*) display a diurnal pattern of organic acid formation. During the day (Figure 6-5) the total organic acid content of the leaves decreases and the pH of the leaf cell sap increases whereas during the night the organic acid content increases and the pH of the leaf cell sap decreases. Accompanying these changes in leaf organic acid content are changes in storage carbohydrate. During the day storage carbohydrates (starch, glucan) increase in leaves and they decrease during the night. Similar patterns of leaf organic acid content and storage carbohydrate content are found in about 20 families of flowering plants, including the Cactaceae, Agavaceae, Orchidaceae, Portulacaceae, and Crassulaceae. All such plants are referred to as crassulacean acid metabolism (CAM) plants.

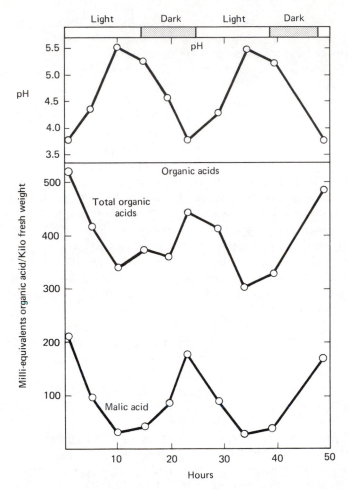

Figure 6-5 Diurnal changes in pH, total organic acids, and malic acid in leaves of *Bryophyllum calycinum* in excised leaves cultured in water during alternating periods of light and darkness. The changes in acidity are due mostly to changes in concentration of malic acid. (After Vickery, 1952; Figure 1.)

     In addition to their organic acid metabolism, CAM plants share other attributes. The leaves—and frequently stems and petioles—are fleshy or succulent. The cells in these tissues contain large water-filled vacuoles, but the cells also contain chloroplasts and other organelles. As far as is known, all CAM plants possess the succulent habit. Certain succulent plants (e.g., halophytes, such as *Salicornia* and *Borrichia*), however, do not carry out crassulacean acid metabolism. The cells of these succulents contain large water-filled vacuoles but lack enzymes of the C_4 cycle. Thus an important attribute of CAM plants is the presence of leaf or stem tissues composed of mesophyll cells containing chloroplasts and vacuoles. Also, there is no well-defined vascular tissue enclosed in bundle sheath cells, as noted in C_4 plants.

     The diurnal pattern of organic acid content in CAM plants is accompanied by a diurnal pattern in the opening and closing of stomata on leaf and stem surfaces. The stomata are open during the night and closed during the day. This pattern of stomatal opening and closing by CAM plants is the reverse of that displayed by other plants, where the stomata are open during the day and closed at night. The CAM plant pattern of stomatal behavior is an advantage in the conservation of plant water. With the stomata closed during the day, the rate of transpiration is reduced. Most CAM plants are native to arid or xeric habitats where maintenance of internal plant water is a serious problem.

The diurnal change in leaf organic acid content shown in Figure 6-5 results almost entirely from the appearance of malic acid in the dark and its disappearance in the light. Malic acid is formed by the carboxylation of phosphoenolpyruvic acid through the action of the enzyme PEPCase and the subsequent conversion of oxaloacetic acid to malic acid, as illustrated in reactions (6-8) and (6-9). The cells in the leaves and stems of CAM plants also contain RuPBCase and other enzymes of the reductive pentose phosphate cycle. CO_2 fixed in malic acid during the dark via PEPCase ultimately is converted to sucrose, starch (or other glucans), and other compounds in the light. A scheme for these dark and light reactions is shown in Figure 6-6.

It should be noted that CO_2 assimilation in both C_4 plants and CAM plants involves the reductive pentose phosphate cycle (C_3) and the C_4-dicarboxylic acid cycle. In C_4 plants the two cycles are separated in different leaf cells: the C_4 enzymes in leaf mesophyll cells and the C_3 enzymes in bundle sheath cells. In CAM plants, on the other hand, both pathways are found in the leaf mesophyll cells, but the operation of the two pathways is separated in time. In the dark, PEPCase and the C_4 enzymes are active and malic acid accumulates. In the light, RuBPCase and the C_3 enzymes are active with CO_2 being supplied by the decarboxylation of stored malic acid. Because stomata are closed during the day, very little CO_2 from outside the leaf can enter the C_3 cycle.

ROLES OF ATP AND NADPH IN CARBON DIOXIDE ASSIMILATION

The biochemical reactions involved in the incorporation of CO_2 into cellular material are "dark" reactions that do not require light as a source of energy. ATP and reduced NADP are necessary, however. Both compounds are formed during the degradative breakdown of foodstuffs (sugar, fats, proteins) through respiratory processes, as was shown in Chapter 4. The fact that the metabolic pathways of CO_2 assimilation and carbohydrate breakdown both involve ATP/ADP and NAD/NADP suggests that a comparative study of the two processes might indicate some common features. Figure 6-7 illustrates some of the features of the two processes.

During respiration (left-hand side of Figure 6-7) sugar is broken down (oxidized) to carbon dioxide and water. Sugar is metabolized by enzymic reactions through a series of intermediates, some of which involve interconversions of ATP/ADP and NAD^+/NADH. In photosynthesis (right-hand side of Figure 6-7) the initial reaction is the excitation of a chlorophyll molecule following the absorption of a photon of radiant energy. The excited chlorophyll molecule then interacts through a number of

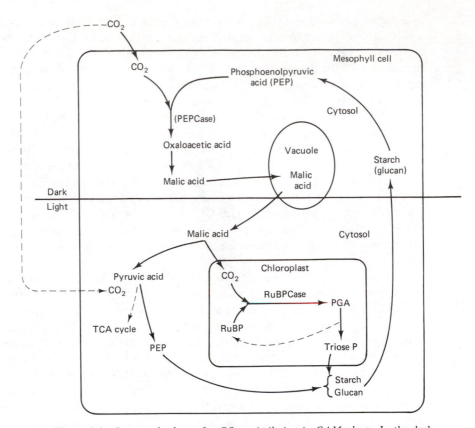

Figure 6-6 *Suggested scheme for CO_2 assimilation in CAM plants.* In the dark, storage carbohydrate (starch, glucan) is metabolized to phosphoenolpyruvic acid, which is then carboxylated to form oxaloacetic acid and eventually malic acid. Malic acid accumulates in the vacuole. In the light, the malic acid is exported from the vacuole to the cytoplasm, where it is decarboxylated to pyruvic acid and CO_2. The CO_2 is metabolized via the reductive pentose phosphate pathway (RuBPCase) to storage carbohydrate. Pyruvic acid also may enter this pathway and be converted to storage carbohydrate. All these reactions occur in the leaf or stem mesophyll cells.

intermediates to form ATP and NADPH, which subsequently participate in the reduction of CO_2 and the formation of carbohydrates and other plant constituents.

Many intermediates in respiration and photosynthesis are identical and some enzymes seem similar. A great deal of useful information about the detailed nature of respiration and photosynthesis has been learned from studies of their comparative biochemistry.

LIGHT (PHOTO) PHASE OF PHOTOSYNTHESIS

The absorption of radiant energy by green leaves is due to the presence of several pigments—chlorophyll *a*, chlorophyll *b*, and carotenoids. The contribution of these

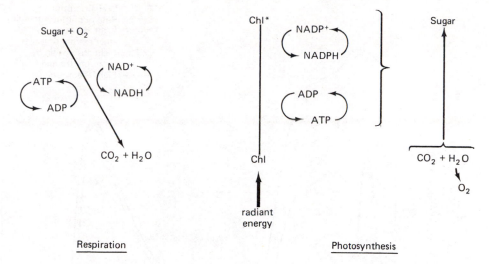

Figure 6-7 Schematic representation of intermediates in respiration and photosynthesis. In the breakdown (oxidation) of sugar to CO_2 and H_2O during respiration, interconversions of ATP/ADP and NAD^+/NADH occur at various steps. In photosynthesis chlorophyll molecules are excited (chl*) through the absorption of photons of radiant energy. The excited chlorophyll, through intermediate reactions, interacts with ADP/ATP and $NADP^+$/NADPH so that CO_2 is assimilated and converted to sugar. Oxygen is released during photosynthesis. Many intermediates between sugar and CO_2 and H_2O in respiration are similar to the intermediates involved in the conversion of CO_2 and H_2O to sugar in photosynthesis.

pigments to the absorption of radiant energy in a leaf of *Elodea densa*, an aquatic dicot, is shown in Figure 6-8. Chlorophylls *a* and *b* account for the absorption of red light (600 to 700 nm) and blue light (400 to 500 nm). The carotenoids also absorb in the blue region of the spectrum. Not much light absorption occurs in the green region (500 to 600 nm). It is also shown in Figure 6-8 that the rate of photosynthesis in the *Elodea* leaf follows closely the curve of radiant energy absorption by the intact leaf. The close relationship between the action spectrum for photosynthesis and the absorption spectra of the chlorophylls and carotenoids indicates that the pigments are involved in photosynthesis.

Excitation of Chlorophyll

Figure 6-8 shows that chlorophyll *a*, chlorophyll *b*, and the carotenoids all absorb radiant energy. It is not clear, however, from the data given whether each of these pigments participates directly in some photochemical reaction or whether some pigments might function only as collectors of radiant energy. The energy collected by such pigments might then be passed on to a pigment (or pigments) closely coupled with the conversion of radiant energy to chemical energy. Information on this subject has come from studying pigment excitation by varying wavelengths of radiant energy.

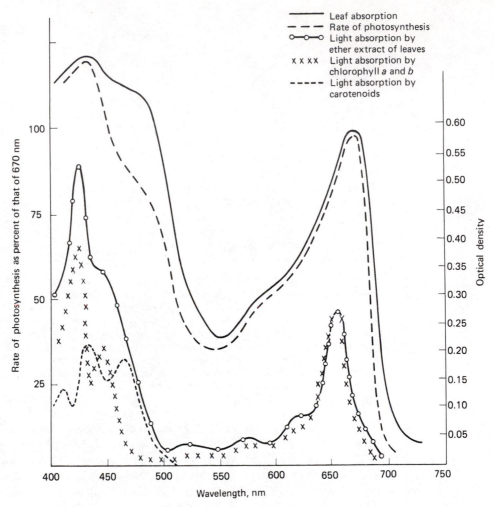

Figure 6-8 Comparison of action spectrum for the rate of photosynthesis by *Elodea densa* leaf with absorption spectra of the intact leaf and with extracted pigments. (courtesy M. H. Hommersand.)

Details of pigment excitation were discussed in Chapter 5 (Figure 5-5). Recall that when a photon strikes a pigment, an electron within the pigment molecule is raised to a more energetic energy level; such a pigment is said to be in the excited state. Depending on the immediate environment of the excited pigment and the chemical nature of the neighboring molecules, the excited molecule may return to ground state by several different processes: heat dissipation, emission of radiant energy (fluorescence and phosphorescence), transfer to a neighboring pigment by resonance transfer, or by participating in a chemical reaction with a neighboring molecule. These processes are summarized in Figure 6-9 with respect to the absorption of red (660 nm) and blue (400 nm) light by chlorophyll *a*.

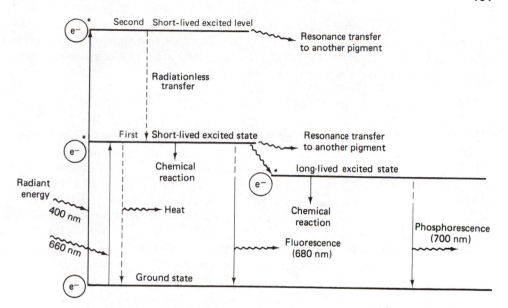

Figure 6-9   Energy levels for excited chlorophyll following absorption of radiant
energy at either 400- or 660-nm wavelength. If electrons in the excited state interact
with appropriate molecules, chemical reactions, such as ATP and NADPH formation,
occur. (See Figure 5-5 for further details of the excited state.)

Red light is less energetic than blue light, and following the absorption of red
light (660 nm) chlorophyll attains the first excited level. The lifetime of the excited
molecule is quite short, often of the order of 10^{-10} to 10^{-8} s, and unless the energy is
transferred to another molecule (as in a chemical reaction), excited chlorophyll returns
to ground state through the loss of energy by the emission of heat or light. The rapid
emission of light following chlorophyll excitation is referred to as fluorescence.

Under certain conditions some of the energy of the first excited level of
chlorophyll may be used to effect the production of a long-lived excited molecule with
lifetimes of the order of 10^{-2} to 10^{-1} s. Such long-lived excited molecules have a much
greater probability of interacting with neighboring molecules and participating in
photochemical reactions. If the energy of these long-lived excited molecules is emitted
as light, the process is referred to as phosphorescence. The major difference between
fluorescence and phosphorescence is that fluorescence occurs rapidly after the
molecule is excited whereas light emission by phosphorescence may be delayed. Both
the short-lived and long-lived energetic states of chlorophyll *a* have been observed
following the absorption of red light by green plants.

Blue light (400 nm) is also used in photosynthesis, as is shown in Figure 6-9.
Following the absorption of blue light, chlorophyll *a* attains the more energetic second
excited state (Figure 6-9). It will be recalled that a photon of blue light is about twice as
energetic as a photon of red light. It is thought, however, that the electron in the second
excited level returns to the first excited level through some sort of radiationless transfer

and that subsequent reactions are as described for the absorption of red light. This situation is believed to occur because chlorophyll *a* fluoresces or phosphoresces at wavelengths characteristic of the first excited state regardless of the wavelengths of radiant energy used to excite the molecule. Thus the energy available for photo-chemical reactions through the excitation of chlorophyll *a* is that equivalent to the energy of a photon of red light (660 nm).

Considerable evidence supports the notion that chlorophyll *a* is closely coupled to reactions concerned with the conversion of radiant energy to chemical energy. More will be said of these reactions shortly. The involvement of chlorophyll *b* and the carotenoids (also some other pigments found in algae) appears less direct. As mentioned in Chapter 5, these pigments apparently function as collectors of radiant energy, which is passed on to chlorophyll *a*. Several kinds of evidence support such a view. It was seen, for example, that the carotenoids absorb radiant energy in the blue region of the spectrum. If a wavelength of light strongly absorbed by a carotenoid but not chlorophyll *a* is presented to a leaf, it is observed that radiant energy is emitted by fluorescence at wavelengths characteristic of chlorophyll *a* fluorescence, not of carotenoid fluorescence. Similarly, if chlorophyll *b* is excited by a wavelength of radiant energy not absorbed by chlorophyll *a*, chlorophyll *a* fluorescence is observed. not chlorophyll *b* fluorescence. It is believed that the energy of excited chlorophyll *b* or carotenoids is transferred to chlorophyll *a* by a process known as resonance transfer.

Coupling of Photophysiology to the Chemical Phases of Photosynthesis

Although chlorophyll is clearly involved in the transformation of radiant energy to chemical energy, details of intermediates and pathways are poorly understood. Three major products are formed as a result of the absorption of radiant energy by chloroplast pigments—oxygen, ATP, and NADPH. Oxygen is evolved during photosynthesis, but the nature of the reactions is largely unknown at this time. Considerably more is known of the pathways leading to the formation of ATP and NADPH, the molecules necessary for the reduction of carbon dioxide to carbohydrate.

The Hill reaction [reaction (6-5)], a process involving the transfer of electrons in the light from water to an electron acceptor accompanied by oxygen evolution, provides a model reaction for the formation of NADPH. Several groups of workers demonstrated that illuminated chloroplasts will form NADPH and oxygen as follows:

$$2\ H_2O + 2\ NADP^+ \xrightarrow[\text{chloroplasts}]{\text{radiant energy}} 2\ NADPH + O_2 + 2\ H^+ \qquad (6\text{-}11)$$

Here $NADP^+$ is the natural hydrogen acceptor replacing $[Fe(CN)_6]^{-3}$ in reaction (6-5). If reaction (6-11) is coupled with a reaction requiring NADPH, $NADP^+$ is regenerated and made available for accepting additional electrons from water.

As described in Chapter 4, ATP is generated during the oxidation of glucose to carbon dioxide and water (also note Figure 6-7). Most of the ATP is synthesized in mitochondria during the transport of electrons through the electron transport chain by a process known as oxidative phosphorylation. In 1954 D. I. Arnon and coworkers demonstrated that illuminated chloroplasts generate ATP.

$$\text{ADP + inorganic phosphate (P}i) \xrightarrow[\text{chloroplasts}]{\text{radiant energy}} \text{ATP} \qquad (6\text{-}12)$$

Arnon called this reaction photosynthetic phosphorylation or photophosphorylation to distinguish it from the generation of ATP by oxidative phosphorylation in mitochondria. Chloroplasts also were shown to generate both ATP and NADPH when illuminated by the following reaction scheme.

$$2 \text{ ADP} + 4 \text{ H}_2\text{O} + 2 \text{ P}i + 2 \text{ NADP}^+ \xrightarrow[\text{chloroplasts}]{\text{radiant energy}}$$
$$2 \text{ ATP} + 2 \text{ NADPH} + 2 \text{ H}_2\text{O} + \text{O}_2 + 2 \text{ H}^+ \qquad (6\text{-}13)$$

Thus illuminated chloroplasts are capable of generating both ATP and NADPH, the substances required for the conversion of CO_2 and H_2O to carbohydrates.

As written, reactions (6-12) and (6-13) do not show the nature of the intermediates or the relationship of the reactions to the excited chlorophyll molecule. Several lines of evidence suggest that the two reactions are closely related and that the formation of ATP and NADPH is linked to a flow of electrons from excited chlorophyll through a series of carriers.

REACTION SCHEME FOR ATP AND NADPH FORMATION

It is believed that the formation of ATP and NADPH following the excitation of chlorophyll occurs by the reaction pathway shown in Figure 6-10. In this scheme electrons from water pass through several intermediates before they are raised to a high enough energy level to reduce $NADP^+$. The energy necessary to raise the electrons to the high energy level is provided by radiant energy acting on appropriate pigment molecules. The generation of ATP (photophosphorylation) also occurs during the flow of electrons (e^-) and protons [H^+] through the intermediates shown in Figure 6-10.

Two pigment systems are shown—photosystem I (PS I) and photosystem II (PS II). Evidence in support of the involvement of two photoevents is derived from several sources. In action spectrum studies R. Emerson noted that radiant energy of wavelengths in excess of 685 nm is inefficiently used in photosynthesis, although chlorophyll a absorbed such wavelengths. Efficiency of long wavelengths (685 nm or longer) is enhanced if light of short wavelengths (less than 680 nm) is simultaneously supplied. These results were interpreted as indicating two separate photosystems, one reacting to long wavelengths and the other to short wavelengths.

Further evidence in support of the idea that two photoacts are involved comes from experiments where a partial physical separation of the photosystems has been achieved. Particles rich in chlorophyll a_1, for example have been isolated from chloroplasts. If a suitable source of electrons is supplied, these particles will form

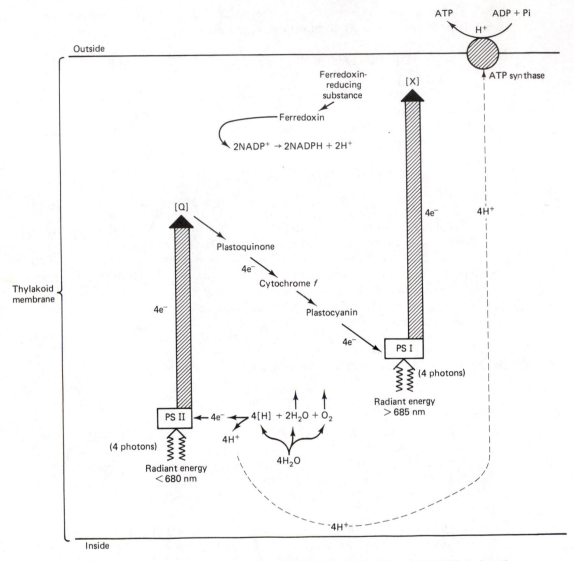

Figure 6-10 Reaction scheme for the formation of ATP and NADPH during photosynthesis. Two pigment systems, photosystem I (PS I) and photosystem II (PS II), play a role in photosynthetic energy transport, photophosphorylation, and formation of NADPH. The various components are organized within the thylakoid membrane. PS I is composed of chlorophyll a_I, chlorophyll b, and carotenes. Radiant energy at wavelengths longer than 685 nm excites PS I. Photosystem II contains chlorophyll a_{II}, chlorophyll b, and xanthophylls. PS II is excited by radiant energy wavelengths shorter than 680 nm. Electrons (e^-) extracted from H_2O at PS II pass through a series of intermediates [Q], plastoquinone, plastocyanin, cytochrome f, [X], ferredoxin-reducing substance, ferredoxin, and ferredoxin-NADP reductase to produce NADPH. Protons (H^+) also supplied by H_2O molecules at PS II are passed through several intermediates (indicated by dashed line) and finally react with ATP synthase to produce ATP.

NADPH when irradiated by light of wavelengths longer than 685 nm, but no O_2 is evolved. Similarly, particles rich in chlorophyll a_{II} with an absorption maximum around 680 nm have been isolated from chloroplasts. If such particles are irradiated with wavelengths of 680 nm or less, O_2 evolution occurs, but no NADPH is formed.

The pigment composition of the two photosystems is not definitely known. Chlorophyll a is present in both systems, but chlorophyll a can exist in several forms, depending on the nature of its association with protein in a chlorophyll-protein complex. PS I contains a form of chlorophyll a known as chlorophyll a_I, chlorophyll b, and carotenes.

Photosystem II contains chlorophyll b, some xanthophylls, and a form of chlorophyll a referred to as chlorophyll a_{II}. Radiant energy of wavelengths shorter than 680 nm absorbed by PS II can drive both photosystems because of resonance transfer of radiant energy between PS II and PS I. Under these conditions, or if the plant is irradiated with natural daylight, the complete photosynthetic process occurs— O_2 evolution, ATP synthesis, and formation of NADPH.

It is seen in Figure 6-10 that electrons extracted from water are driven to [X] by two photochemical reactions mediated by pigments in photosystems I and II. The two photosystems are connected by several intermediates—plastoquinone, cytochrome f, and plastocyanin. A substance designated as [Q] is believed to be the primary electron acceptor from photosystem II. As electrons from [Q] move toward PS I, the intermediates are alternately reduced and oxidized. [Q] is a form of quinone.

Electrons from PS II arrive at PS I at a somewhat lower energy level than those at state [Q]. A second photochemical event, driven by long-wavelength radiation, raises the electrons to the energetic level of [X], a very strong reducing agent. Reduced [X] then reduces a compound referred to as ferredoxin-reducing substance, which, in turn, reduces ferredoxin. Reduced ferredoxin next reduces oxidized NADP ($NADP^+$) to form reduced NADP (NADPH).

ATP generation following excitation of PS I and PS II is also shown in Figure 6-10. It is believed that photophosphorylation in chloroplasts is similar to the mechanism of ATP generation (oxidative phosphorylation) in mitochondria. The underlying mechanism involves the unidirectional flow of protons [H^+] through a membrane-oriented enzyme, ATP synthase. Details of this model—the chemiosmotic model—were discussed in Chapter 4 (pp. 119–122). In brief, ATP is generated by the reaction of protons [H^+] with the enzyme ATP synthase. As shown in Figure 6-10, the various components of PS I, PS II, and the interconnecting molecules are organized within the thylakoid membrane. ATP synthase is situated at the outside surface of the membrane. Electrons (e^-) and protons [H^+] formed during the light excitation of PS I and PS II pass through the various chemical compounds, resulting in O_2 evolution and formation of ATP and NADPH.

For each electron moved through the complete system, two photons are required, one at PS I and one at PS II. The formation of a molecule of O_2 involves the removal of four electrons from four water molecules as shown.

$$4\ H_2O \xrightarrow[\text{energy}]{\text{radiant}} O_2 + 4\ (H^+ + e^-) + 2\ H_2O \qquad (6\text{-}14)$$

The least understood part of the reaction sequences shown in Figure 6-10 are those involved in the evolution of O_2 following excitation of PS II. A number of intermediates are involved, but little is known of their identity or mechanisms of action. Manganese and chloride ions are involved at some point in the oxygen evolution pathway.

For the four electrons removed from water and passed through PS I and PS II, 8 photons are necessary (4 photons at PS I and 4 photons at PS II in Figure 6-10). These 8 photons generate a molecule of O_2 and 2 molecules each of ATP and NADPH [reaction (6-13)]. It will be recalled, however, that in the reductive pentose phosphate pathway (Figure 6-2) 3 molecules of ATP are required in the assimilation of one CO_2 molecule. It is assumed that an additional 2 photons may be sufficient to provide the extra ATP, making the total photon requirement 10; 10 photons to reduce one molecule of CO_2 to the level of a carbohydrate, CH_2O.

Mention was made earlier (p. 161) that the theoretical maximum efficiency for converting electromagnetic radiation to chemical energy was in the range of 25 to 30%. The basis for such values depends on the relationship between incoming radiant energy and stored chemical energy. The photon requirement mentioned in the preceding paragraph can be used in calculating the theoretical maximum efficiency.

If 10 photons are required in the reduction of one CO_2 molecule to the level of carbohydrate (CH_2O), the energy available is 1680 kJ, for the energy of one photon of red light (660 nm) is 168 kJ per mole of photons. (The energy of photons of other wavelengths might be used, but in this calculation it is assumed that red light is utilized most efficiently.) How much energy is in a mole of CH_2O? This factor is calculated as follows. If a mole of glucose ($6 \times CH_2O$) is oxidized to CO_2 and H_2O, 2768 kJ are released. Then one-sixth of this quantity, or 478 kJ, would be released per mole of CH_2O. If 478 kJ are released on the oxidation of CH_2O, at least this amount of energy is necessary to synthesize a mole of CH_2O. Then if 1680 kJ of photon energy are available and 478 kJ are necessary to form a mole of CH_2O, the efficiency of energy conversion is 28% (478/1680).

Light and Enzyme Regulation

The preceding section on the light (photo) phase of photosynthesis described the role of radiant energy in supplying ATP and NADPH used in the assimilation of CO_2. Both molecules are generated continuously (Figure 6-10) in the light by interactions of the two photosystems—PS I and PS II. In the absence of light, CO_2 assimilation by the reductive pentose phosphate (RPP) pathway diminishes or ceases, but the enzymes are still present. If ATP or NADPH is available from other sources—from respiratory pathways (glycolysis, Krebs cycle, or oxidative phosphorylation)—the RPP pathway can continue to function if CO_2 is available. It would be costly, however, to plant metabolic processes by diverting respiratory ATP and NADPH from protein and lipid synthesis and other anabolic processes. Studies have shown that some of the enzymes in the RPP pathway of CO_2 assimilation are photoregulated—activated in the light and deactivated in the absence of light. Thus the RPP pathway is inoperative in darkness and ATP and NADPH generated by respiratory processes are available for anabolic processes.

Similarly, the enzymes involved in the breakdown of starch to triose phosphate are regulated by light. In the presence of light, starch degradation is promoted. These various reactions are illustrated in Figure 6-11. It should be emphasized that the metabolic reactions shown in Figure 6-11 represent only a fraction of those occurring in leaf cells. Clearly, however, the metabolic events are of crucial importance in the assimilation of CO_2 and the formation of starch and sucrose.

In the reductive pentose phosphate (RPP) pathway (Figure 6-11), five enzymes are activated by light. They are

1. ribulose bisphosphate carboxylase
2. NADP-linked glyceraldehyde-3-P-dehydrogenase
3. fructose bisphosphate phosphatase
4. sedoheptulose bisphosphate phosphatase
5. phosphoribulokinase

The RPP pathway synthesizes triose phosphate (an equilibrium mixture of dihydroxyacetone phosphate and glyceraldehyde phosphate, Figure 6-2), part of which recycles to form ribulose bisphosphate (RuBP) and the balance of which is used either in starch or sucrose synthesis. Also, in the light ATP and NADPH are generated by the chloroplast thylakoids.

Several enzymes participate in the synthesis of starch from triose phosphate, none of which is photoregulated. Starch synthesis, however, is regulated by ADP-glucose pyrophosphorylase [(7) in Figure 6-11], an enzyme whose activity in enhanced by glyceric acid-3-phosphate and decreased by high concentrations of inorganic phosphate (Pi). From Figure 6-11 it is seen that when the RPP pathway is operating in the light, levels of PGA [formed by enzyme (1), see Figure 6-2] are high. The elevated level of PGA increases the activity of enzyme (7) and starch is synthesized. Moreover, the activity of enzyme (7) is enhanced by low levels of Pi maintained in light by ATP formation and the synthesis of phosphorylated intermediates in the RPP pathway.

If triose phosphate is exported to the cytosol, where sucrose is synthesized, an active transport system in the inner chloroplast membrane exchanges Pi for triose phosphate. The elevated level of Pi in the chloroplast is sufficient to diminish the activity of enzyme (7) and shut down starch synthesis.

What happens to the reaction sequences of Figure 6-11 in darkness? Less is known of these dark reactions than of the light reactions, but the following events seem reasonable. The cyclic reactions of the RPP pathway will cease in the absence of light because the activities of enzymes (1), (2), (3), (4), and (5) are decreased. In addition, the levels of ATP and NADPH fall in darkness and inorganic phosphate (Pi) increases. As noted, enzyme (7) in the starch synthesis pathway is diminished under high levels of Pi. In the dark, starch synthesis ceases.

Activities of enzymes (6) in the starch degradation pathway are increased in

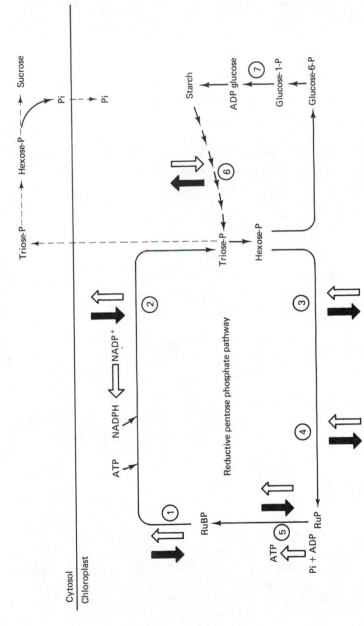

Figure 6-11 Effect of light or its absence on metabolic reactions in the chloroplast and cytosol of leaf cells. All metabolic intermediates are not shown, but the enzymes that show light (dark) modulation are identified. The heavy arrows indicate light effects: ⇧ light promoted, ⇩ light inhibited, ▶ dark promoted, ◀ dark inhibited. Enzymes are as follows: ① ribulose-1,5-P_2 carboxylase; ② NADP-linked glyceraldehyde-3-P dehydrogenase; ③ fructose-1,6-P_2 phosphatase, ④ sedoheptulose-1,7-P_2 phosphatase; ⑤ ribulose-5-P kinase; ⑥ enzymes of starch degradation; ⑦ ADP-glucose pyrophosphorylase. The reactions are discussed in the text.

darkness, leading to triose phosphate formation. Triose phosphate is exported from the chloroplast to the cytosol, where sucrose synthesis occurs. The export of triose phosphate is accompanied by the import of Pi into the chloroplast, reinforcing the inhibition of starch formation via ADP-glucose pyrophosphorylase [enzyme (7)].

THE CHLOROPLAST AS THE UNIT OF PHOTOSYNTHESIS

From the nature of light and dark reactions in photosynthesis it is reasonable to assume that the various processes must occur in the chloroplast. Early studies with isolated chloroplasts, however, did not support such a conclusion. It was found that chloroplasts ceased to function when removed from leaf cells and it was believed that photosynthesis did not occur unless the chloroplasts remained a part of the organized cellular structure. When R. Hill found that isolated chloroplasts evolved oxygen when illuminated, new interest was aroused in the photosynthetic capacity of chloroplasts.

In 1954 Arnon demonstrated that chloroplasts isolated from spinach leaves assimilate CO_2 when illuminated. Using ^{14}C-labeled carbon dioxide, it was shown that ^{14}C-labeled starch is formed by metabolic pathways identical to those followed by algal cells or intact leaves. A physical separation of the light and dark reactions of photosynthesis was achieved with isolated chloroplasts. Chloroplasts are composed of two fractions, grana and stroma, enclosed in a double-layered membrane (Chapter 2). The grana contain the photosynthetic pigments whereas the stroma is pigment free and is composed of soluble protein (enzymes) and other metabolically active molecules. Arnon prepared the chloroplasts so that the chloroplast membrane was broken, releasing the grana and stroma. This preparation was illuminated in the absence of carbon dioxide, after which the grana were removed by centrifugation. Carbon dioxide labeled with ^{14}C was then added to the chlorophyll-free stroma solution in the dark, whereupon the stroma preparation was found to contain ^{14}C-labeled intermediates of the photosynthetic carbon reductive cycle and labeled starch. If, instead of using illuminated grana to provide the "reducing power" for the assimilation of carbon dioxide, ATP and NADPH are added to the stroma in the absence of light, the labeled intermediates are also formed. These and similar experiments strongly support the idea that the chloroplast is the unit of photosynthesis.

In early experiments with isolated chloroplasts the rates of photosynthesis were low, usually less than 5% of the rate of intact cells. In later studies rates of photosynthesis of isolated chloroplasts as high as 80 to 90% the rate of intact leaves have been achieved. These high rates have been achieved through improved isolation procedures. Unless properly handled, soluble cofactors and intermediates rapidly leak out of isolated chloroplasts.

There is no doubt that the chloroplast is the unit of structure and function in photosynthesis. The complex reactions of the light phase of photosynthesis occur on the membrane surfaces of thylakoids (grana), leading to the formation of ATP and NADPH. These molecules, in turn, drive the assimilation of CO_2 by way of ribulose bisphosphate and the reductive pentose phosphate pathway. The end product, triose phosphate, is converted to starch, which accumulates within the chloroplast. A diagrammatic representation of the reactions occurring in the grana and stroma is shown in Figure 6-12. Stromal enzymes are capable of forming a variety of molecules used in the formation of amino acids, protein, lipids, ribonucleic acid, and deoxyribonucleic acid. The chloroplast is a remarkably self-contained unit of metabolism.

Although the chloroplast is capable of synthesizing a variety of molecules necessary for the maintenance and formation of new chloroplasts, plant growth depends on the export of carbon from chloroplasts into cells, tissues, and organs. The major end products of CO_2 assimilation in chloroplasts—starch and hexose phosphate—are not exported across chloroplast membranes. The major export molecule is triose phosphate (glyceraldehyde phosphate $\rightleftharpoons$ dihydroxyacetone phosphate). The central role of triose phosphate in cellular metabolism has been discussed in

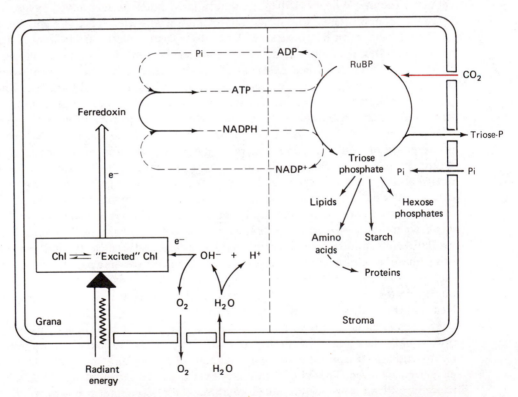

Figure 6-12   The chloroplast as the unit of photosynthesis. The photochemical event occurs in the grana, where excited chlorophyll reduces ferredoxin. Formation of ATP and NADPH also occurs in the grana by dark enzymic reactions. Enzymes concerned in the reduction of CO_2 and formation of starch, lipids, amino acids, etc. are present in the stroma. If carbon is exported from the chloroplast as triose phosphate, inorganic phosphate (Pi) must be imported into the chloroplast to maintain high rates of CO_2 fixation.

Chapter 3. The exit of triose phosphate from the chloroplast, however, results in a decrease in the organic phosphate and eventually the inorganic phosphate (Pi) content of the chloroplast. If the chloroplast is to be maintained at a high level of biosynthetic activity (and triose phosphate export), inorganic phosphate (Pi) must be supplied.

FACTORS INFLUENCING PHOTOSYNTHESIS

From what has been said about the structure of the photosynthetic apparatus and of the complexity of the photosynthetic process, it should be clear that many different kinds of conditions may modify the photosynthetic activity of a plant. Of the elements known to be essential for plant growth (Chapter 8), magnesium, iron, copper, chloride, manganese, and phosphorus are very closely tied with key reactions of photosynthesis and a deficiency in any of them would have serious consequences on photosynthetic activity.

Air pollutants are known to decrease photosynthetic activity in plants. At high concentrations, certain gases react with cellular constituents and cause the death of the affected tissue. If the stomatal apparatus is damaged, the entrance of CO_2 into the leaf is prevented. Sublethal concentrations of certain gases may temporarily inhibit photosynthetic activity without causing permanent damage. The general types of air pollutants that are known to have an effect on plant growth are ozone, hydrogen fluoride, sulfur dioxide, and oxidants produced by the action of sunlight on hydrocarbons and oxides of nitrogen. The oxidants are complex in nature and are sometimes referred to as *chemical smog*.

Moisture stress also modifies photosynthetic activity. Under conditions of high rates of transpiration the leaves may temporarily wilt and close their stomata. At such times, entry of CO_2 is reduced and the rate of photosynthesis will drop. Moisture stress may also have a more direct effect on photosynthetic activity by upsetting the organization of enzyme systems. Little is known about how these moisture stresses affect photosynthesis, but they contribute substantially to decreased photosynthetic activity and lowered crop yield.

Temperature also has a pronounced effect on photosynthetic activity. It will be recalled that studies (Figure 6-1) of the interaction of temperature, light intensity, and carbon dioxide concentration demonstrate that two different kinds of reactions were involved in photosynthesis—a series of chemical reactions and a series of physical reactions. Temperature also influences other physiological processes in the plant. Respiratory processes, for example, are speeded up at high temperatures. Of course, these processes—like the photosynthetic processes—have an upper temperature limit and prolonged exposure to elevated temperatures may permanently damage the enzyme systems.

Carbon dioxide concentration has a marked influence on the rate of photosynthesis. Natural air contains, on the average, 0.03% carbon dioxide and it was shown in Figure 6-1 that the rate of photosynthesis could be increased severalfold by increasing the carbon dioxide concentration. Although feasible under laboratory conditions or even in greenhouses, it is not possible to increase markedly the amount of CO_2 in the air above a field of wheat or in a forest. Certain conditions, however, might improve the availability of CO_2 to crop surfaces. The density of the crop and the height of the canopy will influence the mixing of the carbon dioxide within the immediate vicinity of the leaves. Soil organic matter, together with a vigorous population of soil microorganisms, may increase the concentration of CO_2 at the soil level. Under appropriate conditions of air turbulence this CO_2 may become more readily available to the plants.

PHOTOSYNTHESIS OF CROP PLANTS

Earlier the productivities of a number of crop plants and natural plant communities were discussed. Data were given indicating that plants growing under ordinary agricultural conditions, where moisture and nutrients are limiting, do not convert more than 0.1 to 0.3% of the usable radiant energy into plant organic matter. Under conditions of intensive agriculture, where adequate moisture and nutrients were provided and where modern land management practices were followed, crop plants convert between 2 and 3% of the usable radiant energy into plant material. Under special conditions efficiencies as high as 6 to 10% have been reported. It is obvious that there is a wide spread between crop productivity under various agricultural systems and that large increases in productivity can be achieved by supplying adequate water (irrigation) and nutrients (fertilizer) and by following good land management procedures. In areas where modern agricultural practices are not now followed a ten-to twentyfold increase in crop productivity can be achieved by using information now known to agronomists and soil scientists. Most of this agricultural information is directly applicable to crop production in the temperate region of the world; less information is available regarding crop production in the tropics and subtropics.

An Estimate of Potential Crop Productivity

Assuming that soil moisture, inorganic nutrients, and pests do not limit plant growth, what levels of plant productivity can be expected? This question was posed earlier in the chapter, but it was pointed out that before such an estimate can be made, something of the nature of the chemical and physical aspects of photosynthesis must be known. Such information is now available. The rate of photosynthesis (the amount of CO_2 assimilated per unit of leaf area per unit of time) depends on the intensity of radiant energy, the carbon dioxide concentration, and the temperature. An idea of the quantum requirement of photosynthesis is also required. If it is assumed that the quantum requirement is 10 (i.e., the energy of 10 photons is necessary to convert a molecule of CO_2 to CH_2O), that there is a CO_2 concentration of 0.03% in the air, and that the temperature is about 20° C, Loomis and Williams (1963) calculated that there is a net production of 3.3 μg of CH_2O per joule of incident radiant energy.

This value of 3.3 μg of CH_2O per joule of incident radiation establishes the maximum (or potential) amount of carbohydrate that can be produced by the green plant. The calculations take into consideration physical and biochemical aspects of the photosynthetic apparatus. It should be emphasized that the value is only an estimate and is based on data from a relatively few plants. The order of magnitude appears to be correct, however, and calculations based on different assumptions have arrived at a similar value.

What does the production of 3.3 μg of CH_2O per joule of incident radiation mean in terms of potential net plant productivity? To determine potential net productivity of plants at a particular location, only the amount of daily solar radiation need be known. Such information is available from meteorological data gathered by weather stations and an estimate of total solar radiation at several locations in the United States is given in Table 6-2. This table gives the mean solar radiation for the period June 1 to September 8 (100 days), a period that encompasses the growing season of maize, tobacco, soybeans, and other agronomic crops.

Several points are worthy of note in Table 6-2. Northern latitudes (such as Annette, Alaska) have less radiant energy available for crop growth than the southern latitudes. The air and soil temperatures are also considerably lower at the northern latitudes and there are greater differences between day and night temperatures. At comparable southerly latitudes there are also striking differences in the amounts of available solar radiation. Washington, D.C., and Davis, California, for instance, are located at 39° N latitude, but Davis has roughly 20% more incident solar radiation. This fact is largely due to the greater number of cloud-free days at Davis compared with Washington, D.C. Solar radiation values of Ames, Iowa, are characteristic of many areas in the Midwest where maize and other cereal grains are grown.

Table 6-2 shows that a value of 21 MJ m^{-2} day^{-1} is a reasonable approximation of the solar radiation available for plant growth during the growing season in the major

TABLE 6-2 SOLAR RADIATION RECEIVED AT VARIOUS U.S. LOCATIONS FOR 100 DAYS, JUNE 1 TO SEPTEMBER 8, 1960

Location	Mean solar radiation MJ m^{-2} day^{-1}	Days with radiation exceeding 21 MJ m^{-2} day^{-1}
Annette, Alaska (62° N)	16.4	29
Ames, Iowa (42° N)	20.8	66
Washington, D.C. (39° N)	21.9	61
Shreveport, Louisiana (33° N)	22.4	70
Spokane, Washington (48° N)	26.0	79
Davis, California (39° N)	28.5	99
Albuquerque, New Mexico (35° N)	28.5	94

After Loomis and Williams, 1963.

crop areas of the United States. With this figure and a value of 3.3 μg of CH_2O per joule of incident radiation, a potential net productivity of 69 g of CH_2O per square meter per day (21 MJ m^{-2} day^{-1} $\times$ 3.3 μg CH_2O) can be achieved. How does such a value compare with actual crop yields? Some comparisons are shown in Table 6-3.

In this table it should be noted that daily net productivity is listed. This is not grain yield but the rate of above-ground dry matter accumulation and it is determined by measuring the amount of dry matter accumulated during the growing season. This value is then divided by the number of days in the growing season to give the daily net productivity rate, the values given in Table 6-3. The rates are low early in the growing season, when a complete plant cover is not present, and low toward the end of the growing season, when many leaves have died or are dehydrated. The daily net productivity is used in this way to provide a basis for comparing crop performance with the potential net productivity calculated above.

The daily net productivity of four major agronomic plants and two forage plants is given. For wheat and rice, the world average productivity is 3 to 4% of the potential net productivity. Because they are average values, net productivity in many countries with low rainfall, no fertilizer, poor seed, and lack of pest control will be about 10% of these values, or 0.3 to 0.4% of potential productivity. Under intensive agriculture practices the actual productivities of wheat and rice are about 12% of the potential productivity.

The daily net productivities of sugarcane, maize, Napier grass, and sorghum are two to three times greater than for wheat and rice. Under intensive cultural practices

TABLE 6-3 COMPARISONS OF POTENTIAL PRODUCTIVITY WITH OBSERVED PRODUCTIVITY

Plants	Daily net productivity[a] (g m^{-2})	Percentage of potential productivity (69 g m^{-2})
Wheat, world average	2.3	3.3
Wheat, intensive agriculture	8.3	11.8
Rice, world average	2.7	3.8
Rice, intensive agriculture	8.0	11.4
Sugarcane, world average	4.7	6.7
Sugarcane, intensive agriculture	18.4	26.3
Maize	27.0[b]	38.5
Napier grass (*Pennisetum purpureum*)	26.0	37.0
Sorghum (*Sorghum vulgare*)	22.0	31.4

[a] Above-ground dry matter includes grain, leaves, and stems.

[b] At high plant densities and with no environmental limitations, maize has shown net productivity values up to 50 g m^{-2} for brief periods of time.

and high plant densities these plants have achieved net productivities in excess of 50% of potential productivity. These plants all fix carbon dioxide by the C_4-dicarboxylic acid pathway described earlier.

What Limits Plant Productivity?

As indicated at the beginning of the chapter, the efficiency of energy conversion in plants—the ratio of energy stored as organic substance to the amount of energy available from the sun—varies from between 0.1 and 1.0% under conventional agricultural practices to between 6 and 10% under conditions of intensive agriculture. It has been pointed out that there is a maximum value of energy conversion of about 30% imposed because of the nature of the photosynthetic process. Between the low values of much of traditional agriculture and those of intensive agriculture, however, there is a wide gap. The question was asked: Why are the values of energy conversion in plants so low? It was noted earlier that the very low levels are due mainly to poor agricultural practices. The lack of fertilizer (nitrogen, phosphorus, potassium), deficiency of water, hot or cold temperatures, low-quality seeds, inadequate pest control, and inefficient harvesting and storage practices all contribute to low crop productivity.

Where efficient agriculture practices are followed, as in North America, western Europe, Australia, and New Zealand, agricultural productivity is much greater. The term *green revolution* has been used to denote such a style of intensive agriculture. A major feature of intensive agriculture is the use of new and improved strains and varieties of crop plants, particularly maize, wheat, rice, and sorghum. Plant breeders have modified the form and shape of these crops such that they are usually shorter (dwarf habit) and display their leaves in an upright position near the top of the plant. These features enable more leaves to be exposed to incoming radiant energy for longer periods of time during the day.

Not only does the dwarf habit alter the light environment of the crop, but it also has an influence on the way in which the plants respond to chemical fertilizer. The productivity of most plants can be raised through the application of fertilizer, especially nitrogen. With many plants, particularly members of the grass family (wheat, rice, oats, maize), however, heavy fertilizer applications lead to the formation of tall, spindly plants. Such plants are easily blown over by windstorms during the growing cycle. If it happens before harvest, the production of grain is greatly reduced. Dwarf plants, on the other hand, have short, robust stems and this pattern is not modified by heavy applications of fertilizer. These plants respond to fertilizers by increases in grain yield. In fact, maximum productivity of the new varieties of maize, wheat, and rice is achieved only when the plants receive optimum applications of fertilizer. Moreover, the new varieties must have access to ample supplies of water (irrigation) to achieve high crop yields.

Since the 1960s the techniques of intensive, modern agriculture have been introduced in many parts of the world where agricultural productivity is low. In many instances, dramatic increases in crop yield have been achieved, thereby leading to the coining of the term green revolution to denote the successes in improving food production. As noted, the green revolution style of agriculture requires a number of inputs by the producer: New or improved crop varieties, high-quality seeds, heavy fertilizer applications, access to ample water, use of pesticides (herbicides, insecticides, fungicides, etc.), machinery, access to large enough crop-growing areas to take advantage of the use of mechanization, and access to capital and marketing facilities. There is no question that intensive agriculture leads to increased crop yields, but much of the success has been achieved in temperate parts of the world, in the so-called developed countries, where there is access to all the inputs mentioned. In tropical and subtropical parts of the world the success of intensive agriculture has been less spectacular. If all the practices are followed, crop yields have been greatly increased— rice in southeast Asia and wheat in Mexico, Pakistan, India, and parts of Africa—but the continued success of the green revolution depends on continuous inputs of fertilizer, pesticides, and other energy-using practices.

Energy Uses in Agriculture

It is easy to see that intensive agriculture requires the use of gasoline or diesel fuel for running tractors, harvesters, and other farm machinery. Yet these energy needs are but a small fraction of the total energy requirements necessary in modern agricultural practices. David Pimentel and colleagues estimated the energy requirements for the production of a hectare of maize. The results shown in Table 6-4 enumerate the inputs of labor, machinery, fuel, fertilizer, pesticide, and other factors for 1945 and 1970. In 1945 most of the maize land was planted to open-pollinated varieties and the agricultural practices did not require large inputs of energy. The green revolution in maize production in the United States came during the 1950s when new and improved hybrid maize varieties and intensive agricultural practices were introduced. Although the input of human labor decreased from 57 hours in 1945 to 22 hours in 1970, the average yields per hectare increased from 2140 kg in 1945 to 5100 kg in 1970 through the application of fertilizer, pesticides, and other production factors characteristic of intensive agriculture.

In the last column of Table 6-4 the various agricultural practices in 1970 have been converted to comparable energy equivalents as megajoules (10^6 joules or MJ). The energy in 1 L of gasoline, for instance, is equivalent to 40 MJ and the energy of 206 L of gasoline (1970) is equal to 8200 MJ. The other energy inputs have been calculated from a variety of sources by Pimentel and colleagues. The total energy subsidy for the production of 5100 kg of maize is approximately 30,000 MJ, an amount of energy equivalent to 750 liters of gasoline. In 1974 some 120,000 million kg of maize were harvested in the United States. Clearly a tremendous amount of energy goes into agricultural production. It is estimated that 12 to 13% of the total energy budget in this country goes into the food system, which includes growing crops, harvesting, feeding animals and poultry, processing, transportation, meal preparation, and so on. Another large energy user is the consumer who drives between home and the store. As the cost of fuel (coal, natural gas, and petroleum) increases, the cost of producing food

TABLE 6-4   ENERGY INPUT IN THE PRODUCTION OF 1 HECTARE
OF MAIZE

	1945	1970	Energy equivalents 1970 (MJ)
Human labor (h)	57	22	50
Machinery (MJ)	1863	4360	4360
Gasoline (liter)	140	206	8264
Nitrogen (kg)	8	126	9758
Phosphorus (kg)	8	35	487
Potassium (kg)	6	67	727
Seeds (kg)	11	21	705
Irrigation (MJ)	196	351	352
Insecticides (kg)	0	1	114
Herbicides (kg)	0	1	114
Drying (MJ)	103	1242	1244
Electricity (MJ)	331	3208	3214
Transport (MJ)	207	724	726
Total energy input			30,042 MJ
Maize yield (kg)	2140	5100	

Note: 1 cal = 4.19 joule; 3.8 L = 1 gal; 1 hectare (H) = 2.47 acres.
Maize yield of 63 kg H^{-1} = 1 bushel ac^{-1}; 1 bushel = 25.5 kg.
Calculated from Pimentel et al., 1973.

increases. In countries where crop productivity is low because of the lack of fertilizer,
pesticides, etc., it is both difficult and costly to increase yields.

Another feature of the energy subsidy in agriculture is worthy of note. In 1945,
before intensive agricultural practices were initiated, the ratio of energy output (energy
value of maize) to energy input was 3.7. In 1970 this ratio was 2.8. Even though maize
yields increased over this period, 25% more energy had to be expended to achieve the
increase.

The energy subsidy in maize production does not include the solar energy
received by the crop. Table 6-2 showed that the amount of solar energy falling on the
earth's surface in the maize belt (Ames, Iowa) is about 21 $MJ\,m^{-2}\,day^{-1}$. With a growing
season of 100 days, this amounts to about 21,000 MJ per hectare. Of this amount of
solar energy, the plant converts about 0.4% to grain. The grain weight of maize is about
35% of the total dry weight of the plant (grain + stalk + leaves + roots). Thus the
maize plant converts about 1.2% of the incident solar energy into plant material
through the photosynthetic process. Compared to the amount of solar energy available

for crop growth, the amount of energy used in farming practices is rather small. Solar energy, however, is free and the supply is inexhaustible. Fossil fuel, on the other hand, is expensive and the supply is diminishing.

Not all plants require as large an energy input in crop production as maize. Soybeans, for example, require 10 to 20% less of an energy subsidy than maize because of a lower fertilizer requirement, primarily nitrogen fertilizer. Soybeans are legumes and bacteria in their roots fix atmospheric nitrogen, which becomes available to the plant (Chapter 9). Nevertheless, much of the carbon assimilated by soybeans must be used to support symbiotic nitrogen fixation in the roots and crop yields of soybeans are considerably less than that of maize.

At the other end of the energy scale are agricultural practices in which large inputs of human labor are used. Wetland or paddy rice as grown in Indonesia, China, and Burma returns about 20 times more energy, in the form of grain, than is put into the crop. Fertilizers, pesticides, and fuel are not used. Rice is also grown as an energy intensive crop, however. Somewhere between the two extremes of human-labor-intensive and energy-intensive schemes of agricultural production a balance must be struck if adequate supplies of food are to be produced for an increasing human population.

REFERENCES

ANDERSON, L. E. 1979. Interaction between photochemistry and activity of enzymes. Pages 271–281 in M. Gibbs and E. Latzko, eds. *Photosynthesis II: Photosynthetic Carbon Metabolism and Related Processes. Encyclopedia of Plant Physiology*, new series, vol. 6. Springer-Verlag, New York.

BUCHANAN, B. B. 1980. Role of light in the regulation of chloroplast enzymes. *Ann. Rev. Plant Physiol.* 31:341–374.

BURRIS, R. H., and C. C. BLACK, eds. 1976. *CO₂ Metabolism and Plant Productivity*. University Park Press, Baltimore.

GIBBS, M., and E. LATZKO, eds. 1979. *Photosynthesis II: Photosynthetic Carbon Metabolism and Related Processes. Encyclopedia of Plant Physiology*, new series, vol. 6. Springer-Verlag, New York.

HATCH, M. D., C. B. OSMOND, and R. O. SLATYER, eds. 1971. *Photosynthesis and Photorespiration*. John Wiley & Sons (Interscience), New York.

JENSEN, R. G. 1980. Biochemistry of the chloroplast. Pages 273–313 in N. E. Tolbert, ed. *The Plant Cell*, vol. I. P. K. Stumpf and E. E. Conn, eds. *The Biochemistry of Plants, a Comprehensive Treatis*. Academic Press, New York.

KELLY, G. J., E. LATZKO, and M. GIBBS. 1976. Regulatory aspects of photosynthetic carbon metabolism. *Ann. Rev. Plant Physiol.* 27:181–205.

KLUGE, M., and I. P. TING. 1978. *Crassulacean Acid Metabolism (CAM): Analysis of an Ecological Adaptation*. Springer-Verlag, Berlin–New York.

LILLEY, R. McC., and D. A. WALKER. 1979. Studies with reconstituted chloroplast system. Pages 41–53 in M. Gibbs and E. Latzko, eds., *Photosynthesis II. Encyclopedia of Plant Physiology*, new series, vol. 6. Springer-Verlag, New York.

Loomis, R. S., and W. A. Williams. 1963. Maximum crop productivity, an estimate. *Crop Sci.* 3:67–72.

Pimentel, D., L. E. Hurd, A. C. Bellott, M. J. Forster, I. N. Oka, O. D. Sholes, and R. J. Whitman. 1973. Food production and the energy crisis. *Science* 182:443–449.

Siegelman, H. W., and G. Hind, eds. 1978. *Photosynthetic Carbon Assimilation*. Plenum Press, New York.

Ting, I. P., and M. Gibbs, eds. 1982. *Crassulacean Acid Metabolism*. American Society of Plant Physiologists, Rockville, Md.

Vickery, H. B. 1951. The behavior of isocitric acid in excised leaves of *Bryophyllum calycinum* during culture in alternating light and darkness. *Plant Physiol.* 27:9–17.

Walker, D. A. 1974. Some characteristics of a primary carboxylating mechanism. Pages 7–26 *in* J. B. Pridham, ed., *Plant Carbohydrate Biochemistry*. Phytochemical Society Symposia Series No. 10. Academic Press, London–New York.

Wildman, S. G., and J. Bonner. 1947. The proteins of green leaves. I. Isolation, enzymic properties, and auxin content of spinach cytoplasmic proteins. *Arch. Biochem.* 14:381–413.

Wildman, S. G., and P. Kwanyuen. 1978. Fraction I protein and other products from tobacco for food. Pages 1–18 *in* H. W. Siegelman and G. Hind, eds. *Photosynthetic Carbon Assimilation*. Plenum Press, New York.

7

Photorespiration

It was pointed out in Chapter 6 that the first product of CO_2 fixation in photosynthesis in most higher terrestrial plant species is a three-carbon compound called 3-phosphoglyceric acid. Plant species with this pattern of photosynthesis are called C_3 plants. On the other hand, it will be recalled (Chapter 6) that C_4 species (e.g., sugarcane, maize) are those in which the first product of CO_2 fixation in photosynthesis is oxaloacetic acid, a four-carbon dicarboxylic acid.

Here we will see that C_3 plants are distinguished from C_4 plants by several anatomical, physiological, and biochemical characteristics. Moreover, illuminated leaves of C_3 plants respire away considerable quantities of certain of the early products of photosynthesis, much more than leaves of C_4 plants. This process, called photorespiration, is the main subject of this chapter.[1]

RESPIRATION OF CHLOROPHYLLOUS TISSUES IN THE LIGHT

Everyone who has taken a course in introductory botany knows that respiration and photosynthesis are antagonistic processes. Release of respiratory CO_2 to the atmosphere, whether by root or shoot cells, serves to cancel out a portion of the CO_2 fixed in photosynthesis. A growing plant gains in dry weight to the extent that photosynthesis exceeds respiration.

---

[1] Leaves of CAM plants (Chapter 6) are able to carry on photorespiration, but this matter will not be considered separately.

The opposing natures of photosynthesis and respiration are illustrated by the empirical equation

$$CO_2 + 2\,H_2O \underset{\text{respiration}}{\overset{\text{photosynthesis}}{\rightleftarrows}} (CH_2O) + O_2 + H_2O \qquad (7\text{-}1)$$

where (CH_2O) represents carbohydrate (e.g., glucose). (The reason why an extra molecule of water is added to both sides of reaction (7-1) was considered on pp. 86–87.)

One way to measure the rate of photosynthesis is to place a leaf in a glass chamber and measure the rate of disappearance of CO_2 from the chamber while the leaf is exposed to light. But each and every living plant cell, whether green or not, respires continuously, a fact stressed in Chapter 4. The release of respired CO_2 has the effect of reducing the rate at which CO_2 disappears from the chamber. In fact, the rate of disappearance of CO_2 from the illuminated chamber is a measure of *net* (or "apparent") photosynthesis. To obtain the value for *total* (or "gross" or "true" photosynthesis) requires that the respiratory rate of CO_2 release from a leaf be added to the value of net photosynthesis of the leaf, as shown in the following relation.

$$\text{Rate of total photosynthesis} = \text{rate of net photosynthesis}$$
$$+ \text{ rate of respiration in the light} \qquad (7\text{-}2)$$

Until about 20 years ago the conventional way to calculate the rate of total photosynthesis of a green leaf was to measure the rate of net photosynthesis and to correct the value by the rate of respiration *in the dark*. In this procedure it was assumed that the rate of respiration of a green leaf is the same in the light as in the dark. Today, however, it is well documented that rates of respiration in the light and dark are almost the same only in a few plant species (i.e., C_4 plants). It is recognized that respiratory CO_2 release (and also respiratory O_2 uptake) in chlorophyllous tissues of most species of higher plants proceeds at much higher rates in the light than the dark. This process, which occurs at a relatively rapid rate in C_3 plants but only to a slight extent in C_4 plants, is called photorespiration. Photorespiration is defined rigorously later (p. 209). For the present, it need only be pointed out that photorespiration occurs *simultaneously* with photosynthesis; photorespiration decreases net photosynthesis because a portion of the CO_2 fixed in photosynthesis escapes from the leaf shortly (probably less than a minute) after it is fixed and so is not available for plant growth.

Net photosynthesis is measured more frequently than total photosynthesis—both in laboratory experiments and in the field—not only because it does not require a separate measurement for respiration in the light (i.e., photorespiration) but also because it is a useful measurement in comparative studies. Net photosynthesis is usually expressed in terms of CO_2 uptake on a leaf area basis because the leaf is the main photosynthetic organ. The range of net photosynthesis is generally from 15 to 35 mg CO_2 dm^{-2} leaf surface h^{-1} for C_3 plants and 40 to 80 mg CO_2 dm^{-2} leaf surface h^{-1} for C_4 plants, in full sunlight and at the natural concentrations of O_2 (21%) and CO_2 (0.033%) in the external atmosphere.

The existence of photorespiration was first demonstrated unequivocally by the American plant physiologist Decker in studies beginning in the mid-1950s; he and his associates were the first to use the term photorespiration (Decker and Tio, 1959).

Photorespiration is superimposed on respiratory activities already considered in Chapter 4 (e.g., TCA cycle and the respiratory chain). These latter components of respiration are usually referred to collectively as *dark* respiration even though they also occur at about the same rate in the light. (Whether leaf tissue is held in the dark or light, dark respiration rates are only about 10% of photorespiratory rates.)

Several methods for measuring the rate of photorespiration (together with the continuously occurring dark respiration) have been developed. Only two are mentioned here, primarily to illustrate the general procedure. (Such experiments are usually carried out under optimum or near-optimum conditions of light and water and nutrient supply.) In one method CO_2-free air is passed over an illuminated leaf enclosed in a glass chamber and then through an infrared gas analyzer, an instrument that measures the concentration of CO_2 in the airstream after it has passed over the leaf. The rate of CO_2 evolution from the leaf can be calculated from the measured CO_2 concentration of the airstream and its flow rate. In this way, an estimate of the rate of photorespiration can be made. This method, however, inevitably underestimates the rate of photorespiration because an unknown fraction of photorespired CO_2 released within leaf cells is refixed in photosynthesis even before it escapes from the leaf into the chamber.

In a more accurate method of measuring photorespiratory rate (Ludwig and Canvin, 1971), an attempt is made to avoid the internal photosynthetic refixation of photorespired CO_2 mentioned above. After steady-state photosynthesis in an illuminated leaf in a chamber is achieved in air, radioactive carbon dioxide ($^{14}CO_2$) is introduced into the chamber. [Note: Carbon dioxide in air exists mostly in the *nonradioactive* form ($^{12}CO_2$), see p. 206]. The disappearance of $^{12}CO_2$ from the chamber (measured with an infrared gas analyzer sensitive to $^{12}CO_2$ but not $^{14}CO_2$) and also the disappearance $^{14}CO_2$ (measured with an electrometer) are determined simultaneously within seconds after $^{14}CO_2$ is introduced into the leaf chamber. From these measurements the rate of photorespiration can be calculated.

The rate of photorespiration of many different plants has been determined by different methods. The process of simultaneous measurements of $^{14}CO_2$ and $^{12}CO_2$ uptake described indicates that the rate of photorespiration at 15°C in C_3 plants is about 25% of the rate of net photosynthesis, at concentrations of CO_2 and O_2 naturally present in the atmosphere (Bravdo and Canvin, 1979). [Higher estimates of photorespiratory rates in plants have been reported (Gerbaud and André, 1979) when another method—one that measures O_2 uptake that accompanies photorespiration—was used.] Regardless of the magnitude of photorespiration rates, whether 25% of the rate of net photosynthesis or higher, photorespiratory loss of CO_2 during photosynthesis in leaves of C_3 plants is an important factor in explaining the lower capacity of C_3 plants for net photosynthesis compared to C_4 plants. (As already indicated, photorespiratory rates in leaves of C_4 plants are almost negligible.)

The difference in rates of net photosynthesis between C_3 and C_4 plants (see p. 201) and the existence of high rates of photorespiration in C_3 plants and very low rates in C_4 plants have persuaded plant physiologists in many laboratories throughout the world to study the characteristics that distinguish these two sets of plants. These characteristics are considered in the following section. Later, the biochemistry of photorespiration in C_3 plants and the reasons why photorespiratory rates are very low in C_4 plants are discussed.

THE C_4 SYNDROME

Of the approximately 300,000 species of flowering plants, about 1000 have been reported to be C_4 plants. Most are grasses of tropical origin. Well-known examples are maize, sorghum, sugarcane, Italian millet, crabgrass, and Bermuda grass. The first four are crop plants used since primitive times as sources of feed grains, sweet juices, and forages; crabgrass is a troublesome weed in cultivated lawns; Bermuda grass is an important pasture grass in the southern United States. Also included among C_4 plants are a few dicot species [e.g., *Amaranthus* (pigweed) and a few species of *Atriplex* (a genus of annuals and perennials that grows in the southwestern United States)]. C_3 species, on the other hand, include such crop plants as peanut, soybean, cotton, sugar beet, potato, tobacco, and spinach; cereal grains such as wheat, oats, barley, rye, and rice; and all evergreen and deciduous tree species.

The set of characteristics that distinguishes C_4 plants from C_3 plants is often referred to collectively as the C_4 *syndrome*. One feature was mentioned earlier: There is little or no photorespiratory release of fixed carbon in C_4 plants.

Kranz leaf anatomy   One distinguishing feature of leaves of C_4 plants is the arrangement of chloroplast-containing (i.e., chlorophyllous) cells around vascular bundles. The cells which border the vascular bundles, referred to collectively as the *bundle sheath*, consist of one or more layers of large, thick-walled cylindrical cells with a high concentration of chloroplasts. Bundle sheath cells are surrounded, in turn, by one or more wreathlike layers of mesophyll cells that also contain numerous chloroplasts but usually of a smaller size than those in bundle sheath cells (see maize, Figure 7-1). Because bundle sheath cells cover the sides of vascular bundles as well as their ends, substances that move between mesophyll cells and vascular tissue must pass through bundle sheath cells. This anatomical arrangement of cells is called Kranz type (German, *Kranz*, wreath); the word comes from the wreath of mesophyll cells around bundle sheath cells.

Bundle sheath cells in leaves of C_3 plants have few or no chloroplasts (see pear, Figure 7-1) or are indistinguishable from other chlorophyllous mesophyll cells (see tobacco, Figure 7-1). Moreover, chlorophyllous cells of typical C_3 dicots (see tobacco and pear, Figure 7-1) are arranged in one or more rather compact layers in the region next to the upper epidermis and are known as *palisade* parenchyma; those in the lower part of the leaf are irregular in shape, arranged much less compactly with numerous intercellular spaces, and are called *spongy* parenchyma.

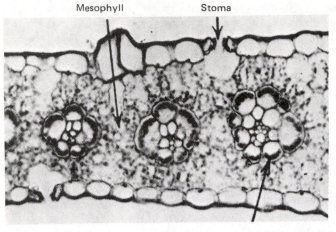

Mesophyll   Stoma

Bundle sheath

Maize

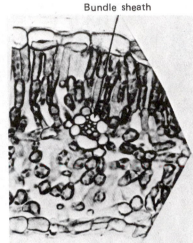

Bundle sheath

Pear

Figure 7-1 Transections of leaves of maize (C_4), pear (C_3), and tobacco (C_3), stained to show chloroplasts. Note the very high concentration of chloroplasts in the bundle sheath cells of maize and their absence in bundle sheath cells of pear. In tobacco, bundle sheath cells are not distinguished in size, shape, or number of chloroplasts from other mesophyll cells in the spongy parenchyma. (After K. Esau, *Plant Anatomy*. John Wiley & Sons, New York, 1953. Plates 59B, 61A, and 63A.

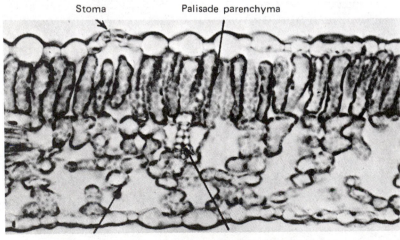

Stoma   Palisade parenchyma

Spongy parenchyma   Vascular bundle

Tobacco

204

An additional feature of the internal leaf anatomy of C_4 plants is the very small number (only four or five) of chlorophyllous mesophyll cells that intervene between bundle sheath cells of laterally adjacent vascular bundles; in C_3 plants there are at least four and often many more chlorophyllous mesophyll cells between laterally adjacent vascular bundles (compare maize with tobacco and pear in Figure 7-1). Furthermore, in C_4 species no chlorophyllous mesophyll cell is separated from the nearest bundle sheath cell by more than one other chlorophyllous mesophyll cell (see maize, Figure 7-1); in C_3 plants there are many more than one (see tobacco and pear, Figure 7-1).

The chloroplasts in each of these two cell types are characterized by their own complement of enzymes. Chloroplasts of mesophyll cells in leaves of C_4 plants, for example, lack ribulose bisphosphate carboxylase/oxygenase, as we will see below; this enzyme is present only in bundle sheath cells of C_4 plants. Another example concerns nitrite reductase; it will be seen (pp. 261–262) that nitrite reductase is present in chloroplasts of mesophyll cells of C_4 plants but not in chloroplasts of bundle sheath cells.

C_4 Dicarboxylic acids as initial products of CO_2 fixation   A major feature of C_4 photosynthesis is a compartmentation of carbon metabolism between mesophyll cells and bundle sheath cells, as described in Chapter 6. Atmospheric CO_2 (as bicarbonate ion) is fixed initially in a reaction with phosphoenolpyruvic acid (PEP) in mesophyll cells; this carboxylation reaction is catalyzed by the enzyme PEP carboxylase (Eqs. 4–33 and 6–8) and leads to the formation of C_4 dicarboxylic acids.[2] These C_4 acids are transported to adjoining bundle sheath cells, where decarboxylation occurs: CO_2 released in bundle sheath cells is refixed by ribulose bisphosphate carboxylase in the photosynthetic carbon reduction cycle (reductive pentose phosphate pathway) in bundle sheath cells. Finally, the three-carbon compound formed in the decarboxylation reaction in bundle sheath cells is transported back to mesophyll cells, where it is regenerated into the initial carboxylation substrate (i.e., phosphoenolpyruvic acid). More will be said about this sequence of reactions later in this chapter (pp. 215–216).

Low CO_2 compensation point   The CO_2 compensation point (or CO_2 compensation concentration) is measured by placing a shoot or a leaf (or a portion of a leaf) in a small, airtight glass chamber initially filled with air. Upon illumination with bright light the CO_2 concentration in the chamber (monitored with a CO_2-measuring device,

---

[2]C_4 mesophyll cells do not contain the enzyme ribulose bisphosphate carboxylase and therefore do not fix CO_2 via the photosynthetic carbon reduction cycle; all the enzymes of this cycle are present only in bundle sheath cells in C_4 plants. However, C_4 mesophyll cells contain NADP-linked triose phosphate dehydrogenase. Moreover, a portion of the 3-phosphoglyceric acid formed in bundle sheath cells (from CO_2 fixation via the photosynthetic carbon reduction cycle) is transported back to mesophyll cells, where it is reduced. Thus CO_2 entry into the photosynthetic carbon reduction cycle in C_4 plants occurs only in bundle sheath cells, but reduction of 3-phosphoglyceric acid occurs in bundle sheath cells *and* mesophyll cells (see Outlaw, Jr., et al., 1981, and references cited there).

such as an infrared analyzer) will decrease because the initial rate of CO_2 fixation in photosynthesis is greater than the rate of production of respiratory CO_2. In a short time (perhaps an hour or two) a concentration is reached that will remain constant. Starting with the natural concentration of CO_2 in air (approximately 0.033% or 330 ppm CO_2 by volume), the concentration of CO_2 in the chamber will be reduced to a range between about 1 to 5 and 60 ppm CO_2—the final CO_2 concentration will depend on the plant species. This equilibrium CO_2 concentration is called the *carbon dioxide compensation point*. The same equilibrium CO_2 concentration will be reached if the chamber is filled initially with air free of CO_2; in this case, there will be a net flow of CO_2 out of the leaf into the chamber. (*Note*: In photorespiration studies CO_2 concentration in air is usually expressed either in ppm by volume or in microliters per liter of air. Thus 330 ppm CO_2 equals 330 μL of CO_2 per liter). The O_2 concentration in the chamber ($= 21\%$ O_2) will remain substantially unchanged in an experiment of this type.

At the CO_2 compensation point, CO_2 released in respiration is exactly *compensated* by CO_2 fixed in photosynthesis. The rate of fixation of CO_2 in photosynthesis *just balances* the rate of CO_2 released in respiration. Therefore net photosynthesis is zero at the CO_2 compensation point [see Eq. 7-2)]. Only at external CO_2 concentrations higher than the CO_2 compensation point is there a net fixation of CO_2.

Measurements have been made of the CO_2 compensation points of literally hundreds of plant species during the past several years. When measured at about 25° C and 21% O_2, C_3 plants have CO_2 compensation points ranging from about 40 to 60 ppm CO_2; those of C_4 plants range from about 1 to 5 ppm CO_2. So it can be said that there is very little leakage of CO_2 into the external atmosphere from C_4 plants.

Low degree of discrimination against ^{13}C   The most unambiguous method for distinguishing between mature C_4 and C_3 terrestrial plants is to determine the $^{13}C/^{12}C$ ratio in a tissue sample (e.g., a leaf). ^{13}C and ^{12}C are the two stable (i.e., nonradioactive) isotopes of carbon; ^{12}C has a far greater natural abundance than ^{13}C. In combination with oxygen atoms, carbon in the atmosphere is present as both $^{12}CO_2$ and $^{13}CO_2$, each of which is absorbed in photosynthesis.[3] The natural abundance of $^{12}CO_2$ in the atmosphere is 98.8% whereas that of $^{13}CO_2$ is 1.1%, assuming no pollution input into the atmosphere.

It should be emphasized at the outset that photosynthesis in *all* terrestrial plants, whether C_3 or C_4, is accompanied by fractionation of the stable carbon isotopes in CO_2. This discrimination is against $^{13}CO_2$ (mass 45) and favors fixation of $^{12}CO_2$ (mass 44): CO_2 of lower mass (44) is assimilated in photosynthesis to a slightly greater

---

[3]The radioactive isotope of carbon (^{14}C) is also present in the atmosphere in combination with oxygen atoms but only to the extent of one part $^{14}CO_2$ to 10^{12} parts of $^{12}CO_2$ and $^{13}CO_2$. It is not known whether photosynthesis discriminates in favor of or against $^{14}CO_2$ at these very low concentrations. Incidentally, the (very slight) degree of radioactivity that occurs naturally in plant material is due to the fixation of $^{14}CO_2$ in photosynthesis and subsequent assimilation of fixed ^{14}C carbon, and also the absorption of trace amounts of radioactive mineral elements.

extent than CO_2 of higher mass (45). Therefore the ratio of ^{13}C to ^{12}C in carbon-containing constituents is slightly lower in tissues of terrestrial plants than in the external atmosphere. (*Note*: ^{13}C discrimination can be detected only from a limitless source of CO_2, such as the earth's atmosphere, not from a small glass container as in Figure 7-6.) The slightly slower rate of movement of $^{13}CO_2$ (i.e., the heavier species) in the gas phase in stomatal pores and intercellular spaces in leaves—as compared to that of $^{12}CO_2$ (i.e., the lighter species)—is the major factor which accounts for the discrimination against $^{13}CO_2$ in photosynthesis.

Isotopic compositions are expressed in $\delta^{13}C$ units, defined as follows:

$$\delta^{13}C \text{ (in parts per thousand)} = \left[\frac{(^{13}C/^{12}C)_{sample}}{(^{13}C/^{12}C)_{standard}} - 1 \right] \times 1000 \qquad (7\text{-}3)$$

where the standard ($^{13}C/^{12}C = .01124$) is a carbonate fossil found in South Carolina. The symbol for parts per thousand (parts per mil) is $^0/_{00}$. $^{13}C/^{12}C$ ratios in carbon-containing samples (e.g., a plant leaf) can be determined accurately by special mass spectrometric techniques.

Not only do all terrestrial plants discriminate against $^{13}CO_2$, as indicated, but there is also a difference in discrimination against ^{13}C in photosynthesis in C_3 and C_4 plants: C_4 plants have a lower degree of discrimination against ^{13}C than do C_3 plants. Therefore *C_4 plants are more enriched in ^{13}C than C_3 plants*; $^{13}C/^{12}C$ ratios in carbon-containing compounds in C_4 plants are higher than those in C_3 plants. In terms of Eq. (7-3), $\delta^{13}C$ values in C_4 plants are higher (i.e., less negative) than those in C_3 plants. $\delta^{13}C$ values in different C_4 plants range from -8 to $-18^0/_{00}$ (average $= -14^0/_{00}$) whereas those in different C_3 plants range from -22 to $-36^0/_{00}$ (average $= -28^0/_{00}$). (Note: In non-industrial areas, atmospheric CO_2 has a $\delta^{13}C$ value of about $-6.5^0/_{00}$.)

An interesting consequence of the carbon isotope fractionation effect is that carbon materials in sugarcane (a C_4 plant) have higher $\delta^{13}C$ values than those in sugar beet (a C_3 plant). Thus table sugar (i.e., purified sucrose) manufactured from each of these two plant species can be distinguished by carbon isotope ratio analysis even though they are chemically identical.

The reasons why C_4 plants have a lower degree of discrimination against ^{13}C than C_3 plants have begun to be understood only recently. They are complex. Space limitations preclude a discussion here.[4]

Insensitivity of net photosynthesis to O_2 concentration at and below natural levels   When the concentration of O_2 in the external atmosphere varies upward from very low levels to 21% (the concentration naturally present in the atmosphere), net photosynthesis in a C_4 plant remains substantially unchanged. This situation is shown in Figure 7-2, where CO_2 concentration is held at 300 ppm and light conditions are optimal. On the other hand, the rate of net photosynthesis in a C_3 plant is considerably reduced at 21% O_2, compared to the rate at about 2% O_2 (see Figure 7-2); in fact, the

[4]The interested reader may wish to consult O'Leary, 1981, and O'Leary, et al., 1981, for details.

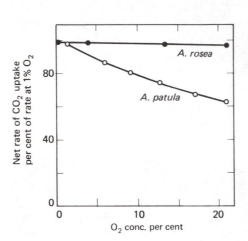

Figure 7-2   The effect of O_2 concentration in the external atmosphere on the rate of net photosynthetic CO_2 uptake for a C_3 plant (*Atriplex patula*) and a C_4 plant (*Atriplex rosea*). Leaf temperature was 25° C and CO_2 concentration was 0.03%; near-optimal light intensities were used for each plant. [*Note*: The rate of net photosynthesis is shown in terms of the rate at 1% O_2 (see vertical axis); when O_2 concentration in the atmosphere external to the leaves is 1%, however, the percentage of O_2 in the chloroplasts would be slightly higher than 1% because O_2 is produced in the light reaction in photosynthesis.] (After O. Björkman, E. Gauhl, and M. A. Nobs), *Carnegie Institution of Washington Year Book No. 68*, 1969, Figure 39.)

rate of net photosynthesis continues to decrease as the concentration of O_2 becomes higher than 21%, but this fact is not shown in Figure 7-2.[5] (In such an experiment, the external concentration of O_2 in the plant chamber at the low limit is usually held at about 2% O_2. Below this level dark respiration would be suppressed because there is an insufficient quantity of O_2; above this level dark respiration is saturated with O_2.)

This inhibitory effect of 21% O_2 in C_3 plants is produced rapidly and is fully reversible; that is, it disappears completely when the O_2 concentration decreases from 21 to about 2%. The German biochemist O. Warburg discovered the inhibition of photosynthesis by O_2 in 1920 and it is now referred to as the Warburg effect. Only recently has it been realized that the Warburg effect involves a competition between CO_2 and O_2 for the reactive site on the enzyme that fixes CO_2 in the photosynthetic carbon reduction cycle (reductive pentose phosphate pathway)—namely, ribulose bisphosphate (RuBP) carboxylase. This enzyme reacts not only with CO_2 but also with O_2, as we will see. Thus O_2 is a competitive inhibitor with respect to CO_2 in the carboxylase reaction [see Eq. (7-4)] and CO_2 is a competitive inhibitor with respect to O_2 in the oxygenase reaction [see Eq. (7-5)]. In other words, both CO_2 and O_2 compete for RuBP at the same reaction site on the same enzyme.

In addition to the competitive inhibition by O_2 of the carboxylase activity of RuBP carboxylase, there is a second reason why O_2 inhibits photosynthesis. The oxygenase reaction [Eq. (7-5)] leads to the synthesis of glycolic acid. This acid can be regarded as the substrate of photorespiration because its metabolism leads to the production of photorespiratory CO_2, as we will see below.

---

[5]Plants held at levels of O_2 higher than 21% are likely to suffer deleterious oxidative effects unrelated to photorespiration.

BIOCHEMISTRY OF PHOTORESPIRATION IN C$_3$ PLANTS

Photorespiration is the uptake of O$_2$ and the release of CO$_2$ in light and results from the biosynthesis of glycolic acid in chloroplasts and subsequent metabolism of glycolic acid in the same leaf cell. To comprehend this definition more fully, we must first consider the *biosynthesis* of glycolic acid (CH$_2$OHCOOH), the two-carbon organic acid known as the *substrate of photorespiration* (because the carboxyl group of glycolic acid is the source of photorespiratory CO$_2$). Relatively large amounts of glycolic acid are synthesized in the light in leaves of C$_3$ plants. (Illuminated leaves of C$_4$ plants synthesize only negligible amounts of glycolic acid.) We will see that the synthesis and subsequent metabolism of glycolic acid in a green leaf of a C$_3$ plant involve release of CO$_2$ and uptake of O$_2$. It is precisely this CO$_2$ release and O$_2$ uptake associated with glycolic acid metabolism that explains the photorespiratory gas exchange in C$_3$ plants.

Let us consider the factors that affect the biosynthesis of glycolic acid in C$_3$ plants. Glycolic acid is synthesized in green leaves in the light but not in the dark. Its synthesis in illuminated leaves has been shown to depend on the operation of the photosynthetic carbon reduction cycle (i.e., reductive pentose phosphate cycle); recall that this is the only pathway of CO$_2$ assimilation in photosynthesis in C$_3$ plants. Moreover, the concentration of CO$_2$ in the external atmosphere (when the concentration of O$_2$ is held at 21%) has a marked effect on the synthesis of glycolic acid. Glycolic acid is synthesized rapidly and accumulates in relatively large amounts in C$_3$ plants held at the natural concentration of CO$_2$ in air (= approximately 330 ppm). But an increase in the concentration of CO$_2$ in the external atmosphere, even only to 1000 ppm, results in a significant reduction in the amount of glycolic acid synthesized.

The concentration of O$_2$ in the external atmosphere (when the concentration of CO$_2$ is held at 330 ppm) also has a marked effect on the synthesis of glycolic acid in plants. The effect of O$_2$, however, is opposite that of CO$_2$. An increase in O$_2$ concentration in the external atmosphere will increase the amount of glycolic acid synthesized in an illuminated leaf. Thus only small amounts of glycolic acid are formed when a C$_3$ plant is held in a chamber at about 2% O$_2$. But glycolic acid is synthesized rapidly and accumulates in relatively large amounts in C$_3$ plants held in air with the natural concentration of O$_2$ (= 21%).

This competition between O$_2$ and CO$_2$, with O$_2$ promoting glycolic acid synthesis and inhibiting photosynthesis and CO$_2$ inhibiting glycolic acid synthesis and promoting photosynthesis, has led to the widely accepted view that glycolic acid is synthesized in vivo as a result of the oxygenation of a member of the photosynthetic carbon reduction cycle—ribulose-1, 5-bisphosphate (RuBP). In fact, the enzyme RuBP carboxylase, which catalyzes the addition of CO$_2$ to RuBP, has also been shown to catalyze the addition of O$_2$ to RuBP. Thus RuBP carboxylase is now called RuBP carboxylase/oxygenase. (This enzyme is the most abundant protein in the biological world; it constitutes as much as one-quarter of the total protein of the leaf of a C$_3$ plant.) Let us compare these two reactions catalyzed by RuBP carboxylase/oxygenase, both of which are irreversible.

The carboxylation reaction has already been considered (pp. 168–169), but it will be useful to repeat it here. As shown in Eq. (7-4), CO_2 is added to carbon atom 2 of RuBP. Together with this addition of CO_2, there is a shifting in the positions of certain protons and valence electrons associated with carbon atoms 2 and 3. Then the bond between carbon atoms 2 and 3 in the postulated intermediate (in brackets) is broken and the elements of water are added across the broken bond so that two molecules of 3-phosphoglyceric acid are produced.

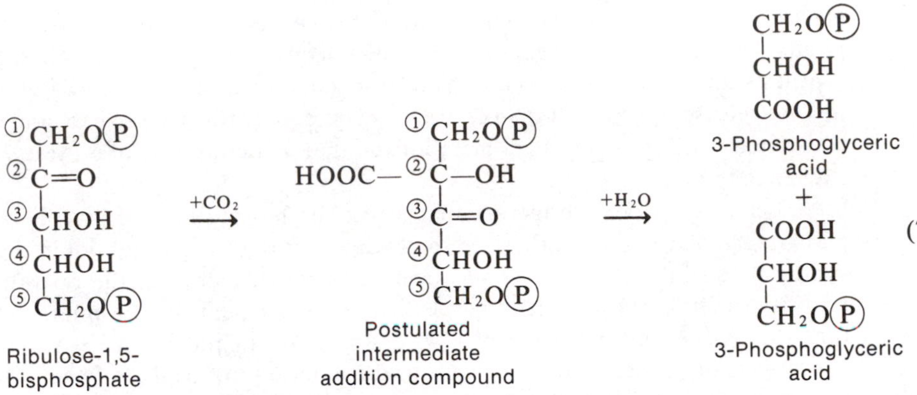

$$
\begin{array}{ccc}
\begin{array}{l}
①\;CH_2O\,\textcircled{P} \\
②\;C=O \\
③\;CHOH \\
④\;CHOH \\
⑤\;CH_2O\,\textcircled{P}
\end{array}
&
\xrightarrow{+CO_2}
&
\begin{array}{l}
①\;CH_2O\,\textcircled{P} \\
HOOC-②\,C-OH \\
③\;C=O \\
④\;CHOH \\
⑤\;CH_2O\,\textcircled{P}
\end{array}
\end{array}
$$

Ribulose-1,5-bisphosphate

Postulated intermediate addition compound

$$
\xrightarrow{+H_2O}
$$

$$
\begin{array}{l}
CH_2O\,\textcircled{P} \\
CHOH \\
COOH
\end{array}
$$

3-Phosphoglyceric acid

+

$$
\begin{array}{l}
COOH \\
CHOH \\
CH_2O\,\textcircled{P}
\end{array}
$$

3-Phosphoglyceric acid

(7-4)

RuBP carboxylase is not only a carboxylating enzyme, as shown in reaction (7-4), but also catalyzes the addition of O_2 to RuBP [see reaction (7-5)]. Although the details of the reaction mechanism are not known precisely at present (consult Johal and Chollet, 1980, and Lorimer, 1981, for further details), O_2 is added to carbon atom 2 of RuBP; then the bond between carbon atoms 2 and 3 is broken. One molecule each of 2-phosphoglycolic acid and 3-phosphoglyceric acid is produced. The overall oxygenation reaction (sometimes referred to as the oxygenolytic cleavage of RuBP) is

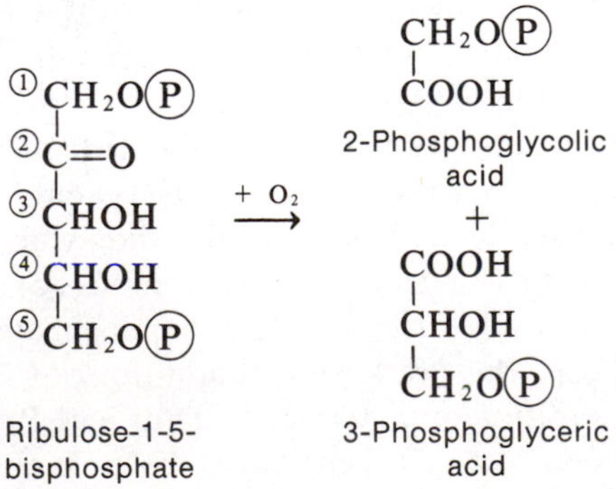

$$
\begin{array}{l}
①\;CH_2O\,\textcircled{P} \\
②\;C=O \\
③\;CHOH \\
④\;CHOH \\
⑤\;CH_2O\,\textcircled{P}
\end{array}
\xrightarrow{+\,O_2}
$$

Ribulose-1-5-bisphosphate

$$
\begin{array}{l}
CH_2O\,\textcircled{P} \\
COOH
\end{array}
$$

2-Phosphoglycolic acid

+

$$
\begin{array}{l}
COOH \\
CHOH \\
CH_2O\,\textcircled{P}
\end{array}
$$

3-Phosphoglyceric acid

(7-5)

In in vitro experiments in which O_2 was labeled with the stable oxygen isotope (^{18}O) and supplied to RuBP and the purified enzyme it was found (Lorimer et al., 1973) that labeled oxygen is incorporated as one of the oxygen atoms of the carboxyl group of 2-phosphoglycolic acid but not in 3-phosphoglyceric acid.

When 2-phosphoglycolic acid [one of the products of reaction (7-5)] undergoes loss of its phosphate group in a subsequent dephosphorylation reaction, glycolic acid is formed. Such a dephosphorylation reaction is catalyzed by an enzyme known as a phosphatase.

$$
\begin{array}{c}
\mathrm{CH_2O\,\textcircled{P}} \\
| \\
\mathrm{COOH}
\end{array}
+ \mathrm{H_2O} \longrightarrow
\begin{array}{c}
\mathrm{CH_2OH} \\
| \\
\mathrm{COOH}
\end{array}
+ \mathrm{HO\,\textcircled{P}}
\qquad (7.6)
$$

2-Phosphoglycolic   Glycolic   Phosphoric
acid   acid   acid

The ability of the enzyme RuBP carboxylase/oxygenase to catalyze both the carboxylation and oxygenation of RuBP supports the prevailing view that the biosynthesis of glycolic acid is a result of the oxygenation of RuBP.[6] This view accommodates the known antagonistic effects of O_2 and CO_2 on the biosynthesis of glycolic acid. With increasing concentration of CO_2 in the external atmosphere, reaction (7-4) is favored over reaction (7-5) and so less glycolic acid would be synthesized. But with increasing concentration of O_2 in the external atmosphere, reaction (7-5) is favored over (7-4) and therefore more gylcolic acid would by synthesized. In effect, the competition between O_2 and CO_2 for RuBP at the reaction site on the enzyme RuBP carboxylase/oxygenase determines the relative rates of photorespiration and photosynthesis in C_3 plants. Low CO_2 or high O_2 concentrations in the external atmosphere favor oxygenation and hence glycolic acid synthesis and photorespiration whereas high CO_2 or low O_2 concentrations in the external atmosphere favor carboxylation and hence photosynthesis.

Let us turn to the metabolic reactions that result in the release of photorespiratory CO_2 from glycolic acid. First, glycolic acid is transported out of chloroplasts and into peroxisomes. Peroxisomes are organelles that have a close spatial relationship to chloroplasts and also to mitochondria, as shown in Figure 7-3. They are especially numerous in mesophyll cells of C_3 plants. (In C_4 plants peroxisomes are present mostly in bundle sheath cells rather than mesophyll cells.) Peroxisomes have been isolated and their enzyme composition studied intensively. Two enzymes always present in peroxisomes were mentioned in Chapter 4 (Eqs. 4–45 and 4–46)— glycolic acid oxidase and catalase.

The metabolic sequence of reactions involved in the metabolism of glycolic acid is shown in Figure 7-4. [Note that the sequence starts with *two* molecules of glycolic acid, produced by *two* turns of the photosynthetic carbon reduction cycle; the reason becomes clear in the following paragraph; see Eq. 7-9.] A peroxisomal enzyme oxidizes glycolic acid to glyoxylic acid; O_2 is a reactant (see Eq. 4–45). Then two molecules of glyoxylic acid are converted to two molecules of glycine, a two-carbon amino acid.

---

[6] Other pathways for the biosynthesis of glycolic acid may exist but account for the formation of only negligible quantities of glycolic acid. For further discussion, consult Chollet and Ogren, 1975; Chollet, 1977; and Tolbert, 1980.

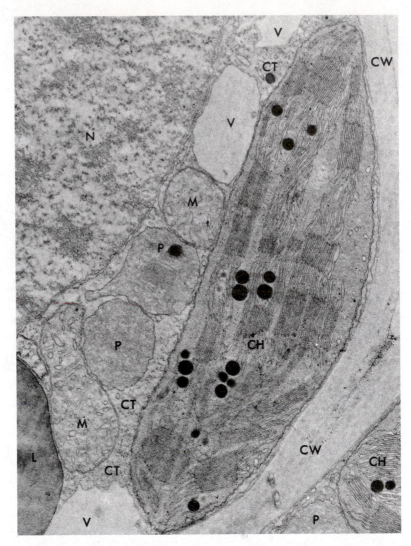

Figure 7-3 An electron photomicrograph of a mesophyll cell in a leaf of orchard grass (*Dactylis glomerata*), a C_3 plant. ×15,000. Note the close spatial relationship of the large chloroplast to peroxisomes and mitochondria. Also note the presence of a crystal of catalase in one of the peroxisomes. The dark spots on the right-hand side of the photomicrograph are organophilic granules. Abbreviations: CH, chloroplast; P, peroxisome; M, mitochondrion; V, vacuole; CT, cytoplasm; L, lipid body; CW, cell wall; N, nucleus. (Courtesy Dr. J. H. Hilliard, Cambrian Process, Ltd., Mississauga, Ontario, Canada.)

$$
\begin{array}{c}
\text{COOH} \\
|\\
\text{CHO}
\end{array}
\ +\
\begin{array}{c}
\text{COOH} \\
|\\
\text{CHNH}_2 \\
|\\
\text{CH}_2 \\
|\\
\text{CH}_2 \\
|\\
\text{COOH}
\end{array}
\xrightarrow[\text{aminotransferase}]{\text{glutamic acid-glyoxylic acid}}
\begin{array}{c}
\text{COOH} \\
|\\
\text{CH}_2\text{NH}_2
\end{array}
\ +\
\begin{array}{c}
\text{COOH} \\
|\\
\text{C}{=}\text{O} \\
|\\
\text{CH}_2 \\
|\\
\text{CH}_2 \\
|\\
\text{COOH}
\end{array}
\qquad (7\text{-}7)
$$

Glyoxylic acid Glutamic acid Glycine α-Ketoglutaric acid

$$
\begin{array}{c}
\text{COOH} \\
|\\
\text{CHO}
\end{array}
\ +\
\begin{array}{c}
\text{CH}_2\text{OH} \\
|\\
\text{CHNH}_2 \\
|\\
\text{COOH}
\end{array}
\xrightarrow[\text{aminotransferase}]{\text{serine-glyoxylic acid}}
\begin{array}{c}
\text{COOH} \\
|\\
\text{CH}_2\text{NH}_2
\end{array}
\ +\
\begin{array}{c}
\text{CH}_2\text{OH} \\
|\\
\text{C}{=}\text{O} \\
|\\
\text{COOH}
\end{array}
\qquad (7\text{-}8)
$$

Glyoxylic acid Serine Glycine Hydroxypyruvic acid

These two reactions are similar insofar as an amino group ($—NH_2$) is transferred from an amino acid [glutamic acid in Eq. 7-7 and serine in Eq. 7-8] to the carbonyl position ($= CO$) of glyoxylic acid. Such reactions are known as *transamination* reactions. Enzymes that catalyze them are called aminotransferases or transaminases. Transamination reactions are discussed again on page 264.

Next, glycine is transported out of peroxisomes into mitochondria (see Figure 7-4). There two glycine molecules react to produce one molecule each of serine (the three-carbon amino acid mentioned in the preceding paragraph,[7] CO_2, and NH_3. This oxidative decarboxylation reaction is complex and involves more than one enzyme (Tolbert, 1980; Journet et al., 1981). The overall reaction is

$$
2\ \text{Glycine} + H_2O + NAD^+ \longrightarrow \text{serine} + CO_2 + NH_3 + NADH[8] \qquad (7\text{-}9)
$$

NH_3 released by mitochondria in Eq. 7-9 is assimilated by chloroplasts within the same cell; this matter is considered in Chapter 9. [The NH_3 assimilatory reactions, and the movements from one organelle to another of certain reactants and products, e.g., glutamic acid and α-ketoglutaric acid in Eq. 7-7, are not shown in Figure 7-4 to avoid overcomplicating the diagram.] CO_2 released in Eq. 7-9 is derived from the carboxyl carbon of glycine and hence from the carboxyl carbon of glycolic acid. It is for this reason that glycolic acid is called the *substrate of photorespiration*.[9]

[7] The two amino acids glycine and serine accumulate in relatively large amounts (Tolbert, 1980) and thus may be utilized for protein synthesis.

[8] The mechanism by which reducing power produced in the mitochondrion in Eq. 7-9 is "shuttled" to the peroxisome (for the reduction of hydroxypyruvic acid to glyceric acid, see Figure 7-4) is considered in Chapter 11 (see Figure 11-5 and related text discussion).

[9] Although past suggestions have claimed that photorespiratory CO_2 released from the leaf may arise, at least in part, from reactions other than Eq. 7-9 (e.g., Grodzinski, 1979; Oliver, 1979), recently it was shown (Somerville and Ogren, 1981) that glycine decarboxylation [Eq. 7-9] is the sole site of release of photorespiratory CO_2 under normal physiological conditions.

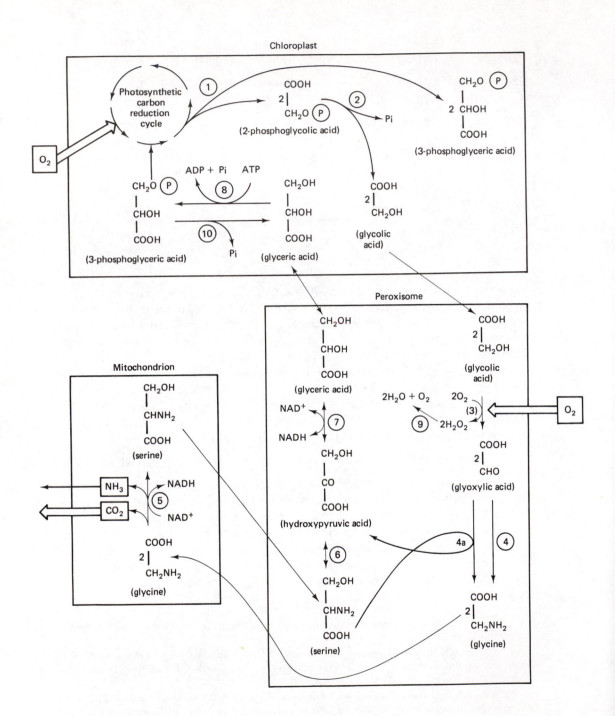

214

Serine produced in Eq. 7-9 is transported out of mitochondria and into peroxisomes (see Figure 7-4), where it is converted successively to hydroxypyruvic acid and glyceric acid. Finally, glyceric acid is transported out of peroxisomes into chloroplasts, where it is phosphorylated by ATP to 3-phosphoglyceric acid.

To sum up, two molecules of phosphoglycolic acid produced by the oxygenase reaction [Eq. 7-5] are converted by reactions shown in Figure 7-4 to one molecule each of CO_2 and 3-phosphoglyceric acid. 3-Phosphoglyceric acid, whether produced by the oxygenase reaction [Eq. 7-5] or metabolism of phosphoglycolic acid (Figure 7-4), becomes part of the 3-phosphoglyceric acid pool of the photosynthetic carbon reduction cycle.

A great deal of experimental evidence has been collected during the past decade in support of the reaction sequence shown in Figure 7-4. But its cyclic nature was understood only recently. To emphasize that the pathway of glycolic acid metabolism shown in Figure 7-4 is a *cyclic* series of reactions, the term *photorespiratory carbon oxidation cycle* has been applied (Andrews and Lorimer, 1978; Berry et al., 1978; Tolbert, 1980). [The reaction sequence shown in Figure 7-4 was formerly known as the glycolate pathway (glycolic acid to serine) and the glycerate pathway (serine to glyceric acid); the glycolate pathway is irreversible, the glycerate pathway reversible (see arrows, Figure 7-4).]

PHOTORESPIRATION IN C_4 PLANTS

To understand why C_4 plants have low rates of photorespiration and high rates of net photosynthesis, it is necessary to emphasize the concerted activity of mesophyll and bundle sheath cells in the fixation of CO_2 in photosynthesis in these plants. This subject was discussed in Chapter 6 (pp. 172–175) and is reviewed briefly here. Photosynthesis in leaves of C_4 plants is characterized by initial fixation of atmospheric CO_2 (as

---

Figure 7-4 A schematic representation of photorespiratory metabolism in a mesophyll cell of a leaf of a C_3 plant. Glycolic acid, the substrate of photorespiration, is synthesized in chloroplasts and metabolized in peroxisomes and mitochondria via the *photorespiratory carbon oxidation cycle* depicted here. (See text for details.) Arrows with double lines indicate sites of release of photorespiratory CO_2 and uptake of photorespiratory O_2. Not shown are the absorption of CO_2 in the photosynthetic carbon reduction cycle in chloroplasts and the evolution of O_2 in the photosynthetic light reactions in chloroplasts. The enzymes are (1) RuBP carboxylase/oxygenase; (2) phosphoglycolic acid phosphatase; (3) glycolic acid oxidase; (4) glutamic acid-glyoxylic acid aminotransferase; (4a) serine-glyoxylic acid aminotransferase; (5) serine-synthesizing enzyme complex; (6) serine-glyoxylic acid aminotransferase; (7) NADH-hydroxypyruvic acid reductase; (8) glyceric acid kinase; (9) catalase; (10) 3-phosphoglyceric acid phosphatase.

In this diagram 3-phosphoglyceric acid (3PGA) is shown *outside* the photosynthetic carbon reduction cycle but actually *is* a member of that cycle (see Figure 6-2 and related text discussion). Of course, 3PGA produced by the photosynthetic carbon reduction cycle mixes with 3PGA produced by the photorespiratory carbon oxidation cycle.

The symbol (P) is used to designate the phosphate group ($-H_2PO_3$), whereas Pi denotes orthophosphoric acid (H_3PO_4). [*Note*: The formulas of the several acids in the figure are shown in the un-ionized form but exist in plant cells as ions—for example, glycolate, glyoxylate, hydroxypyruvate, orthophosphate ($H_2PO_4^-$).]

bicarbonate ion) in mesophyll cells by phosphoenolpyruvic acid (PEP) (see Eqs. 4–33 and 6-8), subsequent transport of C_4 acids out of mesophyll cells into bundle sheath cells, where they are decarboxylated, and fixation of released CO_2 in the photosynthetic carbon reduction cycle in bundle sheath cells; the three-carbon compound that results from decarboxylation of C_4 acids is transported back to mesophyll cells, to begin the cycle again—that is, fixation of atmospheric CO_2 in mesophyll cell by PEP, and so forth.

This pattern of movement of CO_2—from the external atmosphere to mesophyll cells and then to chloroplasts in bundle sheath cells—serves to concentrate CO_2 in bundle sheath cells; in fact, the movement of CO_2 is often called a CO_2 *pump*. Thus the CO_2 to O_2 ratio in bundle sheath cells is increased far above what exists in the external atmosphere. Because CO_2 and O_2 act competitively as substrates for RuBP carboxylase/oxygenase (see pp. 210–211), the effect of the higher CO_2 concentration in bundle sheath cells is to reduce the RuBP oxygenase-catalyzed formation of glycolic acid, thereby decreasing the quantity of glycolic acid available for photorespiratory oxidation to CO_2. Moreover, an elevated CO_2 to O_2 ratio in bundle sheath cells permits CO_2 to compete more effectively with O_2 for RuBP carboxylase/oxygenase and thereby decreases the inhibitory effect of O_2 on photosynthesis in bundle sheath cells (see Figure 7-2 and related text discussion). Finally, any photorespired CO_2 release in bundle sheath cells would be refixed by PEP carboxylase in surrounding mesophyll cells (see Figure 7-1) before it could escape from the leaf.

The foregoing model, which emphasizes that mesophyll cells in C_4 plants function as a biochemical pump, transporting CO_2 from the external atmosphere to the site of its fixation by RuBP in bundle sheath cells, accounts for the low rate of photorespiration and high rate of net photosynthesis in C_4 plants.

OTHER FEATURES OF C₄ PLANTS

In addition to very low rates of photorespiration and high rates of net photosynthesis, C_4 plants have special features not mentioned earlier in this chapter. One is a lower internal concentration of CO_2 in illuminated leaves of C_4 plants as compared to that in C_3 plants. Experimental values reported for C_4 plants are about 100 ppm whereas those for C_3 plants are about 200 ppm (O'Leary, 1981).

C_4 plants have the ability to absorb CO_2 more efficiently than C_3 plants from the external atmosphere. This is shown in Figure 7-5. These curves are from experiments in which the responses of leaves to different concentrations of CO_2 were measured at 21 per cent O_2 and saturating light intensity. In the C_3 plant in Figure 7-5, saturation of the rate of CO_2 fixation begins to occur at a CO_2 concentration of about 500 ppm. In contrast, the C_4 plant in Figure 7-5 begins to reach saturation of net photosynthesis at much lower CO_2 concentrations. (The CO_2 compensation point—the point at which the curve crosses the horizontal axis in Figure 7-5—is almost zero for the C_4 plant; for the C_3 plant, the CO_2 compensation point is about 50 ppm CO_2.) Figure 7-5 shows that the concentration of CO_2 normally present in the external atmosphere (330 ppm) is saturating for the C_4 plant but not the C_3 plant. In essence, C_4 plants are more efficient

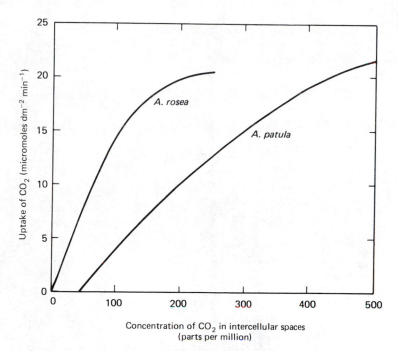

Figure 7-5 The effect of concentration of CO_2 in intercellular spaces on net photosynthesis in a C_3 plant (*Atriplex patula*) and a C_4 plant (*Atriplex rosea*). Measurements were made in 21% O_2 and at near-optimal light intensities for each plant. CO_2 concentrations on the horizontal axis (abscissa) have been adjusted to compensate for diffusion gradients that develop between the external atmosphere and the CO_2 fixation sites inside the leaf; so the responses plotted represent what would have been obtained if there were no diffusion barrier. (*Note*: To convert μ mol of CO_2 $dm^{-2} min^{-1}$ on the vertical axis to mg of CO_2 $dm^{-2} h^{-1}$, use the factor, 20 μ mol of CO_2 $dm^{-2} min^{-1}$ = 52.8 mg of CO_2 $dm^{-2} h^{-1}$.) (After O. Bjorkman and J. Berry, *Sci. Amer.* 229 (October 1973), 85.)

than C_3 plants in absorbing CO_2 from an external atmosphere of low CO_2 concentration.

Another way to demonstrate that C_4 plants are more efficient in absorbing CO_2 from the external atmosphere, under conditions of optimum light, is an experiment that requires little equipment. A C_3 and C_4 plant are placed together in a small airtight glass chamber initially filled with air (see Figure 7-6). When the plants are illuminated, the concentration of CO_2 in the glass chamber will be lowered within an hour or two to the CO_2 compensation point characteristic of the C_4 plant. But this concentration will be much lower than the CO_2 compensation point of the C_3 plant (see pp. 205–206). Thus the C_3 plant will die in a few days when it respires away its food reserves whereas the C_4 plant will continue to survive (see Figure 7-6).

The greater efficiency of CO_2 uptake from the external atmosphere by C_4 plants is due not only to the CO_2 pump in mesophyll cells mentioned earlier (pp. 215–216),

Figure 7-6 The ability of seedlings of maize (a C_4 plant) to survive in an atmosphere that is lethal to soybean (a C_3 plant) can be demonstrated easily. Two seedlings each of maize and soybean were grown on a laboratory bench from seed for 14 days in vermiculite wetted with tap water. Then the cotyledons of the soybean seedlings and the cotyledons and endosperms of the maize seedlings were excised so as to remove the food reserves. The seedlings were replanted in vermiculite wetted with Hoagland's nutrient solution (see pp. 229–230) and sealed in an airtight glass container. Supplemental lighting was provided during both day and night hours. At the end of a week the soybean plants were dead, but the maize plants were still alive. The photograph on the left was taken while the glass container was still intact. The photograph on the right is a closeup view taken a few minutes after the glass container was broken open. Compare the thrifty appearance of the maize seedlings to that of the dead soybean seedlings, the dried and shriveled leaves of which fell away from the stems when the glass container was broken open.

but also to a reduced CO_2 partial pressure during photosynthesis (see p. 216). By fixing CO_2 into C_4 acids in mesophyll cells through the agency of a carboxylating enzyme present in large amount (i.e., PEP carboxylase)—an enzyme with a high turnover rate, and one not subject to competition between CO_2 and O_2 (as is RuBP carboxylase, see pp. 210–211)—and liberating this combined CO_2 in adjacent bundle sheath cells, the CO_2 partial pressure in mesophyll cells becomes much reduced. As a result, the CO_2

gradient from the external atmosphere to the initial CO_2 fixation sites in the leaf is steeper for C_4 than C_3 plants. It is this steeper CO_2 gradient that gives C_4 plants a greater ability, compared to C_3 plants, to scrub CO_2 from the external atmosphere.

The steeper CO_2 gradient in C_4 plants during photosynthesis gives them a greater ability to use water efficiently. Because the rate of transpiration at any given stomatal aperture can be expected to be the same for C_3 and C_4 plants (assuming other factors, such as water content of the plants, to be the same), the steeper CO_2 gradient in C_4 plants results in the assimilation of more carbon per unit of water transpired than in C_3 plants. Stated another way, C_4 plants transpire less water per unit of dry matter production than do C_3 plants.

The temperature optimum for net photosynthesis in C_3 plants is generally in a range somewhat below 25° whereas that for C_4 plants is in the range of 30 to 35°C. At these higher temperatures, the rate of net photosynthesis in C_4 plants may be more than twice that in a C_3 plant.

These features of C_4 plants—their greater efficiency in utilizing low levels of CO_2 in the external environment and in the use of water and their greater rates of net photosynthesis at high temperatures—underscore an important generality: C_4 plants that grow under natural conditions are better adapted than C_3 plants to life in environments associated with hot, arid habitats, such as deserts, grasslands, and other subtropical regions of the world, where the season growth coincides with intense solar radiation, high temperatures, and a limited supply of water. Most C_4 plants are native to xeric habitats.

REGULATION OF PHOTORESPIRATION

One goal of current research is to find ways to decrease photorespiration and the associated O_2 inhibition of photosynthesis in C_3 plants and thereby increase their photosynthetic productivity. We saw earlier that it can be accomplished by increasing the concentration of CO_2 in the external atmosphere—the effect would be to enhance the carboxylating activity of RuBP carboxylase/oxygenase [see reaction (7-4)] and decrease its oxygenating activity [see reaction (7-5)]. It has been found that the optimum concentration for plant growth of C_3 plants at 21% O_2 ranges between 1000 and 1200 ppm CO_2; too high a concentration of CO_2 must be avoided because stomatal closure will result (p. 339). Enrichment of the external atmosphere with CO_2 is feasible at present only in glasshouse culture of plants. Another method of increasing photosynthetic productivity in C_3 plants—but one that is clearly impractical on a large scale—is to reduce the O_2 concentration in the atmosphere surrounding the leaves from 21% O_2 to 1 or 2% O_2 (see Figure 7-2 and related discussion).

One way to learn how photorespiration in C_3 plants can be decreased is to find and study naturally occurring plants with reduced rates of photorespiration. The best-documented example of such a plant is *Panicum milioides*, a member of the grass family; other species also have been identified recently. These plant species, called *C_3–C_4 intermediates* (e.g., see Rathnam and Chollet, 1980c), have rates of photorespiration and CO_2 compensation points (equal to about 15 ppm CO_2) that are intermediate between C_3 and C_4 plants. Study of photosynthetic carbon metabolism in such plants should aid in understanding the relationships between C_3 and C_4 plants and thus improve photosynthetic efficiency of C_3 plants (Morgan and Brown, 1979).

Another way to decrease photorespiration in C_3 plants is by applying chemicals to plant tissues. Studies of the effects of chemicals that inhibit synthesis of glycolic acid in leaf disks and protoplasts have been pursued (Oliver and Zelitch, 1977; Servaites and Ogren, 1977). The possibility exists that chemicals will be found that, when sprayed on whole plants at one or another stage of growth, might reduce photorespiration and increase photosynthetic productivity (Zelitch, 1979).

A third way to reduce photosynthetic inefficiency due to photorespiration is by genetic means. In one approach, suspensions of plant protoplasts are exposed to ultraviolet radiation (to increase the mutation rate) and desirable mutants are identified and isolated. Then cells are grown in solutions of known inhibitors of glycolic acid metabolism, the object being to alter photorespiratory metabolism. Cells that survive are regenerated into fully grown higher plants by standard techniques (pp. 537–538). From these whole plants cells are isolated and subjected again to ultraviolet radiation and the cycle of treatment is repeated. By testing resultant mutant lines for stability and genetic transmission of desired traits, it is hoped to determine whether specific biochemical mutations that affect the metabolism of glycolic acid are also effective in decreasing photorespiration and in increasing net photosynthesis (Zelitch, 1979).

Another genetic approach is to identify whole plant mutants—either naturally occurring or induced—that have low rates of photorespiration. A recent report (Rathnam and Chollet, 1980a) shows promising results. Two cultivars of ryegrass (a C_3 plant), diploid and tetraploid, produced by treatment of seedlings with a mutagen (i.e., a substance with the ability to induce a mutation) have been found to differ in their rates of photorespiratory glycolic acid production; moreover, the lower rate of photorespiration in one of the cultivars (tetraploid) was correlated with a higher affinity for CO_2 by RuBP carboxylase extracted from the two cultivars. This finding may be significant because modification of the carboxylation activity of this critical enzyme relative to its oxygenating activity may prove a practical means of enhancing crop yields of C_3 plants in conventional plant-breeding programs.

Perhaps it might be useful to point out here that a high yield of one or another crop is not always a desirable objective. The *quality* of certain low-yield crops such as barley, grapes, and potatoes—for instance, the malting quality of barley used for the production of beers, the vintage characteristics of grapes, and the crispening quality of potatoes destined for use in potato chips—is by no means associated with high yields.

All efforts to reduce photorespiration in C_3 plants in order to stimulate their rates of photosynthesis are based on the fact that C_4 plants have a higher potential for net photosynthesis than C_3 plants. In this view, photorespiration seems functionally redundant in C_3 plants. Unfortunately, plant physiologists are still quite ignorant of the function of photorespiration. Certain current hypotheses hold that photorespiration is a necessary protective mechanism against oxidative photodestruction

(Heber and Krause, 1980; Foyer and Hall, 1980) of labile chloroplasts in C_3 plants. By consuming O_2, photorespiration helps to prevent the buildup of a hyperoxidative state in chloroplasts of C_3 plants. Free radicals of oxygen gas are thought to be produced when reduced components of photosystem I react with photosynthetically produced O_2; these free radicals are extremely reactive and, by reacting with membrane components, destroy them. If not for photorespiration, oxygen free radicals could reach very high—and very destructive—levels in chloroplasts. Assuming the correctness of these current hypotheses regarding the beneficial effect of photorespiration, reduction of photorespiration in C_3 plants may not be desirable even if possible.

REFERENCES

ANDREWS, T. J., and G. H. LORIMER. 1978. Photorespiration—still unavoidable? *FEBS Lett.* 9:1–9.

BERGMAN, A., P. GARDESTRÖM, and I. ERICSON. 1981. Release and refixation of ammonia during photorespiration. *Physiol. Plant.* 53:528–532.

BERRY, J. A. 1975. Adaption of photosynthetic process to stress. *Science* 188:644–650.

BERRY, J. A., C. B. OSMOND, and G. H. LORIMER. 1978. Fixation of $^{18}O_2$ during photorespiration. *Plant. Physiol.* 62:954–967.

BROWN, W. V. 1977. The Kranz syndrome and its subtypes in grass systematics. *Memoirs Torrey Bot. Club* 23(3):1–97.

BRAVDO, B., and D. CANVIN. 1979. Effect of carbon dioxide on photorespiration. *Plant Physiol.* 63:399–401.

CHOLLET, R. 1977. The biochemistry of photorespiration. *Trends Biochem. Sciences* 2:155–159.

CHOLLET, R., and W. L. OGREN. 1975. Regulation of photorespiration in C_3 and C_4 species. *Bot. Rev.* 41:137–179.

DECKER, J. P., and M. A. TIO. 1959. Photosynthetic surges in coffee seedlings. *Jr. Agri. Puerto Rico.* 43:50–55.

DESJARDINS, R. L., E. J. BRACH, P. ALVO, and P. H. SCHUEPP. 1982. Aircraft monitoring of surface carbon dioxide exchange. *Science* 26:733–735.

EHLERINGER, J., and O. BJÖRKMAN. 1977. Quantum yields for CO_2 uptake in C_3 and C_4 plants; dependence on temperature, CO_2, and O_2 concentration. *Plant Physiol.* 59:86–90.

FOYER, C. H., and D. O. HALL. 1980. Oxygen metabolism in the active chloroplast. *Trends Biochem. Sciences* 5:188–191.

GERBAUD, A., and M. ANDRÉ. 1979. Photosynthesis and photorespiration in whole plants of wheat. *Plant Physiol.* 64:735–738.

GRODZINSKI, B. 1979. A study of formate production and oxidation in leaf peroxisomes during photorespiration. *Plant Physiol.* 63:289–293.

HARRIS, G. C., and A. I. STERN. 1978. Stoichiometry of the ribulose-1, 5-bisphosphate oxygenase reaction. *Jr. Expt. Bot.* 29:561–566.

HEBER, U., and G. H. KRAUSE. 1980. What is the physiological role of photorespiration? *Trends Biochem. Sciences* 5:32–34.

JOHAL, S., and R. CHOLLET. 1980. Ribulose-1,5-bisphosphate carboxylase/oxygenase: enzymic, physiocochemical and nutritional properties. *What's New Plant Physiol.* 11:45–48.

JOURNET, E.-P., M. NEUBURGER, and R. DOUCE. 1981. Role of glutamate-oxaloacetate transaminase and malate dehydrogenase in the regeneration of NAD^+ for glycine oxidation by spinach leaf mitochondria. *Plant Physiol.* 67:467–469.

LORIMER, G. H., T. J. ANDREWS, and N. E. TOLBERT. 1973. Ribulose diphosphate oxygenase. II. Further proof of reaction products and mechanism of action. *Biochemistry* 12:18–23.

LORIMER, G.H. 1981. The carboxylation and oxygenation of ribulose-1,5-bisphosphate: the primary events in photosynthesis and photorespiration. *Ann. Rev. Plant Physiol.* 32:349–383.

LUDWIG, L. J., and D. T. CANVIN. 1971. An open gas-exchange system for the simultaneous measurement of the CO_2 and $^{14}CO_2$ fluxes from leaves. *Can. Jr. Bot.* 49:1299–1313.

MARZOLA, D. L., and D. P. BARTHOLOMEW. 1979. Photosynthetic pathway and biomass energy production. *Science* 205:555–559.

MIFLIN, B. J., and P. J. LEA. 1980. Ammonia assimilation. Pages 169–202 *in* P. K. Stumpf and E. E. Conn, eds., *The Biochemistry of Plants, a Comprehensive Treatise*, vol. 5. Academic Press, New York.

MORGAN, J. A., and R. H. BROWN. 1979. Photosynthesis in grass species differing in carbon dioxide fixation pathways. *Plant Physiol.* 64:257–262.

O'LEARY, M. H. 1981. Carbon isotope fractionation in plants. *Phytochemistry* 20:553–567.

O'LEARY, M. H., J. E. RIFE, and J. D. Slater. 1981. Kinetic and isotope effect studies of maize phosphoenol pyruvate carboxylase. *Biochemistry* 20:7308–7314.

OLIVER, D. J., and I. ZELITCH. 1977. Increasing photosynthesis by inhibiting photorespiration with glyoxylate. *Science* 196:1450–1451.

OLIVER, D. J. 1979. Mechanism of decarboxylation of glycine and glycolate by isolated soybean cells. *Plant Physiol.* 64:1048–1052.

OUTLAW, JR., W. H., B. C. MAYNE, V. E. ZENGER, and J. MANCHESTER. 1981. Presence of both photosystems in guard cells of *Vicia faba* L. Implications for environmental signal processing. *Plant Physiol.* 67:12–16.

RAGHAVENDRA, A.S., and V. S. R. DAS. 1978. The occurrence of C_4-photosynthesis: a supplementary list of C_4 plants reported during late 1974–mid 1977. *Photosynthetica* 12:200–208.

RATHNAM, C. K. M., and R. CHOLLET. 1980a. Photosynthetic and photorespiratory carbon metabolism in mesophyll protoplasts and chloroplasts isolated from isogenic diploid and tetraploid cultivars of ryegrass (*Lolium perenne* L.) *Plant Physiol.* 65:489–494.

RATHNAM, C., and R. CHOLLET. 1980b. Regulation of photorespiration. *Curr. Adv. Plant Sci.* 12(11):38.1–38.19.

RATHNAM, C. K. M., and R. CHOLLET. 1980c. Photosynthetic carbon metabolism in C_4 plants and C_3-C_4 intermediate species. *Progress Biochem.* 6:1–48.

SERVAITES, J. C., and W. L. OGREN. 1977. Chemical inhibition of the glycolate pathway in leaf cells. *Plant Physiol.* 60:461–466.

SINGH, M., W. L. OGREN, and M. WIDHOLD. 1974. Photosynthetic characteristics of several C_3 and C_4 plant species grown under different light intensities. *Crop. Sci.* 14:563–566.

SOMERVILLE, C. R., and W. L. OGREN. 1981. Photorespiration-deficient mutants of *Arabidopsis thaliana* lacking mitochondrial serine transhydroxymethylase activity. *Plant Physiol.* 67:666–671.

SOMERVILLE, C. R., and W. L. OGREN. 1982. Genetic modification of photorespiration. *Trends Biochem. Sciences* 7:171–174.

TOLBERT, N. E. 1980. Photorespiration. Pages 487–523 *in* P. K. Stumpf and E. E. Conn, eds. *The Biochemistry of Plants, a Comprehensive Treatise*, vol. 2. Academic Press, New York.

TOLBERT, N. E. 1981. Metabolic pathways in peroxisomes and glyoxysomes. *Ann. Rev. Biochem.* 50:133–157.

ZELITCH, I. 1975. Improving the efficiency of photosynthesis. *Science* 188:626–633.

ZELITCH, I. 1979. Photosynthesis and plant productivity. *Chem. Eng. New* 57(8):28–48.

8

Mineral Nutrition

For centuries it was realized that roots of terrestrial plants obtain nourishment from soils. The addition of farmyard manures, plant tissue debris, wood ashes, ground bones, composts, and other materials of plant and animal origin to soils was found to increase their productivity long before the nutritional needs of plants were understood.

Only during the first half of the nineteenth century did plant scientists begin to understand that plant growth and development can proceed only when plants are supplied with certain chemical elements referred to as *essential* elements. These elements are absorbed by roots principally as inorganic ions. Inorganic ions in soils are derived mostly from mineral constituents of the soil. The term *mineral nutrient* is generally used to refer to an inorganic ion obtained from the soil and required for plant growth.

In a general sense, mineral nutrition is a broad field concerned with the complex of biosynthetic events by which higher plants produce organic materials from inorganic materials absorbed from their environment. These events include the absorption process itself, the translocation of absorbed mineral nutrients from roots to stems, leaves, and fruits, and, finally, conversion of absorbed (inorganic) nutrients to organic form. The latter subject, assimilation, is considered separately in the next chapter, as are the functions and roles of essential elements. The absorption of inorganic minerals from the soil and their translocation through the plant body are considered under the heading of Transport Processes (Chapter 11).

In this Chapter we examine the needs of higher plants for mineral nutrients.

AUTOTROPHS AND HETEROTROPHS

It has been found useful to classify living organisms on the basis of their nutritional requirements for carbon-containing compounds as either *autotrophs* or *heterotrophs*. Organisms able to utilize CO_2, a relatively simple inorganic carbon-containing compound, as their sole source of carbon are said to be *autotrophic* (literally, self-nourishing). Autotrophs assimilate CO_2 in photosynthesis. In addition to CO_2, nutritional requirements of autotrophs include water, O_2, and several inorganic ions. Except for a few saprophytic and parasitic species that have only a limited ability to carry on photosynthesis, almost all higher plants are autotrophic. Also, most algae—and a few bacteria—have an autotrophic mode of nutrition.

In contrast to autotrophs, many organisms have nutritional requirements that include one or more organic carbon compounds. These organisms are the fungi (including yeasts), most bacteria, and all members of the animal kingdom. Organisms whose nutritional requirements include one or more organic compounds (i.e., carbon-containing compounds with one or more hydrogen atoms and therefore more reduced than CO_2) are called *heterotrophs*. The fact that heterotrophs require an outside supply of one or more organic compounds indicates their dependence on products synthesized by other organisms.

Nutritional requirements of heterotrophs vary considerably. Certain fungi, for example, require only one or two organic nutrients, perhaps glucose and one or another amino acid, in addition to water and a number of inorganic ions. The nutritional requirements of humans are far more complex. Besides water and certain inorganic ions, humans require a bulk source of carbon (usually supplied as carbohydrates and fats), several amino acids (usually supplied in the form of proteins) together with sufficient organic nitrogen to maintain a suitable nitrogen balance in the human body, a variety of vitamins, and a few polyunsaturated fatty acids.

THE THREE MEDIA OF PLANT GROWTH

The nutrients indispensable to the growth and development of autotrophic higher plants are obtained from three sources in the environment: atmosphere, water, and soil.

The atmosphere provides CO_2. All carbon atoms, plus most oxygen atoms, in the dry matter of autotropic higher plants are derived from CO_2, which is assimilated principally in photosynthesis. More specifically, approximately one-third of the oxygen atoms in organic material in higher plants derive from soil water and two-thirds come from CO_2 in the atmosphere. The chemical basis for this conclusion is the reaction responsible for photosynthetic CO_2 fixation in green plants. This reaction involves the addition of one molecule of CO_2 (which contains two atoms of oxygen) and one molecule of H_2O (which contains one atom of oxygen) to RuBP. [See Eqs. (6-7) and (7-4).]

The atmosphere also furnishes O_2, which is the ultimate oxidant in aerobic respiration. In this process, O_2 is reduced to water (Chapter 4). Water produced in respiration (i.e., *respiratory water*) mixes with the water already present in plant cells. In addition to its role in respiration, small amounts of O_2 in the atmosphere are incorporated into certain organic constituents in the process known as *oxygen fixation* (pp. 283–284). Also, some higher plants incorporate N_2 in the atmosphere into organic matter through microorganisms that live in symbiosis with them (Chapter 9). Finally, trace amounts of gaseous ammonia (NH_3) and gaseous sulfur dioxide (SO_2) are sometimes present in the atmosphere; if so, they may be assimilated by some plants.

Water is a second source of nutrition to higher plants. Although the bulk of the water absorbed by roots is lost by transpiration from leaves, a small fraction is used as a reactant in metabolic reactions. Water is incorporated into plant material mainly via hydration and hydrolytic reactions. To give two examples, when fumaric acid is converted to malic acid in the TCA cycle, water is added to fumaric acid (Figure 4-6); when starch is degraded (i.e., hydrolyzed), there is a small gain in dry weight in the products because water is a reactant and thus chemically incorporated into the products (see Figure 4-1). In addition, water is a reactant in the light reaction in photosynthesis (see Chapter 6): Hydrogen atoms in water are incorporated into organic matter (and the oxygen atom in water is released as O_2). Finally, water is incorporated into plant tissue in the photosynthetic process, as pointed out above (p. 225).

The third environmental source of nutrients for autotropic higher plants, the soil, supplies mineral ions derived from the parent rock and decaying plant and animal residues. Mineral ions are absorbed mostly by roots in the soil. Also, certain species of higher plants that grow on the limbs of trees, or on telephone wires, etc., obtain their supply of mineral nutrients from windblown soil and dust particles that alight on their surfaces and then dissolve in rain water or dew. Such plants are called *epiphytes* or *air plants*.

The knowledge that higher plants obtain their nutrients from atmosphere, water, and soil is less than 200 years old. As late as the fourteenth and fifteenth centuries the generally accepted view of plant nutrition, inherited from ancient Greek science, held that the soil is a medium in which "nourishing juices" were prepared. These "predigested foods," as they were called, were thought to be absorbed by roots of higher plants and then incorporated directly and without change into the substance of the plant. Not until the Renaissance was this idea challenged. The German theologician de Cusa proposed that water is transmuted into plant material. His suggestion (published near the end of the fifteenth century) was that water is the source of increase in the weight of a growing plant.

De Cusa's idea was not tested experimentally until the seventeenth century when the Belgium alchemist van Helmont, in one of the first recorded experiments in plant nutrition (published posthumously in 1648), placed a willow shoot weighing 5 lb in 200 lb of dry soil and added water at regular intervals, as needed by the growing plant. After 5 years the fresh weight of the willow tree was 169 lb, 3 oz, while the soil, again dried, was found to have lost 2 oz. Van Helmont did not attach any significance to the loss of 2 oz by the soil. Nor did he recognize that the willow tree gained weight by fixation of CO_2 in photosynthesis and lost water by transpiration. Because he believed that his demonstration showed that only water and no other materials were used by the willow tree, it appeared logical to conclude that all the plant substance was produced from water alone. His conclusion, although erroneous, is understandable in view of the scant knowledge of chemical principles prevailing at that time.

The role of the atmosphere in plant nutrition was understood only when modern chemistry became established in the last quarter of the eighteenth century. At that time chemists learned the elemental composition of relatively simple substances, such as CO_2 and water. The former was found to consist of carbon and oxygen and the latter of hydrogen and oxygen. Also, at that time the proportions in which such elements as carbon, hydrogen, oxygen, and nitrogen combine chemically with one another to form compounds more complex than CO_2 and water were discovered. Thus it was only in the late eighteenth and early nineteenth centuries that plant scientists began to explain the process of photosynthesis in modern terms (see Chapter 6).

Even though it was known in the late 1700s that the addition of certain materials (e.g., manures, urine, sewage effluants) to soils stimulated plant growth, knowledge of the chemical identity of the substances required for plant growth was gained only in the early 1800s. The French plant physiologist de Saussure was the first to state clearly (in a book published in 1804) that plant growth depends on the absorption of nitrogen and certain other (then unidentified) elements by plant roots. Serious study of the identity of mineral nutrients required for plant growth was begun by others who followed de Saussure.

THE WATER CULTURE TECHNIQUE

The growing of higher plants with their roots in dilute solutions of mineral salts instead of soils led to a vastly increased understanding of plant nutrition. The use of the water culture technique for growing plants permits precise control of the supply of nutrient ions in the root environment.

Probably the first recorded use of the water culture technique was by the Englishman Woodward in 1699. He compared the growth of cuttings from mint and other plants in water obtained from distillation, rain, springs, rivers, and sewage effluant. More than a century later, in the early nineteenth century, higher plants were grown with their roots immersed in water solutions consisting of inorganic salts alone, without the addition of either soil or organic matter. By the 1860s the technique of water culture of plants was modernized, especially through the efforts of two German plant physiologists, Sachs and Knop, working independently. The Knop nutrient solution became well known; it consisted of potassium nitrate, calcium nitrate, potassium dihydrogen phosphate, magnesium sulfate, and an iron salt. This solution contained all the mineral elements then considered necessary for plant growth. Because Sachs and Knop made no special efforts to purify the salts used, several mineral elements now known to be required in very small amounts for the growth of plants were added unintentionally.

Figure 8-1 Aerial view of the agricultural research station at Puerto Penasco in the Mexican coastal desert. The white structures are plastic houses in which plants are grown. The desalinization tower is at the right. (After Bazell, 1971.)

The experimental demonstration by Sachs and Knop of the ability of higher plants to grow and develop when supplied only with certain inorganic nutrients—CO_2, O_2, water, and several mineral ions—was a significant scientific achievement. These experiments provided unequivocal evidence that the higher plant fabricates all its cells and tissues from inorganic nutrients alone.

The water culture technique of growing plants with their roots in a recirculating water solution is called *hydroponics*. In a modification of this technique plants are grown with their roots anchored in a solid inert aggregate, such as sand, gravel, or vermiculite (heat-expanded mica), which is wetted with a solution of mineral nutrients. But regardless of whether the rooting medium consists of liquid alone or of a solid aggregate wetted by a liquid, the root system of plants grown in water culture must be provided with an adequate supply of O_2, either by bubbling air around the roots or by frequent replacement of depleted solution with fresh solution.

The water culture technique is used extensively at present to grow plants in the laboratory for experimental purposes. Moreover, it is used to grow certain high-value crops (e.g., tomato and strawberry) in glasshouses out of season in metropolitan areas in the United States. In addition, vegetables are grown in plastic houses in a few coastal desert regions (see Figure 8-1). In these arid land facilities seawater is desalinated and supplied to roots in precisely the amounts required by the growing plants.

There are many successful nutrient solution formulas, each differing slightly in concentrations of individual ions.[1] For several reasons, it is difficult, if not impossible, to designate a "best" nutrient solution. Because roots absorb individual nutrient ions at different rates, the composition of a nutrient solution, no matter what it is initially,

---

[1] More than 150 nutrient solution formulas are cited and described in a monumental work (Hewitt, 1966) that contains much valuable information about nutrient culture techniques, including methods of purifying nutrient salts, preparation of water, cleaning of glassware for nutrient culture work for experimental purposes, and so forth.

undergoes continuous change throughout the period of plant growth. Different species of higher plants have different minimum requirements for each mineral nutrient. Some plants require more boron or iron or zinc or other elements than other plant species. Plants at different stages of growth have different quantitative requirements for nutrient ions. In the early stages of vegetative growth relatively large amounts of protein are produced; consequently, plants require larger amounts of nitrogen than at later stages. The climatic conditions—light, temperature, and so forth—have a considerable influence on the growth rates of plants and hence on their rates of mineral ion utilization. For these reasons, it is evident that no single nutrient solution is superior to others. The "best" solution is the one that the user finds best for his or her purposes.

One nutrient solution that enjoys considerable popularity in the United States was formulated by Hoagland, an American plant physiologist, and his associates. The concentrations of the several mineral nutrients in Hoagland's solution were patterned, at least in part, on the proportions of nutrients found in water extracts of certain fertile California soils and then modified to conform to the approximate proportions in which they were absorbed by tomato plants during trial experiments.

The composition of Hoagland's solution is given in Table 8-1. Stock solutions of four salts are prepared separately. These four stock solutions supply six elements required by plants in relatively large amounts—potassium, calcium, magnesium, nitrogen, phosphorus, and sulfur. In addition to these four stock solutions, a stock solution of an iron salt and a stock solution that supplies several elements required by plants in relatively small amounts are prepared separately. Aliquots are taken from these several stock solutions and are added to water to make the nutrient solution (see Table 8-1).

The final molar concentrations of the four main salts in Hoagland's solution are $KNO_3 = 0.001$ M, $KH_2PO_4 = 0.005$ M, $Ca(NO_3)_2 = 0.005$ M, $MgSO_4 = 0.002$ M. Thus Hoagland's nutrient solution is actually a very dilute solution of mineral salts.

In past years, one problem associated with the growth of plants in water culture was the difficulty of supplying sufficient iron to meet plant needs. The use of soluble inorganic iron salts (ferrous sulfate, ferric sulfate, ferric nitrate, etc.) is accompanied by the formation of iron hydroxides in the nutrient solution, especially if the nutrient solution becomes (even slightly) alkaline. Iron phosphates are formed in nutrient solutions containing phosphate salts. Because iron phosphates and iron hydroxides are not too soluble in water, they will precipitate out of solution. Thus the availability of iron to plant roots is reduced sharply. Iron-deficiency symptoms are likely to appear in plants growing in nutrient solutions in which inorganic sources of iron are used unless iron salts are added regularly to the nutrient solution every few days.

In the early 1900s it was found that the problem of supplying iron to plants growing in nutrient solutions could be avoided, at least for the most part, by the addition of citrate or tartrate of iron to the nutrient solution. Citric and tartaric acids are naturally occurring organic compounds with the ability to react with iron ions and thereby maintain their solubility. [Hoagland used iron tartrate (see Table 8-1).] But in the early 1950s it became common to supply iron to plants in nutrient solution in the

TABLE 8-1 FORMULA FOR HOAGLAND'S SOLUTION

A. Molar stock solutions of each of the following four salts are prepared separately. Each stock solution
   is used in the amounts indicated.

	Milliliters, in a liter of nutrient solution
1 M KH$_2$PO$_4$, potassium dihydrogen phosphate	1
1 M KNO$_3$, potassium nitrate	5
1 M Ca(NO$_3$)$_2$, calcium nitrate	5
1 M MgSO$_4$, magnesium sulfate	2

B. To each liter of nutrient solution prepared in A is added 1 ml of a solution of the following five salts.

	Grams dissolved in 1 liter of water
H$_3$BO$_3$, boric acid	2.86
MnCl$_2$ · 4 H$_2$O, manganese chloride	1.81
ZnSO$_4$ · 7 H$_2$O, zinc sulfate	0.22
CuSO$_4$ · 5 H$_2$O, copper sulfate	0.08
H$_2$MoO$_4$ · H$_2$O, molybdic acid	0.02

C. To each liter of nutrient solution prepared in A, 1 ml of an iron solution is added.

Note: Hoagland recommended the addition of a solution of iron tartrate (0.5%), at the rate of 1 ml liter^{-1}
of nutrient solution about once or twice a week. Since about 1950, however, it has become common
practice to supply iron to nutrient solutions as ferric salts of ethylenediaminetetraacetic acid (see the
text). The addition of 1 ml of a solution of FeEDTA containing 5 mg Fe ml^{-1} to 1 liter of nutrient solution
provides 5 ppm iron by weight. This level of iron is adequate for the growth of most higher plants. For a
report on the best method of preparation of FeEDTA, consult Steiner and van Winden, 1970.
After Hoagland and Arnon, 1938.

form of certain organic compounds that do not occur naturally. Organic compounds
that react specifically with metal ions in solution are now known as *chelating agents*.
The word *chelate*, pronounced "*key-late*" (Greek, *chela*, claw) refers to the clawlike
binding of a metal ion in solution by an organic molecule. One of the best known of the
synthetic (i.e., commercially manufactured) chelating compounds is ethylenediaminetetraacetic acid (EDTA). The chemical formulas of EDTA and its water-soluble
complex with a metal (e.g., with iron) are shown in Figure 8-2.

Another problem regarding growing plants in nutrient solution is the tendency of
the pH of the solution to change while plants are growing in it, perhaps after a few
hours or days. Hoagland's solution has an initial pH of about 5 but undergoes an
increase in pH during plant growth. This pH increase occurs because growing plants,
especially young and vigorous ones, absorb anions (nitrate, phosphate, and sulfate)

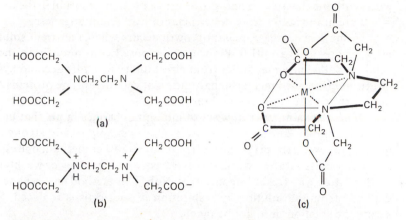

Figure 8-2 Formulas of (a) ethylenediaminetetraacetic acid (EDTA), (b) its ionized form, and (c) its metal complex. The metal atom in the complex is linked to oxygen and nitrogen atoms by six bonds. Two of these six bonds are in the plane of the paper, vertically above and below the metal ion, and four are in a plane perpendicular to the paper. The latter four bonds, shown as dotted lines, are coordinate bonds. (Coordinate bonds are defined in Chapter 9 (pp. 281–282).) The three-dimensional nature of the metal complex is emphasized by the four thin solid lines which surround the metal atom, in a plane perpendicular to the paper. (After Geigy Industrial Chemicals, 1952.)

more rapidly than cations (potassium, calcium, and magnesium) from a nutrient solution. Because anions absorbed by plant cells from a nutrient solution are replaced by hydroxyl ions released by plant cells (or by bicarbonate ions produced in respiration) whereas cations are replaced by hydrogen ions produced metabolically, there will be an excess of base ions (hydroxyl and bicarbonate) in Hoagland's solution after plants have grown in it for a period of time (e.g., see Kirkby and Knight, 1977). This trend toward alkalinity in the nutrient solution will occur with increasing intensity as the buffering system—phosphate ions provide the primary buffering system in Hoagland's solution—becomes progressively depleted as a result of absorption by plant roots.

   Various studies have been carried out on the influence of pH of nutrient solutions on plant growth. Generally it can be said that all species of higher plants tested are capable of healthy growth within a rather wide range of pH, from about pH 5 to 7 (a one hundredfold difference in hydrogen ion concentration).

   A slight decrease in plant growth occurs when a nutrient solution is maintained in the range from about pH 4 to 5 or pH 7 to 9. In these two pH ranges absorption of macronutrient ions is decreased to some extent, perhaps due to competition of hydrogen ions, or hydroxyl ions, for absorption sites on cell membranes of root cells. Moreover, in the range from about pH 7 to 9 precipitation of iron and manganese hydroxides and phosphates (and also calcium phosphate to some extent) will take place. Unless special care is taken to furnish soluble forms of iron and manganese (e.g., with synthetic chelates), plants growing in a nutrient solution in the range of about pH 7 to 9 are apt to suffer from deficiencies of iron and manganese.

A marked decrease in plant growth occurs when a nutrient solution is less than pH 4 or greater than pH 9. Plant roots exposed continuously to these extremes of acidity or alkalinity may suffer from direct injury by hydrogen and hydroxyl ions; the injurious effects will include denaturation of cell membrane proteins and destruction of absorption sites on cell membranes.

There are a number of ways to minimize changes in pH that usually occur in a nutrient solution during plant growth. One is simply to add strong acid or base, as needed, to restore the pH to the desired point. Another method, one especially popular in small-scale laboratory work, involves frequent (e.g., daily or weekly) replacement of used solution with freshly prepared solution. This method has an added advantage: O_2 depleted from the solution by respiration in root cells is replaced. Special steps to aerate the solution need not be taken.

A third method to minimize pH changes that occur in a nutrient solution during plant growth is to supply nitrogen in two inorganic forms: ammonium ion (NH_4^+) and nitrate ion (NO_3^-). [In Hoagland's solution (see Table 8-1) nitrogen is supplied exclusively as nitrate ion.] Ammonium ions absorbed by root cells from a nutrient solution are replaced by hydrogen ions produced metabolically whereas nitrate ions are replaced by hydroxyl ions (or by bicarbonate ions produced in respiration). By supplying both ammonium and nitrate ions, the tendency of a nutrient solution to undergo a change in pH during plant growth will lessen to some extent. Another advantage of the partial substitution of ammonium ions for nitrate ions is that ammonium ions will react with hydroxide ions (released by plant cells when anions are absorbed), forming ammonium hydroxide in the solution. Ammonium hydroxide will act as a buffer because it is only weakly ionized.

ELEMENTAL ANALYSIS OF PLANT TISSUES

The first step in determining the elemental composition of plant tissues is to collect tissue samples (e.g., leaves) from living plants in the field or laboratory. Whether these tissue samples are washed before drying depends largely on the objectives of the analyst. It may be necessary, especially in the determination of some of the micronutrient elements, to wash the tissue samples with a detergent and then rinse with distilled water. This procedure ensures the removal of dust and soil particles and fertilizer and spray residues. The risk involved in washing plant tissues, however, is that some nutrients—for example, potassium—may be leached out of the tissue, at least to some extent.

Because it is more meaningful to express the elemental composition of plant tissues in terms of dry rather than fresh weight, the tissues are dried in an oven. The dry weight of a tissue can be expected to correlate closely to the nutritional status of the plant because dry weight depends on the relative rates of photosynthesis and respiration. The fresh weight of a tissue reflects tissue water content and is influenced considerably by the climatic conditions prevailing when the samples were removed from the plant.

Tissue drying is best accomplished in a forced-draft oven at a temperature of 70 to 80° C. Higher temperatures (e.g., 100° C) are avoided because some of the sulfur- and nitrogen-containing compounds in tissues may volatilize at temperatures higher than 80° C; thus some of these two elements may be lost. The dried tissue is ground to a powder. As a safeguard against contamination, grinders or pulverizers or motor-driven mills with grinding parts made of elements essential to plant growth (e.g., iron, copper) are avoided.

After weighing the powdered tissue sample, an elemental analysis requires that the organic matter in the tissue be decomposed by either *wet digestion* or *dry ashing*. Wet digestion involves digesting plant tissue in boiling, concentrated, reagent-grade acids. Nitric, perchloric, and/or sulfuric acids are generally used. Plant tissue constituents are degraded into small, acid-soluble molecules of both organic and inorganic composition. Dry ashing of plant tissue is carried out in an electric muffle furnace at about 300° C for an hour or two and then at 500 to 550° C for another hour or two. Dry ashing results in the complete oxidation of tissue organic matter: All carbon, hydrogen, and oxygen atoms in the tissue are converted to CO_2 and water, both of which escape into the atmosphere as vapors.[2] The ash residue is dissolved in dilute acid.

Each of these decomposition methods (wet digestion or dry ashing) has its merits and disadvantages. But, in a few instances, one or the other is mandatory. Dry ashing, for example, is required if boron is to be assayed because this element is lost completely when a tissue sample is prepared by wet digestion.

The final step in an elemental analysis of plant tissue is the quantitative determination of the chemical elements in the solution derived from either wet digestion or dry ashing. Until about two decades ago conventional chemical methods were used to determine all these mineral elements. In these methods one or more specific reagent chemicals are added to aliquots of the solution and then titrimetric or colorimetric procedures are followed.[3] At present the atomic absorption spectro-photometer, an instrument not commercially available until about 20 years ago, is widely used.

This device consists of an atomizer (usually a flame), a source of radiation (usually a hollow cathode lamp), a device for dispersing radiation (e.g., a mono-

---

[2]Elemental analyses for carbon, hydrogen, and oxygen in plant tissues are made only infrequently. To analyze for carbon and hydrogen, the dried tissue is oxidized completely to CO_2 and water in a current of O_2 at an elevated temperature. The amounts of these two end products of combustion (i.e., CO_2 and water) are used to calculate the percentages of carbon and hydrogen in the tissue. Elemental analysis of oxygen in plant tissue can be made by pyrolyzing the tissue in a stream of pure nitrogen over heated carbon at a temperature slightly above 1100° C. In this treatment all the oxygen in the pyrolysis products is converted to carbon monoxide, which is then analyzed (usually by iodometric methods) for its oxygen content. For further details, consult Ingram, 1962.

[3]For a concise description of these methods and other matters related to plant analysis, see Johnson and Ulrich, 1959.

chromator), and an electronic processing unit (photomultiplier, amplifier, etc.). The technique is based on the fact that the atoms of every element have the ability to absorb radiation at very narrow wavelength bands, each of which differs for each element. A solution containing the elements to be determined is vaporized in the atomizer. Thus the elements are dissociated from their chemical bonds and are in a condition to absorb radiation. The spectral radiation band characteristic of the element being determined is selected with the monochromator; the vaporized sample absorbs this radiation band and thereby diminishes its intensity. The intensity of the diminished wavelength band is measured and recorded by the electronic processing unit. The concentration of the element in the sample is proportional to its absorbance.

The atomic absorption spectrophotometer is the workhorse of the modern plant and soil analytical laboratory. Only a single solution obtained from a plant tissue is needed to measure the concentration of most mineral elements essential to plant growth. An efficient technician can analyze 20,000 or more sample solutions per year. For technical reasons, nitrogen, sulfur, and phosphorus cannot be determined by atomic absorption spectrophotometry; these three elements are determined in most analytical laboratories by titrimetric or colorimetric methods.

Other even more versatile instruments capable of rapid sequential determination of elements in plant tissues are being developed currently. High on the list is atomic emission spectroscopy (Fassel, 1979). Powdered or liquid samples are introduced into a plasma chamber in which the elements are vaporized and atomized and the free atoms excited and ionized at temperatures much higher than the hottest combustion flames used in atomic absorption spectroscopy. Under these conditions the characteristic spectral lines of the elements are emitted as the excited electrons from the upper energy levels fall back into a lower energy level. The intensities of these spectral lines are measured in electronic processing units.

At least two other methods capable of rapid sequential determination of elements in plant tissues are being investigated today: x-ray emission spectroscopy (Maugh, 1977) and ion chromatography (Wetzel et al., 1981).

Research related to the mineral nutrition of plants, especially to those minerals required in very small amounts, has escalated considerably in recent years. Clearly analytical methods must establish a reliable data source for plant scientists concerned with the mineral aspects of plant nutrition. Knowledge of tissue mineral status and its relation to normal plant growth, as well as the inverse (i.e., the effects of plant growth on tissue mineral composition), should prove invaluable in years to come in the prevention and alleviation of mineral deficiencies in plants.

An elemental analysis of a mature maize plant, including the stem, leaves, fruits, and roots, is shown in Table 8-2. The plant was grown in soil. A total of 14 elements were determined. Several elements present in amounts too small to be measured accurately by the methods used in this study (published in 1938) are grouped together as "undetermined elements" (see the last line in Table 8-2).

Table 8-2 shows that hydrogen is the most abundant element—in terms of *numbers of atoms*—in the dry matter, followed by carbon and oxygen. Compared to these three elements, other elements are present in relatively small numbers.

TABLE 8-2 ELEMENTAL COMPOSITION OF A MAIZE PLANT (*ZEA MAYS*, PRIDE OF SALINE VARIETY), INCLUDING STEM, LEAVES, COB, GRAIN, AND ROOTS, GROWN IN KANSAS IN 1920[a]

Chemical element	Weight (g)	Dry weight (%)	Relative numbers of atoms in the dried tissue
Hydrogen	52.2	6.2	10,000
Carbon	364.2	43.6	5,825
Oxygen	371.4	44.4	4,450
Nitrogen	12.2	1.5	167
Silicon	9.8	1.2	67
Potassium	7.7	0.9	37
Magnesium	1.5	0.2	11.7
Phosphorus	1.7	0.2	10.5
Calcium	1.9	0.2	8.9
Sulfur	1.4	0.2	8.2
Aluminum	0.9	0.1	6.3
Chlorine	1.2	0.1	6.0
Iron	0.7	0.1	2.4
Manganese	0.3	0.04	1.1
Undetermined elements	7.8	–	–

[a]Dry weight (834.9 g) was distributed among the elements as shown in the second column; the dry weight represented 10% of the fresh weight of the plant. From the dry weight data in the second column are calculated the percentages of total dry weight accounted for by each element and the approximate relative numbers of atoms of each element, assuming hydrogen to be 10,000. (The data in the second column were taken from Miller, 1938, p. 284.)

Table 8-2 also shows that carbon and oxygen are the most abundant elements if calculations are made on the basis of *percent of dry weight*. Carbon and oxygen together account for 88% of the dry matter that remains after water is removed from these tissues; hydrogen accounts for slightly more than 6%. All the mineral elements (nitrogen, silicon, potassium, etc.) together contribute slightly less than 6% of the dry weight of the maize plant. In other plants as little as 5% and as much as 15% of the dry weight may consist of mineral elements.

Variations in mineral content of plant tissues are influenced by several factors. One is the kind of plant. Different species of higher plants, and also different varieties of the same species, have different abilities to absorb mineral nutrients even when rooted in the same soil or nutrient solution. The following three well-documented examples illustrate this point.

1. Dicotyledonous plants have been found to absorb more boron than monocots.
2. Legumes usually have a higher calcium content than grasses.
3. Some varieties of soybean are able to absorb greater amounts of iron than other varieties.

Differences in plant species and varieties in their abilities to absorb nutrients may be due simply to differences in the growth habits of root systems. On the other hand, species and varietal differences in the ability to absorb a given nutrient may be due to differences in the activities of specific transport carriers in cell membranes (see p. 323).

Another factor influencing the mineral content of plant tissues is the climatic conditions that prevail during plant growth. A plant grown under high light conditions, for instance, will have a high rate of photosynthesis and will produce a larger food reserve and a larger root system—one more efficient in absorption of nutrients—than a plant grown in low light.

The chemical composition of the medium in which a plant is grown affects the mineral content of plant tissues. If a plant grows on a soil or nutrient solution in which an element is lacking or unavailable, the tissues will have a low content of this element. On the other hand, minerals added intentionally to soils in fertilizer programs will be absorbed by roots; sooner or later analyses of leaves will reflect these additions.

The age of a tissue is another factor that influences its mineral content. A mature leaf is likely to have a higher mineral content than a very young leaf. Also, a mature leaf may have a higher mineral content than an old, senescent leaf. The latter is apt to suffer considerable loss of water-soluble minerals by leaching by rain.

CRITERIA OF ESSENTIALITY OF ELEMENTS

Although only 14 chemical elements are listed in Table 8-2, other more complete analyses have shown that a surprisingly large number can be detected in tissues harvested from plants grown in soils. Most of these elements are present only in trace amounts. About 30 elements can be detected readily by modern analytical techniques and as many as 60 have been reported in one or another tissue.

The presence of an element in a plant is not, of and by itself, a valid basis on which to assess its essentiality to plant life. Not all elements present in plant tissues are required for plant growth. Even though the mechanisms of absorption of molecules and ions by plant cells are selective (see Chapter 11) and plant cells can discriminate among substances presented to them, accumulating some and excluding others, the discrimination process is not perfect.

Of the many elements detected in plant tissues only 16 are essential to all higher plants (see Table 8-3). For each of these essential elements, experimental evidence that the element is required for growth has been obtained. In the absence of each essential element, plants develop deficiency symptoms characteristic of the deficient element and die prematurely.

TABLE 8-3 CHEMICAL ELEMENTS KNOWN TO BE ESSENTIAL FOR GROWTH OF HIGHER PLANTS[a]

		C	H	O	
K					N
Ca					P
Mg					S
		B, Cl, Cu, Fe, Mn, Mo, Zn			

[a] A few chemical elements (e.g., cobalt, selenium, silicon, sodium) have been shown to be essential to only a few plants and therefore are not included in this table. (See p. 240 for further discussion.)

No great experimental difficulties occur in demonstrating the essentiality of several of the essential elements by growth experiments in nutrient culture. This is especially true of *macroelements* (i.e., essential elements that are required in relatively large amounts). The essentiality of macroelements is often demonstrated in introductory botany courses at the high school or university level.[4] In addition to carbon, hydrogen, and oxygen, there are six mineral macroelements: potassium, calcium, magnesium, nitrogen, phosphorus, and sulfur (see Table 8-3). The others in Table 8-3 are *microelements* (i.e., elements required in relatively small amounts).

Demonstrating the essentiality of most microelements entails considerable technical difficulty. These difficulties occur because the requirements of a plant for a microelement may be satisfied by mere traces of incidental contaminants. The unintentional addition to the nutrient solution of elements required in extremely small amounts is difficult to avoid. Impurities are always present in salts used to prepare nutrient solutions; it is impossible to remove completely every atom of a contaminating element. Other possible sources of contamination are the water supply, the container in which the plants are grown, and dust particles in the air.

Besides reducing the level of incidental contamination to a minimum, demonstrating the essentiality of many microelements is difficult because plant species with large cotyledons may carry large reserves of the microelement under investigation. It might be desirable, in experimental work designed to determine the essentiality of an element required in very small quantity, to select a species with a small seed and a high requirement for the element involved. Alternatively, it may be desirable to harvest seeds from experimental plants grown in a nutrient medium subjected to rigorous purification procedures designed to remove the element and then to grow the plants and harvest the second generation of seeds. By repeating this procedure for one or more generations, the content of the element under investigation can be reduced in the harvested seeds and seeds suitable for experimental work can be obtained. Still

[4] Directions for preparing nutrient solutions in which one or another of the mineral macroelements is omitted are given by Hoagland and Arnon, 1938, and Moore, 1973.

another experimental procedure is to remove the cotyledons from the root-shoot axis of ungerminated seed. The excised root-shoot axis devoid of cotyledonary reserves can be grown to the seedling stage by tissue culture techniques (see Chapter 17). Once such a seedling develops sufficient chlorophyllous tissue to be autotrophic, it can be used as experimental material for demonstrating the essentiality of microelements.

Because of the technical difficulties associated with demonstrating the essentiality of elements required in very small amounts, two American plant physiologists, Arnon and Stout, in 1939 suggested the adoption of three criteria of essentiality.

The criteria proposed by Arnon and Stout (1939a) were described by them as follows:

> . . . an element is not considered essential unless (a) a deficiency of it makes it impossible for the plant to complete the vegetative or reproductive stage of its life cycle; (b) such deficiency is specific to the element in question, and can be prevented or corrected only by supplying this element; and (c) the element is directly involved in the nutrition of the plant quite apart from its possible effects in correcting some unfavorable microbiological or chemical condition of the soil or other culture medium.

The experimental procedure for fulfilling Arnon and Stout's criteria involves growing plants from the seedling stage in a nutrient solution to which all elements regarded as essential are added except the one under consideration, which is removed as completely as possible. First, it must be demonstrated that normal growth is impossible in the absence of the element: "Deficiency symptoms" must be produced in the experimental plants. (Deficiency symptoms are disruptions in plant metabolism and growth produced by a less than adequate supply of an essential element and are considered more fully later.) Second, it must be demonstrated that addition of the element under investigation, and no other, to the nutrient solution can correct or prevent the deficiency symptoms. Finally, the element under investigation must be shown to be involved directly in the nutrition of the plant and not to exert its effect indirectly through some type of interaction with another ionic species in the root environment. To fulfill this third criterion, plants with deficiency symptoms must be shown to recover and resume normal growth after a solution of the elements under investigation is introduced into the plant through stems or leaves (e.g., by injection) rather than through roots.

Since their publication in 1939 Arnon and Stout's criteria have exerted considerable influence on the development of ideas relating to the essentiality of elements. Arnon and Stout (1939b) themselves used their criteria to demonstrate for the first time that molybdenum is essential for higher plants.

There has been a growing awareness lately that Arnon and Stout's criteria of essentiality may be too stringent. Chlorine, for example, is listed as an essential element (see Table 8-3), but it has been shown (Broyer et al., 1954) that bromine can substitute for it in some plants, although at much higher concentrations. There can be no question, however, that chlorine is the normally functional, essential halogen for higher plants; chlorine is the most abundant naturally occurring halide in the soil. But strict adherence to Arnon and Stout's criteria—especially the second criterion regarding complete specificity of an element—would mean that chlorine cannot be accepted as an essential element.

One point of view that seems to be gaining favor among plant physiologists is that only the first of Arnon and Stout's criteria is needed to establish the essentiality of an element. (The first criterion, it will be recalled, uses growth as the yardstick of essentiality.) In addition to satisfying Arnon and Stout's first criterion, if the element can be shown to have a role in plant *metabolism*, then it would appear justifiable to declare that the element is essential. It was precisely this reasoning that Wilson and Nicholas (1967) used to establish the essentiality of cobalt in two species of higher plants.

Another suggestion relative to the criteria of essentiality is that some elements might better be called *functional* or *metabolic* elements rather than essential elements (Bollard and Butler, 1966, Nicholas, 1961). This designation is intended merely to indicate that an element is metabolically active; a functional or metabolic element may or may not be essential. Such a designation has the advantage of avoiding labeling an element essential when it is known that another element can replace it. In the chlorine–bromine example mentioned, adoption of this suggestion would dictate that chlorine be designated a functional element rather than an essential element; bromine would also be designated a functional element.

THE ESSENTIAL ELEMENTS

At present 16 chemical elements are known to be essential for the growth of all higher plants. Their chemical symbols are listed in Table 8-3. All except carbon, hydrogen, and oxygen are mineral elements.

Potassium, calcium, and magnesium are grouped together in Table 8-3 because they are present in soils as cations (K^+, Ca^{2+}, Mg^{2+}). Similarly, nitrogen, phosphorus, and sulfur are grouped together because they are normally present in soils as anions (NO_3^-, $H_2PO_4^-$, $SO_4^=$). [Under suitable conditions ammonium ion (NH_4^+) may substitute for nitrate ion (NO_3^-); for further details, see p. 267.] These six elements (potassium, calcium, magnesium, nitrogen, phosphorus, and sulfur) are called macroelements, a term defined on p. 237. The term *macronutrient* (in contrast to macroelement) refers to the chemical form (K^+, Ca^{2+}, Mg^{2+}, NO_3^-, $H_2PO_4^-$, $SO_4^=$) in which a macroelement is presented to plants.

The microelements listed in Table 8-3 are boron, chlorine, copper, iron, manganese, molybdenum, and zinc. Sometimes the microelements are called *minor* or *trace* elements because they are required by plants in only extremely small quantities. But these latter designations are quite unsatisfactory; there is nothing "minor" about the essentiality of the microelements.

Microelements can be presented to plants in different chemical forms, either as inorganic ions or undissociated molecules or organic complexes (e.g., chelates, described earlier). If presented in the inorganic form (e.g., see Hoagland's solution in Table 8-1) the microelements usually are supplied as the following ions: BO_3^{-3} (borate), Cl^- (chloride), Cu^{2+} (cupric), Fe^{2+} (ferrous) or Fe^{3+} (ferric), Mn^{2+} (manganous), $MoO_4^=$ (molybdate), and Zn^{2+} (zinc).

Estimates of the concentrations of the mineral macroelements generally adequate for optimal growth range from a low of about 30 μmol g^{-1} dry weight of tissue for sulfur to a high of about 1000 μmol g^{-1} dry weight of tissue for nitrogen (Epstein, 1965). Estimates of the requirements for carbon, oxygen, and hydrogen range from 30,000 to 60,000 μmol g^{-1} dry matter (Epstein, 1965).

Estimates of the concentrations of microelements required to be present in plant tissues to prevent the appearance of deficiency symptoms (in tomato) range from 7 μmol g^{-1} dry weight of tissue for chlorine to less than 1.4×10^{-3} μmol g^{-1} dry weight of tissue for molybdenum (Broyer et al., 1954). Plant requirements for the other microelements are intermediate between these two extremes.

The essentiality of the elements in Table 8-3 has been established by growth experiments (of the type considered in the preceding section) for many different species of higher plants. The plants tested range from dicots to monocots and from angiosperms to gymnosperms and hence represent a wide variety of taxonomic, anatomic, and physiologic groups. There can be little question that all the elements listed in the table are essential to all higher plants.

In addition to the 16 elements in Table 8-3, other elements have been established as being essential for a few species of higher plants. Sodium is an essential microelement to certain salt-marsh plant species in which CO_2 assimilation in photosynthesis takes place by the C_4 dicarboxylic acid pathway (Brownell and Crossland, 1972). Cobalt is an essential microelement to wheat and one legume species (Wilson and Nicholas, 1967). (More experimental evidence is required before cobalt can be classified as an essential element to all higher plants.) To demonstrate the essentiality of sodium and cobalt, it was necessary to take extraordinary precautions to reduce the levels of these elements in the experimental plants. In the cobalt experiments the cobalt content of the experimental plants was reduced by growing experimental plants from excised embryos instead of whole seeds. In both the sodium and cobalt experiments it was necessary to filter the air in which the experimental plants were grown because the leaves were capable of extracting their total requirement for these elements from dust particles that alighted on their surfaces.

Two other microelements, silicon and selenium, have been shown to be essential to a few higher plants. Silicon is essential to certain grasses and horsetails with naturally high silicon content (Lewin and Reimann, 1969). Selenium is essential for the growth of certain species of *Astragalus*, a genus that grows naturally on seleniferous soils in the western plains of the United States (Shrift, 1969).

Finally, nickel recently has been shown to be an essential component of at least one enzyme, urease (Thauer et al., 1980). For the reaction catalyzed by urease, see Eq. 9-7.

NUTRITIONAL DISORDERS OF PLANTS

Because each essential element performs one or more specific internal roles (see pp. 284–286), a less than adequate supply[5] of an essential element will be accompanied by a set of distinctive metabolic disruptions, including changes in activities of enzymes, rates of metabolic reactions, and concentrations of metabolites. To illustrate some of these internal changes, consider the results obtained in the following mineral deficiency experiments.

Peppermint plants were grown in full nutrient solution and in solutions in which one or another essential element was omitted (see Figure 8-3). When acute visible deficiencies appeared in the plants, leaves were harvested and extracted with ethyl alcohol; this solvent extracts amino acids and amides that are in a free and uncombined form. It can be seen from the figure that 0.087 mg of amino acids and amides was extracted per gram of fresh weight of leaves grown in full nutrient solution, a value that can be regarded as "normal." Leaves of plants grown under conditions of nitrogen deficiency contained very low amounts of alcohol-soluble amino acids and amides (0.019 mg g^{-1} fresh weight of tissue), while those of minus-sulfur plants contained relatively high amounts (1.351 mg g^{-1} fresh weight of tissue). Leaves of plants subject to other mineral deficiencies were intermediate between these two values. There were considerable differences in the relative amounts of different amino acids and amides in the leaves of these plants, as shown by the histograms in Figure 8-3.

In addition to alterations in metabolic patterns such as those illustrated in Figure 8-3, severe deficiencies of individual essential elements also produce a set of characteristic effects in the *external* appearance of leaves, stems, roots, blossoms, and fruits (see Figure 8-4.)[6] Visual symptoms of nutritional deficiency (see Table 8-4) include stunted growth, chlorosis (failure of leaves to produce normal amounts of chlorophyll), mottling of leaves, abnormal curling of leaves, development of abnormal leaf discolorations, development of regions of necrosis (blackening and decay of tissues), premature drying and withering of leaves, and premature senescence of leaves and blossoms.[7]

As a general rule, symptoms of deficiencies are first noticeable in older and lower leaves for elements that are able to move readily from one region to another within the plant. Included among these elements are potassium, nitrogen, and magnesium (see Table 8-4). When the supply of one of them to the plant is less than adequate, the

---

[5] Different varieties of the same species of plant may have different abilities to extract one or another nutrient from a given soil because they can chemically modify the soil close to their roots. For a recent review, especially regarding solubilization of iron by "efficient" varieties of plants, see Olson et al., 1981.

[6] Several pamphlets and books containing excellent color photographs of essential element deficiencies in plants have been published. For example, see (a) Sprague and (b) T. Wallace (1951).

[7] Injurious effects in plants may be produced not only by a deficiency of an essential element but also by an excess. For descriptions of characteristic visual symptoms due to excesses of essential elements, see Altman and Dittmer, 1968.

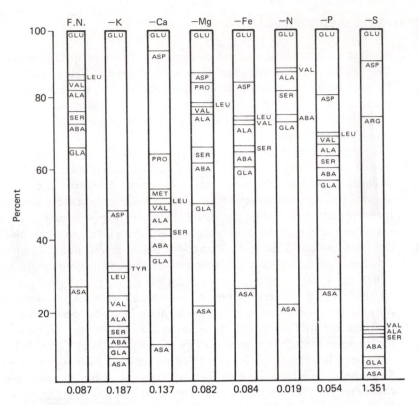

Figure 8-3 Relative composition of the alcohol-soluble amino acids and amides of leaves of peppermint grown in full nutrient (F.N.) solution and in solutions of one or another mineral deficiency. At the base of each histogram is indicated the total quantity of amino acids and amides extractable in alcohol, in milligrams per gram fresh weight of leaf tissue. Key: ASA, aspartic acid; GLA, glutamic acid; SER, serine; ASP, asparagine; ALA, alanine; GLU, glutamine; ARG, arginine; MET, methionine; PRO, proline; VAL, valine; LEU, leucine; TYR, tyrosine; ABA, δ-amino butyric acid. (After Crane and Steward, 1962; Figure 1.)

element is translocated away from older tissues to younger, more metabolically active tissues. On the other hand, deficiencies of such elements as sulfur, iron, boron, manganese, and calcium—elements that are not mobile within the plant—are noticeable first in younger leaves and growing shoot tips (see Table 8-4). These tissues are the first to suffer when the supply of one of these elements is less than adequate because these elements tend to remain in older tissues.

Much experience may sometimes be necessary to diagnose correctly, on the basis of visual symptoms alone, the particular element that is deficient in a plant. Consider a symptom like chlorosis, for example. Although deficiency chloroses, especially those caused by insufficiency of iron or nitrogen, are well known and widely recognized, chlorosis may also be induced in plants by low light levels (etiolation), certain insect or fungal infestations, albinism, and senescence.

Figure 8-4 *Upper*: Lettuce plants grown with their roots in nutrient solutions lacking only nitrogen, or potassium, or phosphorus, and (right) in a nutrient solution containing all essential mineral elements. *Lower*: Left, tomato plants grown with roots in a nutrient solution containing all essential mineral elements except copper. Right, tomato plant grown under the same conditions except that the leaves were sprayed with a solution of a copper salt. (After Hoagland, 1949.)

TABLE 8-4   VISUAL SYMPTOMS OF DEFICIENCY OF ESSENTIAL ELEMENTS[a]

1.	Boron	Terminal leaves necrotic, shed prematurely; internodes of terminal shoots shortened, usually rosetting; apical meristems blacken and die, general breakdown of meristematic tissue; roots short, stubby. Plants dwarfed, stunted; flower development and seed production usually impaired or lacking.
2.	Calcium	Leaves chlorotic, rolled, curled; breakdown of meristematic tissues in stems and roots (death in acute cases); roots poorly developed, lack fiber, and may appear gelatinous. Symptoms appear near growing points of stems and roots. Little or no fruiting.
3.	Chlorine	Wilting of leaf tips, followed by chlorosis, bronzing and necrosis basipetally in areas proximal to the wilting.
4.	Copper	Wilting of terminal shoots, frequently followed by death. Leaf color often faded. Carotene and other pigments reduced.
5.	Iron	Interveinal white chlorosis, appearing first on young leaves; tendency for chlorosis of all aerial parts, often becoming necrotic; in some cases leaves may be completely bleached, margins and tips scorched. Usually has an overall effect.
6.	Magnesium	Mottled chlorosis with veins green and leaf web tissue yellow or white, appearing first on old leaves; severely affected leaves may wilt and shed or may abscise without the wilting stage; brittleness of leaves common; necrosis often occurs.
7.	Manganese	Mottled chlorosis with veins green and leaf web tissue yellow or white, appearing first on young leaves; may spread to old leaves. Stems yellowish green, often hard and woody. Carotene reduced.
8.	Molybdenum	Light yellow chlorosis of leaves; leaf blade may fail to expand.
9.	Nitrogen	In young plants, stunted growth and yellowish green leaves; older leaves light green, followed by yellowing and drying or shedding, often with abundant anthocyanins in veins. Shoots short, thin, growth upright and spindly; flowering reduced. Apple and peach fruit highly colored, developing slowly, small when mature.
10.	Phosphorus	Young plants stunted, leaves dark blue-green sometimes purplish (potato and certain other vegetables have pale green leaves); stems slender; often anthocyanins in veins, and may become necrotic. Meristematic growth ceases in potato. Fruits ripen slowly; plants often dwarfed at maturity.
11.	Potassium	Leaves of potato usually dark blue-green and leaves of monocotyledons pale green or streaked with yellow, with marginal chlorosis and necrosis appearing first on old leaves; usually wrinkled, corrugated, or crinkled between veins.
12.	Sulfur	Leaves light green to yellow, appearing first along veins of young leaves; stems often slender.
13.	Zinc	Leaves chlorotic and necrotic, young growth first affected; rosetting; premature shedding; whitish chlorotic streaks between veins in older leaves and whitening of upper leaves in monocotyledons; chlorosis of lower leaves in dicotyledons.

[a] The descriptions of symptoms are applicable primarily to herbaceous plants. Symptoms of deficiencies of carbon, hydrogen, and oxygen have been omitted intentionally from this table.

Except for chlorine, the data were taken from Altman and Dittmer, 1968, pp. 317–319. The chlorine data were taken from Broyer et al., 1954.

A further complication in correctly diagnosing the element responsible for a given set of visible symptoms of deficiency is the influence of one nutrient in a soil on another in relation to plant growth. Zinc uptake by plants is reduced and zinc-deficiency symptoms may appear when phosphate fertilizers are applied to soils in large amounts or over long periods of time even when adequate amounts of zinc are present in the soil. High concentrations of phosphate in the root environment (and also a high soil pH) are likely to induce the appearance of iron-deficiency symptoms in plants even when adequate amounts of iron are present in the soil. Heavy applications of potassium fertilizers often lead to reduced absorption of magnesium and calcium and to the development of magnesium and calcium deficiencies, especially when the soil has only limited supplies of these latter two elements.

The causes of these and other ion interactions are not the same and not always easy to pinpoint. In some cases, a nutrient ion, especially a micronutrient ion, may become fixed in a soil by an unfavorable management practice and thus rendered unavailable to plants. Visible symptoms of deficiency due to this kind of ion interaction are apt to occur when a nutrient ion is present in a soil in amounts barely adequate to prevent deficiency symptoms. In other cases, deficiency symptoms in plants may be due to suppression of the absorption of one nutrient ion by another, perhaps by competition for the same absorption sites on plasma membranes.

Visible symptoms of deficiency have only limited usefulness in the raising of field crops because they appear only when the deficiency is severe. Furthermore, mineral deficiencies in the field are likely to be multiple. If so, the actual visual symptoms may be altered considerably from those described in Table 8-4 and a correct diagnosis on the basis of visual symptoms alone may be difficult. Also, the problem of ion interactions mentioned earlier tends to complicate diagnosis. Therefore agriculturalists interested in the intensive production of field crops must consider other methods (e.g., soil analyses and leaf analyses) in conjunction with visual symptoms in establishing the causes of nutritional disorders. These methods are considered briefly later in this chapter.

THE SOIL AS A SOURCE OF NUTRIENTS

Soils are complex natural formations on the surfaces of the earth and consist of five main components: mineral matter, organic matter, soil water, soil air, and living organisms. Mineral matter, which consists of inorganic material, ranges in size from rock fragments and large pebbles to minute particles of clay. Organic matter consists of decomposition products of plant and animal life. Mineral matter and organic matter together make up the solid matrix of soils. The pores within the matrix are filled with soil water and soil air. A fifth component of soils is a population of living organisms, including roots of plants, bacteria, fungi, protozoa, and many kinds of animal life such as insects, mites, spiders, and earthworms.

Soils are developed slowly, over many thousands of years, as a result of several geologic processes. These processes include chemical and physical weathering of rocks and minerals, application of tremendous pressures, accumulation and decay of living organisms, and leaching by water of the products of mineral and organic decompositions. Chemical weathering of rocks and mineral particles involves their interaction with atmospheric gases, especially CO_2 and O_2, and with bicarbonate ion and organic acids of biological origin. Fluctuations in temperature, particularly the effects of freezing and thawing, result in the gradual physical disintegration and subdivision of minerals. Eventually some particles are comminuted to dimensions so small as to acquire the properties of colloids. These minute particles constitute the *clay* fraction of the soil.

The remains of plants, animals, and microorganisms are subject to chemical weathering and also to biological degradation by soil microorganisms. Degradation of organic matter is rapid at first and is accompanied by the release of inorganic ions and CO_2. With continuation of the process, which progresses at slower and slower rates, a colloidal carbonaceous residue relatively resistant to further decomposition accumulates. This organic residue is called *humus*. Humus consists largely of dark-brown organic molecules rich in phenolic compounds derived mainly from lignin of plant residues. Humic particles have the dimensions of clay particles and the two mix together intimately in the soil. Finally, all the products of both mineral and organic origin are acted on by water that percolates through the soil. As a result, soluble and dispersible materials are removed from the topmost layers and come to rest in lower layers or are leached into water tables. In addition, the colloidal clay and humus particles remaining in the soil become hydrated to varying degrees.

The most important part of the soil with respect to plant nutrition is the colloidal fraction, which consists of inorganic (clay) and organic (humic) colloidal particles. Individual colloidal particles (micelles) have a large surface area per unit weight and are extremely small, not visible even with the light microscope. They can be observed only with an electron microscope. Most clay micelles have a crystalline structure and consist of silicon and aluminum oxides with oxygen atoms on the surfaces of the micelles. They possess electronegative adsorption sites available for attracting cations, including calcium, magnesium, potassium, aluminum, ammonium, and sodium, as well as hydrogen ions arising from biological activity (see Figure 8-5). Humic micelles (and a few clays) are amorphous rather than crystalline. Like the crystalline clay micelles, the humic colloids are also characterized by the presence of surface charges, but these arise from exposed carboxyl (—COOH) and hydroxyl (—OH) groups from which the hydrogen may disassociate and be replaced by other cations. Thus both clay and humic micelles are surrounded by a swarm of loosely bound cations, each of which is surrounded by a shell of water molecules (see Figure 12-2).

Because clay and humic micelles are predominantly negatively charged, they do not attract anions. When an anion, such as nitrate, is added to a soil as a fertilizer or released into a soil by decomposition of organic matter or by the action of soil microorganisms, it is gradually leached from the soil by percolating water. Nitrate ions that are not assimilated into organic matter by living organisms eventually find their way into rivers and oceans. Cations are adsorbed to clay and humic micelles and so are not likely to be leached away by percolating water.

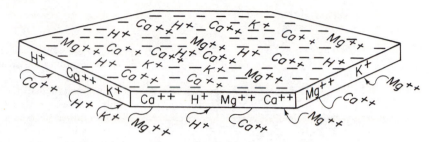

Figure 8-5 Diagrammatic representation of the outside surfaces of a colloidal clay crystal (micelle) showing its sheetlike structure. Note the innumerable negative surface charges and the swarm of adsorbed cations. In this diagram calcium ions are most numerous, but some hydrogen, potassium, magnesium, and sodium ions also are shown. (After Buckman and Brady, 1960; Figure 4.3.)

Cations adsorbed on the surface of soil colloidal particles are capable of exchanging rapidly and reversibly with those in the soil solution. This process is called *cation exchange* (or base exchange).[8] The extent of exchange of one kind of cation by another depends on their relative adsorptive affinities and also on their relative concentrations in the soil solution and on the surfaces of the colloidal micelles.

An exchange reaction between a single micelle and the soil solution can be illustrated by an equation in which two hydrogen ions are represented as being displaced from a soil colloidal micelle by a calcium ion, perhaps added to a soil when it is "limed" with calcium carbonate (limestone):

$$\boxed{\text{Micelle}}\begin{array}{l} H^+ \\ H^+ \end{array} + Ca^{2+} \longrightarrow \boxed{\text{Micelle}}\, Ca^{2+} + 2\,H^+$$

The cation exchange that occurs when finely ground limestone consisting mostly of calcium carbonate is added to an acid soil is diagrammed in Figure 8-6.

The principal immediate source of mineral nutrients to plant roots are ions in the soil solution. This nutrient supply is gradually depleted by absorption of nutrient ions by plant roots and continuously replenished by desorption of exchangeable ions on the clay-humus complex and breakdown of readily decomposable organic debris. These two sources represent reserves that serve to replace—but only at a relatively slow rate—those ions in the soil solution that are absorbed by plant roots. For intensive cultivation of crop plants, application of mineral salts to soils is required.

[8] It is also possible for ions to be exchanged *directly* between root surfaces and soil colloids when they are in intimate contact. This process is called *contact exchange* and does not involve the soil solution; the dimensions involved are 5 nm or less.

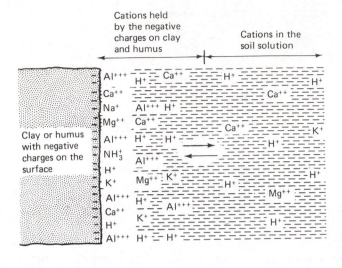

Cations held by the negative charges on clay and humus — Cations in the soil solution

Clay or humus with negative charges on the surface

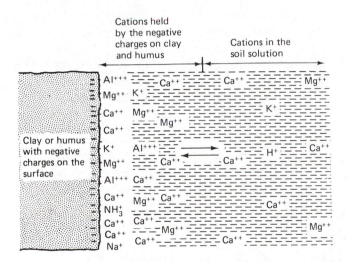

Cations held by the negative charges on clay and humus — Cations in the soil solution

Clay or humus with negative charges on the surface

Figure 8-6 Diagrammatic representation of the role of cation exchange in the conversion of an acid soil to a near-neutral soil. Initially the soil pH was 5.0 (upper diagram). The soil pH was raised to 6.5 (lower diagram) by the addition of finely ground limestone (which consists largely of carbonates of calcium and, to a lesser extent, magnesium). Note that the acid soil (upper diagram) has a larger number of adsorbed aluminum ions than adsorbed hydrogen ions. The preponderance of adsorbed aluminum ions arises from the fact that aluminum hydroxide in clay micelles reacts with hydrogen ions as follows:
$Al(OH)_3 + 3H^+ \rightarrow Al^{3+} + 3H_2O$.
Aluminum ions released by this reaction are strongly adsorbed to soil-colloidal micelles and displace other cations. When an acid soil is limed with calcium carbonate (lower diagram), calcium ions displace many of the adsorbed aluminum ions and hydrogen ions. Desorbed aluminum ions are converted to aluminum hydroxide (by reversal of the above reaction). Desorbed hydrogen ions (as well as hydrogen ions produced by reversal of the above reaction) react with carbonate ions (from calcium carbonate). Because the resulting bicarbonate ions (not shown in the above diagram) are weakly dissociated, hydrogen ions are effectively removed. Thus the pH of the soil increases. (After Worthen and Aldrich, 1956; Figures 110 and 111.)

CHEMICAL FERTILIZERS IN CROP PRODUCTION

One of the important contributions of the nineteenth-century experimental plant physiology to agriculture was the discovery that soil fertility and crop yields could be increased by adding *inorganic* nutrients to soils. Prior to the nineteenth century the common method for increasing crop production was to apply plant and animal debris (manures, composts, etc.) to soils. It was not realized that this treatment returned to the soil only a portion of the nutrients that had been extracted by plants. Another centuries-old agricultural practice was to rotate leguminous crops periodically with other crops. This practice resulted in increased growth of those crops that followed the legumes. The scientific basis for this effect [i.e., the addition of nitrogen to soils by biological nitrogen fixation (see pp. 270–275)], however, was elucidated only in the late nineteenth century.

Only in the early 1800s did agricultural scientists realize that crop plants grow in proportion to the amounts of inorganic nutrients present in soils. This principle was adopted by nineteenth-century agronomists, such as Liebig in Germany, Boussingault in France, and Lawes and Gilbert in England, who emphasized that losses of nutrients to plants could be replaced by additions of inorganic mineral salts to the soil. Since the time of Liebig, Boussingault, and Lawes and Gilbert, fertilizer technology has developed enormously. Today the application of inorganic mineral salts to soils is a basic feature of agricultural practice. Salts of nitrogen, phosphorus, and potassium—the N-P-K in the fertilizer bag—are used in largest amounts because these three elements are most likely to be deficient in most soils in which crops are grown. Without the application of these and other mineral fertilizers to soils, the large crop yields obtained in developed countries throughout the world during the past 50 years or more could not be possible.

Although the mineral nutrients most commonly added to soils as chemical fertilizers are the macronutrient salts, especially those of nitrogen, phosphorus, and potassium, there has been an increasing awareness during the past two or three decades of the need to fertilize certain soils with micronutrients. One result of the intensive use of macronutrient salts in present-day crop production is the relatively rapid utilization of micronutrients by plants; removal of micronutrients from the soil takes place at much greater rates than formerly and deficiencies of micronutrients are more likely to occur. Moreover, some soils may be naturally deficient in certain micronutrients. Copper, zinc, and molybdenum deficiencies occur in some Australian soils; boron and copper are lacking in many peat soils in the United States; zinc and manganese are deficient in soils rich in organic matter, especially those with high pH.

In addition to supplying specific nutrients, another use of inorganic chemical fertilizers is to change the pH of arable soils. [The term *pH* (French, *pouvoir hydrogène*, hydrogen power) is the universally adopted scale of acidity.] Three soil reactions are recognized: neutral, acid, and alkaline. A soil is considered neutral if its pH is in the range of 6.5 to 7.5. Acidic soils have pHs below 6.5 and alkaline soils have pHs above 7.5.

Acidic conditions are found especially in soils in which organic matter accumulates or in regions where precipitation is high. An arable soil with a low pH (e.g., pH 4.5) is likely to have several undesirable features associated with it. As indicated in the legend for Figure 8-6, aluminum hydroxide (and also the hydroxides of iron and manganese) is solubilized at low pH reactions. Thus the concentrations of aluminum, iron, and manganese ions may be increased to the point that they become toxic to plants. Increases in aluminum ions in acid soils will also result in a decrease in available phosphate because insoluble aluminum phosphate will form. Moreover, an acid soil is apt to have a low content of exchangeable calcium and magnesium. These two ions will be desorbed from the clay–humus complex by hydrogen and aluminum ions (see Figure 8-6) and will be deposited into the soil solution, where they may be leached away (at least in part) by percolating water. These and other unsatisfactory soil conditions associated with acidic soils can usually be corrected by neutralizing the soil by liming (see Figure 8-6).

Alkaline soil conditions are generally found in arid and semiarid regions. In contrast to an acid soil, an arable alkaline soil (e.g., one with a pH higher than 7.5) probably has relatively large amounts of calcium and magnesium and little or no ionized aluminum (see Figure 8-6). But the availability of heavy metals, such as copper, iron, manganese, and zinc, will be markedly decreased in alkaline soils; these elements are likely to be fixed as relatively insoluble hydroxides. Furthermore, an alkaline soil is apt to be low in available phosphate because soluble phosphate will react with excess calcium to form insoluble calcium phosphate. These and other unsatisfactory soil conditions associated with alkaline soils can usually be corrected by lowering the soil pH. Elemental powdered sulfur is the chemical fertilizer used most often to render a soil more acidic. Elemental sulfur is absorbed and metabolized by certain micro-organisms present in most soils; during this process sulfate and hydrogen ions are produced.

Although most plants grow best in soils with a neutral or slightly acid reaction, a few species thrive in acid soils. Included are "acid-loving" species, such as cranberry, blueberry, azalea, heather, and rhododendron. On the other hand, a few plant species (e.g., alfalfa, sweet clover) grow best in slightly alkaline soils.

The ability of a plant species to grow better at either an acid or alkaline rather than a neutral reaction is based on the response of the plant to special soil conditions associated with these soil reactions rather than the hydrogen ion concentration itself. Acid soils will be a better growth medium for those plant species that are better able to tolerate the higher concentrations of available iron, manganese, and aluminum usually present in these soils or to absorb sufficient phosphate in spite of its decreased availability.[9] In alkaline soils the availability of calcium is increased whereas the availabilities of copper, iron, manganese, and zinc are reduced. Plant species that require relatively large amounts of calcium are likely to grow well in alkaline soils. Also, certain plant species that grow well in alkaline soils may do so because they have an inherent superior ability to absorb copper, iron, manganese, and zinc when only limited supplies of these elements are available.

In addition to inorganic salts, organic fertilizers (e.g., composts and manures) are occasionally applied to soils. Organic fertilizers are used not because higher plants have specific requirements for organic chemicals but to improve the physical structure of soils by promoting retention of moisture during droughts and better drainage in wet weather. An additional feature of organic fertilizers is their slow breakdown through the activity of soil microbes. Certain breakdown products—organic and inorganic molecules and ions of low molecular weight—are released gradually into the soil solution. Consequently, essential elements are made available slowly to plants, over a

[9] The presence of mycorrhiza on roots of plants growing in natural habitats may result in the increased availability of certain nutrients (e.g., phosphorus) to the plant (see p. 255).

period of weeks or months. In contrast, inorganic fertilizers, being very soluble in water, find their way into the soil solution almost as soon as they are added to soils.

PLANT ANALYSIS AS A GUIDE TO THE NUTRITIONAL STATUS OF PLANTS

The cultivation of plants, especially high-priced field crops, requires that soil nutrients be available in sufficient quantities at appropriate stages of plant development. Unless sufficient amounts of all the essential nutrient ions are present in available forms in soils, crop yields will be limited. Because the amounts of these nutrients are limited in most arable soils under most natural conditions, soil fertility can be controlled by supplying, as fertilizers, those nutrient ions that limit plant growth, provided that their identities are recognized.

One way to increase yields of a crop is to carry out systematic investigations of fertilizer needs. Generally investigations of this type are performed by governmental agencies (e.g., agricultural experiment stations and substations). Data obtained regarding a schedule of fertilizing (i.e., information regarding which nutrient to apply to the soil, when, and how much) can be used by growers in the immediate neighborhood.

In addition, two types of tests are available to evaluate the chemical fertilizer requirements of crop plants: soil analysis and plant analysis.

Soil analysis is a method of assessing the total and available mineral nutrient reserves in a soil by measuring nutrient concentrations in soil samples. Through soil analyses, the fertilizer needs of a soil can be evaluated, provided a relationship between the amount of extractable nutrient obtained through the soil test and crop yield has been established by appropriate field experiments. But the actual amounts of mineral nutrients absorbed by plants from a soil cannot be determined by soil analysis. This type of information can be obtained only by chemical analysis of plant tissues. *Plant analysis* is a method of assessing the nutritional status of plants by measuring mineral nutrient concentrations in tissue samples. Because leaf tissues (especially the petioles of recently matured leaves) are often selected, plant analysis is also called foliar analysis or leaf analysis. The analytical procedures used in soil and plant analyses (e.g., atomic absorption spectrophotometry) were considered briefly on pp. 232–234.

Plant analysis is based on establishing a relationship between the growth of a crop plant and the concentration of a given mineral nutrient in tissue samples. To establish this relationship, plants of a given crop are grown in a soil or nutrient solution in which all nutrients except the one under examination are well supplied. The deficient nutrient is furnished in increasing quantities to different sets of plants. As a result, plant growth is increased until the requirements of the plant for the deficient nutrient are met. After harvesting the different sets of plants and weighing the crop yields, tissue samples are analyzed chemically for the nutrient being examined. Typical results are shown in Figure 8-7.

For each crop and each nutrient, a *critical concentration* exists: Plants that

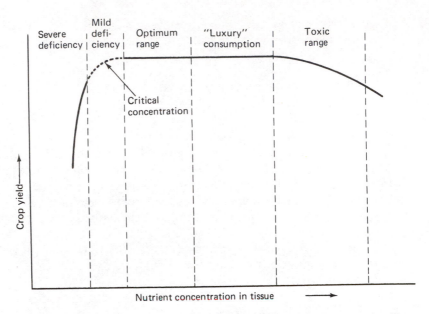

Figure 8-7 A schematic representation of the empirically determined relationship between crop yield and tissue concentration of a nutrient. (See text for details.) (After Smith, 1962; Figure 1.)

contain concentrations slightly lower than the critical concentration are less productive than plants that contain slightly higher concentrations. This critical concentration—the concentration between the *severely deficient zone* and the *optimum range* (see Figure 8-7)—is actually a narrow transition zone rather than an exact point. The *critical concentration* can be defined as the minimal concentration of a nutrient that is present in a plant when maximal growth is achieved.

In practice, several curves like the one in Figure 8-7 are established for a given crop plant, one curve for nitrogen, another for potassium, and a third for phosphorus. Usually these three curves suffice for most crops, but sometimes additional curves for other nutrients may be needed. Once the curves are established, tissue samples of the crop are collected and analyzed at frequent intervals during the growing season. By comparing the concentration of a nutrient in the tissue sample with the concentrations in a curve like the one in Figure 8-7, it is possible to evaluate the nutritional status of the crop plant. Thus plant analysis can serve as a guide to the chemical fertilizer needs of the soil in which the crop is grown. Corrective measures can be taken before potential mineral deficiencies seriously reduce the yield and quality of the crop.

The two methods, plant and soil analysis, are complementary. Each has a role in determining the nutrient situation for a given crop in a given soil. Plant analysis, on the one hand, is based on the premise that the concentration of a nutrient within a plant is the best measure of the ability of the plant to extract the nutrient from the soil. Only the nutritional status of the plant at the time of sampling is revealed, however; the time lag between plant analysis and nutrient absorption from the soil is ignored. Furthermore, when current conditions are corrected, little guidance is provided on what may next prove limiting. Soil analysis, on the other hand, reveals the status of nutrient supply in the soil and tells which nutrients are in low, or high, supply. A soil analysis will often reveal the exact soil conditions that cause a nutritional deficiency, as well as indicating the appropriate corrective measures to be applied.

Plant analysis may also be helpful in diagnosing a mineral nutrient deficiency. The critical concentration in Figure 8-7 represents a threshold concentration below which plants show visual symptoms of deficiency. Thus in cases in which visual symptoms alone are not sufficient to pinpoint a deficient element, plant analysis can be used to verify the specific deficiency.

FOLIAR NUTRITION

In modern agricultural practice various chemicals in solution or aqueous suspension are sprayed onto the foliage of crop plants. Fungicides and insecticides have been applied for almost a century. Also, foliar absorption of plant growth regulators and herbicides is utilized more and more frequently, especially during the last decade or two, with the object of accelerating or retarding and otherwise modifying plant growth and development. Still other foliar sprays include the antitranspirants described on pp. 391–392. In addition, mineral nutrients may be supplied as foliar sprays.

Foliar nutrition is a useful method of fertilizing certain crop plants that can tolerate the aerial spray without damage. Tolerant plants (e.g., orchard trees) normally have a heavy, waxy cuticle layer. The spraying of foliage with nutrients is a method used to deal with special problems that cannot be solved readily by application of chemical fertilizers to the soil.

Foliar nutrition can serve as a means of applying supplemental macronutrients during critical growth periods when it is impractical to apply fertilizers to soils, perhaps because of an unseasonal period of dry weather. This process may offer a remedy in situations in which the time lag between soil application of fertilizers and plant absorption may be too long to satisfy the needs of fast-growing, annual crops during periods of intense growth. Foliar nutrition may also serve as a means of applying micronutrients to crops. Although micronutrient elements can be supplied to plants by addition of chelated salts to the soil, such addition may not be very effective in some cases (perhaps due to special soil conditions that induce hydrolysis or destruction of the chelate) or may not be feasible (e.g., in the case of orchard trees, which generally have deep roots).

The effectiveness of foliar nutrition depends on the ability of nutrients first to penetrate through the outermost plant surface (the cuticle) and the outer cell walls of epidermal cells and then to be transported beyond the epidermal cells. Penetration into leaves may occur through stomata, of course, but no satisfactory evidence presently exists that this is the favored pathway. It is more probable that penetration of nutrients occurs by movement through the cuticle (i.e., the layer of polymerized waxy esters and hydrocarbons that covers the outer surface of epidermal cells of leaves).

Penetration may be facilitated by breaks, fissures, and discontinuities in the cuticular layer. In some plants tiny wax rodlets project outward from the cuticle proper (see Figure 2-15) and prevent the cuticle from being wetted. This difficulty can be partly overcome by using detergents or other substances that reduce the surface tension of the spray solution.

Once having penetrated the cuticle, further penetration of nutrients through the outer epidermal cell wall probably occurs primarily through thread-like, sub-microscopic structures called ectodesmata. *Ectodesmata* have been identified, at least tentatively, as interfibrillar spaces filled with liquid excretion products of protoplasts of epidermal cells; they can be observed, in electron micrographs of epidermal cell transsections, to extend through the outer epidermal cell wall to the inner surface of the cuticle (which they do not perforate). Mineral ions dissolved in water gain access to the protoplasts of epidermal leaf cells by mechanisms similar to those operating in root cells (see Figure 10-3 and related text discussion; also see Chapter 11).

Nutrient uptake by leaves has its counterpart in nutrient loss by leaching. *Leaching* refers to the removal of substances from plant parts by washing with water. Under field conditions rain, snow, dew, mist, and fog are responsible for losses of mineral nutrients from plants. Losses of mineral nutrients for a 24-hour period have been estimated to be as high as 25% of the mineral content of a leaf under certain conditions. Organic substances may also be leached away by rainwater. Sugars, amino acids, plant growth regulators, and organic acids have been detected among leached materials. Older leaves are subject to considerably more leaching than younger ones. Substances leached from above-ground parts may be reabsorbed by roots of the same or nearby plants. Such cycling may be significant in certain ecosystems.

HETEROTROPHIC NUTRITION IN HIGHER PLANTS

Earlier we noted that almost all higher plants have an autotrophic mode of nutrition. Cells and tissues without chlorophyll (and so without the ability to carry on photosynthesis), however, are present in the plant body of all autotrophic higher plants. The root system, for example, is heterotrophic: Roots depend on shoots for a supply of sugars. Meristems and floral parts are heterotropic: Sugars are obtained from nearby leaves. Embryonic root and shoot systems of germinating seedlings are heterotrophic: Sugars are obtained from food reserves in the endosperm or cotyledons. (Although most seeds have adequate food reserves for germination, the tiny, dustlike seeds of orchids are well-known exceptions. These seeds have almost no food reserves and will germinate only when supplied with an external source of sugar.)

Heterotrophic tissues are present even in the *leaves* of some species of higher plants. In variegated leaves chlorophyll is absent in some regions but not in others. Regions in which chlorophyll is absent depend for their supply of sugars on adjoining chlorophyllous regions.

Certain mutant seeds will produce albino seedlings that are totally devoid of chlorophyll. Such seedlings will die when endosperm and cotyledonary food reserves are exhausted. It is possible to grow albino seedlings to maturity, however, by artificially supplying organic nutrients, such as sucrose, through leaf tips (Spoehr, 1942). In effect, such a plant grows by heterotrophic nutrition.

Various higher plant species carry on a heterotrophic mode of nutrition insofar as they obtain part or all of their organic nutriment from outside sources. Some carry on photosynthesis but at rates inadequate to meet all their needs. A few are completely devoid of chlorophyll. Still others are capable of an autotrophic mode of nutrition but carry on a partial type of heterotrophism in their natural habitats. Higher plant species that utilize an outside supply of organic nutriment are generally classified as parasitic or saprophytic.

A heterotrophic higher plant that obtains all or part of its organic nutrient requirements from living tissues of another plant (the host) with which it maintains physical contact is called a *parasite*. In one type of parasitism a higher plant lives in association with another higher plant. Organic substances (e.g., sugars) pass from host plant to parasite through specialized absorptive organs (suckers) that grow out from the parasite and penetrate the host tissues. Examples of such parasites include mistletoes, sandalwoods, broomrapes, figworts, and dodders.

The most widespread type of parasitism in higher plants is the symbiosis with fungi. (A *symbiosis* refers to a partnership of two organisms in which each partner benefits.) Of special concern to the nutrition of higher plants are *mycorrhizae* (singular, *mycorrhiza*) (literally, fungus root); the term designates a symbiotic association between the roots of a higher plant and certain fungal species. The filaments and threads of some mycorrhizal fungi occur as a feltlike sheath or mantle around root cortical cells; other mycorrhizal fungi penetrate into living root cortical cells and fungal threads emanate from the infected root to form a loose network in the adjacent soil.

Mycorrhizal fungi are known to occur naturally on roots of forest trees and food crops. They are probably present on all soil-grown plants. In soils low in available mineral nutrients there is an increased absorption of nutrients by plants infected with mycorrhiza, especially in soils low in available minerals. This increase is usually attributed to increased soil exploration by an increased surface area of the root fungus association. Also, it is possible that the fungus has the ability to dissolve and absorb certain minerals (e.g., phosphorus) that the plant roots cannot readily obtain (Englander, 1981), due perhaps to an unfavorable soil pH [e.g., low levels of available phosphate in very acidic soils (pp. 249–250)]. Higher plants that live naturally in association with mycorrhizal fungi are able to grow very well under laboratory conditions in which the fungus is absent, of course, provided that inorganic nutrients are available in sufficient quantities.

Research in mycorrhizae will probably assume increasing importance in coming years, because the ever-increasing cost of fertilizers will make it impossible to continue to use fertilizer chemicals in a wasteful fashion. Thus it will become imperative to increase the efficiency of nutrient absorption by forest trees and crop plants with judicious management and selection of efficient mycorrhizal fungi.

A few heterotrophic higher plants obtain all or part of their organic nutrients from soils rich in humus; these plants are called *saprophytes*. The coral-root orchid of

European forests and the Indian pipe of the North American forests generally are regarded as examples of saprophytes. The subterranean root tissues of saprophytic vascular plants are associated with fungi; fungal threads and filaments penetrate the forest litter and digest decayed organic matter. Whether the saprophyte obtains organic substances directly from the soil or from the fungus has not yet been clarified.

Another example of heterotrophic nutrition is tissue culture. *Tissue culture* refers to the cultivation of isolated organs, tissues, or cells of a plant on prepared media in sterilized glass containers. This technique is described in Chapter 17.

REFERENCES

ALTMAN, P. L., and D. S. DITTMER, eds. 1968. *Metabolism*. Federation of American Societies for Experimental Biology, Bethesda, Maryland.

ARNON, D. I., and P. R. STOUT. 1939a. The essentiality of certain elements in minute quantity for plants with special reference to copper. *Plant Physiol*. 14:371–375.

ARNON, D. I., and P. R. STOUT. 1939b. Molybdenum as an essential element for green plants. *Plant Physiol*. 14:599–602.

BAZELL, R. J. 1971. Arid land agriculture: Shaikh up in Arizona research. *Science* 171:989–990.

BOLLARD, E. G., and G. W. BUTLER. 1966. Mineral nutrition of plants. *Ann. Rev. Plant Physiol*. 17:77–112.

BROWNELL, P. R., and C. J. CROSSLAND. 1972. The requirement for sodium as a micronutrient by species having the C_4 dicarboxylic photosynthetic pathway. *Plant Physiol*. 49:794–797.

BROYER, T. C., A. B. CARLTON, C. M. JOHNSON, and P. R. STOUT. 1954. Chlorine—a micronutrient element for higher plants. *Plant Physiol*. 29:526–532.

BUCKMAN, H. O., and N. C. BRADY. 1960. *The Nature and Properties of Soils*, 6th ed. The Macmillan Company, New York.

CLARKSON, D. T., and J. B. HANSON. 1980. The mineral nutrition of higher plants. *Ann. Rev. Plant Physiol*. 31:239–298.

CRANE, F. A., and F. C. STEWARD. 1962. Effects of acute deficiency of specified nutrients on *Mentha piperita L.*, Part V. Pages 91–121 *in* Growth, nutrition, and metabolism of *Mentha piperita L.*, parts I–VIII, memoir 379. Cornell University Agricultural Experimental Station, Ithaca, New York.

ENGLANDER, L. 1981. Rhododendron mycorrhizae. Brooklyn Botanic Garden Record/ Plants and Gardens 36(4):24–27.

EPSTEIN, E. 1965. Mineral metabolism. Pages 438–466 *in* J. Bonner and J. E. Varner, eds. *Plant Biochemistry*. Academic Press, New York.

EPSTEIN, E. 1972. *Mineral Nutrition of Plants: Principles and Perspectives*. John Wiley & Sons, New York.

FASSEL, V. A. 1979. Simultaneous or sequential determinations of the elements at all concentration levels—the renaissance of an old approach. *Anal. Chem*. 51:1290A–1308A.

Geigy Industrial Chemicals, Technical Bulletin. 1972. Sequestrene. Geigy Chemical Corporation, Ardsley, New York.

HEWITT, E. J. 1966. Sand and water culture methods used in the study of plant nutrition. Commonwealth Agricultural Bureaux Technical Communication No. 22, rev. 2nd ed. Farnham Royal, England.

HOAGLAND, D. R. 1949. Fertilizers, soil analysis, and plant nutrition. Circular 367. California Agricultural Experiment Station, Berkeley, Calif.

HOAGLAND, D. R., and D. I. ARNON. 1938. The water-culture method for growing plants without soil. *Calif. Agr. Expt. Sta. Cir. 347*. Berkeley, Calif.

INGRAM, G. 1962. *Methods of Organic Elemental Microanalysis*. Van Nostrand Reinhold Company, New York.

JOHNSON, C. M., and A. ULRICH. 1959. Plant analysis and analytical methods. *Calif. Agr. Expt. Bull. 766*. Berkeley, Calif.

KIRKBY, E. A., and A. H. KNIGHT. 1977. Influence of the level of nitrate nutrition on ion uptake and assimilation, organic acid accumulation, and cation-anion balance in whole tomato plants. *Plant Physiol.* 60:349–353.

LEWIN, J., and B. E. F. REIMANN. 1969. Silicon and plant growth. *Ann. Rev. Plant Physiol.* 20:289–304.

MAUGH, T. H. 1977. Electron probe microanalysis: New uses in physiology. *Science* 197:356–358.

MENGEL, K., and E. A. KIRKBY. 1978. *Principles of Plant Nutrition*. International Potash Institute, Berne, Switzerland.

MILLER, E. C. 1938. *Plant Physiology*, 2nd ed. McGraw-Hill Book Company, New York.

MOORE, T. C. 1973. *Research Experiences in Plant Physiology*. A laboratory manual. Springer-Verlag, Inc., New York.

NICHOLAS D. J. D. 1961. Minor mineral nutrients. *Ann. Rev. Plant Physiol.* 12:63–90.

OLSON, R. A., R. B. CLARK, and J. H. BENNETT. 1981. The enhancement of soil fertility by plant roots. *Am. Scient.* 59:378–384.

RAINS, D. W. 1976. Mineral nutrition. Pages 561–597 *in* J. Bonner and J. E. Varner, eds. *Plant Biochemistry*, 3rd ed. Academic Press, New York.

RUEHLE, J. L., and D. H. MARX. 1979. Fiber, food, fuel, and fungal symbionts. *Science* 206:419–422.

SHRIFT, A. 1969. Aspects of selenium metabolism in higher plants. *Ann. Rev. Plant Physiol.* 20:475–494.

SMITH, P. F. 1962. Mineral analysis of plant tissues. *Ann. Rev. Plant Physiol.* 13:81–108.

SPRAGUE, H. B., ed. 1964. *Hunger Signs in Crops, a Symposium*, 3rd ed. Longman, New York.

SPOEHR, H. A. 1942. The culture of albino maize. *Plant Physiol.* 17:397–410.

STEINER, A. A., and H. VAN WINDEN. 1970. Recipe for ferric salts of ethylenediaminetetraacetic acid. *Plant Physiol.* 46:862–863.

THAUER, R. K., G. DIEKERT, and P. SCHONHEIT. 1980. Biological role of nickel. *Trends Biochem. Sciences* 5:304–306.

WALLACE, T. 1951. The diagnosis of mineral deficiencies in plants by visual symptoms. A colour atlas and guide. 3rd ed. H.M. Stationary Office, London.

WETZEL, R., F. C. SMITH, Jr., and E. CATHERS. 1981. Rapid analysis of multiple-ion industrial samples. *Indus. Res. Devel.* 23:152–157.

WILSON S. B., and D. J. D. NICHOLAS. 1967. A cobalt requirement for non-nodulated legumes and for wheat. *Phytochemistry* 6:1057–1066.

WORTHEN, E. L., and S. R. ALDRICH. 1956. *Farm Soils, Their Fertilization and Management*, 5th ed. John Wiley & Sons, New York.

9

Assimilation of Inorganic Nutrients

One of the important principles of plant physiology considered earlier is that a higher plant fabricates all its substance from inorganic nutrients absorbed from its external environment (see Chapter 8). These relatively simple nutrients (CO_2; water; nitrate, sulfate, and phosphate ions; etc.) undergo metabolic transformations to organic plant constituents. This process is referred to as *assimilation* (literally, to make similar).

Our main concern here is the assimilation of inorganic nutrients other than CO_2 and water. CO_2 assimilation was discussed under photosynthesis (Chapter 6). Water is assimilated in a few metabolic reactions: in certain steps of the TCA cycle (Chapter 4) and in hydrolytic reactions (e.g., the breakdown of starch to glucose). In addition, water is a reactant in the light reactions in photosynthesis (Chapter 6). But the bulk of the water absorbed by higher plants is lost in transpiration and therefore not assimilated.

Some inorganic nutrients undergo intensive metabolic transformation during their conversion from inorganic to organic form whereas others do not. Nitrate ion, for example, is first reduced to ammonium ion before conversion to amino acids whereas potassium ion is assimilated simply by relatively weak interionic attractive forces to organic anions already present in plant cells.

In this chapter we are concerned only with the initial steps in the assimilation process—that is, with the metabolic conversion of inorganic nutrients (e.g., nitrate, sulfate, phosphate) to relatively simple organic constituents. The discussion of the assimilation of sulfate ends with its incorporation into the amino acid cysteine. Similarly, the discussion of the assimilation of nitrate ends with its incorporation into glutamic acid, another amino acid. The metabolic conversion of cysteine and glutamic acid to other plant constituents is only of peripheral interest in this chapter.

It seems worthwhile to point out at the outset that a plant cell may not assimilate each and every absorbed nutrient ion. A variable fraction may be deposited in vacuoles and stored there as unassimilated inorganic ions.

Of all the essential elements, nitrogen is most probably in short supply in soils in which crops are grown. Applying nitrogen fertilizers to soils is one of the most important means by which crop yields are increased at the present time. Therefore nitrogen assimilation is considered first.

METABOLIC REDUCTION OF NITRATE

Nitrogen in soils is available to the roots of higher plants mostly as nitrate ion (NO_3^-) and ammonium ion (NH_4^+). Both forms of nitrogen are produced as a result of microbial decomposition of the organic remains of plants and animals. Nitrate ion, being negatively charged, is not bound to clay and humic particles in soils (p. 246). If unabsorbed by plants, nitrate may leach through the soil and into the aquifer (the porous subsurface rock that holds water) or may be washed away into rivers and oceans. Ammonium ion is a cation and so is held by clay and humic particles (p. 246). Thus whether ammonium is produced naturally in soils or introduced intentionally as a fertilizer, little is likely to be lost to the aquifer or the oceans.

Of the two major forms of nitrogen in soils, nitrate rather than ammonia is the principal source of nitrogen to higher plants growing under usual field conditions. In most tillable soils in temperate regions ammonium is transformed by specific soil bacteria into nitric acid. This process is called *nitrification*. Nitrification by soil bacteria is quite rapid: Ammonium fertilizers applied to tillable soils are likely to be transformed to nitric acid (HNO_3) by bacterial action in only a few days. These bacteria excrete nitric acid into the soil, where it dissociates into hydrogen and nitrate ions. Only in poorly drained, unaerated soils, where the specific bacteria that normally convert ammonium to nitrate do not grow well, is ammonium present in relatively large amounts. But even though nitrate is the principal natural source of nitrogen to plants, many plants are able to utilize ammoniacal nitrogen readily, particularly when environmental conditions favor high rates of photosynthesis and vigorous plant growth (see p. 267).

Much of the nitrate absorbed by roots of actively growing higher plants is swept upward in the transpiration stream to leaves. Thus nitrate assimilation in most higher plants occurs mostly in leaves. But whether assimilated in roots or leaves, nitrate is first reduced to ammonia.

The reduction of nitrate to ammonia in higher plants takes place in two sequential reactions. First, nitrate is reduced to nitrite (NO_2^-); in this reaction one atom of oxygen is released as water [see Eq. (9-3)]. Then nitrite is reduced to ammonia; two additional atoms of oxygen are released as water [see Eq. (9-5)]. The reaction sequence from nitrate to ammonia may be summarized as follows:

$$NO_3^- \xrightarrow{\ \ \ \ } NO_2^- \xrightarrow{\ \ \ \ } NH_3 \tag{9-1}$$

Nitrate $\quad H_2O \quad$ Nitrite $\quad 2H_2O \quad$ Ammonia

The first step in this reaction sequence (i.e., the reduction of nitrate to nitrite) is catalyzed by the enzyme nitrate reductase. Nitrate reductase in leaves has been reported present in the cytosol (Hewitt et al., 1976; Beevers and Hageman, 1980). The enzyme is classified as a molybdoflavoprotein because it contains both molybdenum and the flavin coenzyme FAD; each acts as an *electron carrier* (pp. 94–95) in the reaction catalyzed by nitrate reductase [see Eq. 9-2]. The reducing power for the reduction of nitrate to nitrite is supplied by NADH. The source of NADH for nitrate reductase is widely considered to depend mostly on the oxidation of glyceraldehyde-3-phosphate by NAD^+, as indicated in Eq. (4-22) (e.g., see Mann et al., 1978); glyceraldehyde-3-phosphate, it will be recalled, is a member of both the glycolytic pathway (pp. 98–101) and the oxidative pentose phosphate pathway (pp. 101–104).[1] The pathway of electron transfer from NADH proceeds through FAD and molybdenum as follows:

$$\text{NADH} \diagdown \text{FAD} \rightarrow \text{Mo}^{5+} \diagdown \text{NO}_3^- $$
$$\text{NAD}^+ \diagup \text{FADH}_2 \diagup \text{Mo}^{6+} \diagup \text{NO}_2^- \qquad (9\text{-}2)$$

Eq. (9-2) emphasizes the direction of electron flow rather than the actual numbers of hydrogen atoms and electrons transferred from one component to another. Some evidence exists that the nitrate reductase molecule may consist of an additional electron carrier (perhaps a form of cytochrome *b*) that is situated between FAD and molybdenum (Sherrard and Dalling, 1979), but this is not shown in Eq. (9-2).

The overall reaction catalyzed by nitrate reductase [i.e., Eq. (9-2)] may be written

$$NO_3^- + NADH + H^+ \longrightarrow NO_2^- + NAD^+ + H_2O \qquad (9\text{-}3)$$

Because nitrate reductase contains molybdenum, it is apparent that one role of molybdenum in higher plants is to facilitate the assimilation of nitrate. It can be expected that the activity of nitrate reductase will be impaired in plants supplied with less than adequate amounts of molybdenum and therefore reaction (9-3) will not proceed efficiently. In fact, if plants are supplied with nitrate as the sole source of nitrogen *and* if molybdenum is available in less than adequate amounts, then the plants are likely to develop marked symptoms of *nitrogen* deficiency even before they develop symptoms of molybdenum deficiency. These plants will absorb and accumulate nitrate, but they will not be able to assimilate nitrate. Reaction (9-3) becomes a "bottleneck" in the assimilation of nitrate.

The second step in nitrate assimilation is the reduction of nitrite to ammonia (NH_3). In green leaves, the reaction occurs in chloroplasts. The reducing power for nitrite reduction, especially in leaves low in starch reserves, is generally derived directly from light reactions in photosynthesis. That is to say, a portion of the reducing power generated by light reactions in chloroplasts is diverted away from the reduction of CO_2 in photosynthesis and is used instead for the reduction of nitrite to ammonia.

---

[1]NADH for nitrate reduction may also be generated from cytosolic NAD^+ by an oxaloacetic acid-malic acid shuttle system that transports reducing equivalents (in the form of malic acid) from mitochondria into the cytosol (Woo et al., 1980); for further details of this shuttle system, see pp. 326–327.

Alternatively, nitrite reduction to ammonia may occur in chloroplasts in the dark, provided that starch reserves are present in the chloroplasts (Kow et al., 1982). The first type of nitrite reduction will be discussed in the next several paragraphs, the second type will be considered briefly in a later paragraph (see p. 262).

The immediate source of reducing power for nitrite reduction in green leaves is a nonheme iron-sulfur protein of low molecular weight. This substance is known as ferredoxin; it has been isolated and purified. Ferredoxin is one of the components of the photosynthetic electron transfer system (Figure 6-10). Iron-sulfur proteins similar to ferredoxin are also present in the respiratory chain (pp. 111–113).

It will be recalled that ferredoxin is reduced in the light reactions in chloroplasts, and then reduces $NADP^+$ to NADPH in photosynthesis (see Figure 6-10 and related discussion in the text). Reduced ferredoxin also may be used to reduce nitrite to ammonia. The pathway of electron flow from photosynthetic light reactions to nitrite is

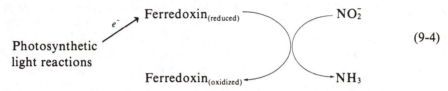

$$(9\text{-}4)$$

The enzyme that catalyzes the reduction of nitrite to ammonia [see Eq. (9-4)] is called nitrite reductase. The reaction catalyzed by this chloroplastic enzyme [i.e., Eq. (9-4)] can be written

$$NO_2^- + 6\,e^- + 7\,H^+ \longrightarrow NH_3 + 2\,H_2O \qquad (9\text{-}5)$$

Here six equivalents of electrons (in the form of reduced ferredoxin) and seven equivalents of hydrogen ion are consumed in the reduction of 1 mole of nitrite to ammonia. Probably a sequence of one or two electron transfers occurs in this reaction and a series of free radicals (none of which is released from the enzyme) is generated during the course of nitrite reduction.

In contrast to leaves, ferredoxin does not participate in nitrite reduction in nongreen tissues (e.g., roots). Instead a still-unidentified electron carrier with properties similar to those of ferredoxin is involved. The reducing power for nitrite reduction in roots is generated in respiratory metabolism, not photosynthetic light reactions.

C_4 plants have been reported to use nitrate more efficiently than C_3 plants. To determine why, the distribution of nitrate reductase and nitrite reductase in C_4 leaf cells was investigated recently (Moore and Black, 1979). It was found that these two enzymes are present only in mesophyll cells, not bundle sheath cells. So it was suggested that nitrate translocated upward in the transpiration stream in xylem tissue moves across leaf bundle sheath cell into mesophyll cells, where it is reduced first to nitrite [Eq. (9-3)] and then to ammonia [Eqs. (9-4) and (9-5)]; in effect, the reducing power supplied in light reactions in mesophyll cell chloroplasts is used for nitrite reduction to ammonia. Because CO_2 reduction in C_4 plants takes place in bundle sheath cells, not mesophyll cells (p. 205), none of the reducing power supplied by light reactions in mesophyll cells needs to be used for CO_2 reduction; that is, reducing power supplied by light reactions in bundle sheath cells is used for CO_2 reduction whereas reducing power supplied by light reactions in mesophyll cells is used for the reduction of nitrite to ammonia. In effect, the greater efficiency of C_4 plants in the use of nitrate, compared to C_3 plants, is due to the spatial separation of nitrate and nitrite reduction from the site of CO_2 reduction.

Recently it was suggested that reducing power derived from the oxidation of hexose phosphate (e.g., glucose-6-phosphate and substances readily converted to glucose-6-phosphate) may be transferred to nitrite with ammonia as the end product (Kow et al., 1982). This type of nitrite reduction may occur *in the dark*, at the expense of starch reserves (if available) that are degraded to glucose-6-phosphate. Oxidation of glucose-6-phosphate in chloroplasts was illustrated earlier (Table 4-3 and related text discussion). The sequence of reactions in Table 4-3 leads to the production of reduced coenzyme (i.e., NADPH), which is able to reduce ferredoxin in the presence of NADP-ferredoxin oxidoreductase. [This enzyme catalyzes a reversible reaction between $NADP^+/NADPH$ and ferredoxin (oxidized)/ferredoxin (reduced).] Then reduced ferredoxin reduces nitrite to ammonia, as shown in Eq. (9-4); but in this type of nitrite reduction, reducing power is derived from the oxidation of glucose-6-phosphate, not from light reactions.

ASSIMILATION OF AMMONIA

Ammonium ion (NH_4^+) and ammonia (NH_3) are conveniently grouped together as ammonia-nitrogen. Both of these inorganic forms of nitrogen are interconvertible in water solution.

$$NH_3 + H_2O \rightleftharpoons NH_4^+ + OH^- \tag{9-6}$$

Ammonia-nitrogen (or more simply, ammonia) may be produced in plant cells in one of several different ways. The reduction of nitrate proceeds through nitrite and results in the formation of ammonia, as already noted in Eq. 9-1. Ammonia is produced in the photorespiratory carbon oxidation cycle (see Figure 7-4 and related discussion in text). Or ammonia may be produced as a result of the metabolic degradation of reserve proteins during the germination of certain seeds. Another way in which ammonia may arise in plant cells is through the absorption and metabolic decomposition of urea (NH_2CONH_2), a substance often used as a nitrogen fertilizer. Decomposition of urea is catalyzed by urease, an enzyme present in many plants.

$$NH_2CONH_2 + H_2O \longrightarrow 2 NH_3 + CO_2 \tag{9-7}$$

Ammonia is also produced in plant cells in which nitrogen fixation occurs. In addition to these several ways of producing ammonia, plants may absorb ammonia directly from a soil solution, perhaps when ammonium salts or anhydrous ammonia are added to the soil as fertilizers.

Ammonia is assimilated rapidly. In most plant tissues only trace amounts are detectable. In fact, ammonia in more than trace amounts is toxic to most plants. Only a few plants—those with very acidic vacuolar contents (e.g., *Begonia* and rhubarb)—can store relatively large amounts of ammonium ion in their vacuoles, as ammonium salts of organic acids.

The assimilation of ammonia in most higher plants occurs primarily into glutamic acid. First, glutamic acid itself is utilized in the synthesis of an amide (i.e., an addition production of an amino acid and ammonia) called glutamine. ATP is required and the enzyme is glutamine synthetase [see reaction (9-8)]. Second, glutamine reacts with α-ketoglutaric acid (a TCA cycle acid) to form glutamic acid (an amino acid), as shown in reaction (9-9). The enzyme is glutamic acid synthase. A reducing agent is required. In green leaves reduced ferredoxin generated in photosynthetic light reactions is used and reaction (9-9) occurs in chloroplasts. In roots NADPH is the reducing agent and the reaction occurs in (colorless) plastids.

The two reactions [(9-8) and (9-9)] and the summary reaction (9-10) are

$$
\begin{array}{c}
\text{COOH} \\
|\\
\text{CHNH}_2 \\
|\\
\text{CH}_2 \\
|\\
\text{CH}_2 \\
|\\
\text{COOH}
\end{array}
+ \text{NH}_3 + \text{ATP} \xrightarrow[\text{synthetase}]{\text{glutamine}}
\begin{array}{c}
\text{COOH} \\
|\\
\text{CHNH}_2 \\
|\\
\text{CH}_2 \\
|\\
\text{CH}_2 \\
|\\
\text{CO(NH}_2)
\end{array}
\begin{array}{c}
+ \text{ADP} \\
+ \text{Pi}
\end{array}
\qquad (9\text{-}8)
$$

Glutamic acid   Glutamine

$$
\text{Glutamine} +
\begin{array}{c}
\text{COOH} \\
|\\
\text{C}=\text{O} \\
|\\
\text{CH}_2 \\
|\\
\text{CH}_2 \\
|\\
\text{COOH}
\end{array}
\xrightarrow[\text{glutamic acid synthase}]{+ (2\,\text{H})} 2\ \text{glutamic acid}
\qquad (9\text{-}9)
$$

α-Ketoglutaric acid

$$(9\text{-}10)$$

The net result of Eq. (9-10) is the synthesis of glutamic acid (from ammonia and α-ketoglutaric acid) through the agency of glutamine. In Eq. (9-10) glutamic acid is both the NH_3 acceptor and the product of NH_3 assimilation. In summary, the sequence of assimilation of ammonia in most higher plants is $NH_3 \longrightarrow$ glutamine $\longrightarrow$ glutamic acid.

Only trace amounts of ammonia are translocated upward in the transpiration stream in the xylem even when ammonia fertilizer is applied to soils. Instead most of the ammonia absorbed by roots is likely to be incorporated in root cells into glutamine [Eq. (9-8)]; glutamine is translocated in the xylem to leaves, where Eq. (9-9) occurs. In effect, the two steps of ammonia assimilation [i.e., Eqs. (9-8) and (9-9)] may be separated in space.

Glutamic acid (i.e., the product of ammonia assimilation) may be used as a starting material for the biosynthesis of amino acids other than glutamic acid and for the biosynthesis of nucleic acids (Miflin and Lea, 1977). The first step in these biosyntheses is a *transamination* reaction. In a transamination reaction (catalyzed by enzymes known as *aminotransferases* or *transaminases*) an amino group ($-NH_2$) is transferred from an amino donor compound to the carbonyl position ($=CO$) of an amino acceptor compound. An amino acid is usually the donor compound and a 2-keto acid is the amino acceptor. A transamination reaction results in the formation of a new amino acid and the keto acid analog of the amino acid that served originally as the amino donor. Most transamination reactions are reversible [except two already considered—Eqs. (7-7) and (7-8). A well-known example of a transamination reaction is that involving glutamic acid and oxaloacetic acid (a member of the TCA cycle).

$$
\begin{array}{c}
\text{COOH} \\
| \\
\text{C}=\text{O} \\
| \\
\text{CH}_2 \\
| \\
\text{COOH}
\end{array}
+ \text{glutamic acid}
\xrightleftharpoons[\text{acid aminotransferase}]{\text{glutamic acid-oxaloacetic}}
\begin{array}{c}
\text{COOH} \\
| \\
\text{CHNH}_2 \\
| \\
\text{CH}_2 \\
| \\
\text{COOH}
\end{array}
+ \alpha\text{-ketoglutaric acid} \quad (9\text{-}11)
$$

Oxaloacetic acid   Aspartic acid

The nitrogen-containing product of this reaction, aspartic acid, is a four-carbon amino acid.

In conclusion, a few words should be said about asparagine. Asparagine, like glutamine (p. 263), is an amide present in plant cells. It is synthesized from glutamine and aspartic acid, as shown in reaction 9-12. (The synthesis of aspartic acid was shown in reaction 9-11.)

$$
\text{Glutamine} + \text{aspartic acid} + \text{ATP}
\xrightarrow[\text{synthetase}]{\text{asparagine}}
\begin{array}{c}
\text{COOH} \\
| \\
\text{CHNH}_2 \\
| \\
\text{CH}_2 \\
| \\
\text{CONH}_2
\end{array}
\begin{array}{l}
+ \text{glutamic acid} \\
\\
+ \text{AMP} \\
\\
+ \text{PPi}
\end{array}
\quad (9\text{-}12)
$$

Asparagine

Here AMP denotes adenosine monophosphate (see p. 55) and PPi denotes pyrophosphoric acid ($H_4P_2O_7$), an inorganic diphosphoric acid. Reaction (9-12), together with reaction (9-8), is another pathway for the assimilation of NH_3 in addition to that shown in reaction (9-10). This second pathway of NH_3 assimilation can be summarized as follows:

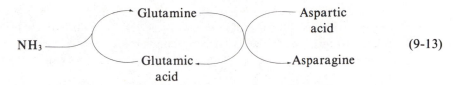

$$(9\text{-}13)$$

NH_3 assimilation into asparagine [reaction (9-13)] is the dominant pathway in legume root nodules that fix atmospheric N_2; on the other hand, NH_3 assimilation into glutamic acid [reaction (9-13)] is the dominant pathway in most higher plants (Miflin and Lea, 1976a and b).

The two amides, glutamine and asparagine, are apt to accumulate in tissues of healthy plants at times when relatively large amounts of ammonia are supplied as soil fertilizers to plants. In this regard, glutamine and asparagine are metabolic "reservoirs" for the temporary storage of excess ammonia. Also, these two amides are important in the translocation of nitrogen through the plant in both xylem and phloem. As noted (pp. 263–264), most plants, when fertilized with ammonia, assimilate it largely into glutamine in roots; glutamine (and to a lesser extent asparagine and amino acids, such as glutamic acid) is translocated in the xylem to leaves. Alternatively, when plants absorb nitrate from soils, nitrate is translocated upward to the xylem to leaves. Leaf cells reduce nitrate to ammonia [Eq. (9-1)] and assimilate ammonia into glutamine [Eq. (9-10)] and asparagine [Eq. (9-13)]. These two amides, together with smaller amounts of amino acids, such as glutamic acid, are the forms in which nitrogen is translocated in the phloem to other parts of the plant.

EFFECTS OF NITROGEN ASSIMILATION
ON CARBOHYDRATE UTILIZATION

Although nitrogen is required by plants in larger amounts than other essential mineral elements (see p. 240), many soils are deficient in available nitrogen. Thus nitrogen fertilizers generally are better than all others with respect to their ability to increase crop yields. The inorganic nitrogen fertilizers most often applied to soils are various chemical forms of nitrate-nitrogen (e.g., potassium nitrate) and ammonia-nitrogen (e.g., anhydrous ammonia, ammonium nitrate).

One important factor in the use of nitrogen fertilizers is their effect on the utilization of carbohydrates in plants. In general, these effects can be summarized as follows. When nitrogen fertilizers are supplied in large amounts, the level of carbohydrates will decrease. But when the supply of nitrogen is curtailed (e.g., by withholding nitrogen fertilizers), carbohydrate levels in plants will increase.

This interdependence of nitrogen supply and carbohydrate utilization can be readily understood in terms of the metabolic requirements of nitrogen assimilation considered earlier. When nitrogen is incorporated into amino acids [see Eq. (9-10)], intermediates of the TCA cycle are consumed. But the continued operation of the TCA cycle requires that these intermediates be replenished. Replenishment occurs at the expense of the utilization of carbohydrates and their derivatives [see Eq. (4-33) and related discussion]. These latter compounds represent the "carbon skeletons" for the biosynthesis of amino acids. Thus the assimilation of inorganic nitrogen represents a drain on carbohydrate reserves.

There is still a second metabolic reason why nitrogen assimilation affects the utilization of carbohydrates. We saw earlier that reducing power is required for the reduction of nitrate and nitrite ions; reducing power is also required for the assimilation of ammonia. If reducing power is generated in respiration, then it will be at the expense of the oxidation of carbohydrates and their derivatives. If reducing power is derived from the light reactions in photosynthesis, then the total photosynthetic machinery will be deprived to some extent of the opportunity to reduce CO_2 to carbohydrates. In effect, the need for reducing power in the assimilation of inorganic nitrogen represents a drain on already existing carbohydrates and/or carbohydrates that might otherwise be synthesized.

The fact that the assimilation of inorganic nitrogen intensifies the utilization of carbohydrate reserves in plants has an important effect on nitrogen fertilizer practices in the production of certain crops. To give a specific example, consider the nitrogen fertilizer needs of celery. Because celery stalks are most edible when soft, it is desirable that the cells be thin walled rather than thick walled. This type of growth can be promoted by application of nitrogen fertilizers. The added nitrogen is incorporated into amino acids, which are subsequently metabolized to proteins. Thus carbohydrate reserves in the celery stalks are depleted. Consequently, the cell walls (predominantly carbohydrates and their derivatives) will be thin rather than thick and the celery stalks will be soft and edible.

To give a specific example of when *not* to fertilize a crop with nitrogen, consider the sugar beet. This is a crop grown in temperate regions; the time from planting to harvest is about 6 months. At harvest time the roots (beets) are sent to the mill, where the sugar is extracted. To produce beets with high sugar content, heavy applications of nitrogen fertilizers must be avoided during the period of several weeks prior to harvest. If heavy applications of nitrogen were made at this late date, their assimilation into amino acids and proteins would result in a decrease in sugar content at harvest time.

Nitrogen fertilizers have a greater effect on shoot growth than root growth. When heavy applications of nitrogen fertilizers are applied to field crops, the leaves are likely to engage in a more intensive utilization of carbohydrates than the roots; the bulk of nitrogen applied to roots [whether nitrate in the case of nitrate fertilizers or glutamine produced by the root system from ammonia fertilizers (see pp. 263–264)] is translocated rapidly via the transpiration system out of roots and into leaves, where carbohydrates are manufactured in photosynthesis. Assimilation of nitrate into amino acids (or conversion of glutamine into amino acids) results in a drain on carbohydrates reserves in leaves. This drain will be reflected in a decrease in the amount of sugars available for translocation via the phloem to the roots. Because the supply of sugars from leaves is a major factor in limiting growth of the root system of a plant, the growth of the shoot system (i.e., leaves and stems) will be favored over that of the root system when heavy applications of nitrogen fertilizers are applied to soils. Thus the *shoot-root ratio* of a plant (i.e., the weight of the shoot divided by the weight of the root) will increase when nitrogen fertilizer supplies are abundant.

The fact that the shoot-root ratio can be controlled, at least partly, by nitrogen supply affects nitrogen fertilizer programs. To give only one example, consider the sugar beet again. In the production of this crop it is desirable to promote the elaboration of photosynthetic tissue early in the season. But late in the season it is desirable to restrict the growth of leaf tissue and promote the growth of root tissue. These objectives can be realized by supplying nitrogen fertilizers at planting time and perhaps once again, a month or two later, and by withholding nitrogen fertilizers near the end of the growing season.

A question often considered is whether nitrate or ammonia should be used in a fertilizer program. To answer this question requires an evaluation of several different factors. One concerns the carbohydrate status of the plants.

The level of carbohydrate is apt to be low when plants are grown under conditions of prolonged periods of cool, cloudy days (as might occur during the fall or winter seasons in temperate regions). Under these conditions nitrate is recommended as a fertilizer rather than ammonia because nitrate-nitrogen is assimilated less intensively than ammonia. The assimilation of ammonia in plants with already low carbohydrate levels will result in an even further depletion of carbohydrate reserves. When supplied with ammonia, plants with low carbohydrate levels are likely to become thin walled, soft, and succulent. Excessively large amounts of ammonia supplied as a fertilizer to plants with low carbohydrate levels may even provoke symptoms of ammonia toxicity due to the accumulation of unassimilated ammonia ions. On the other hand, nitrate is a "safer" source of nitrogen for plants growing under conditions that lead to low levels of carbohydrate. Most plants have the ability to store unassimilated nitrate ion in their cell vacuoles without injurious effects.

Ammonia-nitrogen is a satisfactory source of nitrogen only for plants growing under conditions conducive to high rates of photosynthesis—for example, warm, sunny days in the summertime in temperate regions. These growth conditions favor a high carbohydrate status in plants. Assimilation of ammonia-nitrogen by plants growing under these conditions will be rapid and without deleterious effects.

Not only the carbohydrate status of the plants but also the possible effects on the soil must be considered when deciding whether to use ammonia-nitrogen or nitrate-nitrogen as a fertilizer. When applied to soils, ammonia-nitrogen is converted to nitric acid by soil microorganisms, a fact already noted (p. 259). Soil acidity will probably increase. Whether this is an advantage or disadvantage depends on the kind of soil and the intensity of application of ammonia-nitrogen.

Quite apart from the different effects of ammonia-nitrogen and nitrate-nitrogen on plants and soil, the cost factor may also require consideration in a nitrogen fertilizer program. Ammonia-nitrogen (e.g., anhydrous ammonia) has become increasingly popular in recent years as a fertilizer for the intensive cultivation of field crops because it is often somewhat less expensive than other forms of commercially available inorganic nitrogen.

GENERAL ASPECTS OF NITROGEN FIXATION

Molecular nitrogen (N_2) is present as a gas in the atmosphere to the extent of 79% by volume. It is relatively inert chemically. Under suitable industrial conditions or in special biological systems, however, gaseous N_2 is able to react chemically with other substances. Whether the chemical union of N_2 with other compounds occurs industrially or naturally (e.g., as a result of lightning strokes or biologically), the process is referred to as *nitrogen fixation* or *dinitrogen fixation* (the prefix "di" emphasizes that *two* atoms of nitrogen undergo reaction).

At elevated temperatures (about 500°C) and high pressure (about 350 bars) N_2 combines chemically with H_2 to form ammonia. This discovery was made by the German chemist Haber and his associates in the early twentieth century. Ammonia produced by the Haber process is the starting material for the manufacture of a wide variety of industrial and agricultural chemicals. Other forms of fixed nitrogen (e.g., oxides, cyanides, nitrides, and cyanamides) are produced industrially only to a limited extent. A total of about 50 million metric tons of N_2 are fixed annually throughout the world by industrial methods.

In addition to the fixation of N_2 by industrial methods, a larger quantity of N_2, estimated at about 200 million metric tons annually, is fixed by natural processes. Perhaps 5% of this natural fixation occurs as a result of lightning strokes. There are about 44,000 thunderstorms over the earth each day. More than 8 million bright strokes of lightning hit the earth each day; in addition, there are several lightning bolts in the clouds for every one that strikes the earth. Each of these lightning strokes causes water vapor, O_2, and N_2 in the atmosphere to combine to form fixed nitrogen (e.g., nitric acid), which is brought down by rain to the surface of the earth. Trace amounts of nitric acid are also produced in the stratosphere by photochemical reactions of gases (e.g., ozone and nitrogen oxides) found there. Nitric acid produced in the stratosphere diffuses into the upper atmosphere and is eventually brought to the surface of the earth by rain, just as in the case of nitric acid produced by lightning strokes.

The remaining 95% of natural nitrogen fixation occurs through certain microorganisms. This biological nitrogen fixation results in the synthesis of ammonia. Biological N_2 fixation has an extremely important role in nitrogen economy in agricultural practice. It is widely acknowledged that the low levels of available nitrogen usually present in most tillable soils (e.g., nitrate) cannot be overcome on a worldwide scale by commercial fertilizers alone. Nitrogen supply to crop plants must be met in part by biological N_2 fixation. Thus the study of biological N_2 fixation is an important aspect of agricultural science. The potential benefits of biological N_2 fixation can be utilized fully in agriculture only if a comprehensive knowledge of the various factors influencing the process is available. This knowledge will make it possible to discover means of intensifying the fixation of N_2 through biological systems.

The amount of N_2 removed from the atmosphere by biological N_2 fixation is closely balanced by the amount returned to the atmosphere by soil microorganisms that convert organic nitrogen to gaseous N_2. Several steps are involved (see Figure 9-1). First, ammonia produced by biological N_2 fixation is assimilated by plants and

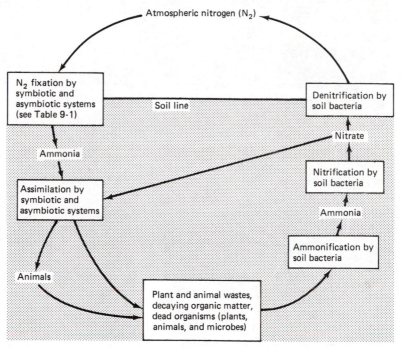

Figure 9-1 The biological nitrogen cycle in terrestrial environments. (See text for details.)

microorganisms into amino acids, proteins, and other nitrogenous products (see left side of Figure 9-1). Some plants may be consumed by animals. When animal wastes and dead plants, animals, and microorganisms are subjected to decay, complex organic nitrogenous compounds undergo decomposition into a number of simpler compounds (e.g., amino acids). Then by a process called *ammonification*, a group of soil bacteria, together with certain fungi, convert amino nitrogen to ammonia. The ammonia thus formed reacts with other chemicals present in the soil (e.g., CO_2 and water) to form ammonium salts (e.g., ammonium carbonate). The next step in the sequence is called *nitrification* (a process mentioned on p. 259). Nitrification is carried out by two kinds of soil bacteria: Bacteria of the *Nitrosomonas* group oxidize ammonia of ammonium salts to nitrite (NO_2^-); bacteria of the *Nitrobacter* group oxidize nitrite to nitrate (NO_3^-), the most readily utilized of all inorganic nitrogen compounds by higher plants. Finally, nitrates in the soil are reduced by certain soil bacteria to N_2 and volatile nitrogen oxides, which escape into the atmosphere. This process is called *denitrification* (see right side of Figure 9-1). Denitrifying bacteria are especially active under anaerobic conditions in wet soils with high organic matter content. Denitrifying bacteria produce not only N_2 but also small amounts of N_2O (nitrous oxide). (The production of N_2O by denitrifying bacteria is not shown in Figure 9-1.) Not only denitrification of nitrate may occur in soils, but soil nitrate may also be lost to rivers and oceans (not shown in Figure 9-1).

The cycle of fixation of gaseous N_2, assimilation of fixed nitrogen by plants and then animals, and return of gaseous N_2 to the atmosphere by denitrifying bacteria is called the *biological nitrogen cycle* (Figure 9-1).

BIOLOGICAL N$_2$ FIXATION

Biological N$_2$ fixation is carried out by two main types of microorganisms: those that are "free living" or asymbiotic (i.e., capable of independent existence) and those that live in symbiosis with other plants. (*Note*: Symbiosis was defined on p. 255.)

The asymbiotic nitrogen fixers can be classified into three groups: aerobic bacteria (mainly of the *Azotobacter* type), anaerobic bacteria (especially those of the genus *Clostridium*), and blue green algae (see Table 9-1). Although the bacteria in the first two groups are found in many soils throughout the world, actually they contribute substantially to the nitrogen content of the soil only under very special soil conditions: a copious supply of decayed plant tissues and high water content. The blue green algae (also known as cyanobacteria) that fix N$_2$ generally consist of chains of cells in long filaments. Occasional cells in the chain are larger than others, have thick cell walls, and are colorless. These cells are known as *heterocysts*—only heterocysts are capable of fixing N$_2$. About 40 species of blue green algae are capable of fixing nitrogen. They seem mainly important in wet tropical soils (e.g., rice fields). Blue green algae use the energy of sunlight for nitrogen fixation, but the mechanism by which light energy is used is only partially understood (Postgate, 1978; Burris, 1978).

Symbiotic N$_2$-fixing systems occur in many vascular plants (Table 9-1). Nearly 200 legumes cultivated as crop or horticultural plants are hosts to N$_2$-fixing microorganisms. In another symbiotic system the host plant is a small water fern *Azolla*. *Azolla* houses colonies of blue green algae that fix N$_2$. This unique symbiotic association is especially important in rice fields; fixed nitrogen (i.e., NH$_3$) leaks out of the fern plant and thereby supplies the nitrogen needs of the rice plant. Also, certain lichens growing on the surfaces of trunks and branches of forest trees are able to fix N$_2$ (see Table 9-1). N$_2$ fixed by lichens may be leached down to the forest floor by rain. In addition, weathering and aging of the bark will cause the lichen to fall to the forest floor, where it decomposes and releases fixed nitrogen to trees.

Symbiotic N$_2$-fixing microorganisms that infect higher plants usually live in small, knoblike protuberances known as *nodules* (Figure 9-2). Nodules are present mostly in roots of host plants. In a few cases, however, nodules occur on leaves of host plants (Table 9-1).

Symbiotic N$_2$-fixing systems, especially those in which root nodules are formed, contribute far more to the nitrogen economy of natural communities and to the fertility of soils than the asymbiotic systems. When calculations are made on a unit of cell material, nitrogen-fixing root nodules can fix 100 to 200 times more N$_2$ than free-living microorganisms, primarily because of the ability of nodules to continue to fix N$_2$ for long periods of time, perhaps for 30 to 40 days. During this time fixed nitrogen[2] is translocated continually (in the transpiration stream of the xylem) from root nodules to other tissues, where it is used in the synthesis of cells and tissues of the host plant. Asymbiotic systems, on the other hand, generally fix no more N$_2$ than can

---

[2]Asparagine rather than glutamine (see p. 264) is the predominant form for translocation of nitrogen in many leguminous plants that are fixing N$_2$ (Miflin and Lea, 1976a and b).

TABLE 9-1 MAJOR NITROGEN-FIXING BIOLOGICAL SYSTEMS

I. Free-living (asymbiotic) microorganisms

Bacteria

Aerobic	Several species of the genus *Azotobacter*
Anaerobic, nonphotosynthetic	Several species of the genus *Clostridium*; also a few other genera
Anaerobic, photosynthetic	Several species of the genus *Rhodospirillum*; also a few other genera
Blue green algae	Several species of the genera *Nostoc* and *Anabaena*

II. Symbiotic systems

Host	Microorganism	Comments
Vascular plants		
Angiosperms		
Certain *leguminous* plants	Several species of bacteria of the genus *Rhizobium*	Nodules located on roots
Certain *nonleguminous* plants, especially trees and shrubs, examples of which are alder (*Alnus* sp.), Australian pine (*Casuarina* sp.), southern wax myrtle (*Myrica* sp.), bitterbrush (*Purshia tridentata*), Western mountain mahogany (*Cercocarpus* sp.)	Several species of actinomycetes, a group of multinuclear, filamentous bacteria	Nodules located on roots
A few tropical plants (e.g., *Psychotria* sp.)	Several bacterial genera (e.g., *Klebsiella*)	Nodules located on leaves
Gymnosperms		
Certain cycads	Certain unidentified species of blue green algae	Nodules located on roots; symbiotic association not well documented
Ferns		
Azolla, a genus of small water ferns	*Anabaena* sp., a blue green alga	Alga grows in pockets within leaves
Nonvascular plants		
Lichens	Fungi and algae	These lichens grow on surfaces of trunks and branches of forest trees
Certain fungi	*Nostoc* sp., a blue green alga	

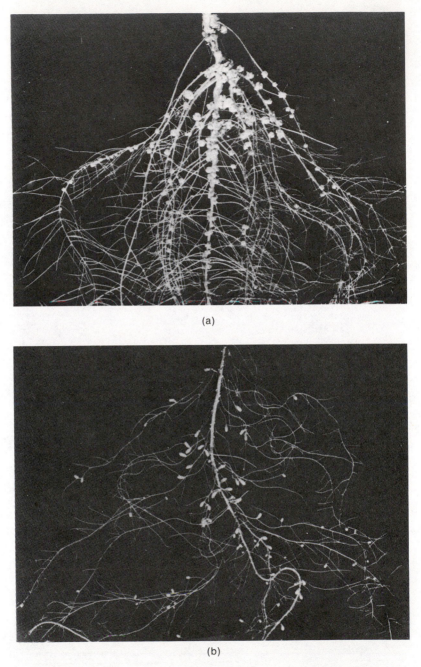

(a)

(b)

Figure 9-2 The typical external appearance of effective legume root nodules of (a) soybean and (b) red clover. ×1 approximately. (After Stewart, 1966; Plate 1.)

be used by the microorganisms themselves. This fixed nitrogen becomes available to higher plants only on the death and decay of the free-living microorganisms that fixed it.

From the agricultural point of view the most important of the symbiotic N_2-fixing systems are those in which the host plant is a member of the family Leguminosae (Table 9-1). Many but not all species of this family engage in symbiotic N_2 fixation. Well-known examples of N_2-fixing leguminous crop plants are peanut, pea, bean, lucerne, soybean, clover, and alfalfa. The infecting microorganisms are several species of bacteria of the genus *Rhizobium*. (Legumes appear to be unique among crop plants in their ability to satisfy their demand for nitrogen *either* by symbiotic fixation of atmosphere N_2 *or* through absorption and assimilation of inorganic nitrogen from the soil.) Also very important, at least in natural ecosystems, is the symbiotic N_2 fixation that occurs in more than 200 species of nonleguminous plants in several families of dicots (Table 9-1). The infecting microorganisms in nonleguminous plants are species of actinomycetes, a group of multinuclear filamentous bacteria. These micro-organisms, like *Rhizobium*, grow in root nodules. Most nonleguminous host plants are trees and shrubs. The term *actinorhizal* is used to describe the association of actinomycetes with host plants.

Not shown in Table 9-1 are the recently discovered N_2-fixing bacteria that live on, in, or under the mucilaginous sheath that covers the root surface of grasses (Vose and Ruschel, 1981). The best known of these so-called associative bacteria are in the genus *Azospirillum*. The *Azospirillum*-grass associations studied to date exhibit only small and variable N_2-fixing activity, but their existence has led to increased interest in the possibility that they may be used eventually to provide nitrogen for important cereal crops, such as maize and wheat.

The discussion of N_2 fixation in plants in the following pages concerns mostly nodulated legumes because they are the primary N_2-fixing plants in agriculture.

The beneficial effects of legumes on soil fertility were realized even in ancient times. The Greek philosophers wrote that leguminous plants should be sown not so much for their yield as for the benefit to the crops that follow them. But it was only in the nineteenth century that the reason legumes are beneficial in crop rotation was discovered.

In the early decades of the nineteenth century the French agronomist Boussingault established that the residues of leguminous plants enrich the soil by furnishing nitrogen. He suspected—but could not prove—that the nitrogen was derived from the atmosphere. It was not until 1888 that Hellriegel and Wilfarth, two German plant physiologists, demonstrated conclusively that the ability of leguminous plants to assimilate N_2 is due to the presence of root nodules formed by infection with certain soil microorganisms. Also, in 1888, the German bacteriologist Beijerinck isolated a pure culture of bacteria (now classified in the genus *Rhizobium*) from root nodules of a legume; moreover, he showed that the application of *Rhizobium* to the roots of an uninfected plant induced the formation of nodules.

Symbiotic N_2 fixation occurs only in legume nodules containing viable bacteria. Leguminous plants themselves do not have the ability to fix N_2 and grow very well

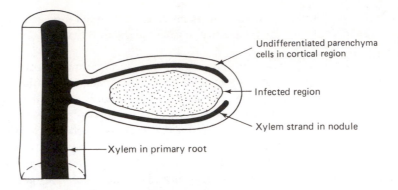

Figure 9-3   Radial longitudinal section through a mature legume nodule showing the central infected region. Also shown are the xylem of the primary root and its connection to two xylem strands in the nodule.

without the bacterium, provided that a source of fixed nitrogen (e.g., nitrate) is supplied. In the symbiotic association of legume host and invading *Rhizobium* the bacterium obtains carbon-containing substances from the host and the host obtains fixed nitrogen through the agency of the bacterium. Thus the symbiotic association of the host and the bacterium is mutually beneficial to both organisms.

The nodules of legumes are small (typically about 2 to 3 mm in apparent diameter), spherical to club shaped, tumorlike structures that are formed laterally on roots (Figure 9-2).

The events in nodule development are complex. Prior to the entrance of rhizobial cells into host tissue, contact between them stimulates the root to secrete growth-promoting substances and organic compounds (including certain plant proteins called lectins) into the soil; lectins are thought to interact with specific rhizobial cells that colonize plant roots and to recognize and permit the entrance of the correct type of rhizobia. Initial infection by rhizobial cells occurs in root hairs, which undergo characteristic curling.[3] Then threads of a mucilaginous substance in which rhizobial cells are embedded grow from root hairs to the root cortical region. There these *infection threads*, as they are called, penetrate into some of the outer cortical cells, which are stimulated to increase the DNA content of their nuclei. At the same time, cells in the inner cortex are stimulated to divide, possibly as a result of secretion of growth-promoting substances by infected host cells. A mass of host cells, some infected, some not, develop into a young nodule. Vascular tissue of the root establishes continuity with newly differentiated vascular tissue in the nodule (Figure 9-3). Bacterial cells inside infected host cells multiply rapidly and are transformed into swollen forms called bacteroids (Figure 9-4). Enzymes necessary for N_2 fixation are present in bacteroids.

[3]For a new hypothesis concerning interactions of cell surface components and enzymes of rhizobial cells and legume root hairs, see Hubbell, D. H., 1981.

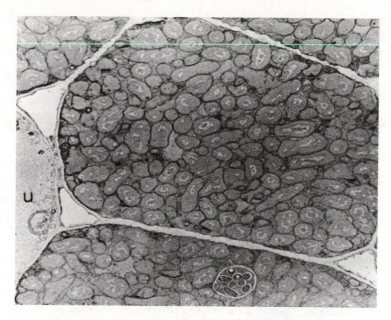

Figure 9-4 Transection of a nodule of a clover plant showing mature infected cells filled with bacteroids. The letter "u" indicates an uninvaded cell. ×3000. (After Stewart, 1966; Plate 2.)

Bacteroids also have a full complement of enzymes (e.g., TCA cycle, oxidative phosphorylation) which enable them to convert photosynthate obtained from the host plant to ammonia. Ammonia is excreted into the cortical tissue of the nodule, converted there to organic form (e.g., asparagine, see footnote 2, p. 270), and then translocated via the xylem (see Figure 9-3) to the shoot system of the host plant.

Bacteroids carry on aerobic respiration; thus they require O_2. But the enzyme system that fixes N_2 in bacteroids is very sensitive to O_2. Excess O_2 in vivo inhibits N_2 fixation. (O_2 applied in vitro to a purified N_2-fixing enzyme preparation produces irreversible damage.)

Protection of the N_2-fixing enzyme system in bacteroids from the deleterious effects of O_2 depends primarily on the following mechanism. Host cells that contain bacteroids produce a reddish protein called *leghemoglobin* (i.e., the hemoglobin of legumes); sliced sections of nodules are pinkish in color. Leghemoglobin is similar to hemoglobin, the O_2 carrier in mammalian red blood cells, insofar as it combines reversibly with O_2, it is the only hemoglobinlike protein present in higher plants. Leghemoglobin has a high affinity for O_2 at the low O_2 tensions that exist in the interior of the nodule and releases O_2 at even lower concentrations within the bacteroid, where O_2 is accepted by a bacteroid respiratory chain oxidase; this has a high affinity for O_2. Thus leghemoglobin delivers O_2 to bacteroids at concentrations necessary for their aerobic respiration but harmless to their N_2-fixing enzyme system.

BIOCHEMISTRY OF NITROGEN FIXATION

Prior to the 1960s little was known about the biochemistry and enzymology of N_2 fixation. Cell-free enzymes capable of fixing N_2 were prepared for the first time in the early 1960s. The first success was with the anaerobic bacterium *Clostridium*. Later, enzymes capable of N_2 fixation were extracted from other free-living nitrogen-fixing microorganisms, including the aerobic bacterium *Azotobacter*. Still later, N_2-fixing enzymes were extracted from bacteroids in legume nodules. Cell-free N_2-fixing enzymes have also been extracted from actinorhizal root nodules.

The enzyme that catalyzes the reduction of N_2 is called *nitrogenase*. Ammonia is the end product of the reaction. The reaction catalyzed by nitrogenase is

$$N_2 + 6\,H^+ + 6\,e^- \longrightarrow 2\,NH_3 \qquad (9\text{-}14)$$

No transient or stable intermediates of this reaction have been detected. Once formed, ammonia is excreted from bacteroids and incorporated into the amides and amino acids considered earlier.

Nitrogenase actually consists of two protein components: One is a nonheme iron protein, the second contains both iron and molybdenum. The first reacts with ATP and reduces the second; the second reduces N_2 to NH_3.

The reduction of N_2 to NH_3 by nitrogenase in bacteroids depends on a continuous supply of both ATP and reduced substrate capable of donating hydrogen atoms (i.e., protons and electrons) to N_2 [see Eq. (9-14)]. ATP is required for the reason mentioned in the next paragraph; it is generated from ADP and inorganic phosphate in bacteroids in their respiratory chain system, as described in Chapter 4. Reduced substrate is obtained from photosynthate supplied by the host plant. Sucrose is the major form of translocation of photosynthate from leaves of higher plants (p. 332), but it is probably not utilized as such by bacteroids in legume nodules. Instead sucrose is probably first hydrolyzed by the enzyme invertase to glucose and fructose. Although carbohydrate metabolism in bacteroids is not well understood at present, it is known that they contain glycolytic enzymes. Thus glucose-6-phosphate has been suggested as a substrate for the nitrogenase reaction (Yoch, 1979). [Recall from Chapter 4 that both glucose and fructose are readily converted to glucose-6-phosphate (e.g., see Figure 4-1 and 4-3).]

Not all details of electron transport to nitrogenase have been elucidated yet. Even so it is possible to present a hypothetical electron transport system (see Figure 9-5). Glucose-6-phosphate is assumed to be the reduced substrate. NADPH (generated in the glucose-6-phosphate dehydrogenase reaction; see Figure 4-4), together with ferredoxin, are shown in Figure 9-5 to be electron carriers. Finally, ATP is shown in Figure 9-5 to interact with the nonheme iron protein component of nitrogenase; it has been suggested that this interaction produces a conformational change in this nonheme iron protein, converting it to a powerful ("superreduced") reductant, one capable of transferring electrons to the iron-molybdenum protein component of nitrogenase, which, in turn, reduces N_2 to NH_3 (e.g., see Evans and Barber, 1977; Yoch, 1979). In in vitro experiments it has been shown that at least 4 molecules of ATP

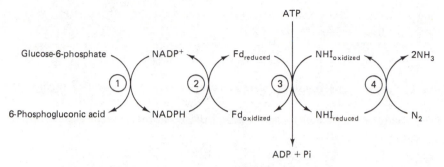

Figure 9-5 A postulated pathway of electron transport in N_2 fixation in legume nodule bacteroids. (See text for details.) Abbreviations: Fd, ferredoxin; NHI, nonheme iron. The numerals refer to (1) glucose-6-phosphate dehydrogenase; (2) NADPH-ferredoxin oxidoreductase; (3) the iron-containing protein component of nitrogenase; (4) the iron-molybdenum protein component of nitrogenase.

are hydrolyzed for each pair of electrons transferred to N_2. Thus reduction of 1 molecule of N_2 to 2 molecules of NH_3 requires at least 12 molecules of ATP because 6 electrons are required per molecule of N_2 reduced [see Eq. (9-14)]. [Reaction (9-14) may be modified by adding 12 ATP to the left side and 12 ADP + 12 Pi to the right side.]

Several other substrates in addition to N_2 can be reduced by the nitrogenase complex. One is gaseous acetylene ($HC{\equiv}CH$), a nonphysiological substance that is reduced to ethylene ($H_2C{=}CH_2$). Because ethylene can be detected with great sensitivity by gas chromatography (a small, portable, and relatively inexpensive instrument), the acetylene–ethylene assay for nitrogenase activity is sometimes used in the field to evaluate the effects of different environmental factors and cultural practices on N_2 fixation (Hardy et al., 1968). In this way, the standard method of measuring rates of N_2 fixation in samples of nodules, soils, and so forth [i.e., with ^{15}N (the heavy isotope of nitrogen) and analysis of ^{15}N in nitrogen-containing compounds by means of a mass spectrometer, a relatively heavy instrument not easily transported from place to place] can be avoided.

The proton is also a substrate for nitrogenase. In fact, a fraction of the electron flow through nitrogenase—in some cases, a third or more—may not be utilized in the reduction of N_2; instead electrons may be used for the reduction of protons.

$$2\ H^+ + 2\ e^- \xrightarrow{\text{nitrogenase}} H_2 \qquad (9\text{-}15)$$

Thus there may be an evolution of gaseous H_2 by bacteroids during N_2 fixation. This H_2 evolution represents a waste of ATP and reducing power.

Certain strains of *Rhizobium* have been found to produce nodules that lose little or no H_2 during periods when nitrogenase is operative. These strains are more efficient in N_2 fixation because they possess a mechanism whereby H_2 evolved by nitrogenase is recycled. The first step in the recycling process is the uptake of H_2, catalyzed by the enzyme hydrogenase in bacteroids.

$$H_2 \xrightarrow{\text{hydrogenase}} 2\,H^+ + 2\,e^- \qquad (9\text{-}16)$$

Subsequent steps involve the reutilization of protons and electrons produced in reaction (9-16) by the electron carriers shown in Figure 9-5. Much current research in N_2 fixation is devoted to determining the combinations of *Rhizobium* strains and legume cultivars that produce nodules that do not evolve H_2.

ASSIMILATION OF SULFATE

Higher plants obtain their supply of sulfur principally by uptake of sulfate ion ($SO_4^=$) by roots. Although trace amounts of gaseous sulfur dioxide (SO_2) are present in the atmosphere and may be absorbed and assimilated by leaves, it has been shown that this substance is converted to sulfate ion before it is assimilated. Thus the assimilation of sulfur dioxide and the assimilation of sulfate are one and the same process.

Most of the sulfate absorbed by roots is carried upward in the transpiration stream to leaves; activation, reduction, and assimilation of sulfate take place in chloroplasts in reactions that depend on light for a supply of ATP and reduced ferredoxin.[4] Although sulfate reduction and assimilation may occur in nonphotosynthetic tissue, this is only of minor importance in higher plants.

The assimilation of sulfate ion begins with its *activation*. The activating reaction is shown in Figure 9-6. Sulfate ion reacts with ATP to yield adenosine-5′-phosphosulfate (designated by the symbol APS); this reaction is catalyzed by an enzyme known as ATP-sulfurylase.

The reduction of sulfate (in the form of APS) takes place first to sulfite ($SO_3^=$). In this reaction APS reacts with a not yet completely identified sulfur-containing reducing agent (designated here as R—SH)

$$APS + R\text{—}SH \longrightarrow R\text{—}S\text{—}SO_3^= + AMP \qquad (9\text{-}17)$$

where AMP is adenosine monophosphate.

Reduced ferredoxin, which is produced in photosynthetic light reactions, reduces sulfite (in the form of R—S—$SO_3^=$) to sulfide (in the form of R—S—$S^=$). The reaction may be written

$$[SO_3^=] + 6\,H^+ + 6\,e^- \longrightarrow [S^=] + 3\,H_2O \qquad (9\text{-}18)$$

Here six equivalents of electrons (supplied by reduced ferredoxin) are consumed in the

---

[4]It was suggested recently that electrons derived from the oxidation of hexose phosphate can be used for sulfate reduction in chloroplasts (Kow et al., 1982); this type of sulfate reduction may occur *in the dark*, Thus, sulfate reduction appears to be analogous to nitrite reduction, which may occur in the light by virtue of reducing power generated by light reactions in chloroplasts (pp. 260–261), and also in the dark at the expense of starch reserves which are degraded to glucose-6-phosphate (p. 262).

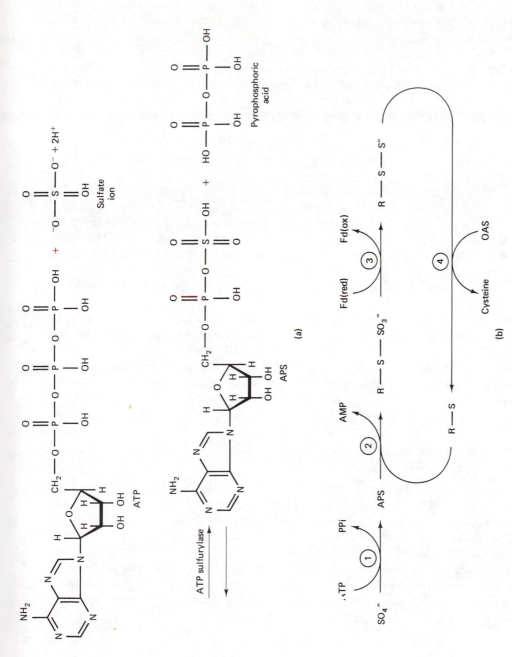

Figure 9-6 (a) The activation of sulfate ion. APS is designated *activated sulfate*. Sulfate undergoes reduction and assimilation only after it is activated. (See text for details.) (b) Summary of the sulfate assimilatory pathway in higher plants. The enzymes are 1, ATP sulfurylase; 2, APS sulfotransferase; 3, thiosulfonate reductase; 4, cysteine synthase. Abbreviations: PPi, pyrophosphoric acid; APS, adenosine-5'-phosphosulfate; AMP, adenosine monophosphate; Fd (red) and Fd (ox), reduced and oxidized ferredoxin; OAS, *O*-acetyl serine.

reduction of one mole of sulfite to sulfide. The latter two components are enclosed in brackets in reaction (9-18), simply to indicate they are in the "bound" forms ($R—S—SO_3^=$ and $R—S—S^=$).

Sulfide produced in reaction (9-18) is incorporated into cysteine, a sulfur-containing amino acid. The compound that reacts with sulfide (in the form of $R—S—S^=$) is a derivation of the amino acid serine—namely, O-acetyl serine. [The designation O indicates that the point of attachment of the acetyl group to serine is at the oxygen atom of serine, as shown in reaction (9-19). The reaction may be written

$$
\begin{array}{l}
CH_2OCH_2COOH \\
| \\
CHNH_2 \\
| \\
COOH
\end{array}
\;+\; [S^=] + 2\,H^+ \;\longrightarrow\;
\begin{array}{l}
CH_2SH \\
| \\
CHNH_2 \\
| \\
COOH
\end{array}
+ CH_3COOH + H_2O
$$

$$\text{O-acetyl serine} \qquad\qquad\qquad \text{Cysteine} \quad \text{Acetic acid} \qquad (9\text{-}19)$$

The amino acid cysteine [produces in reaction (9-19)] is the starting point for the biosynthesis of a wide variety of sulfur-containing plant constituents. Moreover, cysteine is the form in which sulfur in translocated to other parts of the plant in the phloem.

A summary of the sulfate assimilatory pathway in higher plants is shown in Figure 9-6.

A recent investigation of the distribution of ATP sulfurylase in leaf cells of C_4 plants showed that this enzyme is present in chloroplasts of bundle sheath cells, not mesophyll cells (Gerwick and Black, 1980). This finding suggests that sulfur assimilation in leaves in C_4 plants occurs in bundle sheath cells rather than mesophyll cells and contrasts with the site of nitrate assimilation; nitrate assimilation in leaves of C_4 plants occurs in mesophyll cells, not bundle sheath cells (pp. 261–262).

ASSIMILATION OF PHOSPHATE

Phosphate exists in soils and water culture solutions in the form of ions of phosphoric acid (H_3PO_4). $HPO_4^=$ is the predominant ionic form in most soils, but $H_2PO_4^-$ is also present, especially in acid soils. Because some of the phosphate absorbed by roots of higher plants is swept upward in the transpiration stream to leaves, we can expect phosphate to be assimilated in leaves as well as roots.

It is widely assumed that phosphate is assimilated in plant cells primarily by incorporation into ATP. The assimilation of inorganic phosphate corresponds to the process of oxidative phosphorylation (pp. 117–118). So assimilatory reaction may be written

$$ADP + Pi \longrightarrow ATP + H_2O \qquad (9\text{-}20)$$

where ADP is adenosine diphosphate, Pi is phosphoric acid, and ATP is adenosine triphosphate.

In addition to the assimilation of phosphate in oxidative phosphorylation, it is probable that some of the phosphate absorbed by a higher plant is assimilated in green leaves in the light during photosynthetic phosphorylation [Eq. (6-12)]. In this case, reaction (9-20) occurs in conjunction with photosynthetic light reactions in chloroplasts.

Phosphate assimilated into ATP is rapidly transferred by subsequent metabolic reactions into a wide variety of phosphorylated plant constituents, including sugar phosphates, phospholipids, and nucleotides.

Besides its incorporation into ATP, inorganic phosphate may also be assimilated in other ways. In one of the reactions of glycolysis, for instance, phosphate is incorporated into 1,3-bisphosphoglyceric acid [see reaction (4-22)].

ASSIMILATION OF CATIONS

We saw earlier that macronutrient anions, especially nitrate and sulfate, undergo drastic metabolic transformations during their assimilation. In contrast, macronutrient cations (potassium, calcium, and magnesium) and micronutrient cations (cobalt, copper, iron, manganese, sodium, and zinc) can be said to be assimilated either by weak interionic attractive forces of dissolved organic or inorganic anions[5] or by formation of coordinate bonds.

Most of the potassium and part of the calcium and magnesium ions absorbed by plant cells are attracted to anions in solution through their electric fields of force. To give one example, a potassium ion interacts with one or another of various organic anions in solution, including malate ion. Malate ion is produced in plant cells when malic acid (one of the intermediates of the TCA cycle) dissociates. Potassium ion can be said to be assimilated, in this example, by attraction to malate ion.

$$
\begin{array}{c}
\begin{array}{l}
COO^- \\
| \\
CHOH \\
| \\
CH_2 \\
| \\
COO^-
\end{array}
\quad + 2\,K^+ \longrightarrow
\begin{array}{l}
COOK \\
| \\
CHOH \\
| \\
CH_2 \\
| \\
COOK
\end{array}
\end{array}
\qquad (9\text{-}21)
$$

Malate ion

The assimilation of metallic micronutrient cations (e.g., Co, Cu, Fe, Mo, Mn, Zn) takes place through the formation of coordinate bonds. Coordination bond formation also accounts for assimilation of the fraction of macronutrient ions (calcium, magnesium, and potassium) not attracted to anions in solution by the weak ionic interactions mentioned. In a *coordinate* bond an atom or group of atoms with an unshared pair of electrons shares that pair of electrons with another atom or group of atoms. Two examples of coordinate bond formation are shown in Figure 9-7. In the first example, ammonia gas is bubbled through a solution containing cupric ion

---

[5]These attractive forces of dissolved ions of opposite sign for each other are of the Debye–Hückel type.

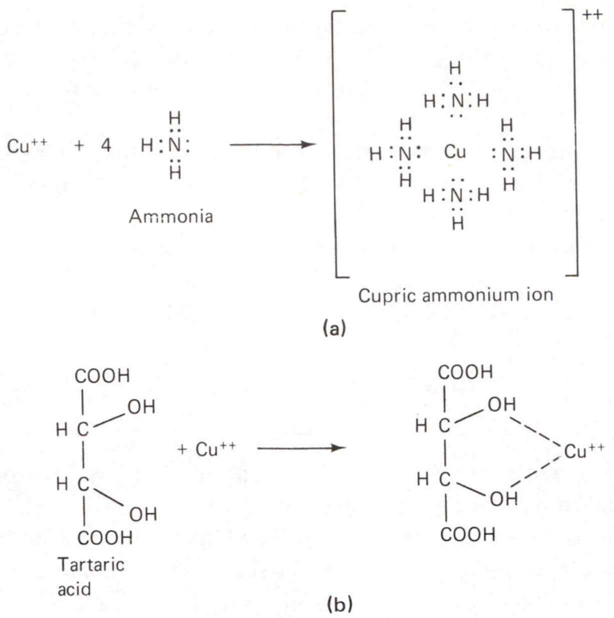

$$\text{Cu}^{++} \; + \; 4 \quad \text{H}:\underset{\overset{\cdot\cdot}{\text{H}}}{\overset{\text{H}}{\text{N}}}: \longrightarrow \;\; \left[\text{H}:\underset{\text{H}}{\overset{\text{H}}{\text{N}}}: \; \text{Cu} \; :\underset{\text{H}}{\overset{\text{H}}{\text{N}}}:\text{H} \right]^{++}$$

Ammonia

Cupric ammonium ion

(a)

Figure 9-7 Two examples of coordinate bonds. (a) In the cupric ammonium ion, nitrogen atoms in four ammonia molecules donate unshared pairs of electrons to cupric ion. Electrons in the valence orbitals in nitrogen atoms are shown as dots. (b) In the tartaric acid-cupric ion complex, two oxygen atoms in the hydroxyl groups of tartaric acid donate unshared pairs of electrons to cupric ion. Coordinate bonds are shown as dashed lines.

(b)

Tartaric acid $+ \text{Cu}^{++} \longrightarrow$

(Cu^{2+}). In the electron dot representation of the ammonia molecule [see Figure 9-7(a)], only the valence electrons are shown; the lone electron pair is available for coordinate bonding. The product of the reaction is cupric ammonium ion $[\text{Cu(NH}_3)_4^{2+}]$. The copper in cupric ammonium ion is surrounded by four ammonia molecules, each of which shares its lone pair of electrons with copper. Thus there are four coordinate bonds in cupric ammonium ion.

Not only nitrogen but also oxygen and sulfur atoms are able to form coordinate bonds with metal atoms, as can be seen in Figure 9-7(b). Tartaric acid forms a complex ion with cupric ion. In this case, two coordinate bonds are formed when unshared pairs of electrons are donated to copper by oxygen atoms in the hydroxyl groups of tartaric acid.

Another example of coordinate bond formation was the chelate compound shown in Figure 8-2(c).

A large number of substances in plant cells is capable of forming coordinate bonds with metals, including about one-quarter of all enzymes in plant cells, plus noncatalytic compounds; the biologically active forms are called *metallobiomolecules* (Ibers and Holm, 1980). Metallobiomolecules contain one or more metallic elements (e.g., Cu, Fe, Mg, Mo, Mn, Zn) in coordination complex. The metal-containing sites are the loci of enzyme catalysis in bond-making and bond-breaking reactions [e.g., phosphatases (Mg, Zn, Cu)], light-dependent oxidation-reduction systems [e.g., chlorophyll (Mg), photosystem II (Mn, Mg)], and electron carriers [e.g., cytochromes (Fe), iron-sulfur proteins (Fe)]. Examples of metallobiomolecules encountered in this chapter include ferredoxin, an iron-sulfur protein that functions as an electron carrier in electron transfer chains [Eq. (9-4)], and multielectron oxidoreductases that catalyze substrate transformations involving the transfer of up to six electrons: nitrite reductase (Fe) [Eq. (9-5)], nitrogenase (Fe, Mo) [Eq. (9-13)], and sulfite reductase (Fe) [Eq. (9-17)]. [Cytochrome oxidase (Fe, Cu) is another multielectron oxidoreductase encountered earlier [Eq. (4-18)].

ASSIMILATION OF MOLECULAR OXYGEN

The role of O_2 in respiration was described in Chapter 4. O_2 moves from the external atmosphere into plant cells and serves as an electron and hydrogen acceptor in reactions catalyzed by enzymes known as *oxidases*. In these respiratory reactions O_2 is reduced to water.

In contrast to oxidases, *oxygenases* catalyze the *addition* of O_2 to organic substrates in reactions generally classified as *oxygen fixation*. There are two types of oxygen fixation reactions. In one, both atoms of O_2 are added to a substrate molecule (S).

$$S + O_2 \longrightarrow SO_2 \tag{9-22}$$

This type of reaction is catalyzed by enzymes known as *dioxygenases*. The enzyme lipoxygenase is an example. Lipoxygenase catalyzes the oxygenation of certain fatty acids to fatty acid hydroperoxides. The natural substrates for the plant enzyme are linoleic and linolenic acids, the two most abundant polyunsaturated fatty acids in plant tissues.

In a second type of oxygen fixation reaction one of the oxygen atoms in O_2 is incorporated into a substrate molecule (S) and the other oxygen atom is reduced to water. Participation of a reducing agent (designed by the symbol AH_2) is required.

$$S + O_2 + AH_2 \longrightarrow SO + H_2O + A \tag{9-23}$$

The reducing agent may be a substance like NADPH or NADH. Enzymes that catalyze reaction (9-23) are called *monooxygenases* (or hydroxylases).

One example of a monooxygenase reaction is the metabolic conversion of the amino acid proline to the amino acid hydroxyproline [Eq. (9-24)]. Because the identity of the natural reducing agent has not been definitely established, only an incomplete reaction can be given at present.

$$\underset{\text{Proline}}{} + \tfrac{1}{2} O_2 \longrightarrow \underset{\text{Hydroxyproline}}{} \tag{9-24}$$

Monooxygenase reactions are involved in metabolic sequences leading to the biosynthesis of complex phenols, sterols, lignin, and so forth. Many details of these biosynthetic pathways, however, remain to be clarified by future research work.

We already encountered an example of an oxygen fixation reaction earlier. Recall that the carboxylating enzyme in the photosynthetic carbon reduction cycle (photosynthetic pentose phosphate pathway) also functions as an oxygenating enzyme [reaction (7-5)]. So this enzyme is called RuBP carboxylase/oxygenase. It is a monooxygenase.

ROLES OF ESSENTIAL ELEMENTS

The discussion of the assimilation of inorganic nutrients by plant cells in preceding sections naturally leads to questions concerning their roles. What are the roles of each of the essential elements? What specific functions do they perform? What are the mechanisms of their action?

The roles of some of the essential elements were emphasized earlier. Carbon and oxygen in CO_2 and hydrogen in water are assimilated in photosynthesis and subsequently metabolized into a variety of plant constituents. Nitrogen, phosphorus, and sulfur also have nutritive roles and are incorporated into plant constituents. O_2 has a nutritive role in oxygen fixation reactions (pp. 283–284). Moreover, O_2 is the final electron and hydrogen acceptor in respiration (see Chapter 4).

Two methods are commonly used in investigating the roles of essential elements. In one, a deficiency of the element is induced by supplying inadequate amounts of it and the results of the deficiency are followed with respect to growth and development, visible deficiency symptoms, and biochemical and physiological abnormalities. Similar kinds of observations are made when the deficient element is resupplied to plants deficient in it. A second method is to discover enzymes that require (or are activated by) the element in question in in vitro experiments.

Using these two methods, considerable research effort has been devoted during the last half century or more to the elucidation of the roles of essential elements, especially the microelements. The literature is voluminous. A summary of information presently available on the roles of most of the essential elements in higher plants is presented in Table 9-2; further details of specific functions of essential elements, many of them listed conveniently in tabular form, may be found by consulting Rains, 1976.

Although one or more roles have been ascribed to most essential elements, knowledge of the roles of boron and chlorine remains obscure at present. Some type of complex between borate ion and carbohydrate is formed and controls the deposition of cellulose micelles in the cell wall. Chlorine appears to be required in O_2 production in light reactions in photosynthesis (see Table 9-2), but not all plant physiologists agree on this point. Little is known today about the roles of cobalt, selenium, silicon, and sodium—elements that have been shown essential to only a few plants (p. 240)—and therefore they are not included in Table 9-2.

Despite the fact that each essential element performs a distinctive set of functions not duplicated completely by other essential elements, some generalizations can be

TABLE 9-2 ROLES OF ESSENTIAL ELEMENTS[a]

Element	Roles(s)
N	Constituent of amino acids, amides, proteins, nucleic acids, nucleotides and coenzymes, hexoseamines, etc.
P	Component of sugar phosphates, nucleic acids, nucleotides, coenzymes, phospholipids, phytic acid, etc. Has key role in reactions in which ATP is involved.
K	Required as a cofactor for 40 or more enzymes. Has a role in stomatal movements. Maintains electroneutrality in plant cells.
S	Component of cysteine, cystine, methionine, and thus proteins. Constituent of lipoic acid, coenzyme A, thiaminepyrophosphate, glutathione, biotin, adenosine-5'-phosphosulfate and 3'-phosphoadenosine-5'-phosphosulfate, and other compounds.
Ca	A constituent of the middle lamella of cell walls as Ca-pectate. Required as a cofactor by some enzymes involved in the hydrolysis of ATP and phospholipids.
Mg	Required nonspecifically by large number of enzymes involved in phosphate transfer. A constituent of the chlorophyll molecule.
Fe	A constituent of cytochromes. A constituent of nonheme iron proteins, which are involved in photosynthesis, N_2 fixation, and respiratory-linked dehydrogenases.
Mn	Required for activity of some dehydrogenases, decarboxylases, kinases, oxidases, peroxidases, and nonspecifically by other divalent, cation-activated enzymes. Required for photosynthetic evolution of O_2.
B	Indirect evidence for involvement of B in carbohydrate transport. Borate forms complexes with certain carbohydrates, but natural borate complexes in plants have not been identified.
Cu	An essential component of ascorbic acid oxidase, tyrosinase, laccase, monoamine oxidase, uricase, cytochrome oxidase, and galactose oxidase. Component of plastocyanin from *Spinacia oleracea*. Deficiency in algae causes decreased activity of cytochrome photooxidase.
Zn	Essential constituent of alcohol dehydrogenase, glutamic dehydrogenase, lactic dehydrogenase, carbonic anhydrase, alkaline phosphatase, carboxypeptidase B, and other enzymes.
Mo	A constituent of nitrate reductase of fungi, bacteria, and higher plants and of xanthine oxidase, aldehyde oxidase from animal sources, and essential for N_2 fixation.
Cl	Required for photosynthetic reactions involved in oxygen evolution.

[a] Carbon, hydrogen, and oxygen have been omitted intentionally.

After Evans and Sorger, 1966 (Table I).

made about their roles. In general, an element is essential to the life of a higher green plant for one or more of the following four reasons.

1. It may perform a nutritive role by being a component of one or more of the major classes of plant constituents.
2. It may have a catalytic role either as an activator of an enzyme or as an integral component of an enzyme.
3. It may function as a noncatalytic "metallobiomolecule."
4. It may function as a free ion.

Several essential elements have a predominantly nutritive role. They are

TABLE 9-3 ELEMENTAL COMPOSITION OF THE SIX MAJOR CLASSES
OF CONSTITUENTS OF LIVING MATTER

Plant constituent	Elements present
Carbohydrates	Carbon, hydrogen, oxygen. (Nitrogen and/or phosphorus are also present in certain derivatives.)
Amino acids	Carbon, hydrogen, oxygen, nitrogen. (Sulfur is also present in some amino acids.)
Nucleotides	Carbon, hydrogen, oxygen, nitrogen, phosphorus.
Porphyrins	Carbon, hydrogen, oxygen, nitrogen. (Magnesium is also present in porphyrins known as chlorophylls.) (Iron is also present in porphyrins known as cytochromes.)
Lipids	Carbon, hydrogen, oxygen. (Phosphorus and/or sulfur are also present in some lipids.)
Enzymes	Carbon, hydrogen, oxygen, nitrogen, phosphorus, sulfur. [A heavy metal (copper, iron, molybdenum, or zinc) is also an integral component of the reactive site in about 150 metalloenzymes.]

components of one or another of the constituents of living matter. The six major classes of constituents of living matter and their elemental compositions are listed in Table 9-3.

Certain essential elements are required for the activation of several enzymes. An activating element is usually a cation (e.g., K^+, Mg^{2+}, Ca^{2+}, Mn^{2+}). Instead of being an integral constituent of an enzyme, the activating ion is linked to the protein portion of the enzyme by electrostatic or coordinate bonds and may be removed from the enzyme by suitable treatment in vitro. In the absence of the activating ion, the enzyme does not perform its catalytic function. The activating ion—sometimes referred to as a *cofactor*—confers on the enzyme certain properties that enable it to perform its catalytic function. The activating ion, for instance, may assist in the attachment of the substrate to the protein portion of the enzyme or it may affect the charge configuration of key reactive groups in the protein portion of the enzyme.

In contrast to activating ions, certain essential elements are specific and integral components of several enzymes. They were considered previously in the discussion of metallobiomolecules (pp. 282–283); also, it was pointed out that some metallobiomolecules have certain *noncatalytic* functions (e.g., as electron carriers and light-dependent oxidation-reduction compounds).

The fourth general role of essential elements is that of a free ion. Potassium is a good example; it is linked by relatively weak interionic attractive forces to organic anions. This function of potassium will be mentioned in the discussion of the opening and closing movements of stomata (pp. 398–399).

REFERENCES

ANDERSON, J. W. 1980. Assimilation of inorganic sulfate into cysteine. Pages 203–223 *in* P. K. Stumpf and E. E. Conn, eds. *The Biochemistry of Plants, A Comprehensive Treatise*, vol. 5, Academic Press, New York.

BEEVERS, L. 1976. *Nitrogen Metabolism in Plants*. American Elsevier Publishing Co., New York.

BEEVERS, L., and R. H. HAGEMAN. 1980. Nitrate and nitrite reduction. Pages 116–168 *in* P. K. Stumpf and E. E. Conn, eds. *The Biochemistry of Plants, a Comprehensive Treatise*, vol. 5. Academic Press, New York.

BRILL, W. J. 1979. Nitrogen fixation: basic to applied. *Am. Scient.* 67:458–466.

BURRIS, R. H. 1976. Nitrogen fixation. Pages 887–908 *in* J. Bonner and J. E. Varner, eds. *Plant Biochemistry*, 3rd ed. Academic Press, New York.

BURRIS, R. H., arranger. 1978. Special issue: future of biological N_2 fixation. *BioScience* 28(9).

CLARKSON, D. T., and J. B. HANSON, 1980. The mineral nutrition of higher plants. *Ann. Rev. Plant Physiol.* 31:239–298.

EPSTEIN, E. 1965. Mineral metabolism. Pages 438–466 *in* J. Bonner and J. E. Varner, eds. *Plant Biochemistry*. Academic Press, New York.

EPSTEIN, E. 1972. *Mineral Nutrition of Plants: Principles and Perspectives*. John Wiley & Sons, New York.

EVANS, H. J., and G. J. SORGER. 1966. Role of mineral elements with emphasis on the univalent cations. *Ann. Rev. Plant Physiol.* 17:47–76.

EVANS, H. J., and L. E. BARBER. 1977. Biological nitrogen fixation for food and fiber production. *Science* 197:332–339.

GERWICK, B. C., and C. C. BLACK. 1980. Initiation of sulfate activation: a variation in C_4 photosynthesis plants. *Science* 209:513–516.

GIVAN, C. V. 1980. Aminotransferases in higher plants. Pages 329–358 *in* P. K. Stumpf and E. E. Conn, eds. *The Biochemistry of Plants, a Comprehensive Treatise*, vol. 5. Academic Press, New York.

HARDY, R. W., R. D. HOLSTEN, E. K. JACKSON, and R. C. BURNS. 1968. The acetylene–ethylene assay for N_2 fixation: laboratory and field evaluation. *Plant Physiol.* 43:1185–1207.

HARDY, R. W. F., and U. D. HAVELKA. 1975. Nitrogen fixation research: a key to world food? *Science* 197:332–339.

HEWITT, E. J., D. P. HUCKLESBY, and B. A. NOTTON. 1976. Nitrate metabolism. Pages 633–681 *in* J. Bonner and J. E. Varner, eds. *Plant Biochemistry*, 3rd ed. Academic Press, New York.

HUBBELL, D. H. 1981. Legume infection by *Rhizobium*: a conceptual approach. BioScience 31:832–837.

IBERS, J. A. and R. H. HOLM. 1980. Modeling coordination sites in metallobiomolecules. *Science* 209:223–235.

JOHNSTON, H. 1972. Newly recognized vital nitrogen cycle. *Proc. Natl. Acad. Sci. U.S.* 69:2369–2372.

KOW, Y. W., D. L. ERBES, and M. GIBBS. 1982. Chloroplast respiration. A means of supplying oxidized pyridine nucleotide for dark chloroplastic metabolism. *Plant Physiol.* 69:442–447.

MANN, A. F., D. P. HUCKLESBY, and E. J. HEWITT. 1978. Sources of reducing power for nitrate reduction in spinach leaves. *Planta* 140:261–263.

MENGEL, K., and E. A. KIRKBY. 1978. *Principles of plant nutrition*. International Potash Institute, Berne, Switzerland.

MIFLIN, B. J., and P. J. LEA. 1976a. The pathway of nitrogen assimilation in plants. *Phytochemistry* 15:873–885.

MIFLIN, B. J., and P. J. LEA. 1976b. The path of ammonia assimilation in the plant kingdom. *Trends Biochem. Sciences* 1:103–106.

MIFLIN, B. J., and P. J. LEA. 1977. Amino acid metabolism. *Ann. Rev. Plant Physiol.* 28:299–329.

MIFLIN, B. J., and P. J. LEA. 1980. Ammonia assimilation. Pages 169–202 *in* P. K. Stumpf and E. E. Conn, eds. *The Biochemistry of Plants, a Comprehensive Treatise*, vol. 5. Academic Press, New York.

MOORE, R., and C. C. BLACK, Jr. 1979. Nitrogen assimilation pathways in leaf mesophyll and bundle sheath cells of C_4 photosynthesis plants formulated from comparative studies with *Digitaria sanguinalis* (L.) Scop. *Plant Physiol.* 64:309–313.

POSTGATE, J. 1978. *Nitrogen Fixation.* University Park Press, Baltimore.

RAINS, D. W. 1976. Mineral nutrition. Pages 561–597 *in* J. Bonner and J. E. Varner, eds. *Plant Biochemistry*, 3rd ed. Academic Press, New York.

SHERRARD, J. H., and M. J. DALLING, 1979. *In vitro* stability of nitrate reductase from wheat leaves. I. Stability of highly purified enzyme and its component activities. *Plant Physiol.* 63:346–353.

SKINNER, K. J. 1976. Nitrogen fixation. *Chem. Eng. News* 54(41):22–35.

STEWART, W. D. P. 1966. *Nitrogen fixation in plants.* The Athlone Press, London.

VOSE, P. B., and A. P. RUSCHEL, eds. 1981. *Associative N_2-fixation*, vols. 1 and 2. CRC Press, Inc., Boca Raton, Florida.

WILSON, L. G., and Z. REUVENY. 1976. Sulfate reduction. Pages 599–632 *in* J. Bonner and J. E. Varner, eds. *Plant Biochemistry*. Academic Press, New York.

WOO, K. C., M. JOKINEN, and D. T. CANVIN. 1980. Reduction of nitrate via a dicarboxylate shuttle in a reconstituted system of supernatant and mitochondria from spinach leaves. *Plant Physiol.* 65:433–436.

YATES, M. G. 1980. Biochemistry of nitrogen fixation. Pages 1–64 *in* P. K. Stumpf and E. E. Conn, eds. *The Biochemistry of Plants, a Comprehensive Treatise*, vol. 5. Academic Press, New York.

YOCH, D. C. 1979. Electron-transport systems coupled to nitrogenase. Pages 605–652 *in* R. W. F. Hardy, F. Bottomley, and R. C. Burns, eds. *A Treatise on Dinitrogen Fixation*, vol. 1. J. Wiley & Sons, New York.

10

Transport Phenomena in Plants: General Considerations

Like all living organisms, higher plants are delineated from their surroundings by their external surfaces. But these boundaries are not impenetrable barriers. Numerous materials gain entrance into plants, some rapidly, others only slowly. Included are water and dissolved inorganic salts in the soil solution, soluble organic substances (e.g., sugars and amino acids) released in soils by decaying plant and animal debris, gases, such as CO_2 used in photosynthesis in leaves and O_2 used in respiration in all plant cells, synthetic substances (e.g., fungicides, insecticides, and herbicides), and such macromolecules as viruses. Nor is the flow of materials directed only into plants. O_2 released in photosynthesis moves out of leaves. CO_2 released in respiration moves out of all plant cells; similarly, ethylene gas (one of the plant growth substances considered in Chapter 14) moves out of plant cells. Small amounts of inorganic ions and organic metabolites are leached out of leaf cells by rain and out of root cells by soil water. Water vapor is evaporated into the external atmosphere as a result of transpiration by leaves.

Transport processes in plants are not restricted to exchanges between the plant body and its surroundings. Transport of substances takes place continually throughout the life of a plant among organelles (plastids, nuclei, mitochondria, etc.) of individual cells, among neighboring cells of a tissue, and from one tissue or organ to another.

Many of the literally hundreds of reactions that occur concurrently in a single growing cell are segregated and compartmentalized by protoplasmic membranes. These membranes include the plasma membrane that surrounds the cytoplasm, the membrane around the vacuole (i.e., the tonoplast), and the membranes that cover cellular organelles. Passage of substances through these cellular membranes is controlled by transport mechanisms located within the membranes themselves. In effect, protoplasmic membranes function not only as barriers but also as "ports" and "pumps." As a result, products of reactions in one part of a cell can be shepherded and delivered as reactants for reactions in other parts of the cell.

In the intact plant body the vascular system of higher plants has as its primary function the transport of substances from one tissue or organ to another tissue or organ. The specialization of leaves for photosynthesis and roots for absorption of water and mineral ions depends on the existence of an efficient interconnecting vascular system that transports photosynthetic products to roots and water and mineral nutrients to leaves.

Transport phenomena unite and coordinate all activities occurring within plants and plant cells. The study of transport phenomena begins the transition from the study of individual biochemical reactions and individual physiological functions to a deeper comprehension and understanding of both cellular and whole plant physiology.

The term *transport* is used throughout this book in a broad sense to refer to transfer of matter from one region to another without regard to mechanism or whether the transfer is spontaneous or nonspontaneous or a membrane is crossed. Similar terms sometimes used by plant physiologists include uptake, release, entry, and exit.

In this chapter several basic principles that apply to all transport processes in higher plants are considered. Included are the differences between spontaneous and nonspontaneous transport processes, diffusion and mass flow, driving force and flux, identification of driving forces in spontaneous transport processes, and the concept of chemical potential, especially as it applies to water in plants.[1] There is also a brief discussion of free space in plant tissues and the pathway of movement of nutrient ions across primary roots is described. One of the major objectives here is to lay the foundations for subjects discussed in the next three chapters.

SPONTANEOUS AND NONSPONTANEOUS TRANSPORT PROCESSES

Experience shows that some changes in nature occur of their own accord under existing conditions. If a cold body is in contact with a hot one, then the latter will lose heat to the cold one until the two are at the same temperature. If one end of a wire is at a higher electrical potential than the other, charge will flow through the wire toward the end with lower potential until electrical potential is everywhere the same throughout the wire. If two vessels at the same temperature and open to the atmosphere—one containing liquid water and the other empty—are placed side by side and connected by a pipe fitted with a valve [Figure 10-1(a)], water will flow into the empty vessel when the valve is opened until the level of water in the two vessels is the same. If two vessels of the same volume and initially the same temperature—one filled with a gaseous substance (e.g., air at a low pressure) and the other evacuated—are separated by a layer

---

[1] It should be emphasized that these matters are considered at the level of a student reader without background in physical chemistry and thermodynamics.

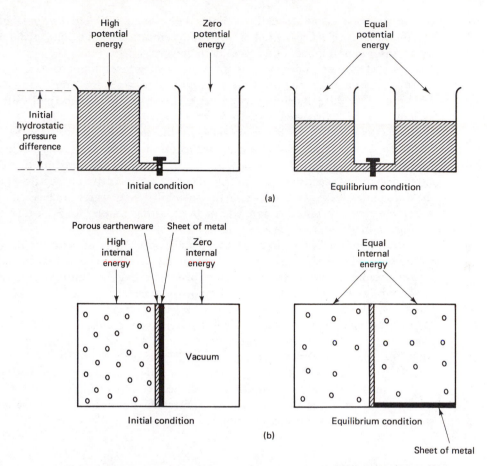

Figure 10-1 Examples of spontaneous transport processes. (See text for details.) In the initial situation in (b), a low pressure of air—much less than atmospheric pressure—is used so as to minimize interactions among the gas molecules.

of porous earthenware (with pore diameters small compared to the distance between the molecules of the gas), as well as by an airtight sheet of metal, an expansion of gas will take place when the metal barrier is permitted to fall until the gas is distributed uniformly throughout the two vessels. These examples illustrate processes that proceed spontaneously without the expenditure of outside effort. They are called *spontaneous* processes.

In each of the spontaneous processes in the preceding systems—a *system* refers to any part of the world isolated by real or conceptual boundaries from the rest of the world (i.e., from its *surroundings*)—one of the properties of the system (temperature, electrical potential, hydrostatic pressure, concentration) is initially different in different *regions* of the system. The change that occurs is in the direction that eliminates regional differences in the property. The process proceeds until uniformity of the property is achieved throughout the system. This *equilibrium* condition, whether thermal, electrical, mechanical, or concentration equilibrium, is characterized by an absence of regional differences in the property.

A system that undergoes spontaneous change loses some of its energy to its surroundings and/or its constituent molecules or ions become arranged more randomly. The difference in energy level between the initial and final states of the system can be "harnessed" to do work. Thus in the first example mentioned, spontaneous flow of heat between bodies at different temperatures can be made to operate a steam engine. In such an engine, heat flows from a heat reservoir into liquid water, which is converted to vapor and, on expansion, drives a piston. In the second example a difference in electrical potential can be used to drive an electrical motor. In the third example [see Figure 10-1(a)] the downhill flow of water can be used to drive a water wheel or a turbine at a dam site. In the fourth example [see Figure 10-1(b)] the expansion of a gas can drive a piston.

In order for a system to undergo a change *opposite* in direction to the one that would proceed spontaneously, work must be done on it. To increase the temperature of a body requires that heat or another form of energy be added to it from the surroundings. Outside work in the surroundings is needed to make electrical potential, initially uniform throughout the length of a wire, higher at one end than the other. Work must be done to restore water in the equilibrium condition in Figure 10-1(a) (right) to the initial condition [Figure 10-1(a), left]. And work is needed to transfer gas molecules uniformly distributed throughout two vessels [Figure 10-1(b), right] to one vessel [Figure 10-1(b), left]. Such examples are called *nonspontaneous* processes.

Some substances are transported spontaneously into and through plants whereas transport of others is nonspontaneous. Diffusion provides a good example of a spontaneous transport process in plants.

TRANSPORT BY DIFFUSION

Whenever a nonequilibrium distribution of molecules or ions of an individual substance exists in a system, movement of the constituent molecules or ions of the substance happens spontaneously. If a can of a volatile substance, such as ethyl ether, is opened in a room, for instance, ether molecules will soon be distributed until their concentration is the same throughout the room. Another example involves placing a small crystal of a water-soluble dye at the bottom of a test tube and then pouring water carefully and gently over the crystal. Dye molecules will dissolve and the color will spread slowly throughout the water, partly because of the movement of dye molecules through the water and partly because of movement of water molecules into a region close to the crystal. Figure 10-1(b) provides still another example.

Assuming that temperature in these three examples is maintained constant throughout the system, net movement of the constituent molecules or ions of the substances initially distributed in a nonequilibrium manner will be from regions of higher concentration of the substance to regions of lower concentration. Regional differences in concentration of the substance will tend to disappear with time. This process, which may occur in gases, liquids, and solids, is diffusion. *Diffusion* is the intermingling of molecules of the same or different species as a result of their random thermal agitation (i.e., inherent kinetic energy). (In Figure 10-1(b) intermingling of gas molecules occurs during and after their initial spontaneous movement into the vacuum.)

Transport of a chemical species by diffusion results from the mobility and internal kinetic energy of component molecules or ions of the diffusing substance. If molecules of other substances are present in the pathway of the diffusing substance, only the rate at which diffusional equilibrium is achieved—not the final equilibrium—is affected.

The Scottish botanist R. Brown was the first to describe a visible manifestation of inherent kinetic energy. In 1828 he found that minute particles suspended in a liquid medium and just large enough to be visible under an optical microscope exhibit a rapid and continually irregular motion, a phenomenon now referred to as *Brownian* movement. This type of motion can be observed by suspending a few specks of dust in a droplet of water on a cover slip and examining them under a microscope. Also, the naturally occurring microcrystals of calcium sulfate in vacuoles of certain green algal cells (of the desmid group) are suitable for observing Brownian movement. (Even mitochondria and other organelles can sometimes be observed to undergo Brownian movement in living cells of higher plants.) Each suspended particle undergoes constant bombardment by vast numbers of small molecules, especially those of water, and so is itself propelled to move. At any instant, however, the numbers of impacts are unequal at different faces of a particle. Thus the suspended particle moves now in one direction and now in another. This *random walk*—the path traversed by a particle that moves in steps in one or more dimension, each step being determined by chance in regard to direction—makes it impossible to predict in advance the direction in which the particle will move in a given interval of time.

The importance of diffusion in plant life cannot be exaggerated. Diffusion is an effective means of transport of matter over very short distances of the order of the thickness of a protoplasmic membrane (e.g., plasma membrane, tonoplast) or a cell wall. Over these short distances, movement by diffusion is very rapid.

TRANSPORT BY MASS FLOW

In contrast to diffusion, which depends on spontaneous movements of individual molecules or ions, another type of transport in plants involves movement of molecules or ions en masse. *Mass flow* is the unidirectional movement of assemblages of molecules or ions either spontaneously or nonspontaneously. Examples of mass flow (also referred to as *bulk flow*) are familiar in everyday life. Water moves through a garden hose by mass flow; the driving force is the difference in hydraulic pressure between the two ends of the hose. Mechanical stirring of water in a vessel gives rise to mass motions of water; the driving force is provided by the stirring action. A third example of mass flow is the convection currents in a vessel of water heated on a stove. The water in the bottom of the vessel becomes hotter—and less dense—than that near the top. So the cool water sinks and the warm water rises toward the top. Consequently, currents are set up within the liquid. The driving force for these mass motions of water is the difference in density of water between the lower and higher levels. Figure 10-1(a) provides still another example of mass flow.

The transport of atmospheric gases to and from the leaves of higher plants depends largely on mass flow. The air in the atmosphere beyond the leaf is agitated constantly, sometimes violently, by winds. In some cases, winds are quite turbulent, with movement in many directions; in other cases, winds are convection currents caused primarily by solar radiation (heating or cooling of masses of air). Vertical movements of masses of air due to heating through the day also help to mix the gases in the atmosphere. Both turbulent and convective mass movements of air ensure that localized depletion of CO_2, O_2, and water vapor do not occur in the vicinity of plants.

The upward longitudinal transport of water and dissolved substances in the xylem elements of higher plants depends on mass flow. The driving force is the difference in hydrostatic pressure in xylem sap between roots and leaves; this driving force develops as a result of the evaporation of water in stomatal cavities of leaves (see p. 400).

Cytoplasmic streaming is another example of mass flow. When the soluble phase of the cytoplasm moves, entrained organelles and inclusions are carried along. This phenomenon is familiar to all students who have completed a course in elementary botany. Rates of linear (translational) movement averaging 25 mm h^{-1} at room temperature have been recorded.

Cytoplasmic streaming in a plant cell is a sure indication that the cell is alive. The motive force depends on the contraction of microfilaments (i.e., elongated protein molecules 5 to 6 nm in diameter) and is linked to the utilization of ATP. The mechanism of the mechanochemical coupling between contractile proteins and ATP, however, is still to be determined.

TRANSPORT ACROSS PROTOPLASMIC MEMBRANES

A special type of transport in higher plants should be emphasized: This is transport *across* protoplasmic membranes (e.g., the plasma membrane, tonoplast, and cyto-membranes, such as those around chloroplasts and mitochondria).

Some uncharged solutes (nonelectrolytes) simply diffuse across protoplasmic membranes because of their solubility in the substances of which the membrane is composed. On the other hand, other solutes (e.g., dissolved ions) are able to cross protoplasmic membranes only by coupling to metabolic reactions. Transport across protoplasmic membranes, whether diffusive or metabolic, is an important field of investigation in plant physiology and is discussed in the next chapter.

DRIVING FORCE AND FLUX

A transport process results from the operation of certain forces acting in such a way as to impart a tendency for movement in a particular direction. These forces are called driving forces. A *driving force* propels matter (or heat or electricity) from one region in space to another.

Several examples of driving forces operative in mass flow have been noted. To illustrate further the nature of driving forces, let us consider the driving force operative in each of the spontaneous processes considered earlier. The driving force for the flow of heat between two bodies at different temperatures is the difference in temperature between them; the greater the difference in temperature, the greater the driving force. The driving force for the flow of electrical current between two ends of a wire is the difference in electrical potential in volts between the two ends; the greater the difference in voltage, the greater the driving force. The driving force for the flow of liquid water between two regions at different hydrostatic pressure [see Figure 10-1(a)] is the difference in hydrostatic pressure between the regions; the greater the difference in hydrostatic pressure, the greater the driving force. The driving force for diffusion of a gaseous substance between two regions at different concentrations [see Figure 10-1(b)] is the difference in concentration of the substance between the two regions; the greater the difference in concentration, the greater the driving force.

In each example the driving force propelling transport from one region to another is the difference in magnitude of a property (e.g., temperature, electrical potential, hydrostatic pressure, concentration) of the transported substance. Also, in each of these examples the driving force is physical in nature. The direction of transport is from the region of greater intensity of the property (i.e., from the region of higher temperature, or higher electrical potential, or higher hydrostatic pressure, or higher concentration of the transported substance) to the region of lesser intensity of the property (i.e., to the region of lower temperature, or lower electrical potential, or lower hydrostatic pressure, or lower concentration of the transported substance).[2]

As indicated in the preceding section, some solutes are propelled across cellular membranes by energy derived from metabolism. In these cases, the driving force is metabolic in nature.

The operation of a driving force, whether physical in nature or metabolic in origin, produces a net flow of matter. The intensity of flow may be evaluated in terms

[2]One of the driving forces for spontaneous transport processes may be an increase in entropy. *Entropy* is a thermodynamic property that refers to the degree of randomness or disorder of a system. An increase in entropy during the course of transport of a substance means that the system becomes more disordered. However, the fact that an increase in entropy may serve as a driving force is of only peripheral concern in this book. Indeed, one of the reasons for introducing the concept of driving force at this point is to lay the foundations for the concept of chemical potential. We will see that the differences in chemical potential of a substance in two regions is a measure of the driving force acting to move the substance from one region to the other. Because a difference in the chemical potential of a substance incorporates a difference in entropy, the subject of entropy is not considered separately in this book.

of flux. *Flux* is the quantity of a substance passing a unit cross-sectional area of surface in a unit of time. To illustrate the mode of expression of flux, consider the following examples. The flux for movement of liquid water from a soil solution into a root might be expressed in microliters of water per square centimeter of root surface per minute ($\mu l \, cm^{-2} \, min^{-1}$). For water vapor released by a leaf during transpiration, the flux might be expressed in milligrams of water vapor per square decimeter of leaf surface per hour ($mg \, dm^{-2} \, h^{-1}$). For an ion that crosses the plasma membrane of a cell, the flux might be expressed in micromoles per square micron of cell membrane surface per second ($\mu mol \, \mu m^{-2} \, s^{-1}$).

A useful relationship between a physical driving force and the flux that it generates is derived on pp. 314–315.

In certain experimental work it may be more convenient to indicate the amount of material absorbed per unit weight of tissue rather than per unit cross-sectional area of surface of tissue. Thus the flux of a solute into a tissue might be expressed in micromoles of solute absorbed per gram of fresh weight of tissue per hour [$\mu mol \, (g \, of \, tissue)^{-1} \, h^{-1}$].

IDENTIFICATION OF DRIVING FORCES IN SPONTANEOUS TRANSPORT PROCESSES

When a substance is transported in plants by one or another of the physical driving forces mentioned, the driving force is readily identified in some cases but not in others.

An example of a readily identifiable driving force is the one that accounts for the movement of water vapor through stomatal pores in leaves, from evaporating sites in stomatal cavities to the external atmosphere. Because water vapor is a gas, water vapor molecules are far apart and do not interact with one another or with molecules of other gases. So the difference in the concentration of water vapor within the stomatal cavities and in the external atmosphere is a measure of its tendency to move spontaneously. This difference in concentration provides a satisfactory measure of the magnitude of the driving force.

Not all driving forces can be identified or their magnitudes ascertained in such a straightforward manner. In the case of the movement of liquid water between two regions of a plant, several complex factors are involved. Unlike water molecules in the vapor state, water molecules in plant cells are in close association with other cellular components and interact with molecules of other substances by intermolecular attractive forces. These attractive forces reduce the ability of water molecules to move from one region to another in the plant, compared to their ability to move in the vapor state.

Still another factor should be considered in evaluating the magnitudes of the total driving force for movement of liquid water in plant cells and tissues. Assuming that temperature is the same throughout a plant and that temperature difference is not a driving force, nevertheless more than one driving force may be acting at the same time to move water from one region to another. One driving force may be due to differences in two regions of a plant in the affinity with which water is held to different cellular constituents. In addition, driving forces due to differences in hydrostatic pressure in different regions of a plant may develop.

In evaluating the total magnitude of the driving force propelling transport of water from one point to another in plants, physiologists use a concept known as *chemical potential*. We will see that the *difference* in *chemical potential* of water in two regions gives a quantitative measure of the total driving force propelling transport of water from one region to the other.

CHANGES IN CHEMICAL POTENTIAL DURING TRANSPORT PROCESSES

It is a well-known fact that each and every mobile chemical species in any system has a tendency to move spontaneously away from a region in which it is present to another region to which the substance has access and has a lesser concentration. The tendency of gaseous substances to move spontaneously from one region to another is shown by their ability to diffuse, expand indefinitely, and exert a pressure when confined within a finite volume. An example of the spontaneous diffusion of a gas into a larger volume was noted in Figure 10-1(b). The tendency to migrate spontaneously is less apparent for liquids and solids than for gases but still is evident in their ability to vaporize. The term *escaping tendency* is a   qualitative apt description of the tendency of a substance to move away from a given region and to pass into another spontaneously.

If the escaping tendency of a substance is greater in one region than another and if no impenetrable barrier exists between the regions, then a net transfer of the substance from the first region to the second will occur. On the other hand, the transfer of the substance from the second region to the first (i.e., in a direction opposite to that of the spontaneous change) requires an expenditure of work. Finally, if no spontaneous net transfer of a given substance occurs between two regions—both accessible to the substance—then an equilibrium exists with respect to transfer of that substance. Therefore the escaping tendency of the specified substance is the same in the two regions.

Naturally a quantitative means of expressing the escaping tendency of a substance in a region of a system is desirable. [For an ion or molecule dissolved in water, the concentration (e.g., mol L^{-1}) relative to that in an arbitrarily designated standard state is not satisfactory; attractive forces between dissolved chemical species and water exist and are difficult to measure.] To resolve this problem, plant physiologists have turned to the science of thermodynamics. Thermodynamics deals with energy and work relations of systems undergoing physical or chemical change.

Thermodynamics tells us that the driving force for the spontaneous transport of a substance from one region to another can be measured accurately in terms of a property known as chemical potential. The *difference in chemical potential* between the initial and final states of the substance in two regions is the driving force for the

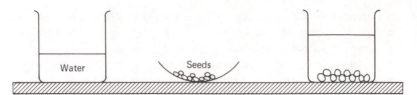

Figure 10-2 Because the chemical potential of water is higher in pure liquid water than in dead, air-dried pea seeds (see the text), a spontaneous transfer of water will take place into the seeds when they are immersed in the liquid. In a few hours the chemical potential of water in the immersed seeds will become equal to that in liquid water and an equilibrium condition with respect to transfer of water will have been reached.

transport of the substance from one region to the other.[3] All the preceding statements on escaping tendency also apply to chemical potential. In the term chemical potential the word *potential* (i.e., what is possible but not necessarily realized) is carried over from analogies to other familiar terms, such as electrical potential and gravitational potential.

To gain a better understanding of chemical potential, let us compare the chemical potential of water in the two systems shown in Figure 10-2. The first system is pure liquid water and the second dead, air-dried pea seeds in which only a small amount of water is present, together with many other plant consitutents. It makes no difference whether a cup or a liter or a swimming pool of water is taken—or a handful or a bushel of dry seeds—because the chemical potential of a substance (in this case, water) is evaluated in terms of a mole of the substance. Because temperature is one of the factors affecting the value of the chemical potential of a substance, we will assume, to simplify the situation in Figure 10-2, that both the beaker of water and the dry seeds are at the same temperature (e.g., 25° C).

As already indicated, water molecules in plant tissues are bound to other plant constituents by adsorptive forces. Therefore the escaping tendency—and the chemical potential—of the water initially present in the dry seeds in Figure 10-2 is *less than* that of pure water.

Because of the lower chemical potential of the water in the dry seeds compared with that in the beaker, there will be a net movement of water molecules into the dry seeds when they are immersed in the water in the beaker (see Figure 10-2). The driving force initially propelling transport of water into the dry seeds will be equal to the difference between the chemical potential of water in liquid water and in the "dry" seeds. As the seeds absorb more and more water, the difference in chemical potential of

---

[3] The *chemical potential* of an individual chemical substance in a system is the free energy of 1 mole (= 6×10^{23} molecules) of the substance. The term free energy in this definition of chemical potential denotes the maximum of useful work potentially available from the spontaneous transport of the substance from one region to another under conditions of constant temperature and constant external pressure. (*Note*: The term chemical potential is also used in a broad sense to denote the tendency of a substance to enter into *any* chemical or physical change; our concern in this book, however, is only with transport from one region to another.)

the water in the two systems—and hence the driving force—will become progressively smaller and smaller. In a few hours, when the seeds become saturated with water, the chemical potential of the water in the seeds will come to be equal to that of the liquid water in the beaker and net transfer of water between the two systems will cease. At this equilibrium point a dynamic interchange of water molecules—rather than a net transfer—will take place between the two systems.

Absolute values of the chemical potential of most substances cannot be measured. Instead the chemical potential is evaluated numerically in terms of differences in two states or regions. A reference state is selected and the chemical potential of a substance is expressed as the difference in chemical potential in the reference state and in the system under consideration. In the case of water, the reference state conventionally designated is that of pure water, at the same temperature (e.g., 25° C) and the same atmospheric pressure as the system under consideration. This reference value is arbitrarily assigned zero. Thus in Figure 10-2 the chemical potential of water in the dry seeds is a negative number because it is less than that of pure water. Its exact value can be determined by methods described on pages 358–364.

Our main concern in this book with respect to chemical potential is the chemical potential of water (pp. 357–358); only brief reference will be made to the chemical potential of solutes and the electrochemical potential of ions in solution.

SOLUTE ENTRY INTO PLANT CELLS

During the passage of a substance from outside a plant cell to its interior—for example, from a soil solution into cortical cells in a primary root—several layers of different kinds of materials are traversed. (See Figure 10-3.) First, there is the unstirred film of water that adheres to the exterior boundary of the cell wall. Generally dissolved molecules and ions penetrate through this layer of water quickly and readily by diffusion.

Next, there is the cell wall. The primary cell wall of a growing plant cell consists mainly of cellulose and noncellulosic polysaccharides. Cellulose is composed of glucose molecules that are linked to one another in straight chains (p. 64). The chains are packed together in regular, parallel, partially crystalline arrays (microfibrils) embedded in an amorphous matrix of noncellulosic polysaccharides. These polysaccharides (often called hemicelluloses) are composed of sugars other than glucose (e.g., xyloglucans). Also, pectic substances, a group of polysaccharides composed in part of polygalacturonic acids, are present in the amorphous matrix.[4] Polygalacturonic acids have weakly acidic carboxyl groups ($-COOH$) that ionize and give rise to fixed negative charges ($-COO^-$) and loosely held hydrogen ions. Thus

---

[4]The *fibrous* materials recommended during the past decade by nutritionists for the human daily diet are the indigestible residues of certain foods (e.g., whole grains, fruits, and vegetables); these residues consist of cellulose, hemicellulose, and pectin in primary cell walls and, in addition, lignin in secondary cell walls. All these substances resist the action of degradative enzymes in the human digestive tract.

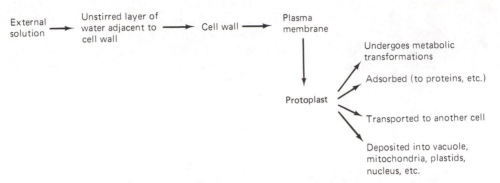

Figure 10-3 To gain entrance into the protoplast of a living cortical cell of a primary root, a substance in a soil or nutrient solution must cross several layers of materials of different nature. These are indicated in the diagram. Not shown here is a layer of mucilaginous material secreted on the outer surface of epidermal cells of some roots, especially the *polysaccharide slime mass* found near root tips. Once having gained entrance into the protoplast, several of the possible fates of entering molecules and ions are indicated in the right-hand column. Note that the arrows indicate entry only. However, exit may also occur; some molecules and ions inside a plant cell may pass out into the external solution.

positively charged ions (e.g., K^+, Mg^{2+}, Ca^{2+}) passing through a plant cell wall will displace the hydrogen ions on the carboxyl groups and will be held by relatively weak interionic attractive forces to the fixed negative charges. These fixed negative charges in plant cell walls are called *cation adsorption sites* or *cation exchange sites*.

Different cations have different affinities for cation exchange sites, depending largely on their charge. A cation with a relatively high adsorptive affinity (e.g., Ca^{2+}) will displace one with a lesser adsorptive affinity (e.g., K^+). The adsorptive affinity of a cation will determine whether it diffuses across a cell wall in a "leapfrog" fashion, by migrating from one negatively charged adsorptive site in the cell wall to another, or diffuses through the water held within the cell wall.

Water forms a large portion of cell walls. It adheres to both cellulosic and noncellulosic cell wall components. Also, relatively large passageways filled with water traverse a primary cell wall. These passageways have been estimated to have diameters of about 400 nm. (In comparison, a water molecule has an apparent diameter of 25 nm.) Thus the primary cell wall has an "open" structure. In fact, except for relatively weak cation attractive forces referred to earlier, a primary cell wall offers little hindrance to the movement of substances across it. Generally water and dissolved nutrient molecules and ions, as well as dissolved metabolites of the size of glucose, sucrose, amino acids, and so forth, diffuse readily across primary cell walls.

Only when a substance reaches the exterior surface of the protoplasm of a plant cell does it encounter an effective barrier. This boundary layer, the plasma membrane, confers on a plant cell the ability to discriminate among different chemical species presented to it, restricting the passage of some but permitting the passage of others (see Chapter 11). Thus the plasma membrane can be said to separate the "interior" of a plant cell from its "outside" environment. Until a substance passes into and becomes part of the protoplasmic matrix, it is outside of a plant cell in the same sense that a morsel of food inside the digestive tract of an animal is topologically outside the animal body.

A substance that enters the protoplast of a plant cell after it crosses the plasma membrane may undergo one of several possible fates, some of which are indicated in Figure 10-3. The substance may be transformed by a metabolic reaction into another substance. Or it may be adsorbed, perhaps to a protein molecule. Or it may be transported to another cell. Or the substance may be transported to another part of the same cell, perhaps across the tonoplast into the vacuole or into such organelles as mitochondria or plastids.

FREE SPACE

The term *free space* refers to the fraction of the volume of a plant tissue readily accessible to diffusion of an externally applied solute dissolved in water. Included in free space are primary cell walls. As noted, primary cell walls offer relatively little hindrance to diffusion of dissolved solutes. The boundary of free space is the plasma membrane because most solutes (e.g., dissolved nutrient ions and sugars) do not diffuse readily through it, but penetrate by active transport (see pp. 317–319).

Free space is a functional concept, the dimensions of which can be measured by physiological experiments described below, and should not be confused with the *apoplast*. The latter is an anatomical designation of the nonliving parts in a plant tissue and so refers to all cell walls in a plant tissue.

The entry of solute into free space can be visualized by immersing a plant tissue into a solution and following the time course of uptake. The solid lines in Figure 10-4 show the time course of uptake of an anion (radioactive sulfate) and a cation (radioactive strontium) by excised root tissues. Both of these solid lines rise very steeply at first, indicating a very rapid uptake of solute immediately on immersion of the tissue in the solutions. After about 30 minutes, the period of rapid absorption ceases and is replaced by a period of slower absorption.

Let us examine the desorption process next. When the plant tissue is removed from the radioactive sulfate solution and placed either in water or a solution of nonradioactive sulfate, most of the radioactive sulfate diffuses out of the tissue. The dashed line (Figure 10-4, upper) shows the time course for exit of radioactive sulfate from the tissue. But not all the radioactive sulfate is removed. About 10 to 15% is retained by the roots even after 60 minutes in water or in a solution of nonradioactive sulfate. This indicates that about 10 to 15% of the sulfate ions in the tissue is not free to diffuse out of the tissue. This fraction of sulfate ions can be presumed to have crossed the plasma membranes of root cells and to have entered the "inner space."

In addition, two strontium fractions—free space strontium and inner space strontium—can be distinguished by desorption experiments. Entrance of cation into free space, however, is characterized by a feature not characteristic of anion entry.

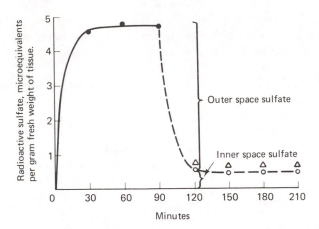

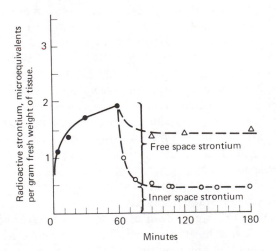

Figure 10-4 Time course of absorption and desorption of radioactive sulfate ($SO_4^=$) and radioactive strontium (Sr^{2+}) by excised roots of barley (*Hordeum vulgare*). *Upper*: Black circles and solid line: roots in solution of radioactive sulfate (K_2SO_4, 20 milliequivalent per liter). Open circles and broken line: roots in water. Open triangles and broken line: roots in solution of nonradioactive sulfate ($CaSO_4$, 20 meq per liter). (After Epstein, 1955; Figure 1.) *Lower*: Black circles and solid line: roots in solution of radioactive strontium ($SrCl_2$, 1 meq per liter). Open triangles and broken line: roots in water. Open circles and broken line: roots in solution of nonradioactive strontium ($SrCl_2$, 1 meq per liter). (For further details, consult the text.) After Epstein and Leggett, 1954 (Figs. 2 and 6).

Some cations in free space will be adsorbed to cation adsorption sites in cell walls. Thus the loss of radioactive strontium from free space into pure water is less than into a solution of nonradioactive strontium (compare the two dashed lines in Figure 10-4, lower). (The lower loss of radioactive strontium into pure water, compared to loss into a solution of nonradioactive strontium, simply reflects the fact that some of the strontium ions in cell walls are adsorbed to cation adsorption sites.) To remove cations adsorbed to cation adsorption sites in cell walls requires exchange of radioactive strontium with other cations (e.g., nonradioactive strontium ions). Water alone cannot displace cations on adsorption sites in cell walls.

Free space in a primary root consists of at least all the region occupied by primary cell walls (i.e., all the apoplast) and intercellular spaces in cortical tissue. It generally extends inward only to the endodermis (see Figure 10-5). Radial and transverse cell walls—but not tangential cell walls—of endodermal cells are impregnated with suberin. *Suberin* is a polymer that contains a phenolic matrix somewhat similar to lignin; within this matrix is embedded a variety of waxes and fatty acids (for further details, see Chapter 3). In the endodermis, suberin takes the form of a bandlike structure, called the Casparian strip, that extends completely around individual endodermal cells (see Figure 10-5). The plasma membrane of an endodermal cell adheres firmly to the Casparian strip. Therefore the Casparian strip acts as an impermeable barrier to the inward radial movement of water and dissolved nutrients through cell walls and so marks the internal boundary of the free space of the primary root. The effectiveness of this endodermal barrier, however, may be reduced in primary roots in which the Casparian strip is poorly developed or broken and/or not continuous. Thus the boundary of free space in a primary root sometimes may extend beyond the endodermis.

The magnitude of free space in a tissue can never be measured directly. Free space must be estimated indirectly from data obtained in experiments like those in Figure 10-4. In his calculation of sulfate free space from data in Figure 10-4, Epstein (1955) used the formula

Free space (in milliliters per gram fresh weight of tissue)

$$= \frac{\text{microequivalents of diffusible sulfate per gram fresh weight}}{\text{ambient sulfate concentration (in milliequivalents per liter)}}$$

where *ambient sulfate concentration* refers to the concentration of external sulfate during the initial absorption period. In applying this formula, it should be assumed that the concentration of sulfate ions in the free space is the same as in the ambient solution when equilibrium is reached.

To focus attention on the approximate nature of estimates of free space in plant tissues, the term apparent free space is often used in place of free space. *Apparent free space* is the apparent volume of the free space in a plant tissue. Estimates of apparent free space range from approximately 5 to 20% of the total volume of fresh plant tissue. These estimates depend on the solute used to make the estimate, plus the age and kind of tissue.

PATHWAY OF MOVEMENT OF NUTRIENT IONS ACROSS THE PRIMARY ROOT

The movement of dissolved nutrient ions across a primary root, from the soil solution to xylem vessels and tracheids in the stele, has been the subject of much research during recent years. This inward movement occurs most rapidly where root hairs reach their maximum length (see Figure 10-6). In this region vessels and tracheids in the xylem are fully mature and, for the most part, have lost their protoplasts and are not alive.

As noted, the single row of endodermal cells that separates the cortex from the stele has a structure in its cell walls—the Casparian strip—that is impermeable to water and dissolved ions and acts as a barrier that prevents inward movement of ions through cell walls into the stelar region.

It is from free space that dissolved nutrient ions in the soil solution are absorbed

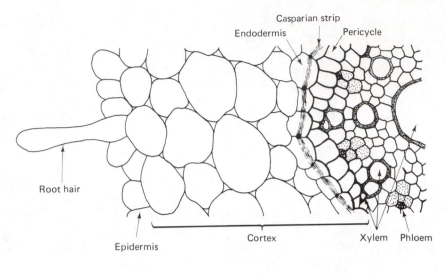

(a)

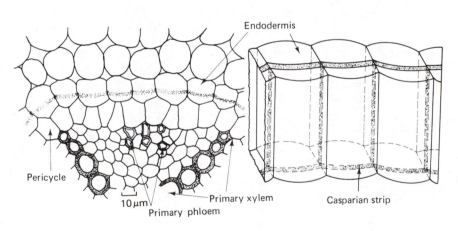

(b)

Figure 10-5   (a) Cross-sectional diagram of part of a primary root of wheat, at the level of root hairs. The Casparian strip in the endodermis is shown as if all the transverse walls are in the plane of the paper. ×330. (After Esau, 1953; Figure 17.13.) (b) Diagram to illustrate structure of endodermis. *Left*: cross section of part of the primary root of morning glory, showing the endodermis in relation to other tissues. Endodermal cells are shown as though all the transverse walls are in the plane of the paper. *Right*: three connected endodermal cells oriented as they are in the left-hand drawing. Note that the Casparian strip occurs in transverse and radial walls but not in tangential walls. (After Esau, 1960; Figure 14.2.)

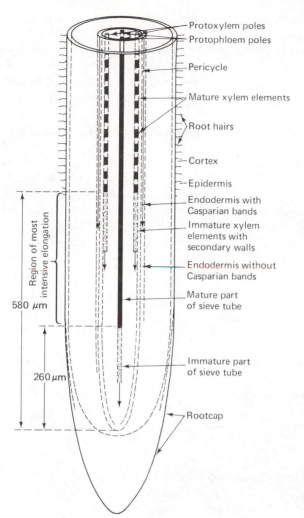

— Protoxylem poles
— Protophloem poles

— Pericycle

— Mature xylem elements

— Root hairs

— Cortex

— Epidermis

— Endodermis with Casparian bands

— Immature xylem elements with secondary walls

— Endodermis without Casparian bands

— Mature part of sieve tube

— Immature part of sieve tube

— Rootcap

Region of most intensive elongation

580 μm

260 μm

Figure 10-6 Diagram of the root tip of tobacco, in longitudinal section. (1 μm = 10^{-4} cm.) (After Esau, 1953; Figure 17.7.)

by root cortical cells. The process involves transport across the plasma membranes of these cells.

In primary roots in which the Casparian strip is intact and effectively continuous around the stelar region, the major pathway of mineral ions across the root is through the protoplasm of root cells, from the protoplast of one cell to that of the next. The protoplasts of all cells of the primary root are interconnected and united by plasmodesmata. Plasmodesmata are threads of cytoplasm that extend through cell walls of adjoining cells (see Chapter 2). This continuum of protoplasts of many cells, together with the plasmodesmata that connect them, is called the *symplasm* or *symplast*. Having entered the symplast, nutrient ions diffuse inward, aided by cytoplasmic streaming and inward flow of water. Because the symplast extends from the cortex into the stele and penetrates the endodermis, clearly movement of nutrient ions from cortex to stele does not require them to be transported across the plasma membrane at the endodermis.

When nutrient ions reach the parenchyma cells of the stele—after having moved successively through the cortex, endodermis, and pericycle—they leave the symplast[5] (by transport across the plasma membrane) and are deposited into xylem tracheids and vessels, where they are translocated upward to leaves in the transpiration stream (see pp. 401–402). Thus movement of nutrient ions across the primary root requires transport across the plasma membrane at two sites: uptake at the plasma membrane of cortical cells and secretion at the plasma membrane of xylem parenchyma cells. [The mechanism of transport across the plasma membrane is considered in Chapter 11.]

It was mentioned that the Casparian strip may not be continuous in all primary roots. Occasional breaks may occur. In these cases, dissolved ions may move across the root, from cortical cells to stele, in cell walls (i.e., in the apoplast) and in films of water that adhere to the outer surfaces of cell walls, without once traversing a plasma membrane. Movement in this pathway is by carriage in a mass flow of water. This second (apoplast) pathway seems only of minor importance in movement of dissolved mineral ions across primary roots, however.

REFERENCES

ANDERSON, W. P. 1976. Transport through roots. Pages 129–156 *in* A. Pirson and M. H. Zimmermann, eds. *Encyclopedia of Plant Physiology*, New Series, vol. 2, part B. Springer-Verlag, Berlin.

EASTMAN, E. D., and G. K. ROLLEFSON. 1947. *Physical Chemistry*. McGraw-Hill Book Company, New York.

EPSTEIN, E. 1955. Passive permeation and active transport of ions in plant cells. *Plant Physiol.* 30:529–534.

EPSTEIN, E., and J. E. LEGGETT. 1954. The absorption of alkaline earth cations by barley roots: kinetics and mechanism. *Am. J. Bot.* 41:785–791.

ESAU, K. 1953. *Plant Anatomy*. John Wiley & Sons, New York.

ESAU, K. 1960. *Anatomy of Seed Plants*. John Wiley & Sons, New York.

LÄUCHLI, A. 1976. Apoplasmic transport in tissues. Pages 3–34 *in* A. Pirson and M. H. Zimmermann, eds. *Encyclopedia of Plant Physiology*, New Series, Vol. 2, part B. Springer-Verlag, Berlin.

NAGAHASI, G., and W. W. THOMSON. 1974. The Casparian strip as a barrier to the movement of lanthanum in corn roots. *Science* 183:670–671.

PITMAN, M. G. 1977. Ion transport into the xylem. *Ann. Rev. Plant Physiol.* 28:71–88.

SPANNER, D. C. 1964. *Introduction to Thermodynamics*. Academic Press, New York.

SPANSWICK, R. M. 1976. Symplasmic transport in tissues. Pages 35–53 *in* A. Pirson and M. H. Zimmermann, eds. *Encyclopedia of Plant Physiology*, New Series, vol. 2, part B. Springer-Verlag, Berlin.

TEPFER, M., and I. E. P. TAYLOR. 1981. The permeability of plant cell walls as measured by gel filtration chromatography. *Science* 213:761–763.

---

[5] After sugars are unloaded from sieve tubes in the phloem, they move across the root from the stele to cortical and epidermal cells mainly in the symplast rather than the apoplast.

11

Transport Processes
in Plant Cells and Tissues

Cells were first examined microscopically three centuries ago. The Englishman Robert Hooke is said to have been the first, in 1685, to use the term *cell* in its biological sense. But only in the middle of the nineteenth century was protoplasm recognized as the material uniquely endowed with the properties of life. And it was only in the latter half of the nineteenth century—about 100 years ago—that studies were begun with the goal of elucidating the properties of protoplasm.

These first studies of the nature of protoplasm dealt primarily with the ability of different substances to penetrate into living cells. From these early investigations it became clear that the protoplast of a plant cell is very permeable to some substances but only slightly permeable to others. It was also recognized that the outermost border region of the protoplast, now called the plasma membrane or plasmalemma, controlled the passage of substances into the protoplasm. Out of these late-nineteenth-century studies arose the concept that some substances gain entrance into individual plant cells by diffusing across the plasma membrane. In later years, especially the twentieth century, it was realized that other substances gain entrance into individual plant cells only by engaging in a biochemical reaction with certain components of the protoplasmic membrane. These and other types of cell membrane transport are considered in this chapter.

Although in most of this discussion we are concerned with transport phenomena in individual cells, near the end of the chapter transport through several layers of cells and also through sieve tubes in phloem tissue is described. The special importance of water in the life of higher plants dictates that its transport into, through, and out of plants be discussed separately (see Chapters 12 and 13).

EARLY STUDIES OF MOVEMENT OF SOLUTES INTO
PLANT CELLS

The systematic study of the ability of plant cells to absorb dissolved substances began in the second half of the nineteenth century. The method used in most of these early investigations was based on the behavior of cells during their recovery from plasmolysis.

Plasmolysis refers to the separation of the living protoplast of a plant cell from the cell wall when the cell is immersed in a solution that withdraws water. To demonstrate plasmolysis, a tissue sample (e.g., a thread of a filamentous alga, or an entire leaf of an aquatic plant, such as *Elodea*, or a thinly sliced section of tissue from an herbaceous plant, or a layer of epidermal cells stripped gently away from a leaf immediately before the demonstration) is immersed in a solution whose solute concentration is higher than the total solute concentration in the cell sap vacuole. Such a solution is said to be *hypertonic* to the cell sap solution. In other words, it has a greater total solute concentration than that in the cell sap solution when the cell is in a condition of zero turgor (see pp. 367–368). Sugars, such as glucose or sucrose, and salts, such as calcium chloride, are often used to prepare plasmolyzing solutions for demonstration purposes because these substances in solution pass slowly through protoplasmic membranes and are not toxic to plant cells.

On immersion of plant cells in a plasmolyzing solution, there is a net movement of water out of the cell sap vacuole and into the solution. [This outward movement of water is a manifestation of osmosis (see pp. 355–357).] Close observation of a single cell under a microscope will reveal that the cell loses its turgor in a short time, sometimes only a few seconds; the vacuole shrinks; the volume of the protoplast decreases; and the protoplast separates away from the cell wall. The space between the protoplast and the cell wall becomes filled with the plasmolyzing solution. Now the cell is said to be plasmolyzed [see Figure 11-1(c) and (d)]. Because only living cells can be plasmolyzed, plasmolysis is a good test of life.

If the plasmolyzed cell remains in certain plasmolyzing solutions, it may recover from plasmolysis. The recovery process will depend on the ability of the dissolved solute in the external plasmolyzing solution to penetrate through the protoplasmic layer and into the vacuole. This entrance of solute into the vacuole will be accompanied by the movement of water (again by osmosis) from the external solution into the vacuole. Entrance of water into the vacuole will cause it to become distended and so the volume of the protoplast will increase.

The rate of deplasmolysis depends on the rate at which a solute in the plasmolyzing solution enters a cell: The more rapidly the solute permeates into the vacuole, the more rapidly will deplasmolysis take place. By making microscopic measurements from time to time of the increasing volume of the protoplast during deplasmolysis, the rate of transfer of a solute from the plasmolyzing solution into the vacuole can be measured.

It is evident that determination of the rate of entrance of a substance into a cell by recovery from plasmolysis is an indirect method. Instead of measuring the entry of the

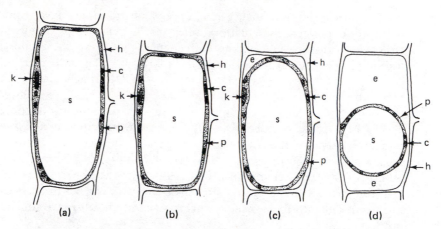

Figure 11-1 Plasmolysis of a plant cell. (a) Longitudinal section of a cell of the cortical parenchyma of a stalk of the immature florescence of *Cephalaria leucantha*. (b) The same cell in 4% solution of potassium nitrate (osmotic potential equals approximately −18 bars). (c) The same cell in 6% solution of potassium nitrate (osmotic potential equals approximately −27 bars). (d) The same cell in 10% solution of potassium nitrate (osmotic potential equals approximately −50 bars). (See text for details.) Abbreviations: h, cell wall; p, protoplasmic layer; k, nucleus; c, chloroplast; s, vacuolar sap; e, potassium nitrate solutions (After de Vries, 1877.) [*Note*: High potassium concentrations have a deleterious effect on protoplasmic membranes. Therefore potassium does not act as a *long-term* plasmolyticum (plasmolyzing agent); after a few minutes exposure to hypertonic potassium solution, potassium moves into cells and deplasmolysis is usually observed.]

substance directly, its rate of entrance into a cell is inferred from the effect that it has on the movement of water. Indeed, were it not for the ability of water to pass into and out of plant cells at a rate far in excess of almost all other substances, the plasmolytic method could not be used to determine the rates of entry of dissolved solutes.

With one or another modification of the plasmolytic method, the permeation of hundreds of different chemical compounds was investigated in the second half of the nineteenth century. Many compounds studied are not normally present in plant cells and are not metabolized to any appreciable extent. The Swiss botanist Overton was one of the best-known investigators in this field of research. At the turn of the century Overton was able to make several broad generalizations relating the chemical properties and structures of compounds and their abilities to pass through protoplasm.

In contrast to the extremely high permeation power of water, Overton found most substances to have lesser ability to permeate protoplasm. At the bottom of the scale Overton listed strongly ionized substances, such as mineral salts dissolved in water.

Of particular interest to Overton were organic nonelectrolytes (alcohols, aldehydes, sugars, etc.). These compounds exist as nonionized molecules in water solution at physiological pH values. Organic nonelectrolytes differ widely in their ability to permeate plant cells. Some permeate rapidly, a few as rapidly as water, whereas others permeate much less rapidly and still others have scarcely detectable permeation powers. Systematic studies of the relation of permeation power to molecular structure led Overton to recognize that lipophilic groups (e.g., —CH_3, —C_2H_5) in a molecule increase its permeation power whereas hydrophilic (i.e., polar) groups, such as hydroxyl (—OH), carboxyl (—COOH), and amino (—NH_2), lower the permeation power.

Overton was able to formulate in broad outline empirical relations between the ability of organic nonelectrolytes to permeate protoplasm and their molecular structures. Hydrocarbons have great permeation power. The monohydric alcohols (e.g., ethanol) permeate rapidly, trihydric alcohols (e.g., glycerol) rather slowly, and mono-, di-, and trisaccharides scarcely at all. Amino acids permeate extremely slowly. Overton stressed that even though cells differ with respect to entry rate of a given nonelectrolyte into their vacuoles, protoplasts of all cells have similar permeability properties.

Realizing that substances with lipophilic groups are more soluble in lipid (fat) solvents (e.g., chloroform, ether, hexane, benzene) than in water, Overton pointed out that a parallelism exists between the ability of a substance to permeate protoplasm and its solubility in lipid solvents. His experimental data, obtained from a wide variety of plant and animal cells, could be summarized as follows. The greater the solubility of a nonelectrolyte in lipid solvents compared to water, the greater its ability to permeate protoplasm.

On the basis of prior work by others, especially the German plant physiologist Pfeffer (who published his findings in 1877), Overton realized that the plasma membrane—the outermost boundary of the protoplast—is the major site of resistance to penetration of substances into cells. So he deduced that the plasma membrane consists largely of fatlike substances (i.e., lipids). Furthermore, Overton emphasized that the permeation of organic nonelectrolytes is a process of "dissolving" in and diffusing through the plasma membrane and differs fundamentally from the mechanism whereby strongly ionized substances gain entrance into cells. The latter process cannot be studied by the plasmolytic method. Other methods, considered later in this chapter, are required.

Finally, Overton did not attempt to explain how ionized salts enter living cells. It will be pointed out below that ions, and also certain organic metabolites (e.g., amino acids, mono-, di-, and trisaccharides), gain entrance into plant cells mostly by active transport processes rather than diffusion.

RECENT STUDIES OF MOVEMENT OF SOLUTES INTO PLANT CELLS

A classical problem of cell physiology is to explain selective permeation of substances through protoplasmic membranes, such as the plasma membrane. Interest in this problem stems partly from the physiological significance of the ability of different substances to penetrate protoplasm and partly from the information that may result regarding the nature of cell membranes.

A landmark in permeability literature is the work of the Finnish plant physiologist Collander and his colleagues, published from the 1930s to 1950s. These plant physiologists studied the abilities of many different organic nonelectrolytes to permeate plant protoplasm but did not use the plasmolytic method. Instead they immersed plant cells in relatively dilute (i.e., nonplasmolyzing) solutions of one or another substance and after a time determined by direct chemical analysis the amount of substance present inside the vacuole. The plants used were members of the family Characeae.

    The Characeae is a family of green algae that grows naturally in fresh or brackish water. Certain cells of some genera (e.g., *Chara* and *Nitella*) are several times longer than wide (Figure 11-2). Most of the volume of these cells consists of vacuole bounded by a thin layer of cytoplasm in which are embedded many nuclei. These cells are often 3 or more cm long and therefore large enough to be handled singly. By snipping the ends of several cells and gently extruding the contents of their vacuoles, a small quantity of

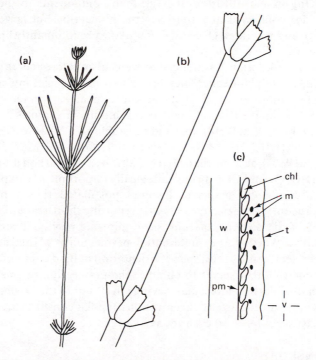

Figure 11-2 (a) A sketch of the apical portion of a plant of the family Characeae. The algal thallus typically is an erect, branched axis consisting of a succession of "nodes" and "internodes." Each node bears a whorl of "branchlets." (b) External view (approximately natural size) of an internodal cell. (c) Section through the cytoplasm of an elongated internodal cell: w, cell wall; chl, chloroplast; m, mitochondria, t, tonoplast; pm, plasma membrane (plasmalemma); v, vacuole. (*Note:* In some species of the genus *Chara* of the family Characeae, the internodal cell is sheathed by a layer of small, vertically elongated cells, one cell in thickness. This ensheathing layer is called the cortex but is not shown in the figure.) (After Briggs et al., 1961; Figure 7.)

vacuolar sap uncontaminated with cytoplasm and sufficiently voluminous to permit direct chemical analysis can be obtained. Because the overall entry of a solute, especially one not metabolized, is from the external solution into the vacuole, insight into the transport process is gained by comparing, during appropriately chosen time intervals, the vacuolar concentration of an entering solute with the concentration of the solute in the external solution.

Collander and his associates found that different organic nonelectrolytes move at different rates from an external solution into the vacuoles of Characean cells. They obtained quantitative data establishing that the permeation of most organic non-electrolytes across Characean cell membranes is governed by their *lipid solubility*. A substance preferentially soluble in lipid solvents has a higher permeation power than one less soluble in lipid solvents.[1] Thus Collander's work with giant algal cells established that organic nonelectrolytes, such as ethanol, propanol, glycerol, and ethylene glycol, to mention only a few, permeate through plasma membranes by dissolving in and diffusing through lipid components in the membrane. In effect, Collander confirmed quantitatively the pioneering but largely qualitative work of Overton and reiterated Overton's view of the predominantly lipid nature of the plasma membrane.

Collander and his associates were also interested in the entrance of small molecules into plant cells. They found that many small molecules, whether polar or nonpolar, move through plant protoplasm at rates much more rapid than can be expected on the basis of their lipid solubility. Examples are ammonia (NH_3, MW 17), water (H_2O, MW 18), hydrogen cyanide (HCN, MW 27), oxygen (O_2, MW 32), hydrazine (NH_2NH_2, MW 32), methanol (CH_3OH, MW 32), hydrogen peroxide (H_2O_2, MW 34), acetonitrile (CH_3CN, MW 41), and carbon dioxide (CO_2, MW 44). (Of these H_2O, O_2, and CO_2 are physiologically important.) To explain these observations, Collander and his associate Bärlund postulated that a plasma membrane acts simultaneously as a *selective solvent*, permitting the passage of substances that dissolve in lipids, and also as a *molecular sieve*, allowing very small molecules to pass through it readily. Whether the preferential permeability of plasma membranes to very small molecules is due to a special orientation of lipid components in the membrane—an orientation that serves to enhance the permeation of small molecules relative to large molecules by a molecular sieving effect—or to the existence of regions of the cell membrane with localized concentrations of polar constituents, or to some other factor, is a matter not resolved even today.

PASSIVE TRANSPORT

The studies of Overton and Collander illustrate a kind of transport that has come to be called passive transport. *Passive transport* refers to transport propelled by physical (i.e., nonmetabolic) driving forces. Examples of these kinds of forces were considered

---

[1]The permeation rate of an organic nonelectrolyte is governed by its *partition coefficient*. This parameter describes quantitatively the solubility of a substance in a lipid solvent compared to its solubility in water.

on p. 295. Thus the driving force for movement of a dissolved chemical species across a protoplasmic membrane, from a region of high to low chemical potential (concentration) of the species, is the difference in chemical potential of the species across the membrane. When the chemical potential of the moving species becomes equal on both sides of the membrane, net movement of the chemical species ceases.

Passive transport has an important role in the life of higher plants. At least three essential nutrients—H_2O, O_2, CO_2—enter and leave individual plant cells by passive transport processes. Also, passive transport accounts for the movement of gas molecules (water vapor, O_2, and CO_2) through stomatal pores. Furthermore, the translocation of water and dissolved solutes from roots to shoots in xylem elements occurs by passive transport.

The membrane (i.e., the sheet or layer of material through which passive transport takes place) in Overton's and Collander's work was the layer of protoplasm between the cell wall and the vacuole, although it was recognized that the major site of resistance to entrance of a substance into a cell is the plasma membrane (i.e., the outermost boundary of the protoplast). In movement of gases through stomatal pores, the "membrane" is a layer of "unstirred" air within the pores (see pp. 387–388). And in the upward flow of sap in xylem vessels in which cross walls are absent, there are no membranes at all.

In contrast to passive transport, active transport (see pp. 317–319) takes place *only* across protoplasmic membranes (e.g., plasma membrane, tonoplast).

PERMEABILITY

Plant membranes differ in their resistance to the movement of substances through them. Primary cell walls of plant cells, for example, offer relatively little resistance to dissolved solutes. On the other hand, preceding material has emphasized that protoplasmic membranes permit rapid passage of some substances but only slow passage of others. *Permeability* is the term used to indicate the degree to which a membrane permits movement of gases, liquids, and dissolved substances across it. A membrane may be highly permeable to one substance, moderately permeable to a second, and only slightly permeable to a third. Thus permeability can be said to denote qualitatively the *conductivity* properties of a membrane. Proper usage of the term permeability dictates that it be reserved for passive transport processes. Occasionally, however, it is used for active transport processes (see pp. 317–319); in these cases, permeability should be designated "active permeability."

Naturally the permeability of protoplasmic membranes is of considerable interest to plant physiologists. The permeability of a protoplasmic membrane depends on the kind and variety of its constituents, its degree of hydration, its state of aggregation, its fluidity, its thickness, and similar factors. Because all plant cells do not have precisely the same permeability properties toward a given substance, clearly the permeability of protoplasmic membranes of different plant cells is different. Even different protoplasmic membranes (e.g., the plasma membrane and the tonoplast) in a single plant cell are known to have somewhat different permeability properties.

The permeability of a given protoplasmic membrane to a given substance is not an invariant property of the membrane. Changes in permeability have been shown to occur by a variety of artificial and natural changes in the environment. Protoplasmic membranes, for example, become highly permeable when high doses of ionizing radiation are applied to cells: Cellular metabolites normally retained within the cell will leak out. Also, the composition of the external solution is known to affect the permeability of cells immersed therein. The absence of calcium ions in the external solution, for instance, increases the tendency of ionic constituents to leak out of root cells. Many other external factors affect permeability of plant cells, including temperature, partial pressures of O_2 and CO_2, availability of water, and the presence of toxic substances.

Not only external factors but also factors operating within a cell cause changes in membrane permeability. The aging of a plant cell is but one example: Senescence is usually accompanied by an increase in permeability. Even transitions from night to day and day to night induce rhythms within the plant that have been shown to be reflected in slight changes in cell permeability.

THE GENERAL TRANSPORT LAW

Flux and driving force are related to one another by an equation known as the general transport law. The general transport law is based on an equation well known to physicists and chemists—Fick's law of diffusion,

$$J = D \frac{dc}{dx} \tag{11-1}$$

where flux J is the quantity of a substance diffusing per second across an area of one square centimeter and dc/dx is the concentration gradient (i.e., the change in concentration c with distance x); D is called the diffusion coefficient and is a constant of proportionality. If the flux is expressed in mol cm^{-2} s^{-1} (see p. 296) and if the concentration gradient dc/dx is expressed in mol cm^{-4} (i.e., concentration c in mol cm^{-3} divided by distance x in cm), then the diffusion coefficient D will be expressed cm^2 s^{-1}.

Equation (11-1) is generally not used as such in transport problems in plant physiology because molecules or ions of a substance move across membranes (e.g., the plasma membrane) whose thicknesses are not always known with precision. But by incorporating dx in Eq. (11-1) into another proportionality constant called the transmission coefficient (i.e. transmisson coefficient = D/dx), Eq. (11-1) can be simplified to

$$J = \text{transmission coefficient} \times \Delta c \tag{11-2}$$

If the units of J and c are the same as in Eq. (11-1), then the transmission coefficient in Eq. (11-2) will be expressed in cm s^{-1}.

In Eq. (11-2), Δc is the difference in concentration of the substance that moves between two regions (e.g., across a membrane). Because a difference in concentration between two regions can be the driving force propelling movement of a substance from one region to another (see Chapter 10), Eq. (11-2) can be generalized as follows:

$$\text{Flux} = \text{transmission coefficient} \times \text{magnitude of the driving force} \quad (11\text{-}3)$$

This equation is known as the *general transport law*. The transmission coefficient is the numerical value of the transmissivity or conductivity of a given membrane to a given substance. Equation (11-3) and its variants [see Eqs. (11-4) and (11-5)] are often applied not only to passive but also to active transport problems.

In the case of a substance moving across a membrane, the transmission coefficient in the general transport law [Eq. (11-3)] is called the *permeability coefficient* or *permeability constant*. Therefore Eq. (11-3) can be written

$$\text{Permeability coefficient} = \frac{\text{flux}}{\text{magnitude of the driving force}} \quad (11\text{-}4)^2$$

This equation states that the permeability coefficient (i.e., the numerical value of the permeability) is the quantity of a substance moving across a unit area of membrane surface per unit of time per unit of driving force.

The permeability coefficient in Eq. (11-4) is actually a measure of the mobility of a substance (i.e., the rate at which a substance crosses a membrane), as the following example illustrates. Assume that the flux of a substance moving across a membrane is measured under conditions of constant temperature and is found to be 0.00001 mol per cm^{-2} of a membrane surface per second. Also assume that the driving force is the difference in concentration of the substance on the two sides of the membrane and is equal to 0.1 mol per liter (L). Then the permeability coefficient is 0.00001 mol cm^{-2} s^{-1} divided by 0.1 mol L^{-1}, or 0.0001 L cm^{-2} s^{-1}. Because 1 liter = 1000 cm^3, the permeability coefficient in this example becomes 0.1 cm s^{-1}. So the dimensions of the permeability coefficient are those of velocity (i.e., mobility).

Another application of the general transport law is in the movement of a gas (e.g., water vapor, CO_2) through stomatal pores. Stomatal apertures change during the course of a day and therefore *restrict*, to varying degrees, the exchange of gases between the mesophyll cells of a leaf and the external atmosphere (see Chapter 13). Thus in considering flux of a gas through stomatal pores, attention is focused on the *resistance* in the pathway rather than on its permeability or conductance. (Resistance is the reciprocal of conductance.) Therefore the general transport law [Eq.

---

[2]Equation (11-4) is subject to certain qualifications. When solute and solvent use the same site in the membrane for their permeation across the membrane, for example, they may interact with one another, perhaps by exerting a "drag" on each other. Thus the permeability coefficient of one substance may be affected by the movement of another. In these interaction cases, Eq. (11-4) is not applicable. Instead, special equations of irreversible thermodynamics must be used.

(11-3)] for movement of water vapor out of a leaf during transpiration can be written

$$\text{Transpirational flux} = \frac{\text{magnitude of the driving force}}{\text{resistance in pathway}} \qquad (11\text{-}5)$$

Equation (11-5) shows that the transpirational flux is inversely proportional to the resistance of the pathway: The smaller the stomatal aperture, the greater the resistance to outward movement of water vapor and hence the less the flux of water vapor. We will encounter Eq. (11-5) again in Chapter 13 [see Eq. (13-1)].

It is pertinent to note that Eq. (11-5) is analogous to Ohm's law, which governs the flow of a current in an electric circuit. Ohm's law states that current flow (I) is proportional to voltage drop (E) and inversely proportional to resistance (R). It is written:

$$I = \frac{E}{R}$$

THE PROTOPLASMIC MEMBRANE

Although the protoplasmic membrane was considered in Chapter 2, a few more words are pertinent here as a prelude to discussion of active transport.

A protoplasmic membrane is a distinct, ultrathin sheet of lipid and protein material approximately 7 nm in thickness. Various protoplasmic membranes are present in plant cells. Protoplasmic membranes that delineate cytoplasmic organelles are either single layered [i.e., consist of a single *unit membrane* (p. 19–20)] or double layered (i.e., consist of two unit membranes layered side by side). Examples of organelles delineated by a single layer of membrane material are leaf peroxisomes and seed glyoxysomes. Also, the membrane at the external surface of the protoplast (the plasma membrane) as well as the one that delineates the vacuole (the tonoplast) are single layered. Nuclei, mitochondria, and chloroplasts are bounded by a double layer of protoplasmic membrane.

Certain cytoplasmic organelles (e.g., mitochondria, chloroplasts) consist of elaborate internal systems of protoplasmic membranes. These membranes are folded and layered in distinctive fashion characteristic of the cytoplasmic organelle (see Figures 2-11 and 2-12 and related text discussion).

Many vital metabolic reactions take place on or in close association with membrane systems (e.g., oxidative phosphorylation in mitochondria, light reactions in chloroplasts). Moreover, protoplasmic membranes constitute a considerable portion, perhaps 50% or more according to some estimates, of the total protoplasmic mass of young plant cells. Thus their importance cannot be exaggerated. The study of their fine structure and chemical composition, and the relation of structure and composition to biochemical events occurring in living cells, is sometimes referred to as membrane physiology.

The concept that the plasma membrane controls the distribution of ions and molecules between a plant cell and its external surroundings is widely accepted. In this view, the plasma membrane acts first as a barrier that separates the cell interior from its environment and, second, as the site of various mechanisms regulating the influx and efflux of matter between cell interior and environment. Similarly, a membrane that delimits a cellular organelle is thought to preserve concentration differences between constituents inside and outside the organelle and to regulate the exchange of matter between the organelle and the surrounding cytoplasm. Such concepts are part of the *membrane theory*, which assumes that the mechanisms controlling the transport of substances across a protoplasmic membrane can be ascribed to the properties of the membrane. These properties include the chemical composition of the membrane and the arrangement of the constituents of which the membrane consists (i.e., the molecular architecture).[3]

In conclusion, it can be said that a protoplasmic membrane is an ultrathin structure that consists of lipid and protein, with water a third important constituent. Various evidence from different sources indicates that a protoplasmic membrane is subject to continual resynthesis and repair. Moreover, it has a dynamic nature, as shown by pinocytosis. Also, in our mind's eye we can visualize "patches" or "islands" of proteinaceous material (both structural and enzymic located here and there, extending from one surface of a membrane to the other and thereby interrupting the continuity of the lipid layer). Many of these protein patches are transport enzymes that function in active transport mechanisms. Other portions of a cell membrane can be visualized as being free of protein. Finally, a protoplasmic membrane should be visualized as a sheet of high electrical resistance that separates two phases of different chemical and pressure potentials.

ACTIVE TRANSPORT ACROSS PROTOPLASMIC MEMBRANES

Although suggestions were made in the late nineteenth century regarding the possible participation of cellular metabolism in the absorption of dissolved ions and molecules that permeate slowly into plant cells, it was not until the 1920s that serious study of this aspect of membrane transport began. The first unambiguous demonstration that the transport of certain substances into plant cells depends on metabolism was provided by the American plant physiologists Hoagland and Davis.

Hoagland and Davis compared the ionic composition of vacuoles of the freshwater alga *Nitella clavata* and the pond water in which these plants grow

---

[3]Opponents of the membrane theory assume that the unequal distribution of ions and molecules between a living cell and its external environment and between cellular organelles and cytoplasmic fluid depends on special properties of the macromolecules of the *entire* protoplasm rather than the membrane itself. For a nontechnical account of this alternative point of view, see P. T. Beale, 1981. The water of life. The Sciences (N.Y. Acad. Sci.) 21(7):6–11. For a comprehensive review, see G. N. Ling. 1969. A new model for the living cell: a summary of the theory and recent experimental evidence in its support. *Internatl. Rev. Cytol.* 26:1–61. Perhaps it is worthy of note that many opponents of the membrane theory are concerned mainly with animal cells, not plant cells.

naturally. *Nitella* plants, like those of *Chara* (Figure 11-2), have giant internodal cells, the vacuolar contents of which can be recovered and analyzed chemically. The pond water in which *Nitella* plants grow is mainly a dilute inorganic solution, comparable in total concentration of electrolytes to a soil solution or a nutrient culture solution (see Table 8-1 and related text discussion).

The vacuolar sap of *Nitella* plants was found to be an inorganic solution consisting primarily of the same inorganic ions as those present in the pond water. But the total ionic concentration of the vacuolar sap was 25 times greater than the ambient pond water. Furthermore, individual ionic concentrations were much higher in the vacuolar sap than in the pond water. Ratios of internal and external concentrations of individual ions are tabulated in Table 11-1. From these data it was evident to Hoagland and Davis that both cations and anions moved into *Nitella* cells against their individual concentration gradients. Because purely physical mechanisms are inadequate to explain these findings, Hoagland and Davis concluded that cellular metabolism is utilized for the transport of the ions in Table 11-1.

The type of transport studied by Hoagland and Davis is now called *active transport*. The term was coined to direct attention to the "active" participation of the metabolic machinery of the cell in the transport of matter across its cellular membranes. A molecular or ionic species is said to cross a protoplasmic membrane by *active transport* if the transport process is linked to metabolism. (The means by which active transport and metabolism are linked are considered on pp. 321–326.) The

TABLE 11-1 RATIOS OF INDIVIDUAL IONIC CONCENTRATIONS IN VACUOLAR SAP OF *Nitella clavata* AND IN THE AMBIENT POND WATER[a]

Ion	Concentration in cell sap divided by concentration in pond water
$(H_2PO_4)^-$	18,050.0
K^+	1,065.0
Cl^-	100.5
Na^+	46.1
$SO_4^=$	25.8
Ca^{2+}	13.17
Mg^{2+}	10.47
NO_3^-	0.0

[a]Calculated from data obtained by Hoagland and Davis, as presented by Osterhout: Osterhout, 1936 (Table II). Although Hoagland and Davis found that concentrations of individual ions in the vacuole varied somewhat from one season of the year to another, the values of individual samples showed only a small deviation from the mean value; so the ratios shown here represent typical values. Of the several ions shown, only nitrate ion did not accumulate in these cells, perhaps because nitrate ions were transformed into other substances (e.g., amino acids) immediately on entry into the cells.

transport of dissolved mineral salts, organic acids, amino acids, and sugars occurs principally by active rather than passive mechanisms.

The mechanism of active transport is discussed below, but a few characteristics of active transport are noted here.

1. Unlike passive transport, active transport of a molecule or ion is not spontaneous and does not proceed toward an equilibrium condition.
2. The driving force for active transport is metabolic in origin rather than physical.
3. Active transport across a protoplasmic membrane may occur from either a lower to higher or higher to lower chemical potential (concentration) of the transported molecule or ion.
4. Active transport processes in protoplasmic membranes are superimposed on their passive transport features.

Several kinds of plant tissues are used to study active transport. Detached segments of root tissues are often favored experimental objects. The usual laboratory procedure is to detach root tips from young seedlings immediately prior to an experiment. The experimental material (referred to as *excised root tips*) normally consists of several centimeters of tip ends of young roots.

In addition to excised root tips, giant algal cells, and leaf and storage tissues are used by many experimentalists. Giant algal cells were considered earlier (see Figure 11-2). Leaf tissues cut into thin disks (of a thickness of about 300 μm) are sometimes used as experimental material.

Also, thin disks cut from mature storage organs (e.g., roots of carrot or beet and tubers of potato or Jerusalem artichoke) will absorb ions rapidly but only after the disks undergo a special treatment. Initially their rate of ion absorption is very low. But when these thin disks (usually of a thickness of about 700 μm) are held in aerated water for several hours—a treatment referred to as *aging*—they acquire (for several days) an enhanced ability to absorb and accumulate ions. The slicing and aeration together stimulate the mature, previously quiescent cells in these tissues into a state of rapid growth. Frequently cell divisions occur in the cells on the cut surfaces. Moreover, "aged" disks exhibit an increased rate of protein synthesis, high metabolic activity, and an increased respiration rate (see p. 89).

A sure test that an individual ionic or molecular species moves by active transport across a protoplasmic membrane is to show that movement occurs from *low* to *high* chemical potential (concentration) of the chemical species. Such a demonstration would provide unequivocal evidence that metabolic work was expended in the transport process. This type of evidence, however, may not be easy to obtain. Nitrate ions absorbed by young, rapidly growing cells, for example, are likely to be transformed quickly on entry into cells into amino acids and proteins. So it may be difficult even to detect nitrate ions in plant cells. It may not be possible, therefore, to decide whether nitrate ion is absorbed with—or against—its chemical potential (concentration) gradient. As another example, consider the absorption of glucose by growing cells low in respiratory reserves. In such cells, glucose is apt to be respired rapidly on entry into the cells. As in the case of nitrate ion, it may be difficult to decide whether glucose actually crosses a protoplasmic membrane with—or against—its chemical potential (concentration) gradient.

In the absence of direct experimental evidence of "uphill" transport, a transport process can be categorized as active if dependence on metabolism is demonstrated. Several techniques are available. Conclusions about whether a given transport process is active, however, are usually made on the basis of results from several rather than only one of these techniques.

One such technique involves comparing the rate of transport of a substance at two temperatures, one $10°C$ higher than the other. For a passive transport process, the rate is increased only slightly when the temperature is raised $10°C$; the average kinetic energy of moving molecules or ions is increased only slightly by a $10°C$ rise in temperature. On the other hand, the rate of an active transport process is influenced by temperature in the same way as the rate of a chemical reaction. That is, an active transport process is linked to one or more metabolic reactions and therefore involves breaking or forming chemical bonds. To break or form bonds, collisions between reacting molecules or ions must exceed a minimum threshold energy—the *energy of activation.* At any given temperature, only a fraction of the molecules or ions of a system have energies greater than the activation energy. but a $10°C$ rise in temperature doubles and sometimes more than doubles the number of high-energy collisions. Therefore a $10°C$ rise in temperature results in an increase, by a factor of 2 or more, in the number of molecules or ions capable of reacting. It is for this reason that a $10°C$ rise in temperature doubles or more than doubles the rate of an active transport process. The *temperature coefficient* of the process—the ratio of the rate at a given temperature to the rate at a temperature $10°C$ lower—is on the order of 2 or 3 for active transport. On the other hand, the temperature coefficient for a passive transport process is only slightly greater than 1.

Other experimental criteria can also be used to demonstrate the existence of metabolic driving forces in active transport. Because active transport is linked to metabolic reactions, the rate of an active transport process is sensitive to the general level of cellular metabolism. If the intensity of cellular metabolism is reduced temporarily by one or another treatment, then the ability of cells to carry on active transport is impaired. One way to reduce the level of metabolism in higher plant cells is to withdraw the supply of O_2 from the cells. It can be done easily—at least for such tissues as roots that do not produce O_2 in photosynthesis—by placing the tissue for a few minutes to an hour or so in an atmosphere of N_2. Whereas active transport processes will stop in the temporary absence of O_2, passive transport processes will proceed regardless of the presence or absence of O_2.

Enzyme poisons can be applied to tissues to determine whether a given transport process is active. The application of cyanide or azide to root tissue, for instance, will inactivate not only the respiratory chain (as pointed out on p. 114) but also the uptake of O_2 and the synthesis of ATP from ADP and Pi; active transport will also cease. Inhibitors of ATP synthesis [e.g., DNP and FCCP (see

p. 119)] also inhibit active transport; we will see below that ATP is required for active transport.

Still another criterion is available to determine whether a given transport process is active. Active transport processes follow "saturation kinetics" (see Figure 11-3 and related discussion).

KINETICS OF ACTIVE TRANSPORT

Many currently accepted ideas regarding the mechanisms of active transport are based on studies of the kinetics of the process. Kinetics deals with measurements of rates of reactions and factors influencing the rates. By measuring the rates at which different substances gain entrance into living cells and the factors that affect the rates, deductions regarding the mechanism can be made. Intact cells are used in the study of transport kinetics. When cells are destroyed, their active transport systems are also destroyed. It is noteworthy, however, that some investigators have succeeded in isolating fragments of cell membranes from cells. These fragments have been used to study various aspects of active transport, especially the chemical composition of transport constituents. This type of research may prove helpful in achieving a more complete understanding of active transport mechanisms.

Typical results of the kinetics of active transport are shown in Figure 11-3. Similar curves have been obtained by many investigators over the past two to three decades. Furthermore, various tests, such as those described in the preceding section, have been used to demonstrate that curves of this type represent active transport. Thus there is no doubt that the data shown in Figure 11-3 depict active transport.

The upper curve in Figure 11-3 shows that the absorption rate of potassium ion is the same whether from a solution of potassium chloride or potassium sulfate. Similarly, the lower curve shows that the rate of absorption of chloride ion is independent of the accompanying cation, whether potassium or calcium. Moreover, Figure 11-3 shows that ions are absorbed by plant cells at rates that vary in a special manner with external concentration of the ion. At very low external concentrations of the ion (less than about 0.05 mM for the ions studied in Figure 11-3), the absorption rate increases sharply with increasing concentration. But with further increase in external concentration (greater than about 0.1 mM for the ions studied in Figure 11-3), the absorption rate approaches a maximal value indicated by the horizontal dashed lines in both the upper and lower curves. In fact, the curves in Figure 11-3 show that the rate of absorption depends on external ion concentration at low external concentrations but becomes independent of external concentration at high concentrations of the ion undergoing active transport.

MECHANISM OF ACTIVE TRANSPORT

Curves of active transport, such as those in Figure 11-3, closely resemble curves in which the rate of an enzyme-catalyzed reaction is plotted against the concentration of the substance whose transformation is being catalyzed (i.e., the substrate of the enzyme

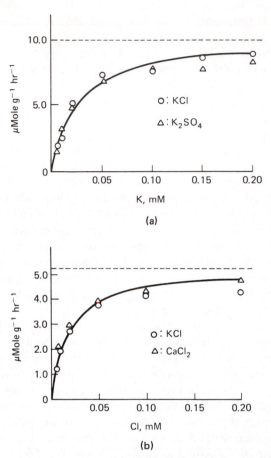

(a)

(b)

Figure 11-3 Fluxes of active transport of a cation (K^+) and an anion (Cl^-) at increasing external concentrations of the ion. Root tips of barley (*Hordeum vulgare*) were used. In these tissues the K^+ ion transport system operates independently of the Cl^- ion transport system. (a) Flux of absorption of K^+ ion (micromoles of K^+ per gram of tissue per hour) at increasing external concentrations (in millimoles) of K^+ ion. The flux of K^+ ion absorption is almost the same from KCl ($\bigcirc$) as from K_2SO_4 ($\triangle$). (b) Flux of absorption of Cl^- ion (micromoles of Cl^- per gram of tissue per hour) at increasing external concentrations (in millimoles) of Cl^- ion. The flux of Cl^- ion absorption is almost the same from KCl ($\bigcirc$) as from $CaCl_2$ ($\triangle$). [*Note*: At high external concentrations of ions, higher than **2.0** mM (i.e., higher than concentrations usually found in most soils), absorption fluxes rise to values far above those shown here. For recent interpretations of this phenomenon, often referred to as the *dual mechanism*, see Epstein, 1976.] (After Epstein, 1965; Figures 11-4 and 11-5.)

reaction).[4] Therefore it is widely assumed that the mechanism of active transport is analogous to the mechanism of catalysis mediated by enzymes. Just as an enzyme molecule first binds a specific ion or molecule and subsequently releases product ions or molecules, so, too, a specialized protoplasmic membrane constituent (i.e., a *carrier*) binds ions or molecules (on one side of the membrane) and releases them (on the other side of the membrane). Carrier molecules, unlike enzyme molecules, however, have not yet been isolated or purified from cells of higher plants.

The "flattening out" of the curves in Figure 11-3 is interpreted to mean that only a limited number of carriers is available on a given protoplasmic membrane. So a maximal rate of substrate absorption is approached at substrate concentrations that "saturate" the carriers. Carriers are visualized as operating in a "turnover" manner, like enzyme molecules, so that a relatively small number of carriers can accommodate the transport of large numbers of specific molecules or ions.

---

[4]The mathematical formulation of the kinetics of an enzyme-catalyzed reaction, commonly called the Michelis–Menton equation, can be found in any elementary biochemistry textbook.

By analogy with enzyme catalysis, the following reaction sequence can be assumed to take place in active transport:

$$M_e + C \longrightarrow MC \longrightarrow M_i + C \qquad (11\text{-}6)$$

The symbol M_e refers to the ion or molecule at the external surface of the protoplasmic membrane, C is the binding compound or carrier in the protoplasmic membrane, MC is the intermediate labile complex, and M_i is the ion or molecule at the internal surface of the protoplasmic membrane.

The stepwise sequence of events in active transport can be summarized as follows. The first step involves binding the substance undergoing transport to a membrane constituent (i.e., the carrier) at an *absorption site* on the outer surface of a protoplasmic membrane (e.g., the plasma membrane). In the second step, the substance is moved across the membrane; this movement may occur because of a conformational change in the carrier when bound to the ion or molecule undergoing transport. Then, in the final step, the carrier-substance complex breaks apart at the inner surface of the protoplasmic membrane. The molecule or ion is released. The carrier, restored to its original conformation, is now ready to engage another ion or molecule at the outer surface of the membrane.

Lately the terms *translocator* and also *transporter* and *porter* have come to be used more frequently than carrier.

A translocator is thought to be a large protein molecule that spans the thickness of the protoplasmic membrane; that is, it is "plugged" through the lipid bilayer of the membrane. Translocators are quite specific with respect to transported ions and molecules: only substances of similar chemical structure are transported by an individual translocator. Specificity of translocators has been demonstrated even in compounds that are mirror images of one another. Plant cells, for example, distinguish between structurally similar compounds, such as D-glucose and L-glucose; the former (an important metabolite) is absorbed, but the latter (a compound not normally metabolized) is not. The difference between D- and L-glucose is that they are optical isomers of one another (i.e., they differ in the spatial arrangement of their atoms, much as the right hand differs from the left (see p. 58).

An active transport system is often called a *pump*. A substance undergoing active transport is said to be "pumped" across a protoplasmic membrane. Various ionic and molecular pumps exist in a protoplasmic membrane, including a proton pump, a pump for K^+, another for Ca^{2+}, still others for Cl^-, $SO_4^=$, glucose, sucrose, and so forth. Some pumps transport a substance from outside to inside a plant cell or cytoplasmic organelle; others operate in the reverse direction; still others transport two substances, either in exchange (i.e., opposite directions) or in the same direction across the membrane.

Many (perhaps all) substances categorized as undergoing active transport do so as a result of generation of a proton motive force across the membrane. The basis of this view is Mitchell's chemiosmotic theory. It will be recalled (see Figure 4-8 and related discussion) that protons are pumped across the inner mitochondrial membrane by the respiratory chain and across the chloroplast thylakoid by light activation of photosystems I and II. These two proton pumps generate a proton motive force across these membranes. The return of protons down their electrochemical gradient, through the enzyme ATPase, drives the formation of ATP from ADP and Pi. Also this proton motive is the source of energy for the active transport of substances across these membranes.

A model for the transport of sucrose across the plasma membrane of a plant cell is shown in Figure 11-4. The membrane contains the enzyme ATPase, which hydrolyzes one molecule of ATP (produced for the most part in mitochondria) and at the same time pumps two protons out of the cell.[5]. Thus the outside becomes positive and the interior negative; the difference in electric potential is about 230 mV. This pump is *electrogenic*, which means that it transports ions—protons in this case—across a protoplasmic membrane by metabolic processes and thereby generates a charge imbalance, or electric potential, across the membrane.[6] Also shown in Figure 11-4 is a sucrose-proton-translocator. The translocator, in its negatively charged form (C^{-1}), combines with a proton and a sucrose molecule (S). Sucrose moves across the membrane, perhaps as a result of a conformational change in the translocator-sucrose-proton complex, as mentioned on p. 323. Now sucrose and the proton are released into the interior of the cell. Thus the translocator, restored to its original conformation

[5] Space does not permit a detailed discussion of the molecular mechanism involved in the transport of two protons for each molecule of ATP hydrolyzed. The interested reader is referred to Mitchell, 1979, and references cited there (see especially Mitchell's Figure 14 and related discussion in reference #66).

[6] The electric potential across a protoplasmic membrane is actually due to three components. In addition to the metabolic (electrogenic) component, there are two (minor) components. One is the diffusion potential, which refers to the slight unequal distribution of cations and anions across the membrane, produced by dissimilar rates of leakage. The other, the boundary potential, is due to an unequal distribution of charges on the inner and outer surfaces of the membrane itself.

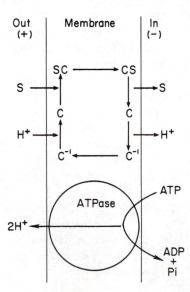

Figure 11-4 A model for sucrose transport across the plasma membrane. (See text for details.) Abbreviations: S, sucrose; C, translocator; SC and CS, complex of sucrose-proton-translocator. (*Note*: Although ATPase is shown to hydrolyze ATP and pump protons out of the cell, actually all ATPases are reversible. Thus a proton gradient from outside to inside, across the plasma membrane, may serve to synthesize ATP from ADP and Pi.) (Courtesy Dr. T. E. Humphreys, University of Florida, Gainesville.)

(C^{-1}), is ready to react with another sucrose molecule and proton at the outer surface of the membrane. In this sucrose-proton model, sucrose will accumulate inside the cell until the difference in chemical potential of protons on the two sides of the plasma membrane (higher outside) becomes equal to the difference in chemical potential of sucrose on the two sides of the membrane (higher inside).

Other ATPases and translocators in the membrane in Figure 11-4 also result in the movement of protons and other ions and molecules in both directions across the membrane. Thus under conditions of steady state the overall outward rate of proton flow across the membrane via ATPases will equal the overall inward rate of proton flow via translocators that move substances inward.

It may be stated as a fundamental principle of plant physiology that the proton motive force created by pumping of protons across protoplasmic membranes [e.g., the plasma membrane and the inner mitochondrial membrane and thylakoid membrane in chloroplasts (see pp. 120–122)] is a basic energy source in cells of higher plants, not only for sucrose transport (Figure 11-4) but also for active transport of other substances. The circulation of protons—pumped in an "uphill" direction via ATPase and coupled to the transport of sucrose and other molecules and ions in their "downhill" return via translocators—enables the metabolic energy released in the hydrolysis of ATP to ADP and Pi to be utilized in active transport processes.[7]

The sucrose-proton translocator in Figure 11-4 is called a *symport*. This term refers to linked movement of two entities across a protoplasmic membrane at the same site in the same direction. (A symport is said to be a *cotransport* system.) Also, two other types of translocators are known to exist in cells of higher plants: uniport and antiport. A *uniport* moves one entity across a protoplasmic membrane at one site. ATPase in Figure 11-4 can be considered an example of a uniport; ATPase pumps protons. An *antiport* refers to the linked movement of two entities in opposite directions at the same site in a protoplasmic membrane. (The manner of coupling of an antiport to ATP hydrolysis, and to the electrogenic proton pump, is not well understood at present.) One example of an antiport (countertransport) is the "phosphate" or "triose phosphate" translocator in the inner membrane of the chloroplast envelope. This translocator catalyzes an exchange of inorganic phosphate, 3-phosphoglycerate, glyceraldehyde-3-phosphate, and dihydroxyacetone phosphate; for each anion entering the chloroplast, one anion leaves the chloroplast. Consequently, the total pool of phosphate in the stromal space of the chloroplast is kept constant. Thus this translocator diverts carbon away from the photosynthetic carbon reduction cycle in chloroplasts only when phosphate is transported into the

---

[7]A few plant physiologists sometimes argue that certain aspects of *active* transport across protoplasmic membranes should be categorized as *passive*: The return flow of protons, together with sucrose molecules, down their gradient via the translocator in Figure 11-4 has been suggested as "passive" transport. The terms *primary* or *direct* (i.e., electrogenic proton pumping via ATPase) and *secondary* or *indirect* (i.e., return flow of protons, together with sucrose molecules in Figure 11-4), however, should be sufficient to distinguish these two aspects of active transport. (For further discussion of primary and secondary transport mechanisms, see Poole, 1978.) Thus the terms active and passive, as used in this book, are convenient and useful because they distinguish two *major* types of transport in higher plants.

chloroplast (i.e., in exchange for exported triose phosphate). Aside from its existence, however, little is known about the factors that control the substances transported by this translocator.

SHUTTLE SYSTEMS

Some metabolites cannot be transported across protoplasmic membranes of certain plant organelles. Here is one well-known example. The inner membrane of the mitochondrion and the inner membrane of the chloroplast envelope cannot be traversed readily by nicotinamide nucleotides (NAD^+/NADH, $NADP^+$/NADPH). (The outer membrane in both organelles is permeable to most small molecules and ions.) Also, the single membrane around the peroxisome and glyoxysome is impermeable to nicotinamide nucleotides. Reducing equivalents from reduced nicotinamide nucleotides, however, can be transferred to suitable acceptor molecules—acceptors capable of traversing the membrane impermeable to nicotinamide nucleotides. The existence of these acceptor molecules is effective in "shuttling" reduced nicotinamide nucleotides across the membranes of these organelles. A *shuttle* system for nicotinamide nucleotides is a cyclic pathway by which reducing (and oxidizing) equivalents are transferred across a protoplasmic membrane impermeable to nicotinamide nucleotides.

As a general rule, a shuttle for nicotinamide nucleotides involves a redox pair of organic acids capable of transversing the protoplasmic membrane. At one end of the shuttle the oxidant of the redox pair is reduced by reduced nicotinamide nucleotide in a reaction catalyzed by an appropriate enzyme. The reductant (reduced organic acid) so formed moves across the protoplasmic membrane, where it is oxidized by an appropriate enzyme, thereby regenerating the oxidant (oxidized organic acid); the reducing equivalents are accepted by oxidized nicotinamide nucleotide. Meanwhile the regenerated oxidant (oxidized organic acid) moves across the protoplasmic membrane, where it can be reduced at the expense of reduced nicotinamide nucleotide. This basic pattern is altered in some cases by additional transamination reactions.

A protoplasmic membrane impermeable to nicotinamide nucleotides has translocators that transport dicarboxylic acid anions (e.g., malate, oxaloacetate, aspartate, glutamate, α-ketoglutarate) across the membrane (see Figure 11/5). But the coupling of these translocators to the hydrolysis of ATP and to the electrogenic proton pump (assuming that they *are* coupled) is not well understood.

Three examples of shuttle systems are presented here. In the first example, reducing equivalents (as malate) are transported out of mitochondria into the cytosol [see Figure 11-5(a)]. The figure shows malic acid oxidized to oxaloacetic acid in the cytosol in the following reaction (from left to right):

$$\text{Malic acid} + NAD^+ \xrightleftharpoons{\text{malic acid dehydrogenase}} \text{oxaloacetic acid} + NADH \quad (11\text{-}7)$$

Oxaloacetate produced in Eq. (11-7) is transported from cytosol into mitochondria,

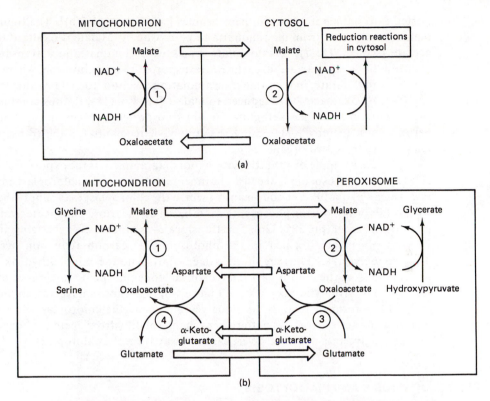

Figure 11-5 Two examples of systems that shuttle $NAD^+/NADH$ across protoplasmic membranes. (See text for details.) Arrows with double lines indicate the shuttle
system. Transport across each membrane is shown at different sites for each organic
acid anion but may be at the same sites(s), depending on the nature of the translocators.
Formulas of malic and oxaloacetic acids are shown in Figure 4-6, those of glutamic,
α-ketoglutaric, and aspartic acids in Eqs. (9-8), (9-9), and (9-11). The enzymes are (1)
and (2), malic acid dehydrogenase; (3) and (4), glutamic acid-oxaloacetic acid
aminotransferase [Eq. (9-11)]. (a) An oxaloacetate-malate shuttle by Woo et al., 1980
suggested to participate in the reduction of nitrate in the cytosol (see p. 260, footnote
1). (A similar shuttle operating between (illuminated) *chloroplasts* and the cytosol, in
the reduction of nitrate, was proposed by Rathnam, 1978.) (b) A malate-aspartate
shuttle suggested by Journet et al., 1981 to participate in the photorespiratory carbon
oxidation cycle [i.e., in the reduction of hydroxypyruvic acid to glyceric acid in
peroxisomes, and the oxidation of glycine to serine in mitochondria (see Figure 7-4 and
related discussion)].

where it is reduced to malate—again by Eq. (11-7) (from right to left)—thus
completing the cycle. [Eq. (11-7), catalyzed by malic acid dehydrogenase, is reversible;
this enzyme, present in both cytosol and mitochondria, was encountered earlier (see
Figure 4-6 and related discussion).]

In the second example, reducing equivalents (as malate) are transported out of
mitochondria into peroxisomes and oxidizing equivalents (as aspartate) are trans-

ported into mitochondria from peroxisomes [see Figure 11-5(b)]. The figure shows malate transported from mitochondria to peroxisomes. Malate is oxidized to oxaloacetate [see Eq. (11-7)]. Oxaloacetate reacts with glutamate, in a transamination reaction, to yield aspartate; aspartate is transported to mitochondria, where it reacts with α-ketoglutarate, in the same transamination reaction, to yield oxaloacetate [see Eq. (9-11)]. Oxaloacetate is reduced to malate [see Eq. (11-7)], thus completing the cycle. (Glutamate and α-ketoglutarate are transported between mitochondria and peroxisomes, thereby replenishing the supply of reactants for the transamination reactions in these two organelles.)

A third example of a shuttle system (not diagrammed) is the export of ATP from chloroplast to cytosol of plant cells. The inner membrane of the chloroplast envelope is impermeable to ATP. Yet one means by which the chloroplast may supply the cytosol with ATP formed during photosynthesis is based on the triose phosphate translocator mentioned earlier (pp. 325–326). This translocator is capable of exchanging dihydroxyaceton phosphate, formed in the photosynthetic carbon reduction cycle, with 3-phosphoglycerate. Thus the shuttle system starts with the exit of dihydroxyacetone phosphate from the chloroplast. Dihydroxyacetone phosphate undergoes oxidation (catalyzed by glycolytic enzymes; see Figure 4-3) to 3-phosphoglycerate in the cytosol and ATP is formed. Now 3-phosphoglycerate enters the chloroplast via the aforementioned translocator; this pattern of exchange of dihydroxyacetone phosphate and 3-phosphoglycerate does not divert carbon away from the chloroplast.

ELECTROOSMOSIS AND PINOCYTOSIS

A transport process for ion movement across a membrane, one which also involves the movement of water, is known as electroosmosis. *Electroosmosis* refers to the movement of the water phase in a colloidal suspension containing two species of charged particles—immobilized or restrained colloidal particles and mobile ions—when an electrical field is applied to the suspension. If the colloidal particles are restrained or immobilized, then only mobile ions move under the influence of the electrical field and carry water molecules adhering to them, thus leading to a movement of water.

For electroosmotic water movement across a protoplasmic membrane, charged protein components of the membrane represent immobile colloidal particles whereas such ions as potassium are mobile. If an electric potential difference exists across the membrane, movement of potassium ions would occur through pores assumed to be present in the membrane. The process may be visualized by referring to Figure 11-6, which shows an electric potential across the membrane and a layer of ions on the pore walls. Ions immediately adjacent to the pore walls are held tightly by intermolecular attractive forces; those in the outer layer are of opposite charge, held only by relatively weak attractive forces. The inner layer of ions is fixed whereas ions in the outer layer are mobile. Such an ionic double layer is always present at a solid-water interface.

Assuming cations (e.g., potassium) are the mobile ions in the ionic double layer in Figure 11-6, a current of potassium ions will move from one end of the pore to the

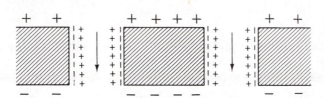

Figure 11-6 Diagram illustrating electroosmosis. A membrane across which an electric potential exists is shown, together with two pores in transection. Also shown are electric charges on the surface of the pore walls. The direction of the transmembrane electric current and also the direction of movement of water by electroosmosis are indicated by vertical arrows. (See text for details.)

other in response to an electromotive force across the membrane. Moreover, each potassium ion will carry several molecules of water with it because an ion in solution has water molecules adhered to it (see Figure 12-2). Thus in electroosmosis a flow of water through the pores of a membrane results from the flow of a current of hydrated ions. Pore diameters must be at least large enough to allow hydrated ions to move readily through them. (A hydrated potassium ion is about 0.5 nm in diameter; the diameter of a water molecule is about 0.25 nm.)

The significance of electroosmosis in plant cells is difficult to assess. Direct experimental evidence is not available at present.

A process known as *pinocytosis* (Greek, *pinein*, to drink; *cytos*, a cell) is a transport phenomenon that depends on changes in the position and structure of the plasma membrane. Materials that penetrate the plasma membrane by pinocytosis become enclosed within tiny vacuolelike structures known as *vesicles*. The sequence of events is illustrated in Figure 11-7. Material exterior to the protoplast adheres to the plasma membrane. Then the plasma membrane undergoes an invagination and engulfs the material within a vesicle. Afterward the vesicle moves into the cell interior. Finally, the vesicular membrane dissolves, and the material is released into the cytoplasm.

Although pinocytosis has been recorded in electron micrographs of plant cells (Nassary and Jones, 1976; Wheeler and Hanchey, 1971), the process is generally believed to have only a minor role in the absorption of water and dissolved mineral nutrients from soils. Pinocytosis, however, provides a possible mechanism to explain

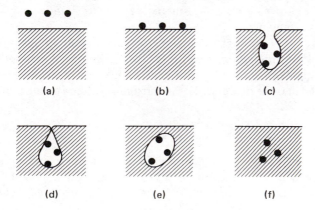

(a) (b) (c)

(d) (e) (f)

Figure 11-7 Diagrammatic representation of pinocytosis. In this series of six drawings three particles (black dots) cross the plasma membrane (solid line) and enter the cytoplasmic matrix (cross hatching) by pinocytosis. (See text for details.) (After Bennett, 1956; Figure 2.)

how plant cells absorb macromolecules, such as virus particles, once they penetrate the cell wall.

TRANSCELLULAR TRANSPORT

The term *transcellular transport* refers to transport of dissolved molecules and ions across a series of a few cells. Transcellular transport is more complex than transport across a single protoplasmic membrane because the substance undergoing transport may cross more than one membrane. Furthermore, such transport is influenced by the features of the individual cells through which transport occurs.

One example of transcellular transport was noted earlier—the movement of dissolved mineral ions across the primary root (pp. 303–306). Other examples of transcellular transport exist in tissues of higher plants, but only one will be mentioned here—phloem loading—mainly as a prelude to the subject of translocation in sieve tubes.

Phloem loading refers to the transport of sugars via interconnecting cells from the site at which synthesis occurs into the phloem. Generally sucrose is the transported sugar; it is synthesized in the cytoplasm of leaf chlorophyllous cells and moves from there to the phloem of minor veins in the leaf. First, sugar is transported across the plasma membrane of a chlorophyllous cell into the water in the apoplast (i.e., the cell wall). Second, it moves through the apoplast of one to several cells to the phloem. Finally, sugar is transported across the plasma membrane of a sieve element of the phloem. The second and third steps of this sequence are called phloem loading. The third step is a sucrose-proton cotransport process and was discussed earlier (see Figure 11-4 and related text discussion).

In some plant species cells adjacent to sieve tubes are specialized to absorb sugars from chlorenchyma cells and transfer them to sieve tubes. These *transfer cells* are modified companion cells. They are characterized by a greatly increased area of their plasma membranes, developed through irregular ingrowths of cell walls, and the production of plasma membrane material that follows the contours of the irregular cell walls. On the side adjoining chlorenchyma cells, transfer cells absorb sugar from the apoplast (see Figure 11-4 and related text discussion). But on the side next to sieve elements, transfer cells have numerous plasmodesmata (see Chapter 2) through which sugar is transported to sieve elements.

The *secretion* of sugar out of the interior of sieve elements into the apoplast (i.e., the cell wall) is called *phloem unloading*. Relatively little is known regarding the molecular mechanism of phloem unloading in higher plants at the present time. However, it may be assumed that the process, like phloem loading (see above), is by an active transport mechanism.

TRANSLOCATION IN SIEVE TUBES

It has long been recognized that sugars manufactured in photosynthesis in leaves of higher plants move in the phloem of petioles and stems to other parts of the plant. The direction of movement is mostly downward (e.g., from leaves to roots) but may also be upward (e.g., from leaves to fruits and buds located above leaves). Longitudinal movement of sugars in the phloem occurs in elongate cells known as *sieve elements* (see below). Sieve elements are joined end to end to form long linear series of cells. The latter is called a *sieve tube*.

Transport of sugars and other substances in sieve tubes is called *translocation*. The term *long-distance transport* is used as well, especially when referring to transport through the phloem that is continuous from leaf cells to root tips, a distance of 100 m or more in tall trees. Both terms, translocation and long-distance transport, are also used in reference to upward transport of water and dissolved mineral salts absorbed from the soil by roots. But the upward movement of water and dissolved substances is a passive transport process that occurs in nonliving xylem cells (see pp. 401–402). In contrast, translocation of sugars and other substances in the phloem is through living sieve elements.

The rate at which sugar molecules are translocated in the phloem has been measured by many investigators. In most experiments a radioactively labeled sugar is applied to a living stem at one point and its arrival at a distant point is timed. Generally measured rates range from 50 to 100 cm h^{-1}, although much higher (and also lower) rates have been recorded. These rates are thousands of times faster than those of diffusion of sugar molecules through water.

The translocation rate is sometimes measured in terms of the amount of sugar translocated per square centimeter of cross-sectional area of phloem tissue per hour. It can be of considerable magnitude. In the pumpkin, for example, a single fruit may attain a dry weight of 4 kg or more within 3 months from the time of pollination. This dry weight increment depends principally on sugars translocated from green leaves via sieve tubes. The amount of sugar translocated per square centimeter of cross-sectional area of phloem tissue per hour is called *specific mass transfer*. The average value of the specific mass transfer of sugars into developing tubers and fruits ranges from about 1 to 6 g cm^{-2} h^{-1}.

It has become customary to refer to regions supplying sugars to the phloem as *sources*. Primary sources are green leaves that have completed their initial phase of expansion growth; sugars produced in these leaves in photosynthesis are exported to other parts of the plant. Also, plant parts in which food materials are stored may serve as sources of sugars for other parts of a plant. Food reserves in the endosperm of germinating seeds, for instance, are metabolized to sugars that are translocated in the developing embryo.

Regions of the plant that utilize sugars translocated in the phloem are referred to as *sinks*. Included are plant parts that carry on little or no photosynthesis—for example, growing points of roots and shoots (buds), storage organs (e.g., fruits and seeds), and tissues in the vegetative body of the plant, both above and below ground, such as stems, rhizomes, and tubers. As a general rule, sinks are supplied mainly by nearby rather than distant sources. It has been estimated, for example, that almost half the carbon in cereal grains, such as wheat, oats, and rye, comes from the flag leaf (i.e., the leaf morphologically just below the mature inflorescence); the remainder is derived from nearby green floral bracts (i.e., glumes), green stem tissue and other leaves. In maize a large portion of the carbon in the grain is derived from the green husk.

Perhaps it might be useful to point out that sinks may occur almost anywhere along the phloem, not just in developing buds, roots, and other tissues. In fact, source and sink may be only a short distance apart; for instance, a source might be one point along the length of a sieve element and the sink another.

More than 90% of the dry weight of sieve tube sap consists of sugars. Although glucose (a monosaccharide) has a central role in respiratory metabolism (see Chapter 4), this sugar and its phosphorylated derivatives are absent from phloem sap. Instead sucrose (a sugar composed of two monosaccharide units, glucose and fructose) is the predominant form in which organic material is translocated from source to sink.

In addition to sucrose, three other sugars have been found in sieve tube sap of some plant species: raffinose (a trisaccharide), stachyose (a tetrasaccharide), and verbascose (a pentasaccharide). Each has a sucrose residue in its structure. Sieve tube sap also contains small amounts of other organic substances, including amino acids, amides, organic acids, sugar alcohols, and plant growth substances, such as auxin, gibberellins, and abscisic acid.

The presence of any given substance in the sieve tube sap does not necessarily mean that it is translocated from a source to a sink. The only sure way to determine whether a substance present in sieve tube sap is translocated is to label it (e.g., with a radioactive tracer) and follow its movement.

The concentration of sucrose in sieve tube sap is quite high, ranging from about 5 to 15% (i.e., 5 to 15 g sucrose per 100 ml of solution). A 10% sucrose solution is about average concentration in the conducting channels of the phloem; this concentration corresponds to a solution about one-fifth saturated at 30° C. But sugar concentration is not the same throughout the length of a sieve tube. It is generally highest near the sites of sugar production, where sugars enter sieve tubes, and falls off near the sites of sugar consumption, where sugars leave the sieve tubes. Moreover, variations in sugar concentration in sieve tubes occur from season to season and from day to day, depending largely on the intensity of the rate of photosynthesis and the prevailing temperature.

It will be useful to describe briefly the experimental evidence demonstrating that translocation of sugars takes place in the phloem. This conclusion is based on results from several different procedures and can be outlined as follows.

1. *Exudation from incision in bark*. When an incision is made in the bark of many deciduous trees, especially in the late summer or early fall after the shoots have completed elongation, there is an exudation of liquid containing sugars in high concentration. By suitable examination, this exudate can be shown to come from sieve tube elements.

2. *Analysis of sap from aphid stylet*. An aphid permitted to feed on a herbaceous plant will seek out the phloem with the stylet portion of its mouthpiece. The stylet terminates in an individual sieve tube (see Figure 11-8). If a feeding aphid is anesthetized by a gentle stream of CO_2 and its body removed with a cutting

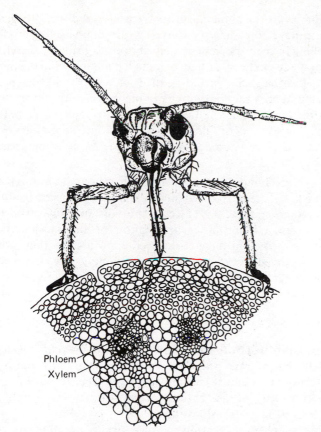

Phloem
Xylem

Figure 11-8 Drawing by Büsgen (1891), showing how the terminal end of the stylet of an aphid has probed a sieve tube. (After Esau, 1961; Figure 15.)

blade, the severed stylet remains in situ as an inert tube connecting the contents of the sieve tube with the exterior. Because hydrostatic pressure exists in the sieve tube—pressures of $+10$ bars have been recorded with aphid stylets (Wright and Fisher, 1980)—sieve tube sap will be extruded through the aphid stylet. Droplets of sap can be collected and analyzed chemically for sugar content.

Severed aphid stylets are used to study other aspects of phloem physiology—for example, measurement of sieve tube membrane potential (Wright and Fisher, 1981).

3. *Effects of girdling*. In the *girdling* or *ringing* technique a broad band of bark completely encircling the stem is removed. Because the bark consists of all the tissue exterior to the cambium, its removal leads to the complete severing of the phloem. The xylem, however, is left intact. Thus girdling interrupts transport in the phloem but has no effect on transport in the xylem. (In laboratory demonstrations precautions must be taken to prevent xylem tissue from drying out in the region of the girdle.)

Because girdling severs the phloem, sugars translocated downward from leaves will accumulate in the phloem tissue above a girdle, often leading in a few weeks to visible swelling in the bark immediately above the girdle.

When a girdle is made on the main stem of a small plant, between the shoot and root, the parts below the girdle, including the root system, will starve and die prematurely (assuming that cambial activity in the region of the girdle does not lead to regeneration of the phloem). But when the main trunk of a large tree is girdled, the tree may survive for considerable periods of time, sometimes even for several years, because considerable food reserves are likely to be present below the girdle.

4. *Autoradiography*. In this technique sugars labeled with radioactive atoms [e.g., tritium (^3H) or carbon (^{14}C)] are introduced into a plant via an incision in a vein of a mature leaf. Thin transverse sections of stem a short distance away from the incision are taken a few minutes after application of the radioactivity. Autoradiographs[8] developed from these tissue sections will show that radioactively labeled sugars moved in the phloem. This technique has shown that sugars move in sieve elements rather than in other phloem cells.

ANATOMY OF SIEVE ELEMENTS

Every physiological process is affected by the form and structure of the cells in which it occurs. This fact is especially true of translocation in the phloem. Any explanation of the mechanism of sugar translocation must be compatible with the structures of the channels of flow. Precise knowledge of the anatomy of sieve elements is essential to understanding the physiology of sugar translocation.

The sieve elements in which translocation of sugar occurs are elongate cells arranged longitudinally end to end. In some plant species the terminal portions of sieve elements are tapered and somewhat wedge shaped; the oblique face of the wedge of one sieve element overlaps that of the adjoining one. Parenchyma cells are closely associated with sieve elements. Fine threads of cytoplasm, called plasmodesmata (see Chapter 2) connect the cytoplasm of sieve elements and contiguous parenchyma cells. Certain parenchyma cells, the *companion cells*, arise from the same parent meristemic cell as the sieve element. That is, a meristematic cell in the vascular cambium divides longitudinally and a daughter cell differentiates into a sieve element whereas the other daughter cell differentiates into one or more companion cells.

During maturation of a sieve element the cell wall undergoes certain distinctive changes. It becomes unusually thickened, takes on an appearance described as "pearllike," and develops clusters of pores in certain regions. Each pore has a single strand of cytoplasm extending through it, connecting the protoplasts of adjoining sieve

---

[8]In the technique of autoradiography an electrosensitive emulsion of silver halide crystallites is pressed tightly against a thin section of tissue. Radiation emitted by radioactive atoms darkens the emulsion (after about 1 to 3 weeks of exposure) and a direct image of the location of the radioactive atoms in the tissue section is obtained.

elements. Perforated cell wall areas are called *sieve areas* [Figure 11-9(a)]. Sieve areas are found in many plant species on the lateral walls of sieve elements; in other plant species sieve areas are localized mostly on the end walls and are called *sieve plates* [Figure 11-9(a), (b), and (c)]. In some species the border of each of the pores that perforates a sieve area becomes impregnated at maturity with callose (a polysaccharide), so that the cytoplasmic strand within it becomes encased in a very thin cylinder of callose [Figure 11-9 (d), (e), and (f)]. The pore itself, however, is not plugged with callose in mature sieve elements. The diameter of a sieve pore ranges from a fraction of a μm to perhaps 1 to 2 μm or even more, depending on the plant species. The total cumulative area of sieve pores in sieve areas is usually a rather large fraction, sometimes 50% or more.

In addition to changes in the cell wall, the protoplast of a sieve element also undergoes marked changes during maturation [Figure 11-9(g) to (k)]. A proteinaceous component, somewhat slimy in texture and referred to as P-protein (P denotes phloem), appears in the cytoplasm of young sieve elements as discrete *slime bodies*. Examination of slime bodies under the electron microscope indicates that they consist of threadlike filaments. Slime bodies tend to fuse with one another during maturation of a sieve element. Later the slime spreads out and disperses throughout the sieve element. Another change in a maturing sieve element is the disintegration of the nucleus. Also, with increasing maturity, the cytoplasm becomes less dense and the vacuolar membrane disappears. The central *vacuolar region* loses its gellike qualities and assumes the consistency of a fluid.

At a still later stage the cytoplasm becomes a thin layer lying next to the cell wall. In some plants a network of membranous material called the endoplasmic reticulum occupies the *lumen* (i.e., the space earlier enclosed by the vacuolar membrane). Moreover, the lumen of mature sieve elements is often traversed by microfibrillar material, presumably consisting of P-protein.

Although general agreement exists on the *gross* anatomical features outlined, there is still considerable disagreement on the exact roles of each feature in the translocation of sugars. Evidence that unequivocally pinpoints the role of each anatomical characteristic in the translocation process is unavailable. These uncertainties are further complicated by the difficulties encountered in elucidating the *fine* structure of sieve elements. Because of its fragility, P-protein undergoes changes during the dehydration, fixation, and embedding procedures necessary for electron microscope examination. Moreover, the sudden release of turgor pressure during cutting is known to result in dislodgement of P-protein and other structures in sieve tubes.

In spite of considerable problems encountered in anatomical studies of sieve elements, the consensus of opinion among specialists in the field of phloem translocation is that a sieve element becomes functional (i.e., capable of translocating organic solutes) when it reaches maturity. At this stage of development the nucleus and tonoplast are lost, the sieve plate becomes fully perforated, sieve pores are not occluded to any appreciable extent either by callose or P-protein, the lumen occupies most of the sieve element, and a thin layer of cytoplasm adheres to the lateral walls.

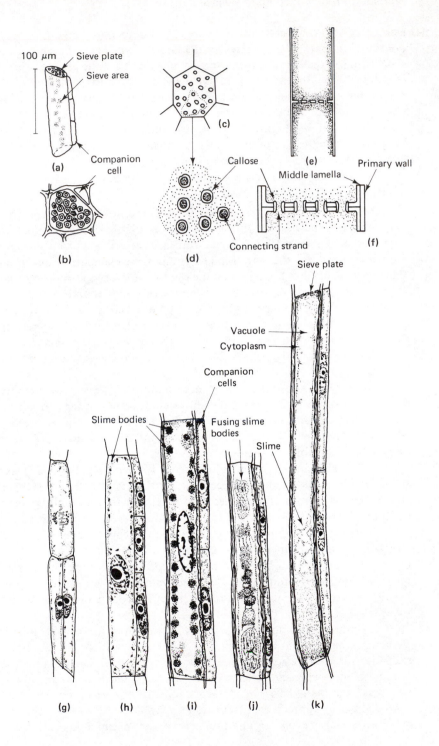

100 μm — Sieve plate

Sieve area

(a)

Companion cell

(b)

(c)

Callose

(d)

Connecting strand

(e)

Middle lamella — Primary wall

(f)

Sieve plate

Vacuole

Cytoplasm

Companion cells

Slime bodies

Fusing slime bodies

Slime

(g) (h) (i) (j) (k)

MECHANISM OF SIEVE TUBE TRANSLOCATION

There is wide agreement that the mechanism of sieve tube translocation is best explained by the *pressure flow* hypothesis.[9] This hypothesis was first proposed by the German forest botanist Münch in the late 1920s. He suggested that the sieve element degenerates as it becomes functional, with most of the cytoplasm being lost except for a very thin layer lining the lateral cell walls. The sieve element at the functional stage was assumed to consist mostly of a lumen filled with sugar solution. Also, it was assumed that sugar solution was able to flow freely through pores in sieve plates, from one sieve element to the next, through the length of the sieve tube.

The principle of the pressure flow hypothesis of phloem translocation is illustrated in the model shown in Figure 11-10. [The figure is somewhat misleading insofar as source and sink are shown as two (relatively large) containers, (a) and (b). Actually, source and sink may be very close together; perhaps the source may be one point along the length of a single sieve element and the sink another, as noted on p. 332.] When sugar is loaded by active transport (pp. 317–319) into the sieve tube at the source, the total concentration of dissolved solutes will become greater than in adjoining cells. Therefore, water will move by osmosis from adjoining cells into the sieve tube. Unloading of sugar by active transport at the sink results in movement of water by osmosis from the sieve tube to adjoining cells. Thus, osmosis into the source and out of the sink establishes the flow of sugar solution in the sieve tube.

In the pressure flow hypothesis, sugar solution is visualized to flow longitudinally in the lumens of sieve elements, from one sieve element to the next, through unoccluded pores in sieve plates. The layer of cytoplasm which lines the vertical walls of the sieve elements has several functions. (1) It provides a differentially permeable membrane, across which a difference in sugar concentration (between apoplast and

---

[9]A few other hypotheses of the mechanism of sieve tube translocation exist. Divergence of opinion stems in large part from difficulties in distinguishing which fine structures in sieve elements are important in translocation. For a brief summary of some of these hypotheses, none of which is widely accepted at present, see Evert, 1982.

Figure 11-9   Anatomy of sieve element: (a) Sieve element in secondary phloem of a dicotyledon, *Robinia pseudoacacia*, in longitudinal view. Note that this sieve element has a length well over 100 μm. (b) Surface view of the sieve plate of the sieve element in (a). (c) and (d) Diagram of a sieve plate in surface view. (e) and (f) Same as (c) and (d) except in sectional view. (g) to (k) Drawings of successive stages in the development of sieve elements in the primary phloem of *Curcurbita*, in longitudinal section. (g) The upper cell is a meristematic phloem cell undergoing division; the lower cell is a meristematic phloem cell immediately following division into a sieve element and a precursor companion cell. (h) A young sieve element with slime bodies beginning to develop, and three companion cells. (i) Nucleus of a sieve element just beginning to disintegrate. Note the thick cell wall in the sieve element. (j) Nucleus is absent and slime bodies partly fused. (k) Mature sieve element with large vacuolar region, thin parietal layer of cytoplasm, and slime (located mainly in the lower portion of the sieve element). Note the protoplast is partly withdrawn from the upper sieve plate but is connected with the lower sieve plate. [(a) to (f): After Esau, 1966; Figures 11.1 and 11.2 (g) to (k): After Esau, 1953; Figure 12.5.]

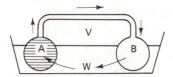

Figure 11-10 Drawing by Münch (1930) of a physical model that illustrates the flow of sugar solution in the phloem. The two containers A and B are fitted with membranes that are permeable to water W but not to sugar molecules. At the start of the demonstration, A is filled with a 10 to 15% solution (to which is added a trace of a dye, such as Congo red) and B with water. Both A and B are immersed in water. Water will move by osmosis from W to A. As a result, the hydrostatic pressure in A becomes higher than in B. This hydrostatic pressure difference between A and B provides the driving force for the movement of the colored sugar solution through the connecting glass tube V into B and for the diffusion of water through the membrane out of B. Münch visualized container A to be analogous to the source and container B to the sink. (After Schumacher, 1967; Figure 18.)

lumen of a sieve element) is maintained; (2) it is the site of active transport of sugar molecules during loading and unloading; (3) it permits osmosis of water to take place across it.

Sugar solution pressures up to 10 bars have been measured in sieve tubes (Wright and Fisher, 1980); in these experiments the aphid stylet technique (see Figure 11-8 and related text discussion) was used. Whether a pressure gradient (i.e., a pressure differential from one point to another along the length of a sieve tube) exists along the length of a sieve tube has not been determined. But if a pressure gradient actually exists along the length of a sieve tube, it would be a consequence of frictional resistance to flow in the sieve tube (Wardlaw, 1974) and thus would be very small and perhaps impossible to measure.

One objection leveled against the pressure flow hypothesis is the phenomenon of *bidirectional* movement (i.e., movement of different substances in opposite directions at the same time). This phenomenon can be demonstrated by applying two different substances at the same time to two different points of the phloem of a stem and following their longitudinal movement along the stem. The critical question is whether bidirectional movement occurs *in a single sieve tube*. Bidirectional movement cannot be expected to occur in a single sieve tube if the mechanism of translocation operates according to the pressure flow hypothesis. Unfortunately, experiments designed to demonstrate bidirectional movement in a single sieve tube are technically very difficult to perform. Some experiments indicate that bidirectional movement may occur in a single sieve tube, whereas other do not (Qureshi and Spanner, 1971).

In summary, the pressure flow hypothesis visualizes a steady, unidirectional, osmotically-generated flow of sugar solution through living sieve elements. Continuous entry of sugar molecules at the source and exit at the sink by active transport, together with the accompanying osmosis of water into the source and out of the sink, maintain the flow of sugar solution along the length of the sieve tube.

REFERENCES

BAKER, D. A., and J. L. HALL. 1973. Pinocytosis, ATPase and ion uptake by plant cells. *New Phytol.* 72:1281–1291.

BENNETT, H. S. 1956. The concepts of membrane flow and membrane vesiculation as mechanisms for active transport and ion pumping. *J. Biophys. Biochem. Cytol.* 2 (suppl.):99–103.

COCKING, E. C. 1977. Uptake of foreign genetic material by plant protoplasts. *Internatl. Rev. Cytol.* 48:323–343.

COLLANDER, R. 1963. Cell membranes: their resistance to penetration and their capacity for transport. Pages 3–102 *in* F. C. Steward, ed. *Plant Physiology, a Treatise*, vol. 2. Academic Press, New York.

DE VRIES, H. 1877. Untersuchungen über die mechanischen Ursachen der Zellstreckung ausgehend von der Einwirkung von Salzlösungen auf den Turgor wachsender Pflanzenzellen. W. Engelmann. Leipzig.

EPSTEIN, E. 1965. Salinity and the pattern of selective ion transport in plants. *Res. Develop. Prog. Rep. No. 161.* Office of Saline Water, U.S. Department of the Interior, Washington, D.C.

EPSTEIN, E. 1976. Kinetics of ion transport and the carrier concept. Pages 70–94 *in* A. Pirson and M. H. Zimmermann, eds. *Encyclopedia of Plant Physiology*, new series, vol. 2, Part B. Springer-Verlag, Berlin.

ESAU, K. 1953. *Plant Anatomy*. John Wiley & Sons., New York.

ESAU, K. 1961. *Plants, Viruses, and Insects*. Harvard University Press, Cambridge, Mass.

ESAU, K. 1966. *Anatomy of Seed Plants*. John Wiley & Sons., New York.

EVERT, R. F. 1982. Sieve-tube structure in relation to function. BioScience 32:789–795.

FENSOM, D. S. 1981. Problems arising from a Münch-type pressure flow mechanism of sugar transport in phloem. *Can. J. Bot.* 59:425–432.

GEIGER, D. R. 1979. Control of partitioning and export of carbon in leaves of higher plants. *Bot. Gaz.* 140:241–248.

GIAQUINTA, R. T. 1980. Translocation of sucrose and oligosaccharides. Pages 271–320 *in* P. K. Stumpf and E. E. Conn, eds. *The Biochemistry of Plants, a Comprehensive Treatise*, vol. 3. Academic Press, New York.

JOURNET, E. -P., M. NEUBERGER, and R. DOUCE. 1981. Role of glutamate-oxalacetate transaminase and malate dehydrogenase in the regeneration of NAD^+ for glycine oxidation by spinach leaf mitochondria. *Plant Physiol.* 67:467–469.

LÜTTGE, U., and N. HIGINBOTHAM. 1979. *Transport in Plants*. Springer-Verlag, Berlin.

MITCHELL, P. 1979. Keilin's respiratory chain concept and its chemiosmotic consequences. *Science* 206:1148–1159.

NASSERY, H., and R. L. JONES. 1976. Salt-induced pinocytosis in barley and bean. *J. Expt. Bot.* 27:358–367.

OSTERHOUT, W. J. V. 1936. Absorption of electrolytes in large plant cells. *Bot. Rev.* 2:283–315.

PATE, J. S., and B. E. S. GUNNING. 1972. Transfer cells. *Ann. Rev. Plant Physiol.* 23:173–196.

POOLE, R. J. 1978. Energy coupling for membrane transport. *Ann. Rev. Plant Physiol.* 29:437–460.

QURESHI, F. A., and D. C. SPANNER. 1971. Unidirectional movement of tracers along the stolon of *Saxifraga sarmentosa*. *Planta* 101:133–146.

RATHNAM, C. K. M. 1978. Malate and dihydroxyacetone phosphate-dependent nitrate reduction in spinach leaf protoplasts. *Plant Physiol.* 62:220–223.

SCHUMACHER, W. 1967. Die Fernleitung der Stoffe im Pflanzenkörper. Pages 61–177 *in* W. Ruhland, ed. *Handbuch der Pflanzenphysiologie*, vol. 13. Springer-Verlag, Berlin.

SPANNER, D. C. 1979. The electroosmotic theory of phloem transport: a final statement. *Plant Cell Environ.* 2:107–121.

SPANSWICK, R. M. 1981. Electrogenic ion pumps. *Ann. Rev. Plant Physiol.* 32:267–289.

STADELMANN, E. J. 1969. Permeability of the plant cell. *Ann. Rev. Plant Physiol.* 20:585–606.

WARDLAW, I. F. 1974. Phloem transport: physical chemical or impossible. *Ann. Rev. Plant Physiol.* 25:515–539.

WHEELER, H., and P. HANCHEY. 1971. Pinocytosis and membrane dilation in uranyl-treated plant roots. *Science* 171:68–71.

WOO, K. C., M. JOKINEN, and D. T. CANVIN. 1980. Reduction of nitrate via a dicarboxylate shuttle in a reconstituted system of supernatant and mitochondria from spinach leaves. *Plant Physiol.* 65:433–436.

WRIGHT, J. P., and D. B. FISHER. 1980. Direct measurement of sieve tube turgor pressure using severed aphid stylets. *Plant Physiol.* 65:1133–1135.

WRIGHT, J. P., and D. B. FISHER. 1981. Measurement of the sieve tube membrane potential. *Plant Physiol.* 67:845–848.

ZIMMERMANN, M. H., and J. A. MILBURN, eds. 1975. Transport in plants I: phloem transport, vol. 1 *in* A. Pirson and M. H. Zimmermann, *Encyclopedia of Plant Physiology*, new series. Springer-Verlag, Berlin.

12

Water as a Plant Constituent

Of all the substances necessary to plant life, water is required in the largest amount. Water is present throughout the plant body, from soil water around roots to the liquid-vapor interface in leaves. The evaporating surface of leaf mesophyll cells mark the discontinuity between water within the plant body and water vapor in the atmosphere. Every individual growing cell is surrounded by and impregnated with water. Water is the most abundant molecular species in actively growing plant cells. Growth rates of higher plants are more sensitive to their water supply and respond more quickly to soil water deficits than to any other factor of the environment. The availability of soil water to plant roots and the demands of the atmosphere for water vapor are among the most important ecological factors governing the distribution of higher plant species on the surface of the earth.

In most cells and tissues of higher plants water constitutes more than 80% of the fresh weight. The water content of some growing cells may rise to 90% or more, but in dormant seeds and buds water content may be 10% or less.

In this chapter the physical and chemical properties of water significant to plant–water relations are described. Also, the concept of water potential is developed. The physiology of plant–water relations, including absorption, translocation, and loss of water, is discussed in the next chapter.

FUNCTIONS OF WATER IN PLANTS

The significance of water to the life of the higher plant can be emphasized best by enumerating some of its functions.

1. Water is a major constituent of protoplasm.

2. Water is the solvent in which mineral nutrients enter a plant from the soil solution. Also, water is the solvent in which mineral nutrients are transported from one part of a cell to another and from cell to cell, tissue to tissue, and organ to organ.

3. Water is the medium in which many metabolic reactions occur.

4. Water is a reactant in a number of metabolic reactions [e.g., certain reactions in the tricarboxylic acid cycle (see Chapter 4)].

5. In photosynthesis the hydrogen atom in the water molecule is incorporated into organic compounds and oxygen atoms are released as O_2 (see Chapter 6).

6. Water imparts turgidity to growing cells and thus maintains their form and structure. In fact, water can be regarded as a material that provides mechanical support and rigidity to nonlignified plant cells.

7. Gain or loss of water from cells and tissues is responsible for a variety of movements of plant parts. Included are the diurnal swelling and shrinking of stomatal guard cells, the nocturnal folding of leaflets of certain plants, the opening and closing of flowers of certain plants at various times of the day, and the sensitivity to touch of leaflets of plants, such as the sensitive plant (*Mimosa pudica*).

8. The elongation phase of cell growth depends on absorption of water.

9. Water is a metabolic end product of respiration (see Chapter 4).

10. More water is absorbed by plants and greater amounts of water are lost (as water vapor) by plants than any other substance.

THE WATER MOLECULE

Water is a substance of unique physical and chemical properties, many of which help to explain its essentiality to living organisms. Several of these properties can be more readily understood if the properties of the water molecule itself are understood.

An electron dot representation of the formation of water from oxygen and hydrogen atoms is shown in Figure 12-1 (upper diagram). The hydrogen atom has one electron; the oxygen atom has eight electrons, but only six (those in the valence orbitals) appear in Figure 12-1. When chemical combination occurs, there is a sharing of valence electrons, in pairs, between the oxygen and hydrogen atoms. In this type of bond, known as the *covalent* bond, each atom contributes one electron; the pair of electrons that constitutes the bond is held jointly by both nuclei. Covalent bonds are very strong and so the water molecule is extremely stable.

A water molecule measures about 0.25 nm from one edge to the other. Because the molecule is not exactly spherical in shape (see Figure 12-1), this overall dimension is referred to as the "apparent" diameter.

Physical and chemical evidence indicates that the hydrogen nuclei and the oxygen nucleus in the water molecule have an unsymmetrical space configuration.

$$\cdot H \quad + \quad \cdot H \quad + \quad :\ddot{O}\cdot \quad \longrightarrow \quad :\ddot{O}:H$$
$$\qquad\qquad\qquad\qquad\qquad\qquad\qquad\qquad H$$

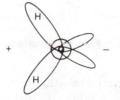

Figure 12-1 Two representations of the distribution of electrons in the water molecule. (See text for details.)

This asymmetry is indicated in the electron dot representation of the water molecule in Figure 12-1 (see upper diagram). The unshared oxygen electrons are positioned on one side of the oxygen nucleus and the two hydrogen nuclei are positioned on the other side. The angle between the lines connecting the hydrogen nuclei with the oxygen nucleus has been found to be 104.5°.

The oxygen nucleus in the water molecule has a greater affinity for electrons than the hydrogen nuclei. Therefore a field of electrical force of a polar character is present around the water molecule. The oxygen side is more negative than the hydrogen side and the water molecule is said to be a *dipole*.

The asymmetrical distribution of charge in the water molecule is shown in the lower diagram in Figure 12-1. Each pair of electrons around the oxygen nucleus (the small circle in the center of the diagram) is represented as a charge cloud in the form of an ellipse. Four ellipses are shown, corresponding to the four pairs of electrons appearing in the upper diagram in Figure 12-1. Each ellipse extends outward as an "arm" from the oxygen nucleus. Two of the charge clouds of electrons contain hydrogen nuclei (each indicated by the letter H). These hydrogen nuclei are responsible for the two positive *point* charges in the water molecule. The other two ellipses, corresponding to the charge clouds of the two pairs of lone electrons, do not contain hydrogen nuclei and are responsible for the two negative point charges. The two latter charge clouds extend above and below the H—O—H plane and are directed away from the hydrogens so that the water molecule takes the form of a tetrahedron in three-dimensional space—the two positive centers of a charge directed toward two of the vertices and the two negative centers of charge toward the other two. It is the asymmetrical distribution of charge that imparts a polarity to the water molecule, indicated by the plus and minus signs in Figure 12-1 (lower diagram).

Because of the dipole character of the water molecule, its positive side is attracted to negative charges and its negative side to positive charges. Therefore water molecules in the immediate neighborhood of an ion in solution will be attracted to the ion. So an ion in solution has a cage of water molecules around it, as shown in Figure 12-2.

The water molecule forms a special kind of bond known as the hydrogen bond. The *hydrogen bond* refers to the attraction of a hydrogen nucleus by two other atoms so that it acts as a link between them. The attractive force of the hydrogen nucleus for

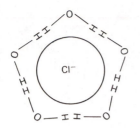

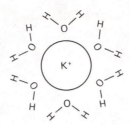

Figure 12-2 A shell of water molecules (hydration shell) will form around ions in solution. Positive ions attract the negatively charged oxygen side of water molecules. Negative ions attract the positively charged hydrogen side of water molecules.

atoms that have a strong affinity for electrons—they are often called "electronegative" and include oxygen and nitrogen—accounts for the hydrogen bond. Several examples of hydrogen bonds between groups of atoms are shown in Figure 12-3.

The strength of the hydrogen bond is much less, perhaps only one-twentieth, that of a covalent bond, such as the O—H bond in the water molecule. Nevertheless, the hydrogen bond is responsible for the attraction of the hydrogen atoms of one water molecule to the oxygen atoms of adjacent molecules in water in the solid (ice) and liquid states. Only in the vapor state are water molecules far enough apart so that they exist as discrete unassociated particles.

To summarize, the dipolar nature of the water molecule and its ability to form hydrogen bonds account for the strong interaction of water molecules with one another and with other molecules and ions in close association with them.

Water molecules in the solid state (ice) are assembled in a symmetrical, hydrogen-bonded lattice structure (Figure 12-4). Each oxygen atom is surrounded in tetrahedronal coordination by other neighboring oxygen atoms in such a fashion that

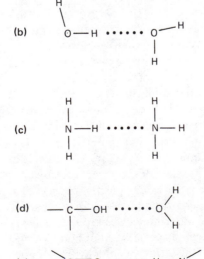

Figure 12-3 Examples of hydrogen bonds, shown as dotted lines: (a) between a group of atoms AH and another group of atoms B; (b) between two water molecules; (c) between two ammonia molecules; (d) between a hydroxyl group and a water molecule; (e) between a carbonyl group and an imino group.

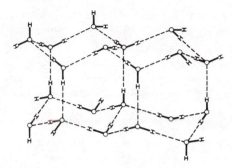

Figure 12-4 A model of the structure of ice. Hydrogen bonds are shown as dashed lines, covalent chemical bonds as solid lines. (After Stillinger, 1980; Figure 1. Copyright 1980 by the American Association for the Advancement of Science.)

oxygen atoms form six-membered puckered rings. Layers of these rings form an extensive three-dimensional network. The structure of this network is said to be "open" because the region within each ring is large enough to accommodate another water molecule but actually is void.

Liquid water also has an extensive hydrogen-bonded structure similar to ice. The detailed features, however, are believed to be short-lived. At any moment about half the water molecules in liquid water have the same orderly structure as a crystal of ice (Figure 12-4). This stable structure extends over long distances in ice, but in the liquid state the icelike clusters are thought to include only a few molecules each and they are constantly forming and disintegrating. The framework collapses and reforms with extreme rapidity, perhaps many billions of times a second. In effect, liquid water is a rapidly fluctuating mixture of hydrogen-bonded aggregates of water molecules. These aggregates have been visualized as "flickering clusters" of varying sizes.

PHYSICAL AND CHEMICAL PROPERTIES OF WATER

Liquid water is a colorless, odorless, tasteless, and highly incompressible substance with several unique properties. One is its peculiar behavior on freezing. Water expands when it freezes whereas almost all other substances contract on cooling and freezing. A given weight of ice is about 9% greater in volume than the same weight of liquid water. Therefore ice is less dense than liquid water and floats. This behavior occurs because water molecules in ice have an open structure (pp. 344–345). When ice melts, there is a partial collapse of its lattice and water molecules occupy some of the spaces that were void.

Other unique properties of water are its high boiling point and high heat of vaporization. A comparison with other hydrides of similar molecular weight (e.g., methane, CH_4, MW 16; ammonia, NH_3 MW 17; hydrogen sulfide, H_2S, MW 34) reveals that water (MW 18) has a much higher boiling point and a much higher heat of vaporization than might be expected on the basis of its molecular weight alone. To cite only one example, hydrogen sulfide has a boiling point of $-62°$ C, and its heat of vaporization (i.e., the number of calories required to convert 1 g of liquid into vapor) is 132 cal g^{-1} at its boiling point. Judging from its molecular weight alone, water would be expected to have a boiling point and a heat of vaporization even lower than that of hydrogen sulfide. But water boils at 100° C. Its heat of vaporization varies from about 540 cal g^{-1} at 100° C to about 580 cal g^{-1} at 25° C.

The high boiling point and high heat of vaporization of water are due to the attraction of water molecules for one another by hydrogen bonding. For a water molecule to escape from the liquid phase into the vapor, it must acquire not only sufficient kinetic energy but also an extra quantity of heat energy to break its hydrogen bonds with other water molecules.

One result of the high heat of vaporization of water is its cooling effect in evaporation. When water vaporizes from a leaf in transpiration, energy in the form of heat is lost from the leaf and the temperature of the leaf is reduced to some extent. The evaporated water containing the heat energy remains in the atmosphere in the form of water vapor until it condenses or precipitates.

Another unique property of water is its great capacity to absorb heat with very little rise in temperature. More heat is required to increase the temperature of 1 g of water by 1° C than to increase the temperature of 1 g of almost any other substance by 1° C. This quantity of heat—the heat required to raise the temperature of 1 g of a substance by 1° C at 15° C (i.e., from 15 to 16° C)—is called the *specific heat*. (The specific heat of water is about four times that of air.) The high specific heat of water is due to the strong attraction of water molecules to one another. Part of the heat energy supplied to water to raise its temperature is required to break its hydrogen bonds. This means that a body of water must absorb relatively large amounts of heat before its temperature rises even slightly. As a corollary, water has a much better storage capacity for heat than does air. To give one example, as much heat is stored in the upper 3 meters of ocean water as in the entire atmosphere stretching between the ocean surface and outer space. The temperature of ocean water, therefore, changes much more slowly than does that of the atmosphere. Larger quantities of heat must be taken from or added to the ocean—than taken from or added to the atmosphere—before a change in water temperature will occur.

The high specific heat of water helps plants maintain relatively stable internal temperatures. When large fluctuations in temperature take place in the external atmosphere, only relatively small changes in temperature occur within the plant—the water in the plant acts as a "cushion." In addition, water tends to maintain plant cells at constant temperature while heat-producing chemical reactions are taking place within them.

The ability of water molecules in the liquid state to cling to one another tenaciously—a property called *cohesion*—explains "surface tension." Molecules of water at the surface of liquid water are more strongly attracted by molecules within the interior of the liquid than by molecules in the air above the liquid. Therefore the liquid–air interface behaves as an elastic skin that contracts to a minimum area. This property of a liquid is known as *surface tension*. Because water molecules are attracted to one another by hydrogen bonding, water has an especially high surface tension, higher than that of most liquids.

Not only do water molecules attract one another by hydrogen bonding but water molecules also hold firmly to the molecules of other substances that contain large numbers of oxygen or nitrogen atoms, such as glass, wood, clay, soil, cellulose, and protein. These two properties (i.e., the ability of water molecules to *cohere* to one another and to *adhere* to molecules of certain other substances) depend on the ability of water molecules to form hydrogen bonds.

The high surface tension of water and the ability of water to adhere to certain other substances account for the capillary action of water. This phenomenon refers to the "spreading out" of liquid water when in contact with such materials as blotting paper, soil, or a plant cell wall. The tiny fissures, pores, and interstices characteristically present in these materials are the spaces through which water moves spontaneously from damp to dry regions.

A striking example of the capillary action of water can be seen in a glass tube of fine bore. Water will rise slowly above the general level of the water in which the tube is dipped. The hydrogen bonding between oxygen atoms in the glass tube and molecules of water at the edge of the water surface serves to raise the edge of the water in the tube. Simultaneously, the hydrogen bonding of water molecules to one another in the water–air interface pulls the water surface taut. The surface, in turn, pulls up the water beneath it and brings more water near the edge. The sequence may be visualized as being repeated over and over again as water climbs upward in the tube. Capillary ascent ceases when the weight of the elevated water inside the tube becomes equal to the surface tension force.

Still another unique property of water is its ability to dissolve more substances and in greater quantity than almost any other liquid. The solvent action of water depends on one or another of at least three types of interaction between water molecules and solute molecules and ions.

The solubility of substances that do not ionize when dissolved in water, especially substances of low molecular weight that contain oxygen and nitrogen atoms in the form of such groups as —OH and —NH_2 (e.g., glucose, fructose, sucrose, undissociated amino acids), depends on hydrogen bonding. The molecules of these substances form hydrogen bonds with water molecules. A different type of solvent action accounts for the solubility of substances with molecules that ionize. These substances owe their solubility to the dipole character of the water molecule. Each ion in solution has a shell of oriented water molecules around it (see Figure 12-2). This shell acts as a field of electrical insulation that decreases the mutual force of attraction between oppositely charged ions and thus serves to keep ions in solution separated from one another. In a third type of solvent action, nonpolar substances (e.g., hydrocarbons) dissolve—but only to a small extent—in water because a strong attractive force exists between water molecules and molecules of all other substances. (Intermolecular attractive forces between molecules of different substances in close proximity to one another arise from small shifts of electrons relative to atomic nuclei and are referred to as van der Waals' forces.) Placing a molecule of a nonpolar substance in water may be likened to making a "hole" in the water structure. An alteration in the structure of water inevitably occurs. Thus the solubility of nonpolar substances is very limited in liquid water.

Although the interaction of solar radiation with certain plant pigments and other constituents has significant effects on plant behavior (see Chapter 5), water itself is hardly affected by solar radiation. Water absorbs ultraviolet and visible radiation to only a slight extent. Absorption is stronger in the infrared, especially at wavelengths greater than about 1.2 μm, but solar radiation is of only relatively low intensity at these longer wavelengths. Interaction of water with these three forms of radiation increases the extent of vibration and rotation of hydrogen and oxygen atoms in water molecules, but this increased motion is dissipated as heat. Thus radiation in the ultraviolet, visible, and infrared regions has relatively little influence on water itself, aside from a slight heating effect. It can be said, in fact, that the heating effect of these forms of radiation on water in plant tissues is not due to their influence on water itself but rather to interaction with plant constituents other than water, followed by conduction of heat to water.

In contrast to the weak interaction of water with ultraviolet, visible, and infrared radiation, water interacts strongly with high-energy radiation, such as x rays, gamma rays, and beta rays. Although these radiations are not usually encountered under natural conditions, plants may be intentionally exposed to them under certain experimental conditions. High-energy radiation is likely to split water molecules into short-lived H and OH radicals. Both radicals are highly reactive chemical species: The former is a strong reducing agent and the latter a strong oxidizing agent. Most H and OH radicals produced in plants exposed to high-energy radiation combine with each other immediately on formation so that water molecules are reformed (thereby nullifying the potentially adverse effects of the radiation). Nevertheless, a few radicals can be expected to collide directly with cellular constituents. It is these *secondary* reactions with H and OH radicals, rather than direct "hits," that account for most of the deleterious effects produced in plants by high-energy radiation. (See also pp. 135–136.)

The chemical properties of water are no less important to life processes than the physical properties. Water may react chemically with a variety of metabolites (organic esters, peptides, proteins, polysaccharides, etc.) to yield hydrolysis products. Also, water dissociates into hydrogen and hydroxyl ions; in pure water, the equilibrium concentration of hydrogen ions and also of hydroxyl ions, is 10^{-7} mol L^{-1}. Hydrogen and hydroxyl ions also are produced by chemical reaction of water molecules with ions of weak acids and bases in aqueous solution. The course of many reactions occurring in living cells may be influenced considerably by hydrogen ion concentration. Control of cellular hydrogen ion concentration by plant constituents (e.g., organic acids) that act as natural buffering systems has primary significance in determining the rates of many metabolic processes.

CELLULAR WATER

Studies of the water in plant cells indicate that a large fraction can be identified with vacuolar water. Vacuolar water is probably quite similar in nature to a dilute salt solution in a beaker except that it is usually subjected to a hydrostatic pressure of several bars (see discussion of pressure potential on p. 367).

In contrast, a small but sizable fraction of the water in plant cells is adsorbed to various plant cell constituents by dipolar and hydrogen bond forces described earlier. Especially important in this regard is the water held in microcapillaries in protoplasmic colloids and cell wall components. Moreover, the surfaces of all protoplasmic membranes can be visualized as encased in a sheet of water molecules one molecule thick; these water molecules can be considered integral components of protoplasmic membranes. Also, the surfaces of cellulose microfibrils and polysaccharide colloids in cell walls can be considered coated with water molecules one molecule thick.

Water molecules that adhere to the surfaces of cellular constituents, such as proteins, protoplasmic membranes, and cellulose microfibrils, can exchange readily with water molecules in other parts of the plant cell.

FACTORS AFFECTING THE CHEMICAL POTENTIAL OF WATER

Recall from Chapter 10 (pp. 297–299) that the best way to express quantitatively the escaping tendency of a substance (i.e., its tendency to move spontaneously from one region to another) is in terms of the difference in chemical potential of the substance between the two regions. It will be useful to restate some of these concepts in a few concise sentences, especially as they apply to water.

If a difference in chemical potential of water exists between two regions, spontaneous transfer of water will take place, provided that the two regions are not separated by a barrier that prevents passage of water. The direction of spontaneous transfer of water will be from the region of *higher* to *lower* chemical potential of water. Furthermore, the difference in chemical potential between the two regions is a unique quantitative measure of the driving force for the spontaneous transfer of water from one region to the other. Spontaneous transfer of water will occur until the chemical potential of water in the two regions is the same. At this point of equilibrium, net transfer of water ceases.

Effect of Temperature

Because the chemical potential of water is a measure of the energy content of water in a system, it can be expected that any factor influencing the energy content of water will influence the chemical potential of water in the system. Temperature is one such factor. As shown in Figure 12-5(a), the chemical potential of water in a system increases with an increase in temperature in the system.

In addition to temperature, two other factors are of considerable concern to plant physiologists. One is the presence of solutes and imbibants in the system. The other is the pressure, or tension, that exists in or on the system. Each is discussed below. Other forces (e.g., the force of gravity) also influence the chemical potential of water, but their effect on plant cells and tissues is small and generally negligible.

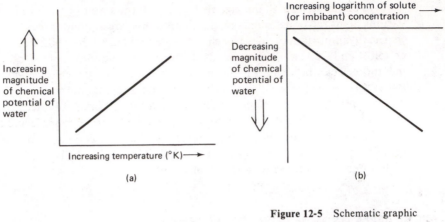

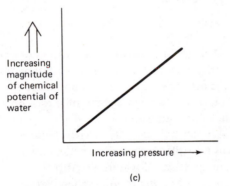

Figure 12-5 Schematic graphic representation of the relation between chemical potential of water in a system and (a) temperature, (b) solute or imbibant concentration, and (c) pressure. In each graph, either temperature, or concentration of solute or imbibant, or pressure is the independent variable, while the other two factors are maintained constant. Although the graphs are drawn as straight lines, deviations from linearity may occur. [In (b) a straight line is obtained only when the logarithm of the solute or imbibant concentration is plotted on the horizontal axis.]

Effect of Imbibants and Solutes

The kind and quantity of constituents other than water in a system considerably influence the chemical potential of water in the system. Water molecules in the near vicinity of protoplasmic constituents and cell wall material are attracted to them, as was noted on pp. 346–347. These attractive forces cause a reduction in the escaping tendency—and hence in the chemical potential—of water in plant cells. The term *imbibant* is used to designate insoluble hydrophilic substances (e.g., cellulose, starch) which absorb water or to which water adheres.

Moreover, the attractive forces that exist between ions and molecules in solution and water molecules (considered on p. 347) results in a lowering in the pressure of water vapor in equilibrium with the solution. This lowering in water vapor pressure is proportional to the total number of dissolved particles in the solution, whether ions or molecules. Furthermore, the lowering in vapor pressure reflects a reduction in the escaping tendency—and hence in the chemical potential—of water in the solution. The influence of a solute or an imbibant on the chemical potential of water is depicted graphically in Figure 12-5(b).

Effect of Pressure or Tension

Water in growing plant cells is usually subjected to pressure greater than that in a beaker. Let us compare the two situations.

The total pressure on the inside surface of a beaker of water consists of the pressure of the liquid and the pressure of the gases of the atmosphere, which press downward on the surface of the liquid. The pressure of the gases of the atmosphere is equal to 1 bar at the surface of the earth. (Although variations in atmospheric pressure occur hourly and daily, they are only on the order of hundredths or thousandths of a bar and are negligible in plant–water relations.) As for the pressure of the liquid, a column of water approximately 10 meters high exerts a hydrostatic pressure of 1 bar at its base.

Next, consider the pressure in the vacuole of a growing plant cell. The contribution of the atmosphere, as in the beaker, is 1 bar. But the cytoplasm and cell wall that surround the vacuole resist expansion and press inward on the vacuole. The intensity of the inwardly directed pressure increases as more and more water is secreted into the vacuole. Consequently, hydrostatic pressure may develop in the solution in the vacuole, sometimes as high as +10 to +20 bars (in excess of atmospheric pressure).

Not only positive hydrostatic pressure but also negative hydrostatic pressure (i.e., tension) may exist in certain plant tissues. Xylem sap in the conducting elements of xylem tissue is likely to be under tension during daytime hours in the growing season, when the rate of loss of water vapor by transpiration from leaves exceeds the rate of absorption of water by roots (see pp. 403–404). Measurements of tensions in xylem sap in trees and shrubs indicate that values on the order of −3 to −5 bars often are reached during the daytime, but even more negative values have been recorded.

The hydrostatic pressure (or tension) in a system that contains water is directly related to the chemical potential of water in the system. A change in hydrostatic pressure or tension results in a proportionate change in the chemical potential of water in the system. This relationship is depicted graphically in Figure 12-5(c).

A Demonstration of Changes in the Chemical Potential of Water in a Solution

It is instructive to compare the change that occurs in the chemical potential of water when a solute is introduced and pressure is applied to the solution. This can be done with the aid of Figure 12-6. Here the chemical potential of water in the liquid can be evaluated indirectly by measuring the pressure of water vapor in equilibrium with the liquid. This procedure is based on the fact that equilibrium vapor pressure reflects the tendency of water molecules to escape from the liquid and therefore is proportional to the chemical potential of water in the liquid. A nonvolatile solute (i.e., sucrose) is used in this demonstration so that vapor pressure will be due to water molecules only. Temperature will be maintained constant at 25° C.

First, water vapor pressure will be measured with a mercury manometer. This is an instrument with two arms, each partially filled with mercury; one arm is evacuated and the other is connected to the chamber [see Figure 12-6(a), (b), and (c)]. The

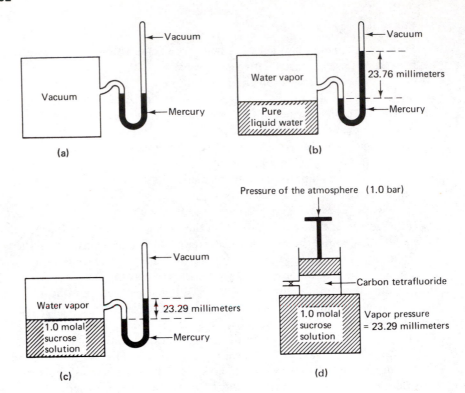

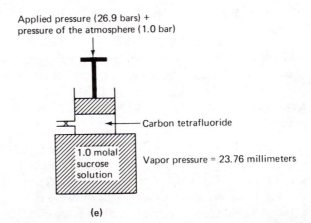

Figure 12-6 Effects of dissolved solute and pressure in excess of atmospheric on the chemical potential of water at 25° C. (See text for details.) The vapor pressure of pure water (= 23.76 mm Hg) was taken from the *Handbook of Chemistry and Physics* (Chemical Rubber Publishing Company, Cleveland). The chemical potential of water in a 1 *m* sucrose solution (= −26.9 bars) was taken from Michel (1972) and is believed to be the best available. From these data the vapor pressure of water in a 1 *m* sucrose solution was calculated by the equation that defines chemical potential of water in terms of vapor pressure (see footnote, p. 357).

difference in height of the mercury in the two arms of the manometer is a measure of vapor pressure in the chamber.

At the start of the demonstration [see Figure 12-6(a)], the chamber is evacuated and so the level of mercury in the two arms of the manometer will be the same. But when pure liquid water is introduced into the chamber [see Figure 12-6(b)], water vapor molecules will escape from the liquid and exert a pressure on the mercury. At equilibrium, the same numbers of water molecules will pass the liquid–vapor interface in each direction in a unit of time and therefore the chemical potential of water will be the same in both the liquid and the vapor phases. As shown in Figure 12-6(b), the difference in height of mercury in the manometer arms will be 23.76 mm.

When sucrose molecules are dissolved in water [see Figure 12-6(c)], there is a lesser tendency for water molecules to escape into the vapor compared to their tendency to escape from pure water. Thus the chemical potential of water in a sucrose solution will be less than that in pure water. And the equilibrium vapor pressure in Figure 12-6(c) will be less than that in Figure 12-6(b). The pressure of water vapor in equilibrium with a 1 m sucrose solution at 25° C is 23.29 mm of mercury [see Figure 12-6(c)]. (Note: The abbreviation for molal is m.)[1]

The *lowering* in chemical potential of water produced by the presence of a *solute* in water can be *overcome* by applying *pressure* to the solution. This can be demonstrated by placing a 1 m sucrose solution in a chamber equipped with a movable, frictionless piston [see Figure 12-6(d) and (e)]. To measure water vapor pressure over the solution when piston pressure is applied, a gas with very low solubility in water (e.g., carbon tetrafluoride) is introduced into the chamber through a side arm. By taking an aliquot of the vapor phase ($CF_4 + H_2O$) slowly (so as not to disturb the equilibrium) and then condensing the water, the mole fraction of water in the vapor phase can be determined. From this information the vapor pressure can be calculated:

Vapor pressure of water = total pressure in the vapor phase × mole fraction of water

At the start of this part of the demonstration (i.e., when only the atmosphere presses down on the sucrose solution), water vapor pressure in the solution [see Figure 12-6(d)] will be exactly the same as in Figure 12-6(c). But by applying pressure in excess of atmospheric on the piston [see Figure 12-6(e)], the hydrostatic pressure against the walls of the chamber is increased and so the chemical potential of water in the sucrose solution is also increased. At 25° C a piston pressure of 26.9 bars (in excess of atmospheric pressure) will be required to raise the chemical potential of water in the sucrose solution so that it becomes equal to that of pure liquid water. Thus the vapor pressures in Figure 12-6(b) and (e) are equal.

[1] Molality is the concentration of a solution expressed as the number of moles of dissolved substance per 1000 grams of water. The molality is more appropriate than the molarity for specification of relationships considered here because it is the ratio of molecules of solute to molecules of water which is important. Molarity—the abbreviation for molar is M—is the concentration of a solution expressed as the number of moles of dissolved substance per liter of water. In very dilute solutions, molality and molarity are practically equal; at higher concentrations the difference between them is significant. (In this book molality is used only in this chapter.)

Later we will see that water vapor pressure (and hence chemical potential of water) can be measured not only by the methods used in Figure 12-6 but also by an instrument known as the thermocouple psychrometer (see Figure 12-9).

Conventions in Plant-Water Relations

One convention adopted in plant–water relations is to specify that constant temperature prevails in the system involved. (The temperature in the demonstration in Figure 12-6 was maintained at 25° C.) In this way, the effect of temperature variations on the chemical potential of water is eliminated. Another convention is to specify that atmospheric pressure is equal to 1 bar.

By specifying conditions of constant temperature and constant atmospheric pressure, no distortion of fundamental principles is involved. Moreover, attention can focus on changes in the chemical potential of water produced by changes that occur within the system itself (e.g., in solute concentration, hydrostatic pressure). So in our discussions of plant–water relations here and in the next chapter it will be understood that conditions of constant temperature and constant atmospheric pressure prevail in the system under consideration unless otherwise specified.

Another convention adopted in plant–water relations is to use the term *water potential* (pp. 357–358) in place of *chemical potential of water*. It may be useful, for future reference, to record the numerical values, in bars, of the water potentials in Figure 12-6 (see Table 12-1). These values were calculated from the vapor pressures shown in Figure 12-6 by solving the equation in footnote 4 (p. 357).

TABLE 12-1   VALUES OF WATER POTENTIAL
IN FIGURE 12-6[a]

	Water potential (bars)
Figure 12-6(b)	0
Figure 12-6(c)	−26.9
Figure 12-6(d)	−26.9
Figure 12-6(e)	0

[a] See text for details.

IMBIBITION AND OSMOSIS

It is customary for plant physiologists to distinguish between two processes of water movement into plant cells. These two processes are imbibition and osmosis. Both are examples of passive transport, as defined earlier (see pp. 312–313). Suggestions have been made at times (e.g., see Cailloux, 1972) that active transport may account, at least in part, for movement of water into plant cells. No unequivocal evidence, however, is available for a pump that transports water actively across protoplasmic membranes.[2]

Imbibition

Imbibition in plant cells refers to the absorption and adsorption of water by insoluble, solid, hydrophilic protoplasmic and cell wall constituents. Water is imbibed as a result of both diffusion and capillary action. (Capillary phenomena participate to the extent which the imbibing material is permeated with minute capillaries.) The direction of water movement is from a region of higher to one of lower chemical potential of water (= water potential), as noted on p. 349. Imbibition was demonstrated earlier (see Figure 10-2 and related text discussion); the imbibing material was dead, air-dried pea seeds. The force responsible for the binding of water molecules to plant solids in these dead, air-dried pea seeds is hydrogen bonding (see pp. 343–344 and also pp. 346–347).

Imbibition is a process that occurs only when solid plant material (e.g., dry wood, dead or living air-dried seeds) comes in contact with water. In the case of living, air-dried seeds wetted with water for the purpose of germination, imbibition occurs during the first several hours. Thereafter water is absorbed by osmosis.

Osmosis

Osmosis refers to the movement of water across a differentially permeable membrane that separates two solutions, the direction of movement being from higher to lower water potential (as in imbibition). A differentially permeable membrane, one that has different permeability properties for different substances, permits the free passage of water molecules but restricts the passage of dissolved solutes. All protoplasmic membranes are differentially permeable. The term differentially permeable is preferred to *semipermeable* even though the latter often appears in the literature of osmosis. The concept of a semipermeable membrane was originally used in the late nineteenth century by the Dutch chemist van't Hoff—who used osmosis to study some properties of dilute solutions—to designate an ideal partition permeable to one component of a mixture but completely impermeable to all others. Such a membrane is probably never encountered in living systems.

To demonstrate osmosis, pure water is placed in a beaker and a concentrated solution of a substance, such as sucrose, is placed in the well of an inverted thistle tube (see Figure 12-7). The sucrose solution is supported by a membrane permeable to water but not to sucrose; a sheet of unglazed cellophane can be used. To start the demonstration, the sucrose solution is immersed in the beaker of water. (This setup is usually called an *osmometer*.) Because the chemical potential of water in pure water in

---

[2]All data regarding the possibility of active transport of water can be interpreted to result from movement of water immediately *following* an osmotic gradient produced by active *solute* movement (Levitt, 1967). This type of water movement is sometimes called "active absorption of water" (Oertli, 1967).

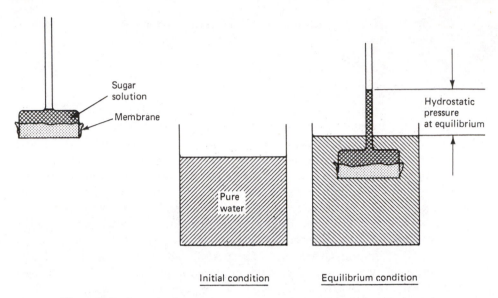

Figure 12-7 Osmosis of water from pure water to a sugar solution. (See text for details.)

the beaker is initially higher than the chemical potential of water in the sucrose solution, there will be a spontaneous net movement of water from the beaker through the differentially permeable membrane into the sucrose solution. As the transport of water proceeds, the height of the sucrose solution in the thistle tube increases. Therefore hydrostatic pressure on the membrane increases. This hydrostatic pressure will tend to press water molecules through the membrane and out of the sucrose solution. When the hydrostatic pressure in the thistle tube increases to the point that the chemical potential of water in the sucrose solution becomes equal to that in pure water in the beaker, further net movement of water through the membrane ceases. Equilibrium with respect to water movement across the membrane has been reached.[3] The equilibrium hydrostatic pressure is called the *osmotic potential* of the solution. (The term osmotic potential takes the place of *osmotic pressure* used in the older literature; see p. 365 for sign conventions for these two terms.)

Water movement by osmosis proceeds rapidly across all plant membranes, whether the membrane is a layer of protoplasm lying between the cell wall and the

---

[3]When this experiment is performed with an *ideal semipermeable* membrane (see p. 355), the equilibrium condition will be final and stationary: The hydrostatic pressure is called the osmotic potential of the solution. With a *differentially permeable* membrane (see p. 355), however, only a transient maximum of hydrostatic pressure will be reached because the solute diffuses gradually out of the solution in the thistle tube through the differentially permeable membrane into the beaker. The ratio of this transient maximum hydrostatic pressure to the final hydrostatic pressure developed by the same solution in an osmometer with a semipermeable membrane (i.e., the osmotic potential of the solution) is called the *reflection coefficient*. This parameter is encountered more and more frequently in current literature (e.g., Palta and Stadelmann, 1980).

vacuole of a plant cell or a protoplasmic membrane, such as the plasma membrane or the tonoplast. In fact, recall that water moves across protoplasmic membranes at much more rapid rates than any other substance (see discussion of plasmolysis and recovery from plasmolysis on pp. 308–309).

The Importance of Osmosis

The importance of osmosis in the absorption of water by cells of higher plants cannot be overestimated. When it is realized that 70% or more of the water in a mature plant cell is vacuolar—and that this water moves across the plasma membrane and vacuolar membrane (tonoplast) and the layer of protoplasm between these two membranes by osmosis—it is evident that osmosis has a much greater quantitative role in water absorption by plant cells than does imbibition.

In addition to its role in the transport of water into plant cells, osmosis has a special significance in certain procedures in experimental plant physiology. The phenomenon of plasmolysis (pp. 308–309) depends on osmosis. As another example, the isolation of cell organelles (e.g., mitochondria) from plant cells depends on the elimination of osmosis. If exposed during isolation procedures to distilled water or to very dilute solutions, these organelles are likely to burst. The rapid entry of water by osmosis into "dead spaces" within these organelles leads to the rupture of the outer membrane. To minimize possible damage of this kind, solutes with no adverse physiological effects (e.g., sucrose) are added intentionally to the isolation medium so that the chemical potential of water in the isolation medium is approximately the same as that in the fluid that bathes the organelles in vivo.

In contrast to the behavior of such organelles as mitochondria, intact plant cells, when placed in distilled water, will expand—because of the absorption of water by osmosis—but will not burst, for cell walls are relatively rigid and stretch but do not rupture during such treatment. The cells, however, are likely to be damaged because small ions and molecules will leak (slowly) out of protoplasts.

WATER POTENTIAL

As noted (p. 354), the term *water potential* is used by plant physiologists in place of *chemical potential of water*. The Greek letter psi (ψ) designates the water potential in a system, whether a plant cell or tissue, or a soil sample in which plants are growing, or a solution in a beaker.[4] Water potential is expressed either in bars (a pressure unit

---

[4]Water potential in a system is defined thermodynamically as the amount of work that becomes available when a mole of water in the system is transferred isothermally and reversibly from pure free water to a point in the system. This work is equivalent to the difference in free energy per unit molal volume of water in the system and in pure water at the same temperature and is related to the vapor pressure of water in the system and in pure water by the thermodynamic equation $\psi = (RT \ln e/e^\circ)/\overline{V}$, where ψ is the water potential in the water in the system, in bars [1 bar = $(10^6$ dynes cm$^{-2})$ = 10^6 ergs cm^{-3}], R is the ideal gas constant (= 8.3×10^7 ergs mol^{-1}deg^{-1}), T is the absolute temperature in $^\circ$K), e is the vapor pressure of water in the system at temperature T, e° is the vapor pressure of pure free water at the same temperature, and $\overline{V}$ is the molal volume of water (= 18.0 cm^3 mol^{-1}).

defined on page 351 and more rigorously in footnote 4 on p. 357) or megapascals (1 Mpa = 10 bars). (Bars will be used in this book rather than megapascals because uniformity of usage of these terms has not yet been achieved in the literature.)

Water potential is a diagnostic tool that enables the plant scientist to assign a precise value to the water status in plant cells and tissues. The lower the water potential in a plant cell or tissue, the greater is its ability to absorb water. Conversely, the higher the water potential, the greater is the ability of the tissue to supply water to other more desiccated cells and tissues. Thus water potential is used to measure *water deficit* (see p. 372) and *water stress* (see p. 409) in plant cells and tissues.

Absolute values of water potential are not measured. Instead measurements are made of the difference between the water potential in a system under investigation (e.g., a plant tissue) and that in a reference state. The reference state is pure liquid water at the same temperature and same atmospheric pressure as the system under investigation. Water potential in the reference state is arbitrarily assigned a value of 0 bar.

Water potential in a plant tissue is always less than 0 bar and hence a negative number.

The range of water potentials in representative plant tissues is illustrated in Figure 12-8. As a general rule, leaves of most plants rooted in well-watered soils are likely to have water potentials between about −2 and −8 bars. With decreasing soil moisture supply, leaf water potential will become more negative than −8 bars and leaf growth rates will decline (for further discussion of this matter, see p. 410). Most plant tissues will cease growth completely (i.e., will not enlarge) when water potential drops to about −15 bars.

Leaves of herbaceous plants may survive for relatively short periods of time—but will not grow—when water potential is less than −15 bars (see Figure 12-8). Generally leaves of herbaceous plants are not likely to recover if the water potential drops below about −20 to −30 bars. In contrast, leaves of desert shrubs have a greater ability to survive, and for longer periods of time, when subjected to low water potentials. As shown in Figure 12-8, leaf water potentials in desert shrubs may be very low under conditions of severe drought, perhaps in the range of −30 to −60 bars. Even lower values, as low as −100 bars, have been recorded. Viable, air-dried seeds are also likely to have very low water potentials (see Figure 12-8), perhaps as low as −60 to −100 bars, or even much lower, depending on the extent of drying and the plant species.

Also shown in Figure 12-8 are values of water potential in a few selected solutions at 25°C; for 1 *m* sucrose solution, see legend to Figure 12-6 and also Table 12-1.

MEASUREMENT OF WATER POTENTIAL

The measurement of water potential in plant cells and tissues is of critical concern in studies of plant–water relations. Several methods are in use today, but none is the last word on the subject. Undoubtedly the need for making rapid and precise measurements of tissue water potential, especially in field studies, will provide the

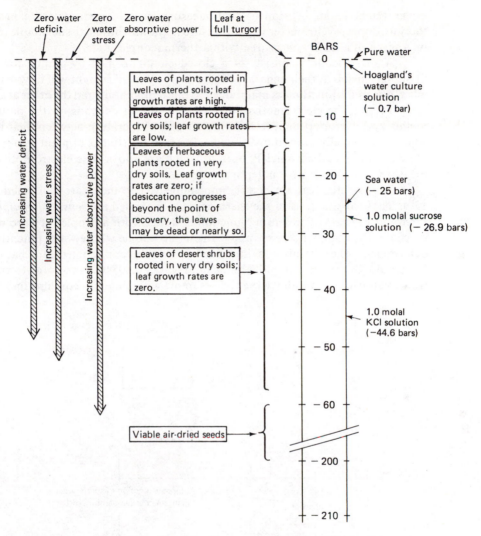

Figure 12-8 The scale of water potential. (See text for details.)

impetus during the next few years for improvement and refinement of existing techniques and for the development of new methods.

Vapor Equilibration Method

In the vapor equilibration method the pressure of water vapor in equilibrium with the water in a tissue sample enclosed in a small chamber is measured. This process is based on factors discussed earlier (see Figure 12-6 and related discussion) except that water vapor pressure is measured with a thermocouple psychrometer. The term *psychro-*

meter refers to an instrument that measures water vapor pressure in air. In a thermocouple psychrometer vapor pressure in the air above a tissue sample (e.g., a leaf or a piece of a leaf) is measured with a thermocouple.

A thermocouple consists of a closed circuit formed by joining together two dissimilar metals at their ends (see Figure 12-9). A small current will flow in the circuit when one of the junctions is maintained at one temperature and the other at a different temperature. The electromotive force of this current depends on the nature of the metals and is proportional to the difference in temperature between their junctions. The metals usually used in instruments designed for studies of plant–water relations are constantan and chromel-P, both of which are alloys. The electromotive force is measured with a sensitive potentiometer.

The essential features of a thermocouple psychrometer are illustrated in Figure 12-9. The two junctions of the thermocouple are called the *sensing junction* and the *reference junction*. The sensing junction is wetted with a droplet of pure water and placed in the chamber directly above the tissue sample whose water potential is to be determined. The chamber is immersed in a constant temperature bath. Precise temperature control, within the limits of at least ±0.005° C, is required to ensure that water vapor in the air above the tissue sample will remain in equilibrium with water

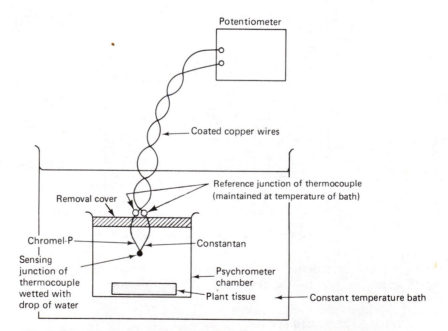

Figure 12-9   The thermocouple psychrometer for measuring water potential in plant tissues is shown schematically. Note the following essential components: (1) a small chamber in which the tissue sample is enclosed; (2) a constant temperature bath; (3) a thermocouple sealed into the removal cover of the psychrometer chamber; (4) a sensitive electrical measuring device capable of measuring very small differences in electromotive force induced in the thermocouple circuit.

inside the sample during a measurement. Also, the constant temperature bath provides a fixed temperature for the reference junction of the thermocouple.

When water evaporates from the wetted sensing junction, it will be cooled (see p. 346). In fact, the rate of evaporation of water from the wetted sensing junction—and hence its temperature—will be determined by the vapor pressure in the chamber. Because of the difference in temperature between the sensing junction and the reference junction, an electromotive force proportional to the difference in temperature is generated in the thermocouple.

Before a thermocouple psychrometer can be used to measure water potential, the electrical output of the instrument must be calibrated against solutions of known water potential. Accurate water potential data for solutions of certain sugars and salts (e.g., sucrose, potassium and sodium chloride) are available in the literature (e.g., see Michel, 1972; Stadelmann, 1966; and Wiebe et al., 1971).

The psychrometer method of measuring water vapor pressure in equilibrium with a tissue sample is generally regarded as the most accurate of all methods of measuring tissue water potential. But because thermocouple psychrometers are subject to errors introduced by temperature fluctuations and so require a bath with very precise temperature control, these devices have been used in the past only in the laboratory. Recently, however, efforts have been made by several investigators to modify the design of thermocouple psychrometers so that they are not subject to temperature-induced errors. A thermocouple psychrometer developed by Hoffman and Rawlins (1972), for example, can measure the water potential of leaves attached to intact plants in the field with a precision of ± 1 bar. Instruments of this type promise to open the way for routine measurements of leaf water potential in investigations carried out under field conditions.

Vapor Immersion Method

In the vapor immersion method, when a plant tissue is placed in an atmosphere in which water vapor is maintained at constant vapor pressure, there is a spontaneous net transfer of water between the tissue and the surrounding atmosphere until equilibrium is reached (see p. 349). The quantity of water transferred depends on the magnitude of the difference in water potential between the tissue and the surrounding atmosphere. If the water potential in the surrounding atmosphere is higher than in the tissue, water moves spontaneously into the tissue and its water content—and also its weight—increases. Conversely, if the water potential in the surrounding atmosphere is lower than in the tissue, water will move spontaneously out of the tissue and its water content—and also its weight—decreases. No transfer of water between the tissue and the surrounding atmosphere will occur if their water potentials are equal. In this case, tissue water content—and also tissue weight—remains unchanged.

In this procedure several representative samples of a tissue are weighed. Then each individual tissue sample is placed in the atmosphere above the surface of one of a series of graded solutions of known water potential, each of which is contained in a closed chamber. Sodium chloride is preferred as a solute because it does not support bacterial growth; stock solutions can be kept almost indefinitely. All chambers are maintained at the same constant temperature. Thus tissue samples are exposed to water vapors of fixed vapor pressures. After a few hours (i.e., when equilibrium with respect to the transfer of water between tissue sample and vapor is attained), each tissue sample is weighed again to determine whether it gained or lost weight. Tissue water potential is equal to the water potential in the solution that produces no change in weight of a tissue sample exposed to its equilibrium vapor.

One disadvantage of this method is that more than one tissue sample must be prepared. Another is the need for precise temperature control. Chambers must be held at one and the same temperature to ensure that no fluctuations in water vapor pressure occur in the vapor phases. In spite of these disadvantages, the vapor immersion method is more accurate than liquid immersion methods (see pp. 362–364) because the solution and the tissue sample do not come in contact. Therefore the solution cannot infiltrate into the intercellular spaces in the tissue sample.

Liquid Immersion Methods

There are two well-known liquid immersion methods. One is known simply as the *liquid immersion* method and the other is the *dye method.*

The liquid immersion method is similar to the vapor immersion method except that each weighed tissue sample is immersed directly into one of a graded series of solutions of known water potential. The volumes of the solutions should be much larger than the volumes of the tissue samples to ensure that the water potential in the solution will remain unchanged during the period of immersion. Sucrose is the solute usually used; most plant cells do not absorb sucrose readily from an external solution. After an hour or two (i.e., when equilibrium with respect to transfer of water between tissue sample and solution is attained), the tissues are removed from the solutions, blotted lightly with filter paper to remove adherent liquid, and weighed. The water potential in the tissue is equal to the water potential in the solution in which the tissue neither gained nor lost weight. (Instead of measuring change in weight, it is also possible to use either tissue volume or length of narrow slices of tissue as indices of gain or loss of water.) The advantage of the liquid immersion method is that not too much time is required. On the other hand, one of the sources of error is the possible infiltration of bathing solution into intercellular spaces in the tissue samples. This may occur to a greater or lesser extent in many tissues and adds to or subtracts from the final weight of the tissue sample. This source of error may be avoided in the vapor immersion method (see pp. 361–362).

In the dye method of measuring tissue water potential several uniform tissue samples (e.g., whole leaves from the same branch of a plant) are selected. Each tissue sample is immersed in one of a series of graded (sucrose) "test" solutions of known water potential and each test solution is colored lightly with a dye, such as methyl orange or methylene blue. A parallel series of *control* solutions, of the same water potentials as the test solutions, is prepared but not colored (see Figure 12-10).

The dye method depends on changes in density in the test solutions. If water is

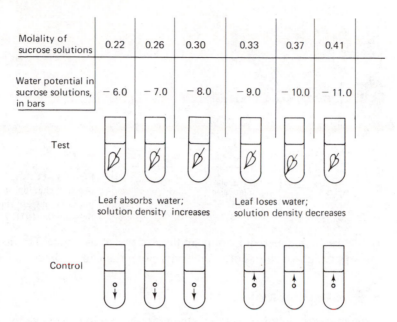

Molality of sucrose solutions	0.22	0.26	0.30	0.33	0.37	0.41
Water potential in sucrose solutions, in bars	− 6.0	− 7.0	− 8.0	− 9.0	− 10.0	− 11.0

Test

Leaf absorbs water; solution density increases   Leaf loses water; solution density decreases

Control

Figure 12-10 The dye method of measuring tissue water potential depends on density changes that take place in a graded series of colored test solutions of known water potential when tissue samples are immersed in them. (See text for details.) In this illustration the measured tissue water potential is −8.5 ± 0.5 bars. (After Knipling, 1967; Figure 1.)

transferred into the tissue during the immersion period, the density of the test solution increases. If water is transferred out of the tissue during the immersion period, the density of the test solution decreases. If the water potential in the tissue sample is the same as that of the test solution, there will be no net water transport between the tissue sample and the test solution during the immersion period. In this case, the density of the test solution is unchanged.

After an hour or two (i.e., when equilibrium with respect to the transfer of water between tissue sample and test solution is attained) the density of each test solution is ascertained as follows. A drop of test solution is placed carefully in the uncolored control solution of corresponding water potential. The drop—made visible by the presence of the dye—will fall in the control solution if there was an increase in the density of the corresponding test solution. Alternatively, the colored drop will rise if there was a decrease in the density of the corresponding test solution. If there was no change in density in the test solution, the colored drop will diffuse away when placed in the corresponding control solution. The water potential in the test solution in which no density change developed during the immersion period is equal to the water potential of the tissue.

There are several advantages to the dye method. No elaborate equipment is required. Measurements can be carried out in the field as well as in the laboratory.

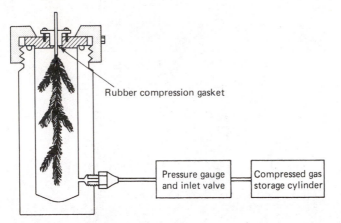
Rubber compression gasket

Pressure gauge and inlet valve | Compressed gas storage cylinder

Figure 12-11 Cross section through a pressure chamber used to measure water potential in leafy shoots. (After Scholander, 1971; Figure 14.)

Also, measurements require only an hour or two to complete. The disadvantage is that only approximate values of tissue water potential can be determined by this method.

Pressure Chamber Method

In the pressure chamber method a living freshly excised, leafy shoot is placed within a chamber able to withstand gas pressures up to 50 or more bars. The cut end of the stem protrudes through a gastight seal (see Figure 12-11). An inert gas (e.g., nitrogen) is admitted into the chamber from a compression tank. When the pressure around the leafy shoot is increased sufficiently so that xylem sap appears at the cut surface, equilibrium with respect to the transfer of water between the cells of the shoot and the xylem conducting system is considered reached. The gas pressure (with a negative sign) just sufficient to force xylem fluid to appear at the exposed cut end of a shoot is taken as being equal to the water potential of the leaf cells.[5]

With the pressure chamber, measurements of water potential can be made in a matter of minutes. No elaborate temperature controls are needed and the apparatus is relatively inexpensive. The process seems well suited for field studies. Stems, however, must be firm enough to withstand sealing without injury. Furthermore, to make accurate measurements of water potential requires that the pressure chamber be calibrated against measurements made with a thermocouple psychrometer.

THE COMPONENTS OF WATER POTENTIAL

One assumption often made to simplify discussions of plant–water relations is to consider that a typical plant cell consists of a cell wall and the cell contents (i.e., protoplast). The latter consists of a vacuole and a layer of cytoplasm (including the plasma membrane) between the vacuole and the cell wall. The cell wall is considered an

---

[5]The theoretical background for determination of water potential with a pressure chamber is considered by Boyer, 1967a.

external envelope that surrounds the protoplast, capable of pressing inward and thereby inducing hydrostatic pressure within the protoplast.

By considering a "typical" plant cell in this manner, it is possible to focus attention on the two major sets of internal factors that contribute to water potential in plant cells and tissues. Also, it is possible to partition the total water potential in plant cells and tissues into two component potentials as follows:

$$\psi_w = \psi_s + \psi_p \qquad (12\text{-}1)^6$$

This equation is equivalent to the statement that the water potential (ψ_w) in a plant cell or tissue is equal to the algebraic sum of the solute potential (ψ_s) due to dissolved solutes (most of which are present in vacuoles), and the pressure potential (ψ_p) due to pressures developed within cells and tissues. Each of these component potentials, as well as water potential itself, is expressed in pressure units of bars. Each component potential is considered separately below.

Solute Potential

The decrease in water potential brought about by substances either ionic or nonionic in solution is referred to as *solute potential* (ψ_s). These dissolved substances are mostly in the vacuole. The terms solute potential and osmotic potential (p. 356) are used interchangeably and take the place of *osmotic pressure* in the older literature. Whereas osmotic pressure was expressed in bars with a positive sign, solute potential (= osmotic potential) is expressed in bars with a negative sign. Therefore when solute potential decreases, it becomes more negative.

Solute potential in a solution depends on the total number of solute particles (molecules or ions) in a solution rather than on their kind or their charge. Thus, for example, because one mole of a salt such as potassium chloride dissociates in solution into two moles of ions, an 0.5 *m* solution of potassium chloride has a solute potential approximately equal to that of a 1.0 *m* sucrose solution. [The solute potential at 25°C in 1.0 *m* potassium chloride is −44.6 bars; in 1.0 *m* sucrose it is −26.9 bars (see Figure 12-8)]. As another example, consider calcium chloride, each mole of which dissociates in solution into 3 moles of ions. An 0.33 *m* solution of calcium chloride has approximately the same solute potential as a 1.0 *m* sucrose solution. In a solution that consists of several different solutes, the solute potential is the sum of the individual solute potentials contributed by each of the solutes. Thus the solute potential in a solution consisting of 0.5 *m* potassium chloride together with 1.0 *m* sucrose is approximately −49 bars (i.e., −22 bars contributed by potassium chloride and −27 bars contributed by sucrose). Seawater has a solute potential of approximately −25 bars (see Figure 12-8); seawater is an ionic solution consisting mainly of sodium chloride (approximately 0.5 *m*), together with small amounts of various ions, including potassium, calcium, magnesium, and sulfate. The solute potential in Hoagland's

---

[6]"Matric potential" (commonly included in this equation in the literature on plant water relations) is absent for reasons discussed by Passioura (1980) and McClendon (1981). Matric potential has significance only for dried plant material (e.g., dry seeds or wood); its magnitude can be measured with a thermocouple psychrometer (see Figure 12-9 and related discussion).

solution, a water culture medium that consists of several inorganic salts (see Table 8-1), is approximately −0.7 bar, as already indicated in Figure 12-8.

The solution phase of plant cells is present mostly in vacuoles. Perhaps the most widely used technique to extract this solution is to freeze and thaw a tissue sample. In many frozen and thawed tissues the solution phase will drain away in response to gravity alone. For other tissues, hand pressure may be sufficient to obtain the solution phase. For still other tissues, especially those of low water content, a hydraulic press may be required.

The pressed liquid consists mainly of a solution of ions and molecules of low molecular weight (e.g., sugars, salts, or inorganic and organic acids). Generally this *cell sap* is considered to be mostly vacuolar contents, although it is recognized that small quantities of other cellular materials may be present. Cell sap obtained by pressing a frozen and thawed tissue sample may contain water from dead cells and intercellular spaces; also, it may contain variable amounts of protoplasm from sheared cells (depending to some extent on the magnitude of the pressure applied to the tissue). Moreover, the possibility exists that the extraction process itself may lead to chemical interactions between vacuolar fluid and the cellular matrix. Nevertheless, the procedure of subjecting frozen and thawed tissues to manual or hydraulic pressure is considered the most direct and most practical method of obtaining a fluid reasonably representative of the solution phase of plant cells.

Several methods are available for measuring solute potential in an extracted cell sap. (In an extracted cell sap, *solute* potential = *water* potential.) One involves the use of the thermocouple psychrometer (see Figure 12-9 and related discussion). A sample of cell sap is used instead of living tissue. A second method utilizes an osmometer, such as diagrammed in Figure 12-7.

In a third method a well-known principle of physical chemistry is used—namely, that a solution freezes at a slightly lower temperature than does pure water. Whereas pure water freezes at 0°C, a 1 *m* sucrose solution [solute potential = −26.9 bars (see Figure 12.8)] freezes at −1.86°C. From this relationship the solute potential in an extracted cell sap can be calculated if the freezing point is known. Sensitive thermometers capable of measuring the freezing point of a solution to tenths and hundredths of a degree are available commercially. Several milliliters of solution must be available, however. This method of measuring solute potential is sometimes called the *cryoscopic method*.

A fourth method enables the investigator to determine the solute potential without extracting the cell sap. In this process, which is based on the phenomenon of plasmolysis (pp. 308–309), a solution is identified that will cause only slight—just barely visible—separation of the protoplast from the cell wall. This condition is known as *incipient plasmolysis*. At incipient plasmolysis the cell wall exerts no pressure on the cell contents (i.e., pressure potential is zero or almost zero). Therefore the *water* potential in the cell is equal to the *solute* potential [see Eq. (12-1)].

The procedure involves immersing different pieces of the same tissue in a series of graded solutions of known water potentials. Usually sucrose solutions are used because plant cells absorb sucrose from an external solution at very low rates, if at all. After the tissue samples have been immersed for about 15 to 30 minutes, the cells in each sample are examined under a microscope. None of the cells in the pieces immersed in dilute solutions will be plasmolyzed whereas all the cells in the pieces immersed in concentrated solutions will be plasmolyzed. Somewhere in the series of graded solutions will be one in which half the cells will be slightly plasmolyzed. This solution is taken to be the one that induces incipient plasmolysis. The water potential in this solution is equal to the solute potential in the cells. It may be necessary, however, to apply a small correction factor. A cell with stretchable elastic cell walls undergoes a decrease in volume when incipient plasmolysis is induced (see Figure 11-1). Thus the cell sap concentration at incipient plasmolysis will be slightly higher than that in the unplasmolyzed condition.

Values of solute potential in plant cells have been found to vary widely from species to species. For leaf tissue of most crop plants grown in temperate regions, the range of solute potential averages from about −10 to −20 bars. But much lower values, on the order of −100 bars or even lower, have been reported for leaves of xerophytic plants grown in desert regions.

Small variations in solute potential in a leaf are likely to occur from day to night. A daily pattern of rise and fall in solute potential is a common occurrence during the growing season. It results from the daily rise and fall in water content in leaves (see pp. 406–408). During the daylight hours, when leaf water content falls, the concentration of solutes in leaf cells increases and therefore solute potential becomes more negative. On the other hand, when leaf water content rises during the night hours, the concentration of solutes in leaf cells decreases and so solute potential becomes less negative.

Pressure Potential

The pressure normally exerted on the contents of a plant cell is due mainly to the tensile strength and elastic stretch of the cell wall. As a result of inwardly directed wall pressure, hydrostatic pressure develops in the vacuole. This hydrostatic pressure is called turgor pressure. *Turgor pressure* is the outwardly directed pressure exerted by the contents of the cell on the cell wall. It is equal in magnitude and opposite in direction to the pressure of the cell wall. In some cases, turgor pressure in a plant cell may arise, at least in part, from pressures of neighboring cells. Mesophyll cells in dicotyledonous leaves, for example, are situated between two sheets of epidermal cells, the cell walls of which are cutinized and relatively stiff and unyielding. Therefore entrance of water into a leaf may be accompanied by compression of mesophyll cells by epidermal tissues.

Whereas solute potential is invariably negative, pressure potential (ψ_p) in plant cells is usually positive. In leaf cells, for example, turgor pressures in herbaceous crop plants may vary from about +3 to +5 bars during the afternoon of a warm summer day to about +15 bars or more during the night hours. This typical diurnal fluctuation in turgor pressure (from about +3 to about +15 bars) corresponds to the typical diurnal fluctuation that occurs in leaf water content (pp. 406–408). [Turgor pressure in guard cells during stomatal opening may be much higher (p. 398).]

On some occasions, pressure potential may be zero or negative. Cells that are plasmolyzed have zero or almost zero turgor pressure. Negative hydrostatic pressure is likely to exist in the water in xylem elements during periods of rapid transpiration (p. 404).

Pressure potential can be measured directly, but some of these direct methods are of only limited usefulness. The resonance frequency method (Burström, 1971) involves fastening one end of a strip of tissue (e.g., a stem segment or an individual root) in a holder and inserting a steel wire into the free end. By using an electromagnet supplied with alternating current of varying frequency, the tissue is induced to vibrate. Its turgor pressure can be determined by measuring the frequency at which the greatest vibration occurs. The method depends on the fact that the rigidity of a plant tissue increases as turgor pressure increases. (Compare, for example, the stiffness of a turgid celery stalk with a wilted one.) In a second method a microcapillary tube is fused at one end. The open end is inserted directly into the vacuole of a plant cell; only cells that are especially large (e.g., giant algal cells) can be used. Using this technique, it was found possible to relate the compression of air in the microcapillary tube to the turgor pressure of an individual call (Green, 1968). In a third procedure the positive pressure of exudate from the xylem in a decapitated plant induced to develop "root pressure" (p. 412) can be measured by attaching a tube equipped with a pressure gauge to the rooted stump. A fourth process uses a microcapillary that is introduced into a plant cell; the latter is connected to a sensitive pressure transducer (Hüsken et al., 1978). Finally, there is the aphid stylet method, which can be used for measuring turgor pressure in sieve tubes only (p. 332–333). Besides these methods (and one or two others not mentioned here), knowledge of the magnirte of pressure potential in plant cells is derived mainly from Eq. 12-1, provided that water potential and solute potential have been measured separately.

ROLES OF TURGOR PRESSURE IN PLANT CELLS

The turgidity of a living plant cell is considerably significant to its physiological well-being. If not for the fact that a plant cell is usually swollen and distended through most of its lifetime, its internal structural integrity could not be maintained. In the absence of turgidity, cellular organelles (e.g., mitochondria, plastids, microbodies) within a cell would not be able to retain the special spatial relationship necessary for normal metabolic functions. Turgor pressure ensures the maintenance of normal rigidity in a plant cell. Indeed, the normally upright position of the shoot of a very young seedling whose cell walls are not yet lignified depends largely on positive turgor pressures developed within individual cells of the shoot.

A second role of turgor pressure is in cell enlargement during cell growth. This type of cell enlargement is accompanied by permanent, *irreversible* stretching of the primary cell wall and should not be confused with *reversible* stretching. The reversible type of stretching of the cell wall is categorized under the term turgor movements.

Although events associated with the permanent, irreversible elongation of the cell wall during growth are complex and still require elucidation, it is generally agreed that the primary cell wall is stretched by the turgor pressure developed within the protoplast. A plant cell grows by the yielding of the cell wall to turgor pressure. Only when the turgor pressure of a cell is high, and above a critical minimum value, does growth occur (Boyer, 1968).

The process of irreversible stretching starts in response to a metabolic event that acts to "soften" the primary cell wall; certain of the chemical bonds in cell wall constituents are weakened or "loosened." Consequently, the balance between inwardly directed pressure of the cell wall and outwardly directed turgor pressure is disturbed. Then the cell wall "yields" to its turgor pressure and stretching takes place. At the same time solute molecules are secreted into the vacuole so that osmotic uptake of water from neighboring cells and tissues is induced. Thus the solute potential of the cell is maintained at its original value. Also, at the same time, newly formed cell wall material is deposited in the cell wall.

The sequence of events during a turgor-driven *pulse* of elongation growth in a cell wall is outlined in Figure 12-12. Subsequent pulses of growth produce further increments of (irreversible) cell wall stretching, together with further increments of osmotic uptake of water and deposition of new cell wall material. The special role of plant growth substances in promoting cell growth is through the control that they exert on "loosening" of cell wall constituents of the cell wall (see Chapter 15).

In contrast to the irreversible increase in cell size during growth, reversible changes in cell volume occur periodically and recurrently in plant cells. Often these reversible changes are categorized as *turgor movements*. Turgor movements depend on the reversible deformability of the cell wall. A familiar example is the drooping of

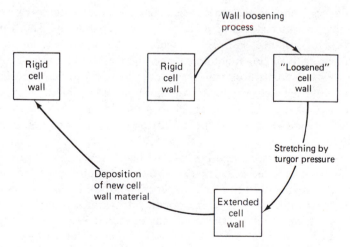

Figure 12-12 Schematic representation of the cycle of postulated events during a single pulse of growth of the cell wall. The cell wall is shown in the rigid condition at the beginning of a pulse of growth and in a "slightly extended" rigid condition at the end of the pulse. (See text for details.)

leaves of some plants during wilting and their recovery of turgidity when water is absorbed again. The opening and closing of stomata is another example of reversible turgor movements (see p. 398). Still other examples (e.g., the folding and unfolding of leaflets and the opening and closing of flowers at certain times of the day) are noticeable in some plant species but not in others.

An especially striking example of turgor movement is the rapid (less than one second) drooping of the leaves and leaflets of the sensitive plant (*Mimosa pudica*) in response to one of several stimuli (e.g., touch, heat). Following the spectacular and sudden drooping of leaves and leaflets, a slow recovery takes place when the plant is left undisturbed. These reversible turgor movements are the result of redistribution of water within a group of specialized cells at the base of petioles and leaflets. This group of cells is called the *pulvinus*.

Although details of the mechanism are still to be verified, an electrical signal is thought to transmit from the stimulated region of the sensitive plant to the pulvinus. This signal induces the cells in the pulvinus to undergo changes in transport properties. As a result, changes in the distribution of potassium ion within the cells of the pulvinus occur, perhaps in a similar manner to the opening and closing of stomata (see pp. 398–399). Thus water is redistributed within the pulvinus and the leaves collapse. The recovery process takes place because the distribution of the potassium ion within the pulvinus—and hence the distribution of water—slowly returns to the normal condition.

CHANGES IN ψ_p AND ψ_w DURING REVERSIBLE CHANGES IN CELL VOLUME

Under environmental conditions usually prevailing in temperate climates, appreciable changes in water status of plant cells occur frequently. The water status of cells in leaves, for instance, is likely to decrease to some extent on a hot summer day and increase again during night hours (see Figures 13-14 and 13-15). Accompanying these changes in water status will be changes in cell volume, especially if the leaves are young and cell walls are elastic and not lignified. Sometimes the volume of a cell in a leaf may change as much as 20 to 40%. Also, there will be changes in cell sap concentration (i.e., solute potential) as well as in turgor pressure (i.e., pressure potential).

These reversible changes are depicted schematically in Figure 12-13 for an idealized vacuolated cell. The two extreme conditions, plotted on the horizontal axis, are *zero turgor* (relative cell volume = 1.0) and *full turgor* [relative cell volume (arbitrary value) = 1.4], corresponding to the maximum turgor pressure that the protoplast can exert. [Zero turgor may be induced *experimentally* by bringing a cell to a state of incipient plasmolysis (pp. 366–367), full turgor by placing a cell in distilled water (p. 357).] Plotted on the vertical axis are arbitrary values in bars. The uppermost curve indicates pressure potential (positive bars) and the lowermost curve solute potential (negative bars). The middle curve is water potential of the cell; at any cell volume the value of the middle curve may be obtained by adding the values of

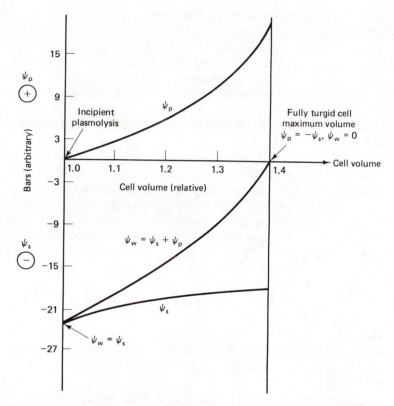

Figure 12-13 Diagrammatic relationship of water potential (ψ_w), solute potential (ψ_s), and pressure potential (ψ_p) in an idealized cell with stretchable elastic walls. (See text for details.) (Adapted from Meidner and Sheriff, 1976; Figure 5.3.)

the upper and lower curves. Figure 12-13 is generally referred to as the Höfler diagram, in honor of the German plant physiologist who first plotted the water potential and its components (vertical axis) against the water content (horizontal axis) in the early twentieth century (McClendon, 1982).

Figure 12-13 shows that a cell at zero turgor (i.e., a flaccid cell) has a solute potential equal to its water potential [see Eq. 12-1]. On the other hand, a cell at full turgor has a water potential equal to 0 bar; its solute and pressure potentials are equal but opposite in sign [see Eq. 12-1].

Under natural growing conditions a cell is usually at a state between zero turgor and full turgor. Zero turgor is approached under natural conditions when a tissue is severly wilted (see pp. 410–411). On the other hand, the cells of a young leaf of a young, thrifty plant growing in well-watered soil will approach full turgor when the leaf reaches maximum water content during the early morning hours. Also, it is often possible to induce a condition of full turgor by placing the petiole of a young, excised leaf in pure water for several hours.

MEASUREMENT OF WATER STATUS BY WATER
CONTENT METHODS

There are two general ways to describe the water status or internal water balance of plants and plant tissues. One is water potential (see Figure 12-8 and related discussion). Water potential is considered by most plant physiologists to be a most useful and significant way to describe the water status of plant tissues. In terms of water potential, *water deficit* exists in a tissue whenever its water potential is less (i.e., more negative) than zero bar.

The second way to describe water status is to measure the quantity of water in a tissue (i.e., its *water content*) and to express it in relation to a selected standard of reference. Three procedures for doing so are described next.

In the fresh weight method the fresh weight of a tissue sample (e.g., a leaf) is measured, preferably at the time when the sample is collected. Then the dry weight is determined by bringing the tissue sample to constant weight in an oven held at approximately 100°C. The water content of the tissue on a *fresh weight basis* is given by

$$\text{Water content in percent} = 100 \times \frac{W_f - W_d}{W_f} \qquad (12\text{-}2)$$

where W_f is the fresh weight of the tissue, W_d is the dry weight, and $W_f - W_d$ is the water content of the tissue. To give an example, let us assume that the fresh weight of a leaf is 1.0 g and its dry weight is 0.2 g. Then the water content on a fresh weight basis is $100 \times (1.0 - 0.2)/1.0 = 80\%$.

To express the water content of a tissue on a fresh weight basis is generally unsatisfactory because the denominator in Eq. (12-2) (i.e., the fresh weight of the tissue) is not likely to remain constant, either with time or different treatments. But despite this drawback fresh weight is useful when it is desired simply to note differences in water content among different plant materials or different plant species. Thus, for example, the water content of fleshy leaves of succulent plants may be over 90% on a fresh weight basis compared to leaves of certain desert shrubs whose water content on a fresh weight basis may be 20% or less.

The second method of expressing water content is on a *dry weight basis.*

$$\text{Water content in percent} = 100 \times \frac{W_f - W_d}{W_d} \qquad (12\text{-}3)$$

Here W_f and W_d have the same meanings as in Eq. (12-2). To use the same example as earlier, the water content of the tissue on a dry weight basis is $100 \times (1.0 - 0.2)/0.2 = 400\%$.

Dry weight is not always a satisfactory base of expression, at least for young, actively growing leaves. The dry weight of a leaf [i.e., the denominator in Eq. (12-3)] is likely to increase with time because of photosynthesis. In fact, all growing tissues, photosynthetic as well as nonphotosynthetic, undergo increases in dry weight over long periods of time. In short-term studies in which comparisons are made over periods not exceeding a few hours, however, this base of expression may be used to advantage. It is possible, by this means, to focus attention on water deficits that develop in tissues during the day, as compared to the night, even in spite of the slight increase in dry weight that may have occurred over the 24-hour experimental period. (For an illustrated example of this method, see Figure 13-14.)

The third method is the relative water content method. *Relative water content—* the water content of a tissue expressed as a percentage of the water content of the fully turgid tissue—is given by

$$\text{Relative water content in percent} = 100 \times \frac{W_f - W_d}{W_t - W_d} \qquad (12\text{-}4)$$

where W_f and W_d have the same meanings as in Eqs. (12-2) and (12-3) and W_t is the weight of the fully turgid tissue.

To determine relative water content, the fresh and dry weights of the tissue are measured as described. In addition, samples of the tissue (e.g., entire leaves or disks punched from leaves) are brought to full turgor, either by enclosing a leaf (with its cut end standing in water) in a moist chamber for a few hours or by floating leaf disks in water. Constant temperature conditions should be maintained when a tissue is brought to full turgor. Again, using the same example and assuming that the weight of the leaf at full turgor is 1.4 g, the relative water content of the leaf is

$$100 \times \frac{1.0 - 0.2}{1.4 - 0.2} = 67\%$$

Relative water content is widely acknowledged to be the most meaningful basis of expression so far devised for measuring the water status of a tissue in terms of the quantity of water in the tissue because the denominator in Eq. (12-4)—the water content of a tissue at full turgor—can be reproduced consistently. In this scale, water deficit exists in a tissue whenever its water content is less than 100%.

Measurements of relative water content can be used to estimate water potential, provided that calibration curves are established. Such curves are illustrated in Figure 12-14 for leaves of three different species.[7] The relationship between relative water content and water potential is not linear. When relative water content drops from 100% to about 90%, there is a large decrease in leaf water potential (more so in *Acacia* than in the other two species), probably due to a rapid reduction in turgor pressure in leaf cells. But as the quantity of water in the tissues continues to decrease and the relative water content drops from about 90% to about 60 to 50%, there is a less rapid decrease in water potential. Finally, when the relative water content becomes less than about 60 to 50%, the curves flatten out (especially in the case of *Acacia* and privet). This "flattening out" occurs because the solutes in the tissues become more and more concentrated. With continuing decrease in relative water content, the water remaining in the tissues is held more and more tightly by the tissue components and so leaf water potential decreases more and more rapidly.

[7]Similar curves have been published recently for maize and sorghum (Acevedo et al., 1979).

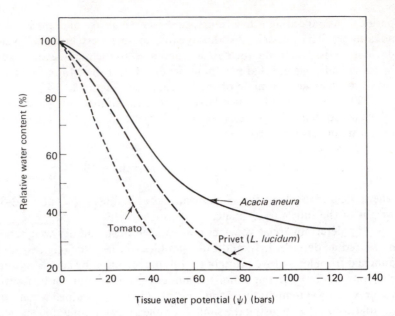

Figure 12-14   The relationship between relative water content and water potential for the leaves of three plant species. (After Slatyer, 1967; Figure 5.3.)

REFERENCES

ACEVEDO, E., E. FERERES, T. C. HSIAO, and D. W. HENDERSON. 1979. Diurnal growth trends, water potential, and osmotic adjustment of maize and sorghum leaves in the field. *Plant Physiol.* 64:476–480.

ANDREWS, F. C. 1976. Colligative properties of simple solutions. *Science* 194:567–571.

BOYER, J. S. 1967a. Leaf water potentials measured with a pressure chamber. *Plant Physiol.* 42:133–137.

BOYER, J. S. 1968. Relationship of water potential to growth of leaves. *Plant Physiol.* 43:1056–1062.

BURSTRÖM, H. G. 1971. Resonance frequency measurements on plant tissues. *Endeavour* 30:87–90.

CAILLOUX, M. 1972. Metabolism and the absorption of water by root hairs. *Can. J. Bot.* 50:557–573.

GREEN, P. B. 1968. Growth physics in *Nitella*: a method for continuous *in vivo* analysis of extensibility based on a micro-manometer technique for turgor pressure. *Plant Physiol.* 43:1169–1184.

HAMMEL, H. T., and P. F. SCHOLANDER, 1976. *Osmosis and Tensile Strength.* Springer-Verlag, Berlin.

HOFFMAN, G. J., and S. L. RAWLINS. 1972. Silver-foil psychrometer for measuring leaf water potential in situ. *Science* 177:802–803.

Hüsken, D., E. Steudle, and U. Zimmermann. 1978. Pressure probe technique for measuring water relations of cells in higher plants. *Plant Physiol.* 61:158–163.

Knipling, E. B. 1967. Measurement of leaf water potential by the dye method. *Ecology* 48:1038–1041.

Kozlowski, T. T., ed. 1968–1981. *Water Deficits and Plant Growth.* 6 vols. Academic Press, New York.

Kramer, P. J. 1969. *Plant and Soil Water Relationships: a Modern Synthesis.* McGraw-Hill Book Company, New York.

Kramer, P. J., E. B. Knipling, and L. N. Miller. 1966. Terminology of cell-water relations. *Science* 153:889–890.

Levitt, J. 1967. Active water transport once more: a reply to J. J. Oertli. *Physiol. Plant.* 20:263–264.

McClendon, J. H. 1981. The balance of forces generated by the water potential in the cell-wall matrix—a model. Amer. J. Bot. 68:1263–1268.

McClendon, J. H. 1982. Water relations curves for plant cells: toward a realistic Höfler diagram for textbooks. *What's New Plant Physiol.* 13:17–20.

Meidner, H., and D. W. Sheriff. 1976. *Water and Plants.* Blackie and Son, Glasgow.

Michel, B. E. 1972. Solute potentials of sucrose solutions. *Plant Physiol.* 50:196–198.

Millar, B. D. 1971a. Improved thermocouple psychrometer for the measurement of plant and soil water potential. I. Thermocouple psychrometry and an improved instrument design. *J. Expt. Bot.* 22:875–890.

Millar, B. D. 1971b. Improved thermocouple psychrometer for the measurement of plant and soil water potential. II. Operation and calibration. *J. Expt. Bot.* 22:891–905.

Oertli, J. J. 1967. Comments on active water transport. *Physiol. Plant* 20:814–818.

Oertli, J. J. 1969. Terminology of plant–water energy relations. *Zeitschr. Pflanzenphysiologie* 61:264–265.

Oertli, J. J. 1971. A whole-system approach to water physiology in plants. *In* L. Chandra, ed. *Advancing Frontiers of Plant Science*, vol. 27, pp. 3–200; vol. 28, 201–283.

O'Leary, J. W. 1970. Can there be a positive water potential in plants? *BioScience* 20:858–859. [Also see J. J. Oertli, Letter to editor, *BioScience* 21:54(1971).]

Palta, J. P., and E. J. Stadelmann. 1980. On simultaneous transport of water and solute through plant cell membranes: evidence for the absence of solvent drag effect and insensitivity of the reflection coefficient. *Physiol. Plant.* 50:83–90.

Passioura, J. B. 1980. The meaning of matric potential. *J. Expt. Bot.* 31:1161–1169.

Ray, P. M. 1960. On the theory of osmotic water movement. *Plant Physiol.* 35:783–795.

Ray, P. M., P. B. Green, and R. Cleland. 1972. Role of turgor in plant cell growth. *Nature* 239:163–164.

Scholander, P. F. 1971. Imbibition and osmosis in plants. Pages 138–147 *in* A. Kramer, ed. Topics in the study of life: The bio source book. Harper & Row, Publishers, Inc., New York.

Scholander, P. F. 1972. Tensile water. *Am. Sci.* 60:584–590.

Shepherd, W. 1973. A simple thermocouple psychrometer for determining tissue water potential and some observed leaf-maturity effects. *J. Expt. Bot.* 24:1003–1013.

Slavik, B. 1974. *Methods of Studying Plant Water Relations.* Springer-Verlag, New York.

SLATYER, R. O. 1967. *Plant–Water Relationships*. Academic Press, New York.

SLATYER, R. O., and S. A. TAYLOR. 1960. Terminology in plant- and soil-water relations. *Nature* 187:922–924.

SPANNER, D. C. 1973. The components of the water potential in plants and soils. *J. Expt. Bot.* 24:816–819.

STADELMANN, E. J. 1966. Evaluation of turgidity, plasmolysis, and deplasmolysis of plant cells. Pages 143–216 *in* D. M. Prescott, ed. *Methods in Cell Physiology*, vol. 2. Academic Press, New York.

STILLINGER, F. H. 1980. Water revisited. *Science* 209:451–457.

TAYLOR, S. A., and R. O. SLATYER. 1961. Proposals for a unified terminology in studies of plant–soil–water relationships. Pages 339–349 *in* Proceedings of a Madrid symposium on plant water relationships, 1959, UNESCO Arid Zone Research Series XVL, UNESCO, Paris.

TYREE, M. T. and H. T. HAMMEL. 1972. The measurement of turgor pressure and the water relations of plants by the pressure-bomb technique. *J. Expt. Bot.* 23:267–282.

WARING, R. H., and B. D. CLEARY. 1967. Plant moisture stress; evaluation by pressure bomb. *Science* 155:1248–1254.

WIEBE, H. H., et al., eds. 1971. Measurement of plant and soil water status. *Utah Agr. Expt. Sta. Bull. 484.* Utah State University, Logan.

13

The Plant in Relation to Water

Unlike animals that move, higher plants cannot satisfy their need for water by migrating to favorable water sources. Their roots anchor them to a fixed terrain. The terrestrial higher plant is interposed between the soil, which is a source of water, and the atmosphere, which is a sink. There is a continual passage of water from soil solution to plant roots, through vascular tissues to leaf cells, and into the external atmosphere. Counteracting the loss of water to the atmosphere is a cuticular layer of waxy material, rather impermeable to water, on the outermost leaf surfaces. Loss of water to the atmosphere is also regulated by stomata in leaf epidermal tissues. Yet even so, the bulk of water absorbed from the soil is lost as water vapor to the atmosphere. Therefore the terrestrial higher plant is faced at all times during growth with the need to maintain within itself a quantity of water adequate to support normal functions. Not only must the plant contend with atmospheric demands for water but it must also extract water from soils in which the water supply is not always unlimited. A balance is struck between the demands of the atmosphere and the ability of leaves to retard water loss by evaporation on the one hand and the ability of roots to absorb water on the other. This compromise, aptly designated by the term water balance, represents the difference between water "income" and "expenditure" by the plant.

Water losses by evaporation from leaves often exceed water gains by absorption by roots. In this situation, a negative water balance, or water deficit, is said to exist within the plant. Small internal water deficits generally are unavoidable for crop plants growing under field conditions and nearly always occur with recurrent regularity during the heat of the day throughout the growing season. Moderate water deficits, whether initiated by low soil moisture or desiccating atmospheric conditions, may impair the plant's ability to carry on one or another physiological process at a normal rate. If sufficiently severe and prolonged, an internal water deficit may even threaten the plant's survival.

WATER LOSS BY TRANSPIRATION

Only a small fraction, generally much less than 1%, of the water absorbed by terrestrial plants is used in metabolic reactions (e.g., hydrolyses). Most water absorbed by roots is lost by transpiration from leaves. The process of *transpiration* refers to the evaporation of water from the aerial portions of the living plant and movement of water vapor into the external atmosphere.

The ratio of transpirational loss of water by a plant to its dry matter production during the growing season (i.e., the *transpiration ratio*) measures the efficiency of water consumption by a plant species: the larger the ratio, the less efficient is the species in its use of water. Transpiration ratios for most crop plants range from 200 to 500 or more; that is, it takes 200 to 500 g of water or more to bring 1 g (dry weight) of a plant to maturity. Clearly higher plants in terrestrial habitats are quite inefficient in their use of water. Yet some plants are more efficient than others. It is well known, for example, that C_4 plants produce two to three more dry matter per unit of water used than C_3 plants.

Water loss by transpiration may take place from any part of a plant exposed to the external atmosphere. It occurs, however, principally from leaves and almost entirely through stomatal pores. Only relatively small quantities of water vapor are transferred to the external atmosphere by evaporation from the cuticle; the cuticle of most leaves is very impermeable to water. So stomatal transpiration far exceeds cuticular transpiration.

As we saw on pp. 345–346, 1 g of water consumes more than 500 cal of heat energy when it vaporizes. Therefore the effect of transpiration is to cool a leaf. Transpiration dissipates energy that would go into heating a leaf and, in fact, abstracts heat energy from a leaf.

The heat demand for vaporization of water in transpiration is met by solar radiation, which also warms the surface of the earth, drives winds and ocean currents, and fuels life processes in the plant through photosynthesis. Solar radiation is supplied to a leaf mainly in three ways:

1. as light (i.e., direct, scattered, or reflected).
2. as thermal radiation (from the atmosphere, ground, or surrounding objects).
3. by warm air currents across a leaf.

Of the total heat absorbed by a leaf, only a small fraction is received by conduction from other plant parts.

The rate of transpiration from a leaf usually exhibits a diurnal cycle. On a typical day in the summer there is a rapid increase in transpiration during the morning hours and a peak is reached in the early afternoon (see Figure 13-1). A decline in the late

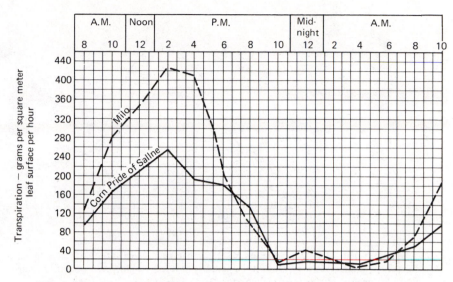

Figure 13-1   The transpiration rates of maize (Pride of Saline) and sorghum (milo) grown under the same conditions in Kansas, showing water loss (in grams per square meter of leaf surface per hour) during the day and night. (After Miller, 1938; Figure 14.)

afternoon and evening follows. During night hours transpiration rates are likely to be very low.

Because transpiration is energized by solar radiation, the diurnal cycle of transpiration shown in Figure 13-1 can be expected to parallel the radiation received at the surface of the earth, as can be seen by comparing Figure 13-1 with the diurnal cycle of solar radiation shown in Figure 13-2.

Figure 13-2 also shows the temperatures of a fully sunlit, exposed leaf and the air around it during the day and night under conditions assumed to be ideal. Nighttime leaf temperature is usually a few degrees below air temperature because leaves lose heat by thermal radiation to the sky and receive relatively little heat from the air around them. In the morning, after the sun rises, a sunlit leaf warms quickly and its temperature rises above the temperature of the air. But at the same time, the stomata, which were closed during the night, will be open; thus the leaf will transpire and lose heat. Also, the leaf will lose heat by thermal reradiation. As a result, the temperature of a sunlit leaf is usually only slightly higher than the air temperature (see Figure 13-2). The temperature of a shaded leaf will also follow a diurnal cycle similar to that shown in Figure 13-2 except that its daytime temperature will probably not exceed the temperature of the air around it.

Transpirational losses of water are often excessive and the large flux of water through the plant does not seem essential to the plant. Some plant physiologists (e.g., Gates, 1968) have suggested, however, that the cooling effect of transpiration helps to keep the temperature of a leaf below lethal levels during periods of abundant solar radiation and may make the difference between healthy and damaged leaf tissue.

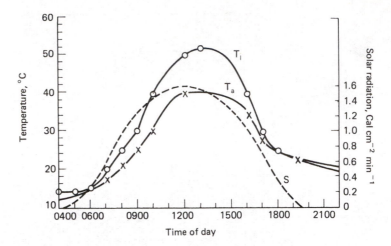

Figure 13-2 The variation in solar radiation received at the surface of the earth (S), air temperature (T_a), and temperature of a fully sunlit, exposed leaf (T_l) during the course of a day and night. (After Gates, 1966; Figure 7.)

Unfortunately, it is not easy to distinguish between the cooling effect of transpiration per se and the cooling effect of thermal reradiation from leaves, especially under conditions existing in nature. Thus the possible beneficial effect of cooling of a leaf by transpiration is not yet completely resolved.

MEASUREMENT OF TRANSPIRATION RATES

The rates at which plants and plant parts lose water by transpiration are naturally of interest to plant scientists. Rates of transpiration for individual plants may be expressed in such units as grams of water vapor per second per plant. When tranpiration rates from leaves are considered, it is appropriate to refer to the *transpirational flux*, which means the quantity of water vapor transpired by a unit area of leaf surface in a unit of time. The units used most often are $g\ m^{-2}\ h^{-1}$ (see Figure 13-1) or $\mu g\ cm^{-2}\ s^{-1}$.

In some field investigations transpiration rates may be expressed most conveniently in terms of a unit of land area—for example, in liters hectare^{-1} day^{-1}. It may be of interest to point out here that about two-thirds of the water falling as precipitation in temperate land areas is returned to the atmosphere by transpiration. This amounts to many centimeters of rain or liters of water on a yearly basis.

Some techniques used to measure transpiration and transpiration rates are relatively simple and require little equipment whereas others can be carried out only with elaborate experimental setups and sophisticated measuring devices.

The most commonly used method, the gravimetric or pot method (also called the lysimeter method) is simple, straightforward, and reliable and may serve in research investigations as well as class demonstrations. A potted, intact plant is selected and the soil around the roots is watered thoroughly. (In long-term experiments it may be necessary to add measured quantities of water through a watering tube sealed into the soil.) Then the pot and the root system are sealed within a container composed of material impervious to water (e.g., a metal sheet constructed to fit over the pot or a sheet of polythene) so that water loss to the atmosphere occurs only from the shoot system and not from the soil. The entire assembly is then weighed periodically—for example, at hourly intervals. The loss in weight represents the quantity of water transpired. Evaluation of the rate of transpiration in terms of a unit of evaporating area (e.g., grams of water vapor per square centimeter of leaf surface per hour) requires that the total leaf area be estimated, bearing in mind whether the leaves of the experimental plant have stomata on one side only or on both sides. Although this process also measures changes in weight due to photosynthesis and respiration during the experimental period, actually such changes are usually very small compared to the loss of weight due to transpiration and may be ignored, especially if the experimental period is short.

In a second procedure, an entire plant or portion of a leafy shoot or single leaf is enclosed in a sealed chamber, usually made of a transparent material, such as Plexiglas. An airstream is forced through the chamber and the water content of the inflowing and outflowing airstream is measured with a humidity sensor (e.g., an electric hygrometer whose electrical resistance varies with humidity, or an infrared analyzer sensitized to water vapor, or a psychrometer). In another version of this technique, air is passed over the transpiring tissue and is then conducted through an absorption vessel containing a water vapor absorbent (e.g., anhydrous calcium chloride or phosphorous pentoxide). At the same time, a stream of air is passed at the same rate directly through a separate water vapor absorption vessel without passing over plant tissue. The difference in the gain in weight of the two absorption vessels represents the quantity of water vapor transpired during the experimental period. With this technique, even the transpiration rates of plant communities rooted in out-of-door habitats have been estimated; in such cases, large transparent plastic tents have been used. Because environmental conditions in glass or plastic chambers differ from those in nature, estimates of transpiration rates under natural conditions can be made only if necessary corrections to the data are applied.

Still another method involves the use of a porometer. A porometer consists of a small cup that covers a portion of a leaf. This instrument measures the rate at which water vapor moves out of a leaf through stomatal pores and can be calibrated to measure either stomatal aperature or transpiration (see pp. 386–387).

Finally, two other methods of measuring transpiration deserve mention even though their use is mainly for demonstration purposes. In one, pieces of absorbent paper are soaked in a dilute (about 3 to 5%) solution of cobalt chloride. After drying in an oven, the paper is stored in a desiccator. When dry, the paper is bright blue; its color changes to pink when exposed to moist air. If a piece of dried paper is transferred quickly from a desiccator to the surface of a leaf and protected from atmospheric moisture with a celluloid or glass cover, a qualitative estimate of the rate of transpiration can be obtained by determining the rate at which the paper changes from blue to pink. A visible change in color usually requires several minutes. Of course, the rate of transpiration is affected by the presence of the paper and may differ somewhat from a leaf surface freely exposed to the external atmosphere. Even so, this technique can be used to reveal relative rates of transpiration.

The second semiquantitative method suitable for class demonstration uses an instrument known as a potometer. In this instrument the rate at which a small air bubble moves in a water column in the horizontal capillary side arm of known diameter is determined (see Figure 13-3). A potometer actually measures the rate of water absorption rather than the rate of transpiration. If conditions are such that the rate of transpiration is not excessive and water absorption does not lag too far behind water loss by transpiration, however, then the rates of water absorption and transpiration will be almost equal. A small excised shoot is the usual experimental material (see Figure 13-3), but whole-tree potometers have now been constructed (Knight et al., 1981).

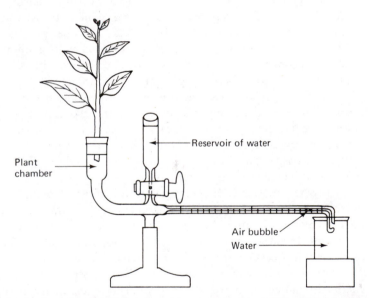

Figure 13-3 The potometer method of measuring transpiration. The stem of a cut shoot is placed in a closed chamber of water. To prevent the formation of air bubbles in open xylem vessels, the stem should be cut under water. The chamber is attached to a graduated capillary tube into which an air bubble is introduced (by lifting the tube momentarily out of the water in the beaker). The rate of travel of the air bubble across the length of the graduated tube can be used to demonstrate the effects of environmental changes on the rate of transpiration. The air bubble can be returned when necessary to the right-hand end of the capillary tube by admitting water through the stopcock. In another type of potometer, the plant chamber is large enough to admit the entire root system of a small plant. (After Kramer, 1959; Figure 9.)

GASEOUS EXCHANGES BETWEEN LEAVES
AND THE EXTERNAL ATMOSPHERE

Exchange of respiratory and photosynthetic gases CO_2 and O_2 between leaves and the external atmosphere and loss of water vapor in transpiration occur primarily through stomata in epiderms. The term *stoma* (Greek, *stoma*, mouth) refers to a pair of guard cells and the pore between them. [Guard cells are cells that flank a stomatal pore (see Figures 13-5, 13-8, 13-9, and 13-10)]. The plural of stoma is *stomata*. (Also, the terms *stomate* and *stomates* are sometimes used in place of stoma and stomata.) Stomata are present not only in the epiderms of leaves but also in certain fruits (e.g., banana, citrus, cucumber, avocado).

A structure known as the *cuticle* is formed—by secretions from epidermal cells—on the outside surfaces of their cell walls in above-ground parts of terrestrial higher plants. (For additional discussion of the cuticle, see Chapter 3.) The width of the cuticle varies from species to species and even from leaf to leaf on the same plant, but as a first approximation can be said to be about the same as the width of the epidermal cell wall. The cuticle consists of *cutin*, a waxlike material of long-chain lipid polymers (mostly esterified hydroxy fatty acids), and provides an almost impervious barrier to the passage of water vapor, CO_2, and O_2. So only a negligible portion of the total gas exchange occurs through the cuticle. Therefore, gas exchange between leaves and the external atmosphere takes place almost entirely through stomatal pores.

Stomata in most plants are open through the day and closed at night. Photomicrographs of open stomata (and a closed stoma) in surface view in monocotyledonous and dicotyledonous leaves are shown in Figures 13-4 and 13-8.

In some plant species the epidermal cells adjoining guard cells differ in shape and size from other epidermal cells and are called *accessory* or *subsidiary* cells. Although present in Figure 13-4 (top), better representations are shown in Figures 13-5, 13-8, and 13-9.

A stomatal pore is generally elliptical in surface view (e.g., see Figure 13-4). In transverse section, the shape of the stomatal pore varies from species to species but may be assumed, at least for simplification, to be that of a cylindrical tube. The length of this tube in most plant species is much greater than its diameter even when the pore is wide open. The actual dimensions of stomatal pores vary from species to species. But, on the average, a stomatal pore measures about 20 μm long and about 10 to 20 μm wide at its widest point when fully open. Although these dimensions may seem small, a stomatal pore is actually an enormous chasm when compared with the dimensions of the gas molecules that move through them. The diameter of a water molecule is about 0.00025 μm; molecules of CO_2 and O_2 are only slightly larger than water molecules. So even when a stoma is almost closed and its opening is only 1 μm, several thousands of these molecules can pass through the pore at the same instant.

The number of stomata per unit area of leaf surface (i.e., the frequency) varies considerably among plant species. Leaves of some plants have many stomata and others few. Also, environmental conditions influence the frequency of stomata. Leaves

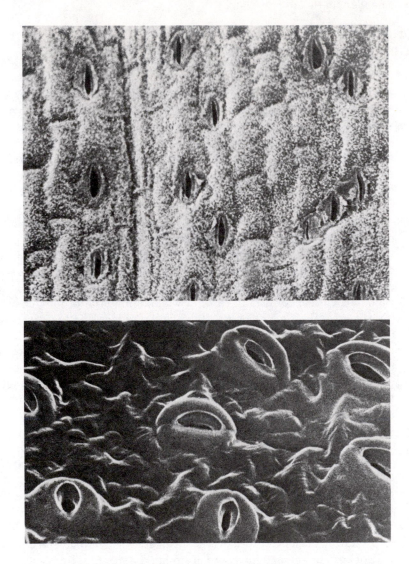

Figure 13-4 Scanning electron photomicrographs of open stomata on the upper surface (×600) of a leaf of maize (*Zea mays*) (top), and the lower surface (×900) of a cucumber leaf (bottom). The stomata on a maize leaf, as in other monocots, are arranged in parallel rows along the length of the leaf; in dicotyledons, such as cucumber, stomata are distributed at random over the leaf surface. These two species have stomata on both upper and lower surfaces; that is, they are amphistomatous (see p. 385). (After Troughton and Donaldson, 1972; Plates 11 and 13.)

that grow in drier environments and under conditions of higher light intensity tend to have smaller and more numerous stomata than those in wet and shaded environments. On the average, stomatal frequencies range from less than 2500 per cm^2 of leaf surface in some plants to 40,000 or more in other plants. Moreover, stomatal frequency varies somewhat from leaf to leaf of the same plant and even in different parts of the same leaf.

The combined pore area of fully open stomata usually amounts to only a small percentage of the total leaf area. In some plant species the combined pore area is not over 2% when both epiderms are considered; in many plant species the combined pore area is much less than 1%.

The relative spacing of stomata (i.e., their distance apart) varies for different plants. This parameter is expressed in terms of the diameter of the wide open pore and ranges from about 5 diameters in some plants to about 15 diameters in others.

The location of stomata, whether on the plane of the leaf surface or buried in pits and whether on one or both surfaces of leaves, varies from species to species. Also, variation occurs in the distribution of stomata. In many species of broad-leaved plants stomata are present on both surfaces of a leaf, usually being more numerous in the lower epidermis. On the other hand, many other plant species, especially dicotyledonous trees, have stomata in the lower epidermis only. Leaves are said to be *hypostomatous* if stomata are present only in the lower epidermis. *Amphistomatous* leaves have stomata in both epiderms.

MEASUREMENT OF STOMATAL APERTURE

Measurement of stomatal aperture (i.e., the width of the pore at its point of widest opening) is a widely used technique in both field and laboratory research on water relations of plants. (Note: The length of a pore is more or less fixed for a leaf of a given plant species.) Stomatal aperture can be measured by several methods. Some can be carried out without injury to stomata; others permit rapid and repeated measurements on selected leaves even while attached to the plant; still others involve serious alteration or destruction of stomatal function. No one single method is best for all plant species or for all investigations. The choice of one or another method depends on its rapidity and limitations and on the nature and properties of the plant material and the stomata.

Perhaps the most direct method of measuring stomatal aperture is by microscopic examination of intact leaves. A micrometer eyepiece enables the observer to calculate the aperture. Unfortunately, direct microscopy is rather time consuming. Moreover, microscopic examination of stomata may not be possible for a plant species in which hairs and scales on the surface of a leaf epidermis obscure the view of the stomatal aperature. Furthermore, it may be difficult to bring the stomatal pore into focus if stomata are recessed in pits and not located on the plant of the leaf surface.

In the stomatal-imprint technique a leaf impression is obtained by spreading a substance, such as silicone rubber, over a leaf surface. The hardening process is hastened by using a catalyst. Then the hardened silicone rubber strip is peeled away from the leaf. To prepare a replica of the microrelief cast, a transparent substance, such as liquid cellulose acetate (i.e., clear fingernail polish), is painted on the silicone rubber impression. When dry, the layer of cellulose acetate (i.e., the "positive" replica of the epidermis) is examined microscopically and the aperture is determined with a micrometer eyepiece. The imprint provides a permanent record. This method, however, cannot be used when stomata are sunk below the plane of the leaf surface.

Another microscopic measuring technique involves stripping pieces of epidermal tissue from a leaf and immediately plunging them into absolute ethyl alcohol. This treatment dehydrates and hardens cell walls and kills the cells, thus fixing the stomata. The epidermal strips are examined later under a microscope. This technique can only be used with plants with easily detachable epiderms.

In addition to microscopic methods, stomatal aperture can be assessed in other ways. In the infiltration method a small drop of a liquid of suitable viscosity and surface tension properties—organic solvents, such as xylene, benzene, commercial kerosene, have been used successfully—is applied to a leaf surface. The extent of liquid infiltration through stomatal pores into the leaf and the accompanying "darkening" of the underlying intercellular spaces is proportional to the width of the stomatal pores. The extent of infiltration can be scored either by eye or by the time (seconds) required for a drop of standard size to infiltrate into the leaf. Although the liquid infiltration technique is a relatively crude procedure, it is rapid and easy to perform and may be useful in certain field studies, especially if properly calibrated (e.g., by direct microscopy or stomatal imprint).

Finally, there are the porometer methods. In these methods a small detachable glass or plastic cup is attached and secured to the surface of a leaf. The rate at which a gas (e.g., air, water vapor) passes through the portion of the leaf epidermis covered by the cup is measured. This rate is expressed in terms of the resistance offered collectively by all the stomata enclosed in the cup; the greater the resistance, the smaller the aperature. Resistance is usually expressed in seconds cm^{-1} (p. 389) and can be converted to aperture by means of a suitable calibration curve.

Porometer cups cover only a small portion of a leaf, usually an area of 2 to 3 cm^2. Because the number of stomata per square centimeter of leaf surface varies from 2000 to 40,000, depending on the plant species, clearly results obtained by porometer methods are representative of very large numbers of stomata.

There are two kinds of porometers, *flow* and *diffusion*. In the flow porometer an inert gas (usually air) is drawn through stomatal pores so that a pressure difference is maintained across the epidermal layer in which the pores are located. A change in stomatal aperture is reflected in a change in the bulk flow of air through the porometer cup. In certain sophisticated versions of flow porometers changes in air flow are detected by specially designed electronic transducers and recorded automatically.

Resistance to air flow in flow porometers is determined not only by the resistance of stomatal pores but also by the intercellular space system in mesophyll tissue. In the case of hypostomatous leaves (i.e., leaves with stomata only on the lower surface), air enters the leaf through the stomata in the area outside the porometer cup, then flows through the air-space system in mesophyll tissue in a plane parallel to the epiderms, and finally flows through the stomata enclosed within the cup. Therefore a correction must be made for the resistance offered by the air space in mesophyll tissue. On the other hand, the resistance offered by mesophyll tissue in amphistomatous leaves (i.e., leaves with stomata on both surfaces) is generally regarded as negligible because air is drawn from one leaf surface to the other and therefore the path of airflow is very short.

In the diffusion porometer the fact that a leaf loses water vapor as a result of transpiration is used. The sensing device in this instrument, housed within the porometer cup, is a miniature hygrometer sensitive to changes in humidity; these changes are measured electrically. In older models of diffusion porometers the interior of the porometer cup is flushed, immediately prior to making a measurement, with air dried by passing it through a dessicant. Then the porometer cup is attached to the surface of a leaf and the time required for the humidity to increase a given amount is recorded with a stopwatch. The porometer cup itself covers the stomata for only a short time, less than a minute. In newer models air in the porometer chamber is stirred with a miniature fan; these models are called "ventilated diffusion porometers" (see Johnson, 1981, and references cited therein). Instruments of this type are commercially available and designed to be used not only for flat-leaf surfaces but also for needlelike leaves of conifers and other nonplanar, irregularly shaped, stomata-bearing structures.

MOVEMENT OF WATER VAPOR THROUGH STOMATAL PORES

The path of outward movement of water vapor in transpiration starts at the sites of evaporation from wet mesophyll cell walls that border intercellular spaces, continues through substomatal cavities and stomatal pores (see Figure 13-5), and then proceeds through a thin layer of water vapor that lies next to the surface of the leaf. It is useful to visualize movement through this pathway as analogous to the flow of an electric current through a circuit: the greater the resistance to flow, the smaller the flow. By visualizing the process in this manner, the pathway of movement can be considered a system of resistances in series. They can be grouped in two categories: (a) the internal resistance and (b) the resistance external to the leaf.

The internal resistance to outward movement of water vapor from a leaf is primarily associated with stomata and can be called the stomatal resistance. Stomatal resistance in an individual stoma depends on several factors, including the shape and size of the substomatal cavity and stomatal pore (see Figure 13-5). The latter is especially important: the smaller the aperture, the greater the resistance to outward movement of water vapor. Stomatal aperture varies with the turgidity of guard cells (see p. 393) and so is a variable resistance. Despite the fact that several factors determine internal resistance, a considerable simplification is possible for a leaf of a given plant species. When attention is directed to a *leaf* rather than one stoma, internal resistance to outward movement of water vapor depends on the number of stomata per square centimeter of leaf surface (i.e., the stomatal frequency): the greater the stomatal frequency, the smaller the internal resistance of the leaf. But stomatal frequency, as well as the geometry of substomatal cavities and stomatal pores, can be regarded as fixed

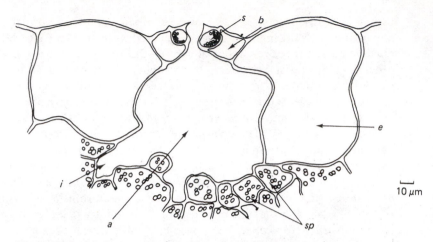

Figure 13-5 Median transverse section through a stoma of *Zebrina pendula* (wandering Jew), to illustrate the nature of the substomatal cavity. The small circles in the mesophyll cells represent chloroplasts. Abbreviations: s, guard cell; b, accessory cell; a, substomatal cavity; i, intercellular space; sp, spongy mesophyll; e, epidermal cell. (After Bange, 1953; Figure 10.)

and invariant for a leaf of a given plant species. Therefore the variation in internal resistance of a leaf is determined chiefly by stomatal aperture.

The resistance external to the leaf, on the other hand, exists because a leaf is sheathed in a *boundary layer* of relatively moist air through which transpired molecules of water vapor move from leaf to atmosphere. This boundary layer is a zone immediately adjacent to the leaf wherein air velocity is modified by the leaf itself (see Figure 13-6).

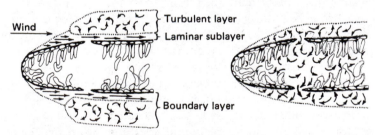

Figure 13-6 Schematic drawings of a leaf boundary layer. On the left, the boundary layer is depicted on the basis of laminar boundary layer theory. The boundary layer consists of two regions: the laminar sublayer not subjected to turbulence (straight arrows) and the turbulent layer (curved arrows). Of course, *diffusion* of water vapor and other gases occurs readily through both regions. In the drawing on the right, the boundary layer is depicted as it may exist in nature. Turbulence is indicated (by curved arrows) both within the leaf and in the turbulent region of the boundary layer. The region enclosed by dotted lines (i.e., those very close to cells inside the leaf and the cuticle outside) is the laminar sublayer but much reduced by thickness. (Also consult footnote 1, p. 389.) (Courtesy Dr. J. B. Shive, Jr., Winthrop College, Rock Hill, S.C.)

By adding the internal (stomatal) resistance to the external (boundary layer) resistance, the total resistance to outward movement of water vapor from a leaf can be determined. This total resistance to outward movement is one of the parameters needed to calculate transpirational flux. Transpirational flux is formulated according to the general transport law. This formulation, it will be recalled (pp. 315–316), relates transpirational flux to the driving force for movement of water vapor out of a leaf[1] and the total resistance to movement in the pathway.

$$\text{Transpirational flux} = \frac{\text{magnitude of the driving force}}{\text{resistance in the pathway}} \qquad (13\text{-}1)$$

The driving force in Eq. (13-1) is the difference in concentration of water vapor between a leaf and the external atmosphere. The total resistance, as we saw, is the sum of the stomatal resistance and the boundary layer resistance. Therefore Eq. (13-1) can be rewritten

$$\text{Transpirational flux} = \frac{C_l - C_a}{R_s + R_a} \qquad (13\text{-}2)$$

where C_l and C_a are the water vapor concentrations in substomatal cavities in the leaf and in the external atmosphere, respectively; R_s is the stomatal resistance; and R_a is the resistance of the boundary layer external to the leaf. Assuming that transpirational flux is given in micrograms of water vapor transpired per square centimeter of leaf surface per second and water vapor concentration in micrograms of water vapor per cubic centimeter, then the units of resistance will be seconds per centimeter.

In a now classic study of the relation of stomatal aperture to transpirational flux, the Dutch plant physiologist Bange showed that stomatal resistance is far more important than boundary layer resistance in determining transpirational flux under natural conditions. Bange's experiments, published in 1953, were carried out with turgid leaves of *Zebrina pendula*, a horticultural specimen plant commonly known as wandering Jew; this species has hypostomatous leaves (i.e., stomata are present only on the under surface of the leaf). A line drawing of a median transverse section through a stoma of *Zebrina pendula* was presented earlier (Figure 13-5).

Bange collected experimental data of transpirational flux from leaves with different stomatal apertures. He measured the loss in weight of excised leaf disks held on an analytical balance, in still air and in moving air. Moving air conditions were generated with a small air fan. Stomatal aperture was measured microscopically. In addition, Bange carried out detailed theoretical analyses of boundary layer diffusive resistance and of expected stomatal diffusive resistance at stomatal apertures varying

---

[1]Although it is widely assumed that gas exchange of water vapor, CO_2, and O_2 between leaf and external atmosphere occurs by diffusion alone, bulk flow also takes place at least to some extent (Shive, 1980). (For the distinction between diffusion and bulk flow, see pp. 292–294.) Bulk flow may be particularly important in the turbulent region of the boundary layer (see Figure 13-6). Within this region, airflow may be even more turbulent than in the surrounding air. Also, bulk flow may occur through a leaf, as shown by the curved arrows in Figure 13-6.

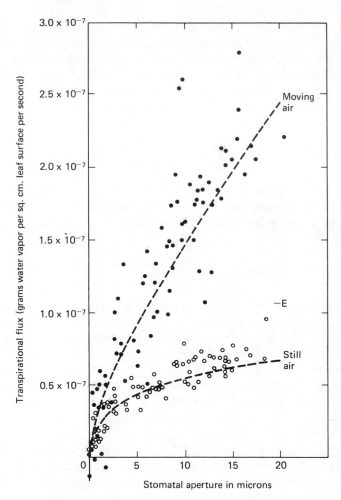

Figure 13-7 The relation between transpiration flux and stomatal aperture in *Zebrina pendula* in still air and moving air. Experimental values in still air are indicated by open circles and those in moving air by black dots. The dashed lines represent theoretical values. Rates of cuticular transpiration in still air and wind were very low (less than 0.03×10^{-7} g of water vapor per square centimeter of leaf surface per second) and are not indicated in the figure. The rate of evaporation of water in still air (E) in these experiments was about 1.0×10^{-7} g cm^{-1} s^{-1}. (After Bange, 1953; Figure 14.)

from completely closed to "wide open." These theoretical analyses were based on microscopic studies of the geometry of substomatal cavities and stomatal pores and on equations derived from diffusion laws. Finally, Bange compared the transpirational flux determined experimentally to that expected theoretically at the same stomatal aperture.

Bange's results are shown graphically in Figure 13-7. It can be seen that close agreement exists between values expected theoretically (dashed lines) and values obtained experimentally (open circles and black dots). Also, Figure 13-7 shows that transpirational flux at any given stomatal aperture is always lower in "still" air than in moving air,[2] for the boundary layer that lies over the surface of a leaf in "still" air restricts transpiration.

[2] Conditions of *absolutely* still air are almost impossible to obtain even in a laboratory.

In "still" air transpirational flux increases linearly with stomatal aperture when apertures are small but levels off when apertures are large (see Figure 13-7). When apertures are small, stomatal resistance [R_s in Eq. (13-2)] has a much greater effect than external resistance [R_a in Eq. (13-2)] in the regulation of the rate of transpiration. When apertures are large, the resistance offered by the boundary layer (R_a) has an increasingly greater role in regulating the rate of transpiration.

The situation changes in moving air. When wind blows over a leaf, even at low velocities, the boundary layer is swept away and virtually eliminated. Thus the external resistance [R_a in Eq. (13-2)] becomes negligible. Therefore, in moving air, stomatal resistance [R_s in Eq. (13-2)] is the only major resistance to outward movement of water vapor. It can be seen in Figure 13-7 that transpirational flux in moving air increases almost linearly with increasing aperture, from closed to "wide open" stomata. The transpiration rate in moving air depends essentially on stomatal aperture alone.

Bange's studies focused attention on the critical role of stomata in transpiration. His demonstration that the rate of transpiration depends primarily on stomatal aperture provided convincing evidence that stomata are the major physiological factor regulating transpiration.

Equation (13-2) can be used to determine transpiration flux in laboratory experiments, such as those carried out by Bange, and also in field studies. In the usual applications of Eq. (13-2), relative humidity inside the leaf (i.e., in the substomatal cavities) is assumed to be 100%. This assumption makes it easy to calculate the water vapor concentration inside the leaf [C_l in Eq. 13-2)] if the leaf temperature is known. The water vapor concentration in air [C_a in Eq. (13-2)] can be calculated from the measured relative humidity and temperature of the air above the leaf. Stomatal resistance [R_s in Eq. (13-2)] is measured with a porometer calibrated to give readings in resistance units (p. 386 and p. 389). Because truly "still" conditions are rarely if ever realized in nature, the boundary layer resistance [R_a in Eq. (13-2)] can be assumed to be zero in field studies. Knowing C_l, C_a, and R_s, transpirational flux can be calculated from Eq. (13-2).

PLANT ANTITRANSPIRANTS

It was pointed out earlier that almost all the water absorbed by plant roots is lost to the atmosphere by transpiration. Water use by plants is very inefficient. To retard transpiration by artificial means—perhaps in the cultivation of high-priced field crops, or in seedling transplantation in nurseries, or in other circumstances—is a desirable objective.

In recent years efforts have been made to find substances that, when applied to plants, will reduce transpiration but will have relatively little effect on CO_2 fixation in photosynthesis or on plant growth. The term *antitranspirant* is used to designate any material applied to plants for the purpose of retarding transpiration. Although the use of antitranspirants to reduce the enormous losses of water by transpiration has only limited application in crop production at present, it can be expected that research in this important aspect of applied plant physiology will accelerate in years to come.

One group of antitranspirants consists of colorless plastics, silicone oils, and low-viscosity waxes that are sprayed on leaves with the object of forming a film permeable to CO_2 and O_2 but not to water. Only limited success has been achieved to date with these sprays. A second group of antitranspirants consists of substances that cause a partial closure of stomata when sprayed on leaves. In some cases, these substances are metabolic inhibitors that prevent the complete opening of stomata when applied at certain concentrations. Such substances, to be useful, should have little effect on leaf cells other than guard cells. Today the fungicide phenylmercuric acetate is one of the more promising stomatal-closing types of antitranspirant. When applied as a foliar spray at a concentration of about 10^{-4} M, phenylmercuric acetate has been found to have little toxic effect on leaves and results in partial closure of stomatal pores for periods of 2 weeks or more. Also, application of the naturally occurring plant growth regulator abscisic acid to the surfaces of leaves is highly effective in causing stomata to close. Effects do not persist for more than a few hours, however, and so this plant growth regulator has no practical value. Still, it may be possible in future to develop analogs of abscisic acid that may be of use in agriculture.

CO_2 is another substance that is a very effective antitranspirant. Raising the concentration of CO_2 only slightly in the air surrounding a leaf—from the natural 0.03% to perhaps 0.05%—is known to induce partial closure of stomata. High concentrations of CO_2 in the air surrounding a leaf must be avoided because stomata will be induced to close completely (see p. 399) and photosynthesis will cease. One advantage of using CO_2 as an antitranspirant is its inhibition of photorespiration (see Chapter 7). At present, however, enrichment of the ambient atmosphere with CO_2 is economically and practically feasible only in glasshouses.

The success of antitranspirants of the stomatal-closing type is based on the fact that partial closure of stomata reduces transpiration relatively more than net photosynthesis. To understand why, it is necessary first to consider Eq. (13-3).

Because CO_2 uptake in photosynthesis takes place primarily through stomata in leaves, the rate of net photosynthesis may be formulated in an equation analogous to Eq. (13-2) for transpiration. The pathway for inward movement of CO_2 through stomata, however, from the external atmosphere to the sites of photosynthesis in chloroplasts, is longer than that for the outward movement of water vapor. In fact, an additional resistance—the resistance encountered by CO_2 molecules in moving from the exterior surfaces of mesophyll cells to the chloroplasts—must be included in the formulation of the rate of net photosynthesis. Thus the rate of net photosynthesis (i.e., the net flux of CO_2 from the external atmosphere to the site of CO_2 fixation in the chloroplasts) can be written

$$\text{Rate of net photosynthesis} = \frac{C_a - C_c}{R_s + R_a + R_c} \qquad (13\text{-}3)$$

where C_a and C_c are the CO_2 concentrations in the external atmosphere and the chloroplasts, respectively; R_s is the stomatal resistance; R_a is the resistance of the boundary layer of CO_2 external to the leaf; and R_c is the resistance to CO_2 movement offered by the liquid phase in the cell walls and cytoplasm of mesophyll cells and by the photosynthetic carboxylation reaction itself. The units used in Eq. (13-3) are usually as follows: rate of net photosynthesis, in nanomoles of CO_2 uptake per square centimeter of leaf surface per second; CO_2 concentrations C_a and C_c, in nanomoles per cubic centimeter; resistances, in seconds per centimeter. Equation (13-3) is applicable only when CO_2 limits the rate of photosynthesis (as often occurs under natural conditions in sunlight), not when other factors (e.g., light, temperature) are limiting.

Although R_c in Eq. (13-3) is difficult to measure accurately, it is known to be of considerable magnitude. Hence the total resistance to inward movement of CO_2 through stomata, from the external atmosphere to the site of CO_2 fixation in chloroplasts, is much greater than the total resistance to outward movement of water vapor in transpiration [compare the resistances in Eqs. (13-2) and (13-3)]. It follows that an antitranspirant that closes stomata partially [i.e., increases R_s in Eqs. (13-2) and (13-3)] will increase the total resistance to movement of water vapor more than to CO_2. (For example, if a given partial stomatal closure increases the total resistance to movement of water vapor by 100%, the total resistance to CO_2 movement may be increased by only 50%.) Thus such an antitranspirant can be expected to reduce transpirational flux more than the rate of net photosynthesis.

STOMATAL MOVEMENTS

It has been recognized for over a century that the movements of stomata—open in the daytime and closed through the night in most plants—are due to the entrance and exit of water into and out of guard cells. The entrance of water into the vacuoles of a pair of guard cells induces the development of a higher turgor pressure within these cells than in adjacent epidermal cells. This increase in turgor pressure in guard cells, relative to that in adjacent epidermal cells, causes changes in their volume and shape so that the size of the stomatal pore increases. Conversely, the exit of water from the vacuoles of a pair of guard cells induces a lower turgor pressure within these cells than in adjacent epidermal cells; as a result, the size of the stomatal pore decreases (see Figure 13-8). Inasmuch as these opening and closing movements depend on the elastic properties of the cell walls and guard cells, they are classified as reversible turgor movements (see pp. 369–370). Stomatal movements have considerable interest not only because of their role in gas exchange but also because they exemplify a basic—but still not completely resolved—problem in cell physiology.

Guard cells have unusually thick walls compared to other leaf cells; in some plants the thickenings are equal throughout but in other unequal (see Figure 13-9). Moreover, cellulose micelles (aggregates of chainlike cellulose molecules) in the cell walls of guard cells are oriented radially, like the reinforcement bands in an automobile tire, rather than laterally and thereby restrict expansion of the cross section of guard cells during stomatal opening. Also, guard cell walls have special elastic properties that enable them to stretch laterally to a considerable extent, more so than other leaf cells. In addition, the cell walls between which the stomatal pore appears during stomatal opening are free and not attached to cell walls of other cells.

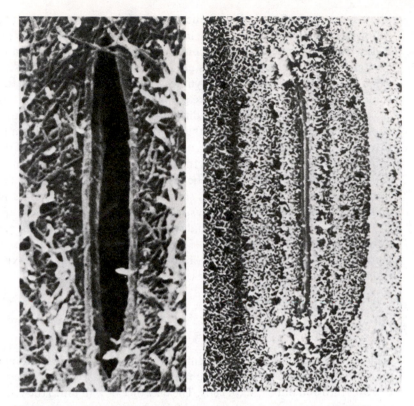

Figure 13-8   Scanning electron photomicrographs of an open stoma (×2800) on the lower surface of a wheat leaf (left) and a closed stoma (×1400) on the upper surface (right). In the latter, the guard cell is indicated by the letter G and the subsidiary cell by the letter S (near top of photo). Wax on the cuticle on the lower leaf surface appears as long rodlets. (After Troughton and Donaldson, 1972; Plates 20 and 21.)

These attributes give guard cell walls the ability to increase or decrease stomatal aperture under the influence of changes in turgor pressure in guard cells.

In guard cells of grasses and sedges, which constitute one distinct type, the two ends of a guard cell are bulbous and thin walled and joined on one side to the corresponding ends of the sister guard cell; the cell walls of the middle portion are strongly thickened, to such an extent that the protoplasm between is reduced to a thin thread (Figure 13-10). When turgor pressure in a pair of guard cells is low, the thickened cell walls (in the middle portion of the cells) touch each other and the stomatal pore is closed. But an influx of water increases turgor pressure within the guard cells and causes the thin-walled ends of the cells to expand. At the same time, the cell walls of the middle portions remain straight but are pulled away from each other so that the size of the stomatal pore increases.

In contrast to grasses and sedges, dicotyledonous plants exhibit considerable

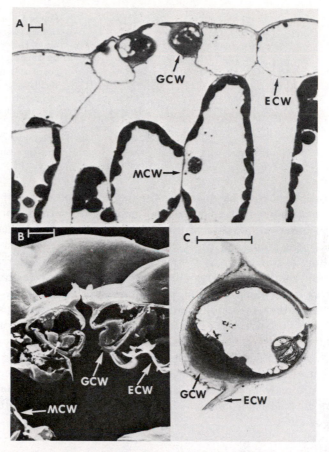

Figure 13-9 Three cross-sectional views through a leaflet of *Vicia faba*, showing the relative thickness of guard cell walls (GCW), epidermal cell walls (ECW), and mesophyll cell walls (MCW). (A) Light micrograph, (B) scanning electron micrograph (C) transmission electron micrograph. The stomatal pore in (C) is to the left of the guard cell. The scale bar in each of the diagrams is 5 μm. Note the cuticular horn on each guard cell. (Courtesy Dr. W. H. Outlaw, Jr., Florida State University, Tallahassee.)

variability in the distribution of cell wall thickenings in guard cells. In one common type, the portion of the cell wall bordering the pore is thicker than the other portions of the cell wall (see Figure 13-9). When water moves into the vacuoles of a pair of guard cells and the turgor pressure within them increases, the thin cell wall (away from the pore) becomes distended and pulls the thickened front wall (facing the pore) with it so that each of the guard cells bends away from the pore and the pore increases in size. Reverse changes occur when the turgor pressure of the guard cells decreases.

Guard cells have a full complement of organelles, such as mitochondria and

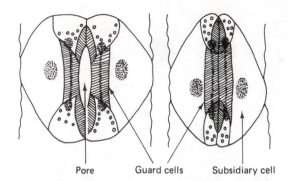

Figure 13-10 Schematic diagram of the surface view of a stoma of a grass (sugar cane). Left, open; right, closed. The thickened portions of the cell walls of guard cells are indicated by hatching. The small circles in the bulbous portions of the guard cells represent chloroplasts. Note that the nucleus of a guard cell appears as two masses connected by a thin thread. Also note the subsidiary cells with their nuclei. (After Esau, 1953; Figure 7.5.)

Pore   Guard cells   Subsidiary cell

chloroplasts. (It is pertinent to note that chloroplasts are absent in other epidermal cells). Also, many droplets of lipid material are normally present in guard cells and often observed in active protoplasmic streaming. Chloroplasts in guard cells have prominent grana and (in most plant species) large accumulations of starch;[3] the amount of starch increases slowly during the night and decreases during the day—in marked contrast to mesophyll cells, in which starch reserves are built up in the daytime, during hours of illumination, but diminish during the night.

Another feature of guard cells is their small size compared to other epidermal cells. This property is thought to give them a special ability to undergo marked changes in turgor pressure. For a given quantity of dissolved solute moving into or out of guard cells, there would be a larger change in osmotic potential—and hence a larger change in turgor pressure upon entry or exit of water—than in other epidermal cells.

Stomata open and close in response to changes in one or another of several biologic and environmental factors. Under natural growing conditions the most important are

1. supply of water to a leaf.
2. level of atmospheric humidity around a leaf.
3. temperature of the leaf.
4. whether the leaf is lighted or darkened.
5. the concentration of CO_2 within the leaf.

Each is discussed in turn here and on pages 398–399.

Water supply to a leaf has a critical role in determining the size of the stomatal pore, especially during periods of water shortage. When leaf water deficits begin to develop, guard cells become less turgid and stomatal aperture decreases. If water shortage continues and leaf water stress becomes severe, stomata usually will close completely and will remain closed. In fact, a deficiency of water within a leaf will override all other opening stimuli. When the water content of a leaf drops below a

---

[3]Starch is absent from guard cells in a few species of plants, including members of the lily, iris, and amaryllis families.

critical value, stomata will close partially or completely even under conditions that are otherwise conducive to opening.

Another factor influencing stomatal movement is the degree of atmospheric humidity. Stomatal aperture decreases as the difference between the vapor pressure of the air surrounding a leaf and the vapor pressure of the substomatal cavity increases. That is, stomata close slightly as the air surrounding a leaf becomes slightly drier. [The moisture on the surfaces of mesophyll cells in the substomatal cavity is assumed to have near-saturated vapor pressure, as already indicated on page 391.] This closing response of stomata to evaporative demand does not result from lowered leaf water content; instead it is due to increased evaporation from the sides of the guard cells facing the stomatal pore, and the inability of guard cells to absorb water from other cells as rapidly as it is lost (Aston, 1976). It is interesting to note, in passing, that most higher plants grow best at a humidity of 70% or higher as a general rule; if atmospheric humidity drops below 50%, plants will probably grow less rapidly because stomata close partially and so the rate of photosynthesis decreases somewhat.

Temperature also influences stomatal movements. The effects of temperature become noticeable especially at the extreme ranges—that is, near 0°C and above about 30 to 35°C. At low temperatures near the freezing point of water, stomatal opening does not occur even when other environmental conditions favor opening. Undoubtedly this situation arises because water transport between guard cells and other epidermal cells occurs slowly at these temperatures. On the other hand, when the temperature of the air becomes higher than about 30 to 35°C—at these air temperatures the temperature of leaves exposed to strong sunlight may be 45°C or even higher (see Figure 13-2)—stomata often close, at least partially. This type of stomatal closure often occurs at midday and is referred to as *midday closure*. The pattern of behavior of midday closure—wide stomatal opening in the morning, partial closure for an hour or two at midday, followed by reopening of stomata in the afternoon—is most probable in many plants exposed to strong sunlight in temperate regions in the summer on hot, sunny days.

The mechanism of midday stomatal closure is not yet completely understood. Quite possibly there is more than one cause. One is the inability of the plant to absorb and replace water as rapidly as it is lost by transpiration. When water potential of leaf cells drops to low values because water uptake by roots cannot keep pace with transpiration, open stomata close. This condition is most likely to occur in full sunshine at midday. Another cause is a decrease in atmospheric humidity around a leaf in full sunshine at midday, as noted on this page.

Midday closure of stomata is usually a transient phenomenon, lasting only an hour or two. When it occurs, leaf temperature rises sharply, leading to a further loss of water vapor from the leaf. As a result, there is a partial loss of leaf turgidity, perhaps even accompanied by transient wilting. Therefore the heat load on the leaf is lessened considerably and the temperature of the leaf decreases accordingly. This decrease in temperature is followed by a recovery in the rate of photosynthesis in the guard cells and a rapid reduction in the CO_2 concentration in substomatal spaces, the effect of which is to promote stomatal opening (see p. 399). Thus stomata open again.

METABOLIC CHANGES IN GUARD CELLS
DURING STOMATAL MOVEMENT

If the water supply to a leaf is adequate and leaf temperatures are not extreme, light induces stomatal opening and darkness stomatal closure. Stomatal movements in most higher plants are *photoactive* (i.e., movements are initiated by light–dark transitions).[4] Thus stomata are open during the day and closed during the night.

Moderate sunlight can result in maximum opening in most temperate plant species. The response time to light varies from species to species; complete opening requires only a few minutes in some species but an hour or more in others. Stomatal opening in light is produced by light absorption in guard cells themselves (Sharkey and Raschke, 1981). The action spectrum that causes stomatal opening shows a strong peak in the blue region, suggesting the existence of a blue-light receptor in guard cells.

Recall that stomata open when turgor pressure in guard cells increases (p. 393). This increase in turgor pressure is an osmotic effect, due to an accumulation of inorganic and organic ions in vacuoles of guard cells, followed by an influx of water. Measurements of ionic concentrations in guard cell vacuoles indicate increases from less than 1 M to more than 2 M during the stomatal opening. Osmotic potential of fully turgid guard cells may be −45 bars and turgor pressure as high as +45 bars (Raschke, 1975). The technique of dissecting single pairs of guard cells from leaves and analyzing them by sensitive quantitative histochemical methods (Outlaw, Jr., 1980; Outlaw, Jr., 1981) has shown potassium as the major cation whereas malate (and citrate to a lesser extent) is a counter anion. (Chloride also functions as a counter anion.) These ions increase in concentration during stomatal opening and decrease during stomatal closing. In addition to these changes in ionic composition of guard cells, starch concentration in guard cell chloroplasts is lower in light than in darkness.

Current hypotheses regarding the metabolic events in guard cells during stomatal opening indicate starch is degraded to phosphorylated hexoses and then to phosphoenolpyruvate via glycolysis (see Figure 4-3). Phosphoenolpyruvate is converted first to oxaloacetate [see Eq. (4-33)] and then to malate and citrate by reactions of the TCA cycle (see Figure 4-6).[5] Thus these two organic anions are produced metabolically. On the other hand, potassium and chloride ions are imported from adjacent epidermal cells.

Guard cell chloroplasts of most plants lack ribulose bisphosphate carboxylase (Madhavan and Smith, 1982) and NADP-linked triose phosphate dehydrogenase (Outlaw, Jr., 1981). Thus the photosynthetic carbon reduction cycle is absent. Guard cells are heterotrophic. The substances utilized for starch formation in guard cell

---

[4]The discussion here is restricted to light-induced (or -maintained or -mediated) stomatal opening. Dark opening, which occurs in CAM plants growing under conditions of water stress, may be by a different mechanism (see footnote 6, p. 399). [*Note*: By increasing the supply of water to CAM plants, stomata will open through the day and close at night, just as in most plants (Nobel and Hartsock, 1979).]

[5]Plants in which starch is absent in guard cells (see footnote 3, p. 396) probably produce malate during stomatal opening from carbohydrate reserves other than starch (e.g., fructosans).

chloroplasts are sugars transported into guard cells from adjacent mesophyll cells (Dittrich and Raschke, 1977). The light-harvesting pigments of both photosystems I and II are present in guard cell chloroplasts (Outlaw, Jr. et al., 1981; Zeiger et al., 1981), however, a fact that suggests guard cells have the ability to carry on photosynthetic phosphorylation. Presumably ATP (produced in photosynthetic and/or oxidative phosphorylation) is needed to sustain the transport of potassium ions across the plasma membrane and tonoplast of guard cells during stomatal opening; but little is known about this translocator system at present.

The mechanism by which light precipitates the biochemical events in guard cells—import of potassium and chloride ions and production of malate and citrate ions from starch—leading to increased turgor pressure and stomatal opening is not understood today. One suggestion (Sharpe and Zeiger, 1981) is that light initiates an electron transport event in protoplasmic membranes in guard cells, thereby leading to the generation of a proton motive force across the membranes, in accord with chemiosmotic principles (see Figure 4-8 and related text discussion). In this view, proton motive force is responsible for an active transport of potassium into the vacuole of the guard cell.

The concentration of CO_2 in the substomatal cavities of a leaf influences stomatal movement. If a leaf is placed in a stream of air free of CO_2, stomata will open whether the leaf is in light or darkness. On the other hand, stomata of most plant species show some tendency to close, even in the light, when exposed to concentrations of CO_2 somewhat above the natural atmospheric level (approximately 330 ppm). In maize, a species whose stomata are especially sensitive to CO_2, stomata that were induced to open in air free of CO_2 can be closed by raising the concentration of CO_2 to about 1000 ppm, irrespective of whether the leaf is in light or darkness. But in other species the concentration of CO_2 required to close stomata is above 2500 ppm.

The mechanism by which CO_2 in the 1000- to 2500-ppm range closes stomata is not understood. One recent suggestion is that CO_2 dissipates the proton gradient across protoplasmic membranes in guard cells, but how it occurs is not known (Sharpe and Zeiger, 1981); when the proton gradient is dissipated, active transport of a potassium ion into guard cells ceases. In fact, it has been suggested that CO_2 accumulates in the leaf in the dark, in the absence of photosynthesis (Sharpe and Zeiger, 1981); in this view, CO_2 is the natural closing agent.[6] Besides this suggestion, little is known about stomatal closure.

Of course, guard cells dispose of accumulated ions during stomatal closure. Potassium ion is transported out of guard cells into subsidiary cells or other epidermal cells. Also, during stomatal closure guard cells dispose of accumulated malate, partly by consumption in the TCA cycle and partly by release to other adjacent cells (Van Kirk and Raschke, 1978).

---

[6]In leaves of CAM plants growing under conditions of water stress, malic acid decarboxylation occurs through the day (pp. 175–177) and internal tissue CO_2 concentration increases dramatically (perhaps up to 2%); it is probably the high CO_2 concentration in leaves of CAM plants during the day that explains their daytime closure. Perhaps low CO_2 concentration in leaves at night accounts for nighttime opening.

ABSORPTION OF WATER BY TRANSPIRING PLANTS

Throughout the period of active growth of higher plants water absorbed by roots passes along the length of the vascular system in specialized water-conducting cells in xylem tissue. This upward translocation of water in the xylem, from roots to leaves, is called the *transpiration stream* because its ascent depends on transpiration. Although the direction of the transpiration stream is normally upward, under certain conditions it may be downward. In some plants in dry soils water that condenses on leaf surfaces during fogs may be absorbed through stomata and conducted via water-conducting xylem elements to various parts of the plant, even the roots (Bornman, 1972; Tibbitts, 1979).

The transpiration stream is initiated by water vaporization at the outermost surfaces of moist cell walls of mesophyll cells that line substomatal spaces. At these evaporation sites cell wall constituents become partially dehydrated and water from nearby, more hydrated cell wall constituents moves into them. It is the adhesive properties of water [see pp. 346–347] that ensure the continued movement of water molecules from hydrated cell wall constituents to evaporation sites.

As water continues to evaporate from evaporation sites, it continues to move into these sites from more hydrated regions in mesophyll cell walls. Now water from the terminal ends of water-conducting xylem elements in leaves begins to move toward evaporation sites. But because of the cohesive properties of water [see pp. 346–347], water in the water-conducting cells of the xylem is "pulled" upward in bulk flow. The driving force is the difference in hydrostatic pressure in xylem sap between roots and leaves. As will be seen (pp. 403–404), continuity in water columns in the xylem is necessary for continued upward flow of the transpiration stream.

In the roots themselves the pull on the water in the water-conducting xylem cells generates a difference in pressure potential between the terminal ends of these cells and the external soil solution. It is this difference in hydrostatic pressure between these two regions that is the driving force for water movement across the root of a transpiring plant.

The proportion of water moving into roots by osmosis compared to "transpirational pull" depends on the transpiration rate. In rapidly transpiring plants osmotic movement of water from the soil solution into roots is negligible. Only in plants with very low rates of transpiration does osmosis have a role in water movement from the soil solution into roots.

Because water absorption by roots of transpiring plants depends on transpiration, clearly solar radiation is the ultimate source of energy for water absorption by transpirational pull. Moreover, the adhesion of water molecules to the evaporating surfaces of mesophyll cell walls and the cohesion of water molecules to one another account for transpirational pull. A useful but perhaps oversimplified point of view is to regard the evaporating surface in mesophyll cell walls as the "hook" from which water "hangs" in a column extending from leaves to roots.

The lifting power of transpiration can be easily demonstrated with little equipment. A leafy shoot is severed from a plant. The operation should be performed under water to ensure that no air bubbles are introduced into water-conducting xylem elements. Then the shoot is attached with rubber tubing to one end of a long glass tube filled with water. The other end of the glass tube is immersed in a vessel of mercury and the whole apparatus is fastened in a vertical position. After a time—the time required depends on the rate of transpiration by the shoot—a column of mercury will be supported in the glass tube, perhaps to a height of 50 cm or more, depending on the care taken to prevent introduction of air bubbles.

The lifting power of transpiration is an essential feature of the *cohesion theory* of the ascent of sap.

THE COHESION THEORY

Studies on the upward translocation of water in plants are among the earliest dealing with plant physiology. How water is raised in tall trees has long drawn the attention of plant physiologists.

It has been recognized for almost 200 years that translocation of water through the length of a plant takes place primarily in the xylem portion of the vascular tissue system. This can be demonstrated by observing the effect of removing a ring of phloem tissue from the stem of a plant. Upward movement of water past the point of girdling will continue almost as well as in a normal plant. Additional evidence that the xylem is the pathway of longitudinal transport of water can be obtained by supplying the root system of a potted plant with isotopically labeled water. After the passage of an hour or two, the label will be detectable in the xylem tissue of the stem but not the phloem.

The cells in the xylem that function in translocation of water are mature, dead tracheids and vessels. A tracheid is an elongate, spindle-shaped cell with tapering ends and pitted cell walls (Figure 13-11). Tracheids overlap each other in longitudinal arrays. A vessel is an elongate, tubular-shaped cell with pitted side walls and perforated end walls (Figure 13-11). In some plant species the perforations in the end walls are small holes; in others there is a single wide hole; in still others the end wall may disintegrate and disappear completely at maturity. Vessels are stacked longitudinally one on another, end to end, to form long, continuous tubes or conduits consisting of an overlapping series of several dozens or even hundreds of individual vessel elements.

Both tracheids and vessels become devoid of protoplasm at maturity, once their cell walls are lignified and rigid. Thus longitudinal movement of water in these cells is not obstructed by living matter. Moreover, the rigid cell wall prevents collapse of these dead cells when tension exists in the transpiration stream.

The fluid translocated in the xylem is a dilute (about 0.1%) solution of mineral ions absorbed from the soil. Dissolved mineral ions are carried along passively in the water in the transpiration stream; when delivered to leaf cells, dissolved ions are absorbed by leaf cells by active transport mechanisms (pp. 321–326) operating at the plasma membranes of these cells. In addition to mineral ions, small amounts of a few organic substances (e.g., amino acids, amides, sugars, cytokinins, abscisic acid) are

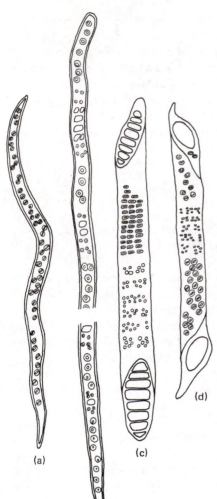

(a)

(b)

(c)

(d)

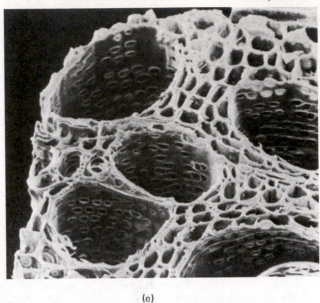

(e)

Figure 13-11 Tracheids [(a) and (b)] and vessels [(c) and (d)] from several different plant species, in side view. Only one-third of the tracheid is shown in (b). Note the different kinds of pits in the side walls of these cells. Also note the different kinds of perforations in the end walls of the vessels, varying from a cluster of several openings to a single opening of different widths. (Tracheids are drawn to scale and to the same scale as are those of vessels.) (After Eames and MacDaniels, 1925; Figures 33 and 35.) (e) A scanning electron micrograph of the cross section of elm wood (enlarged about 430 times), showing five large vessels running longitudinally through the xylem. Each of the small pits, or openings, in the side walls of these vessels has a membrane that permits water to flow through it in a lateral direction, but the membranes are not visible in this micrograph. [After *Sci. Amer.* 245(2), (1981), 61. Courtesy H. J. Miller and D. M. Elgersma, Phyto-pathological Laboratory "Willie Commelin Scholten," Baarn, The Netherlands.]

almost always present. The water solution translocated in the xylem is sometimes referred to as *xylem sap*.

Upward translocation of xylem sap may be quite rapid during the daytime, especially when transpiration rates are high. Linear rates ranging from about 3 to more than 45 m h^{-1} have been recorded, depending on the kind of plant and transpiration rates. The lowest maximum rates are found in evergreen conifers, intermediate rates in deciduous trees and herbaceous plants, and the highest in vines and lianas.

It was pointed out in the preceding section that the ascent of sap in transpiring plants is initiated by the "pull" of transpiration. Liquid water, which fills the nonliving tracheids and vessels of the xylem and is present in continuous, freely mobile columns through the length of the plant, is literally pulled upward en masse.[7] But a matter not considered explicitly in the preceding section is that water in xylem conduits in transpiring plants is in a state of *negative hydrostatic pressure* or *hydrostatic tension*. This tension—a result of the fact that water absorption by roots lags slightly behind water loss by leaves (see Figure 13-13)—exists throughout the continuous water system in the xylem.

The ability of liquid water to sustain a metastable condition of negative pressure has been known for years. At the end of the nineteenth century Dixon, an Irish plant physiologist, realized that the ability of water confined to xylem tracheids and vessels to withstand rupture when subjected to tension is crucial to the ascent of sap. His proposal that the cohesive properties of water (see pp. 346–347) are sufficiently great to maintain unbroken water columns from leaves to roots is known as the *cohesion theory*.

Today it is universally recognized that the tensile strength of water—the force needed to rupture a column of water when it is subjected to a strong, steady pull—is more than adequate to account for the lifting of xylem sap from the base to the leaves of the tallest trees. Values in excess of −260 bars have been reported.

Xylem sap undergoes a gain in gravitational potential during its ascent to the level of leaves—work is done against gravity. Also, work must be done to overcome frictional resistance to the passage of xylem sap through nonliving xylem conduits. The gravitational work has been calculated and found to be almost negligible.[8] The frictional resistance can be measured by forcing water through logs lying in a horizontal position. Depending on the plant species, pressures of only 1 to 3 bars are required to force water through horizontal logs more than 10 m long at rates comparable to those for water being translocated upward in trees. Thus despite the frictional resistance encountered by the passage of water through xylem conduits, the tensile strength of xylem sap (= −260 bars approximately) is far in excess of that required to maintain continuous columns in the xylem of the tallest trees.

Since its introduction, the cohesion theory has been challenged and, as a result, subjected to repeated reexamination and reevaluation. But this theory is almost universally accepted as the only reasonable explanation of how xylem sap can be lifted to the tops of tall trees. Such forces as surface tension or root pressure are completely inadequate. Although surface tension can pull liquids upward in capillary tubes (see pp. 346–347), this force alone cannot explain the ascent of sap in tall trees; capillary rise of xylem sap cannot exceed a few feet in tubes of the dimensions of tracheids and vessels. Root pressure can push water upward from roots, but this force is developed in intact plants only infrequently (see p. 412). Furthermore, root pressures are usually no more than +1 to +2 bars. Therefore root pressure could account for the ascent of sap only to a height of about 20 m. (One bar of pressure is capable of causing a

---

[7] Water is driven passively upward because of the difference between its chemical potential in the roots and the atmosphere external to the leaves, irrespective of whether the mechanism of water movement in the plant is diffusive or bulk flow or any combination of the two (Andrews, 1976).

[8] It has been calculated (Rosenberg, 1954) that the work required for the reversible raising of 1 mole of liquid water through a tube 25 m in length is only 1 cal.

flow of water up to a height of about 10 m.) This is far short of the heights of which water is delivered in tall trees. (Giant redwood trees in California measure 120 m or more, from root tips to the highest leaves.)

Ample experimental evidence in support of the cohesion theory is available today. Most criticisms of past years have been refuted and no serious evidence against it exists at present.[9] A few of the points that seemed to jeopardize the validity of the cohesion theory but are no longer considered critical are as follows:

1. Because tracheids and vessels through which upward transport of water occurs are not living, suggestions have been made that living cells bordering on tracheids and vessels have a role in water movement. Many investigators, however, have established that water will be drawn up to leaves through segments of stems that have been treated so that all cells are killed (e.g., by application of poisons or by scalding).

2. It has been argued that spontaneous bubble formation occurs in xylem sap and, as a result, water columns undergo rupture and fractionation—therefore upward movement of water is thought to cease. Yet the known ability of water molecules to cohere to one another and to adhere to the cell walls of water-conducting elements in the xylem makes the probability of rupture of water columns quite small. Furthermore, even if water columns were to rupture and thereby render xylem elements inoperative, other intact water columns would ensure the continuation of sap ascent. In fact, only a small fraction of the water-conducting elements in the xylem is required, at any one time, to supply adequately all the water lost by transpiration. Preston (1958) found that many (perhaps even most) of the xylem ducts are blocked with gas; freely mobile and continuous sap columns exist only in the most recently differentiated xylem elements in the outermost (youngest) annual growth rings.

3. Although hydrostatic tensions in the xylem sap of transpiring plants were assumed to exist from the time when the cohesion theory was first proposed, techniques for their direct measurement were developed only recently. Using the pressure bomb technique (see p. 364), Scholander and colleagues (1965) were able to demonstrate that hydrostatic tensions exist in xylem sap. Values ranging from −4 bars in forest trees in moist soils to −80 bars in desert shrubs were recorded. Also, they verified that hydrostatic tensions are increasingly negative with increasing tree heights.

MOVEMENT OF WATER ACROSS ROOTS
AND THROUGH LEAVES

The flow of water through a transpiring plant and into the external atmosphere occurs through a sequential series of tissue segments. First, water moves radially across the root, from the soil solution to the water-conducting elements in the root xylem. Then

[9]Consult Plumb and Bridgman, 1973, for a proposed alternative hypothesis and also for technical comments and rebuttals.

water is translocated longitudinally in the water-conducting elements in the xylem of stems and leaves. Next, water moves from the terminal ends of xylem conduits in leaves to the evaporating sites in the cell walls of mesophyll cells lining substomatal spaces. Finally, water vapor moves outward through stomatal pores.

Because these four segments are in series, the flow of water through a plant, from the soil solution to the external atmosphere, is controlled primarily by the segment exerting the greatest resistance. Undoubtedly it is the last segment. The resistance to movement of water vapor through stomata is many times greater than through other portions of the plant. In fact, the size of the stomatal pore is the most important factor controlling the rate of transpiration from leaves (pp. 389–391).

Of the four segments mentioned, the least resistance is encountered by water in its ascent through nonliving, water conducting tracheids and vessels in the xylem. The rapid rate of translocation of water in the xylem (p. 402) is evidence, of and by itself, of relatively small resistance to longitudinal water flow.

Let us compare, at least qualitatively, the relative resistances to movement of water across roots and through leaves.

In young dicotyledonous roots consisting only of primary tissues water is absorbed from the soil solution most rapidly in the region directly behind the root tip, between the elongation zone and the suberized zone farther from the root tip. The zone of most rapid water absorption corresponds to the zone of most rapid absorption of mineral ions (see p. 303). In this zone, tracheids and vessels in the xylem are completely differentiated and dead; also, numerous root hairs are present in the epidermal layer. The ability of root hairs to absorb water was demonstrated (Cailloux, 1972) by direct measurement of water uptake by a single root hair on an intact plant.

The pathway of movement of water across the primary root—from the external soil solution to the water-conducting xylem elements in the stele—is the same as that described for dissolved ions (see pp. 305–306). The presence of water-impermeable suberin in the Casparian strips of endodermal cells forces water to move through the protoplasm of several layers of cells. Thus the major pathway of water movement across the primary root is through the symplasm (i.e., the protoplasmic continuum). Water enters the symplasm in the cortical region of the root and leaves the symplasm at the xylem (see Figure 13-12). Only relatively small quantities of water move across the root exclusively through cell walls and in films along the outer surfaces of cell walls.

In older roots, first the epidermis and then the cortex are sloughed off and secondary tissues laid down. The cell walls of these secondary tissues become impregnated with suberin, the same fatlike substance present in Casparian strips in endodermal tissue in primary roots. Water in the soil solution enters these older roots through breaks and cracks in the suberized secondary tissues, then moves across the protoplasm of parenchyma cells in secondary phloem tissue and in vascular cambium, and finally reaches the water-conducting tracheids and vessels of the xylem. Thus a layer of protoplasm is crossed during the movement of water across older roots, just as in the case of primary roots. Even in those plants in which roots grow in association with mycorrhizae (see p. 255), water must pass through a layer of protoplasm—in this case, the protoplasm of mycorrhizal cells.

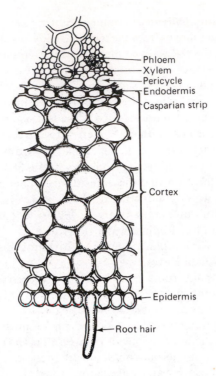

Figure 13-12 Cross-sectional diagram through the primary tissue of the root of a dicotyledonous plant. (After Kramer, 1956; Figure 3.)

Evidence that the protoplasm of root cells offers considerable resistance to water movement can be obtained by attaching a vacuum pump to the stump of a decapitated root system. The intake of water through the root system is increased greatly after the roots are killed—for example, by plunging them in boiling water. Although movement of water is rapid across a plasma membrane and little resistance is offered (see Chapter 11), the resistance to the flow of water across a layer of cytoplasm is considerable.

In contrast to roots, leaves do not have a barrier, such as the Casparian strip. Moreover, vascular strands are distributed in a networklike fashion throughout a leaf so that almost all mesophyll cells that border stomatal cavities are separated from vascular elements by only two or three, or perhaps five, cells. Thus water passing out of the terminal ends of water-conducting xylem cells moves predominantly in the cell walls of only two to five cells before reaching evaporation sites in mesophyll cells that border substomatal cavities. If has been suggested, however, that the resistance to water flow in this short cell wall pathway is relatively high (Aston, 1976).

The resistance to movement of water across roots is generally considered greater than that through leaves. Next to stomatal resistance to movement of water vapor into the atmosphere, roots offer the greatest resistance to water flow through the plant.

WATER CONTENT AND ITS DAILY FLUCTUATION

One of the most important aspects of the water relations of higher plants is their failure to maintain a perfect internal water balance. Fluctuations in the internal water status of a plant occur daily in nature even when a plant is grown with its roots in a soil with high moisture content or in a water culture solution. These fluctuations in water status are due to a difference in rate between water absorption by roots and transpiration by leaves.

In a now classic experiment the American plant physiologist Kramer made simultaneous measurements of water absorption by roots and water loss by transpiration. Several individual potted plants rooted in soil were selected. The soils were wetted to maximum water-holding capacity at the start of the experiment and were maintained at this water content by replacing soil water continually and automatically. Water absorption was determined by measuring the amount of water supplied to the soil. Transpiration was determined by measuring the loss in weight of the plant–soil system.

Figure 13-13 shows that the curves for water absorption and transpiration are almost parallel during both day and night. During the day, however, a plant loses slightly more water by transpiration from leaves than is absorbed by roots. This situation occurs because of the greater resistance offered by roots to the flow of water across them compared to the flow of water through leaves. But during the night hours, when transpiration rates are very low, water absorption is able to "catch up" to transpiration.

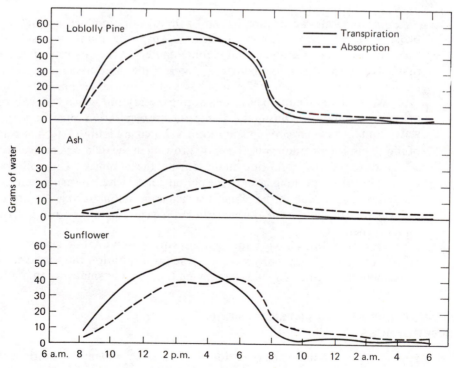

Figure 13-13 The relation of water absorption to transpiration for loblolly pine, green ash, and sunflower. (See text for details.) (After Kramer, 1937; Figure 2.)

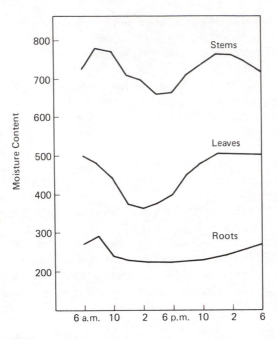

Figure 13-14. Fluctuation in the water content of leaves, roots, and stems of sunflower, on a clear, sunny day and the following night during the summer. The plant material was grown in the field, in the southeastern United States. Water content is expressed on a dry weight basis. (After Wilson et al., 1953; Figure 1.)

Because water absorption by roots and transpiration by leaves are slightly out of phase with one another, the tissues of transpiring plants are subject to daily fluctuations in water content. Water content refers to the quantity of water in a plant tissue as compared to the quantity of water in the same tissue in a selected base of reference. (For discussion of different bases of reference, see pp. 372–373).

Water content is apt to be lowest during the daytime, when water absorption lags behind transpiration. During the night, when transpiration rates are very low, tissue water content will increase. This situation is shown in Figure 13-14. The water content of the stems, leaves, and roots of a plant growing in the field was measured on a clear, sunny day and the following night during the summer. The soil was watered thoroughly on the evening prior to the period during which measurements were made. Water content of each of the tissues reached a minimum during the afternoon hours but was the same at the end of the period of measurement (i.e., at 6 A.M.) as it was 24 hours earlier.

Of course, not all days during a growing season are hot, sunny days. On cool, cloudy days, or during rainy weather, when transpiration rates are very low, tissue water deficits may develop, but their magnitudes will be smaller than in Figure 13-14.

WATER STRESS AND ITS PHYSIOLOGICAL CONSEQUENCES

Higher plants are subjected to various adverse environmental conditions throughout their lifetimes. Included are periods of drought and low soil moisture content, cold temperatures, hot temperatures, desiccating atmospheric conditions, prolonged periods of cloudiness, high wind velocities, excessive soil salinity, air pollution, snow, frost, and so forth. These and other adverse environmental conditions may cause injury to plants, ranging from mild reduction in growth with negligible lasting effects to severe disorders or even death. The term *stress*, as used by plant physiologists, sometimes refers to the external environmental factors (e.g., drought, cold) capable of inducing a potentially injurious effect in a living organism but more often refers to the injurious effect itself (e.g., reduction in photosynthesis, reduction in the rate of growth, stomatal closure, wilting). Which of these two different but related meanings is intended generally is clear from the context.

Naturally growers of plants—farmers, horticulturists, agronomists, orchardists, foresters, etc.—are greatly interested in why stress develops in plants and also in the nature of plant adaptations for withstanding or avoiding the injurious effects of stress. The term *hardiness* generally is used to refer to the innate ability of plants to withstand or avoid injury from stress.

Our present concern is with some of the physiological effects associated with water stress in plants. Much research effort is presently devoted to the study of this subject. To discover the optimum degree of hydration for crop plants—and hence obtain maximum yields with minimum use of water—would have great practical benefits to agriculture.

The most frequent cause of water stress in plants is a suboptimal soil moisture supply, coupled with a rate of transpiration in excess of the rate of absorption of water by roots. But even when a plant is well watered, water stress may develop during the heat of the day in the growing season if water absorption by roots fails to keep pace with transpiration (see Figure 13-13 and related discussion). Water stress may also arise from other causes. When a plant is exposed to *slowly* decreasing temperatures near and below freezing, ice crystal formation begins in *extracellular* regions, where salt concentrations are very low. In this case, water will move out of cells and into the growing ice crystals because ice has a lower water potential than liquid water. With continued gradual loss of cellular water, water stress will become more and more severe. This kind of injury—from which many plants can recover if the temperature increases—probably occurs only infrequently under natural conditions. On the other hand, when a plant is exposed to *rapidly* decreasing temperatures near and below freezing, ice crystals will form *inside* cells. Such an occurrence is almost always lethal because protoplasm is likely to be ruptured. Cell damage under these conditions is irreversible. This type of freezing occurs frequently under natural conditions.

One of the consequences of water stress in a plant is a change in the structure and configuration of enzymes. As a result, enzyme activities change and metabolism rates are affected adversely. It is widely recognized that the leaf tissue content of proline (an amino acid), as well as abscisic acid (a plant growth substance known to induce stomatal closure), increases somewhat during water stress. Unfortunately, the basis of these effects is not yet well understood; they are grouped together under the heading "dehydration of protoplasm." A further complication in the study of plant water stress is the different response of different physiological processes to desiccation. Still another concerns the different responses of different species and varieties to the same degree of desiccation.

To illustrate the complexity of the matter, consider the results of a study by Boyer (1970a). Rates of *leaf enlargement* in sunflower, maize, and soybean were measured at various leaf water potentials. When leaf water potential dropped to −4 bars, leaf enlargement in sunflower was almost completely suppressed. In contrast to the marked sensitivity of sunflower, leaf enlargement in maize and soybean at −4 bars was 20 to 25% of the maximal rates. (Maximal rates of leaf enlargement occurred at about −1.5 to −2.5 bars in all three species.) When water potential dropped below −4 bars, leaf enlargement continued to decrease in maize and soybean and became almost zero at about −15 bars. In a separate study (Boyer, 1970b), it was found (in those species studied) that rates of *net photosynthesis* decreased as water potential decreased. The photosynthetic response to reduced leaf water potential was found to depend in most species on the tendency of stomata to close during desiccation. In a third study (Sung and Krieg, 1979), it was found that the rate of net photosynthesis and also the rate of translocation of photosynthate from leaves decreased as water potential decreased in cotton and sorghum.

The different responses of various physiological processes to a given degree of water deficit sometimes can be used advantageously in cropping practices. To give a well-known example, higher seed yields are obtained in certain crops (e.g., alfalfa) when irrigation practices promote slow, continuous growth during the entire production period; this type of growth can be achieved by applying small amounts of irrigation water at regular intervals. On the other hand, large amounts of irrigation water stimulate vegetative growth and are desired if alfalfa is to be grown for hay.

As a general rule, the adverse physiological effects of severe water stress become clear in most plant species when the water potential drops to about −8 bars. When water potential drops to about −14 to −15 bars, most physiological processes (e.g., enlargement growth, net photosynthesis) reach very low levels or cease entirely. In the range of water potential between −8 and −14 to −15 bars, deleterious effects of water deficit are due primarily to dehydration of protoplasm and to accompanying changes in the spatial relationship of protoplasmic macromolecules.

When desiccation progresses so far that tissue water potential drops below -15 bars, the question of survival becomes important. Yet considerable variability exists among plant species in their ability to tolerate low water potential. Many herbaceous plants can withstand water potentials in the range of −15 to −20 bars for only a day or two. On the other hand, desert shrubs rooted in dry soils may survive for many weeks when their water potential is in the range of −30 to −60 bars. Also, air-dried seeds with water potentials in the range of −60 to −100 bars and below remain viable for months and even years. The range of water potentials in different tissues was indicated in Figure 12-8.

Perhaps the most familiar visible manifestation of water stress is the drooping and sagging of plant tissues, especially leaves, referred to as *wilting*. Wilting is due to a change in elastic properties of cell walls when turgor pressure declines below a certain critical value. Whether wilting actually occurs, however, depends on the amount of supporting tissue. Some leaves with abundant lignified tissue may not wilt even when desiccation proceeds to the point of death.

Older leaves of herbaceous plants usually wilt before younger leaves, due to the tendency for the water remaining in a plant to be redistributed when water stress develops. The younger, more actively growing leaves will hold and attract the available water within the plant more readily than older leaves and therefore the latter are likely to wilt first.

Leaves of some herbaceous plants sometimes sag or droop slightly during the afternoon of a hot summer day and recover again at night even when the plant is growing in a soil with high moisture content. This "temporary" or "transient" wilting is associated with the diurnal fluctuation of water content (see Figure 13-14 and related discussion). On the other hand, "permanent" wilting refers to wilting caused by an actual lack of water in the soil in which a plant is rooted. A plant will not recover from permanent wilting unless water is added to the soil.

These two kinds of wilting can be better appreciated by examining Figure 13-15, in which leaf and soil water potentials are plotted during the period of one drying cycle. During the first several days, when soil moisture availability is high, tissue water potential declines slightly through the day and returns to a high level at night. If wilting occurs during these days, it would be referred to as temporary or transient wilting. But on the eighth and ninth days, leaf water potential does not rise as high as during the first seven days. And on the tenth day (see Figure 13-15) the plant reaches the point of permanent wilting. In most herbaceous plant species leaf water potential is then about −14 to −15 bars.

In conclusion, it may be useful to point out that different plants may have different means of coping with water stress. Some annual plants, for example, are able to *escape* periods of substantial water deficits in semiarid or arid environments by germinating, growing, and producing seed during periods when water is available; in these plants the life cycle is completed before the onset of drought conditions. Other plants are able to *avoid* water stress, perhaps by deep root systems that seek out all available soil water, or by leaves that orient themselves parallel to the sun's rays, so that the absorption of solar radiation is minimized and water loss is reduced. Finally there

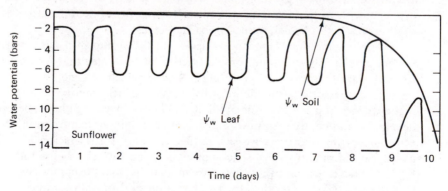

Figure 13-15 Schematic representation of changes in leaf water potential (ψ_w *leaf*) and soil water potential (ψ_w soil) for a sunflower plant rooted in initially wet soil ($\psi_w = 0$). During each day of the 10-day drying cycle, the same evaporative conditions are considered to prevail. The horizontal dashed line at the base of the diagram indicates alternate periods of day and night. After Boyer, 1968 (Fig. 6).

are plants that are able to *tolerate* water stress. Some of them have the ability to carry on a reduced rate of photosynthesis at extremely low levels of leaf water potential, in the range of −30 to −60 bars or even lower. In other plants in this last group, old leaves wither and fall off during drought conditions and newly formed leaves are smaller— and therefore have lower transpiration rates—than the old withered ones.

ROOT PRESSURE AND GUTTATION

Root pressure refers to positive hydrostatic pressure that sometimes develops in the xylem sap of roots. The phenomenon can be demonstrated readily in the laboratory. A short piece of rubber tubing is attached to the stump of a freshly decapitated, young, well-watered herbaceous plant plotted in a well-aerated soil. A glass tube is attached to the rubber tubing and secured in a vertical position. The level of xylem sap in the glass tube will rise slowly in a few hours.

The first recorded demonstration and measurement of root pressure were carried out by the Englishman Stephen Hales, an early experimental plant physiologist (Hales, 1727). Root pressures are usually in the range of +1 to +2 bars, but slightly higher values have been recorded.

Root pressure is an osmotic phenomenon. Transport of mineral ions across a root gives rise to a slightly higher solute concentration in xylem sap than in the external soil solution. The difference in solute potential between the external soil solution and the xylem sap in a root that exhibits root pressure is the driving force for the movement of water across the root.

Root pressures have been found to fluctuate in a rhythm of small amplitude with a period of about 24 hours. This diurnal rhythm persists even if a decapitated plant is placed in a growth chamber in which environmental conditions (e.g., temperature, light, humidity) are held constant. Rhythms in biological processes (e.g., root pressure) are due to rhythms in protoplasmic activity—and these are reflected in changes in the properties of protoplasmic membranes. But the fundamental nature of biological rhythms has yet to be determined (see Chapter 18).

The volume of xylem sap exuded from the stump of a detopped viable root system is relatively small and only a fraction of the water that would be lost by the (intact) plant by transpiration during the same period of time.

Root pressure is generally regarded as having little importance in the life of most higher plants. If it has any role at all, it is only in very young plants, before leaves are developed and before transpiration becomes a dominant feature in the overall water economy of the plant. As we saw earlier (p. 400), transpiration accounts for the absorption of water by most species of higher plants during their adult lives.

The development of root pressure in an intact plant becomes visibly manifested in guttation. *Guttation* refers to the exudation of droplets of liquid water from the margins and tips of leaves. Guttation water is exuded from groups of leaf cells called *hydathodes*. Typically a hydathode is an opening or pore in the leaf epidermis, around which are grouped several thin-walled parenchyma cells. Frequently the pore of a hydathode is an incompletely differentiated stoma incapable of opening and closing movements.

Guttation depends on root pressure. The development of root pressure in a plant leads to positive hydrostatic pressure in the xylem sap throughout the plant. Because water-conducting xylem elements of a vascular bundle terminate in a hydathode, xylem sap is forced to flow through the hydathode. Thus guttation water is exuded from the leaf.

To demonstrate guttation in the laboratory, a potted, well-watered, herbaceous plant, preferably one that is young and growing vigorously, is placed on a flat surface and covered with a bell jar. The rim of the bell jar is sealed (e.g., with petrolatum) to ensure that loss of water by transpiration is negligible. In a few hours guttation water will appear on the leaves.

Guttation water contains small amounts of both organic and inorganic solutes. The presence of organic solutes can be explained by leakage from cells that border on the xylem tissue. The inorganic solutes are mostly those absorbed by roots from the soil solution and carried passively to leaves in the upward-flowing xylem sap.

The concentration of inorganic solutes in guttation water is much smaller than in xylem sap that exudes from the rooted stump of a decapitated plant. This reduction in solute concentration is due to the fact that leaf cells absorb most of the dissolved salts passing through their xylem tissues. The small amounts of inorganic solutes present in guttation water represent the remnants of dissolved salts not removed during their passage through leaves. Thus leaf cells may be said to be very effective but less than perfect demineralizers.

BLEEDING PHENOMENA

Bleeding is a general term that refers to the slow exudation of a watery solution from an incision made in a plant tissue. Generally four major types of bleeding can be distinguished.

First, there is the exudation of xylem sap from an incision made in the root or stem of a young, well-watered plant rooted in a well-aerated soil. This type of bleeding is associated with the development of root pressures and was considered in the preceding section.

The second type of bleeding is attributed to local pressures developed within stems themselves. The most familiar example of this phenomenon is the flow of xylem sap in the sugar maple (*Acer saccharum*), a tree that grows in northeastern United States. When holes about 5 cm deep are drilled into the sapwood in late winter or early spring, before the leaves have expanded, sap will be forced through them. The environmental conditions most favorable for flow exist on days when temperatures alternate above and below freezing; the greatest flow rates are likely to occur during late afternoon. Marked fluctuations in temperature from day to night cause the xylem sap to rise and fall, somewhat in the fashion of mercury in a thermometer. The sap, a solution of about 2 to 3% sugars, is derived from starch stored in ray cells during the previous growing season. Starch is hydrolyzed to sugars during the winter, before the flow of sap begins.

The third type is also a result of local pressures developed in stems themselves. When the stems of certain species of palms, yuccas, and agaves are cut, a phloem sap containing up to 20% sugars is exuded.

In addition to exudations from incisions in xylem or phloem tissues, a fourth type of bleeding occurs when cuts are make in certain tissues of certain plants. A milky sap called latex is secreted in lactiferous ducts in the cortex and pith, as well as in the secondary tissues, of some plant species (e.g., the para rubber tree, members of the milkweed family, certain species of poppy). Also, some species of pine secrete sap into special cells referred to as ducts in stems; when pine sap exudes out of a tree trunk and comes in contact with air, it thickens into a substance called resin.

REFERENCES

ANDREWS, F. C. 1976. Colligative properties of simple solutions. *Science* 194:567–571.

APFEL, R. E. 1972. The tensile strength of liquids. *Sci. Amer.* 227:58–71.

ASTON, M. J. 1976. Variation of stomatal diffusive resistance with ambient humidity in sunflower. *Aust. J. Plant Physiol.* 3:489–501.

BANGE, G. G. J. 1953. On the quantitative explanation of stomatal transpiration. *Acta Bot. Néerl.* 2:254–297.

BORNMAN, C. H. 1972. *Welwitchia mirabilis*: paradox of the Namib Desert. *Endeavour* 31:95–99.

BOYER, J. S. 1968. Relationship of water potential to growth of leaves. *Plant Physiol.* 43:1056–1062.

BOYER, J. S. 1970a. Leaf enlargement and metabolic rates in corn, soybean, and sunflower at various leaf water potentials. *Plant Physiol.* 46:233–235.

BOYER, J. S. 1970b. Differing sensitivity of photosynthesis to low leaf water potentials in corn and soybean. *Plant Physiol.* 46:236–239.

CAILLOUX, M. 1972. Metabolism and the absorption of water by root hairs. *Can. J. Bot.* 50:557–573.

DITTRICH, P., and K. RASCHKE. 1977. Uptake and metabolism of carbohydrates by epidermal tissue. *Planta* 134:83–90.

EAMES, A. J., and L. H. MacDANIELS. 1925. *An Introduction to Plant Anatomy.* McGraw-Hill Book Company, New York.

ESAU, K. 1953. *Plant Anatomy.* John Wiley & Sons, New York.

GATES, D. M. 1966. Transpiration and energy exchange. *Quart. Rev. Biol.* 41:353–364.

GATES, D. M. 1968. Transpiration and leaf temperature. *Ann. Rev. Plant Physiol.* 19:211–238.

HALES, S. 1727. *Vegetable Statics.* W. & J. Innys and T. Woodward, London, Reissued in 1969. History of Science Library, Macdonald and Co., London.

HSIAO, T. C. 1976. Stomatal ion transport. Pages 195–221 *in* A. Pirson and M. H. Zimmermann, eds. *Encyclopedia of Plant Physiology*, new series, vol. 2B. Springer-Verlag, Berlin.

JARVIS, P. G., and T. A. MANSFIELD, eds. 1981. *Stomatal Physiology*. Cambridge University Press, New York.

JOHNSON, J. D. 1981. Two types of ventilated porometers compared on broadleaf and coniferous species. *Plant Physiol.* 68:506–508.

KNIGHT, D. H., T. J. FAHEY, S. W. RUNNING, A. T. HARRISON, and L. L. WALLACE. 1981. Transpiration from 100-yr-old lodgepole pine forests estimated with whole-tree potometers. *Ecology* 62:717–726.

KOZLOWSKI, T. T., ed. 1968–1981. *Water Deficits and Plant Growth*. 6 vols. Academic Press, New York.

KRAMER, P. J. 1937. The relation between rate of transpiration and rate of absorption of water in plants. *Am. J. Bot.* 24:10–15.

KRAMER, P. J. 1956. Roots as absorbing organs. Pages 188–214 *in* W. Ruhland, ed. *Handbuch der Pflanzenphysiologie*, vol. 3. Springer-Verlag, Berlin.

KRAMER, P. J. 1959. Transpiration and the water economy of plants. Pages 188–214 *in* F. C. Steward, ed. *Plant Physiology, a Treatise*, vol. 2. Academic Press, New York.

KRAMER, P. J. 1969. *Plant and Soil Water Relationships: a Modern Synthesis*. McGraw-Hill Book Company, New York.

KRIZEK, D. T., chairman. 1981. Adaptation to water stress in plants. Proceedings of a workshop, 1980. *HortScience* 17(1):1–38.

LEVITT, J. 1956. The physical nature of transpirational pull. *Plant Physiol.* 31:248–250.

MACKAY, J. F. G., and P. E. WEATHERLEY, 1973. The effects of transverse cuts through the stems of transpiring woody plants on water transport and stress in the leaves. *J. Expt. Bot.* 24:15–28.

MADHAVAN, S., and B. N. SMITH. 1982. Localization of ribulose bisphosphate carboxylase in the guard cells by an indirect, immunofluorescence technique. *Plant Physiol.* 69:273–277.

MEIDNER, H., and D. W. SHERIFF. 1976. *Water and Plants*. Blackie, Glasgow.

MILBURN, J. A. 1979. *Water Flow in Plants*. Longman, London.

MILLER, E. C. 1938. *Plant Physiology*, 2nd ed. McGraw-Hill Book Company, New York.

NOBEL, P. S., and T. L. HARTSOCK. 1979. Environmental influences on open stomates of a crassulacean acid metabolism plant, *Agave deserti*. *Plant Physiol.* 63:63–66.

OUTLAW, W. H., JR. 1980. A descriptive evaluation of quantitative histochemical methods based on pyridine nucleotides. *Ann. Rev. Plant Physiol.* 31:299–311.

OUTLAW, W. H., JR. 1981. Unique aspects of carbon metabolism in guard cells of *Vicia faba* L. Pages 4–18 *in* C. Rogers, ed. Proceedings of a Symposium of the Southern Section of the American Society of Plant Physiologists. Published by Houston Baptist University. Houston, Texas.

OUTLAW, W. H., JR., B. C. MAYNE, V. E. ZENGER, and J. MANCHESTER. 1981. Presence of both photosystems in guard cells of *Vicia faba* L. *Plant Physiol.* 67:12–16.

PALEG, L. G., and D. ASPINALL. 1981. *Physiology and Biochemistry of Drought Resistance in Plants*. Academic Press, New York.

PLUMB, R. C., and W. B. BRIDGMAN. 1972. Ascent of sap in trees. *Science* 176:1129–1131. Also see Technical comments: On the ascent of sap. *Science* 179:1248–1250 (1973).

PRESTON, R. D. 1958. Movement of water in higher plants. Pages 257–321 *in* A. Frey–Wyssling, ed. *Deformation and Flow in Biological Systems*. North-Holland Publishing Company, Amsterdam.

RASCHKE, K. 1975. Stomatal action. *Ann. Rev. Plant Physiol.* 26:309–340.

RASCHKE, K. 1979. Movements of stomata. Pages 383–411 *in* W. Haupt and M. E. Feinleib, eds. *Encyclopedia of Plant Physiology*, new series, vol. 7. Springer-Verlag, Berlin.

ROSENBERG, T. 1954. The concept and definition of active transport. *Symp. Soc. Expt. Biol.* 8:27–41.

SCHOLANDER, P. F. 1972. *Tensile water. Am. Scient.* 60:584–590.

SCHOLANDER, P. F., H. T. HAMMEL, E. D. BRADSTREET, and E. A. HEMMINGSEN. 1965. Sap pressure in vascular plants. *Science* 148:339–346.

SHARKEY, T. D., and K. RASCHKE. 1981. Separation and measurement of direct and indirect effects of light on stomata. *Plant Physiol.* 68:33–40.

SHARPE, P. J. H., and E. ZEIGER. 1981. Chemiosmotic hypothesis of ion transport in guard cells. Pages 47–64 *in* C. Rogers, ed. Proceedings of a Symposium of the Southern Section of the American Society of Plant Physiologists. Published by Houston Baptist University. Houston, Texas.

SHIVE, Jr., J. B. 1980. Leaf gas exchange: Does bulk flow occur? *What's New Plant Physiol.* 11:1–4.

SLATYER, R. O. 1967. *Plant-water Relationships.* Academic Press, New York.

SLAVIK, B. 1974. *Methods of Studying Plant Water Relations.* Springer-Verlag, New York.

SMART, R. E., and G. E. BINGHAM. 1974. Rapid estimates of relative water content. *Plant Physiol.* 53:258–260.

SUNG, F. J. M., and D. R. KRIEG. 1979. Relative sensitivity of photosynthetic assimilation and translocation of ^{14}carbon to water stress. *Plant Physiol.* 64:852–856.

THOMAS, D. A. 1975. Stomata. Pages 377–412 *in* D. A. Baker and J. L. Hall, eds. *Ion Transport in Plant Cells and Tissues.* North-Holland Publishing Co., Amsterdam.

TIBBITTS, T. W. 1979. Humidity and plants. *BioScience* 29:358–363.

TROUGHTON, J., and L. A. DONALDSON. 1972. *Probing Plant Structure.* McGraw-Hill Book Company, New York.

VAN KIRK, C. A., and K. RASCHKE. 1978. Release of malate from epidermal strips during stomatal closure. *Plant Physiol.* 61:474–475.

WILSON, C. C., W. R. BOGGESS, and P. J. KRAMER. 1953. Diurnal fluctuations in moisture content of some herbaceous plants. *Am. J. Bot.* 40:97–100.

WU, H., and P. J. H. SHARPE. 1981. Cell wall mechanics of guard cell motion. Pages 19–46 *in* C. Rogers, ed. Proceedings of a Symposium of the Southern Section of the American Society of Plant Physiologists. Published by Houston Baptist University. Houston, Texas.

ZEIGER, E., P. ARMOND, and A. MELIS. 1981. Fluorescence properties of guard cell chloroplasts. *Plant Physiol.* 67:17–20.

14

Plant Growth Substances: Structure and Physiological Effects

Botanists have been interested for years in understanding how plants achieve their distinctive form. During the mid-1800s the famous German plant physiologist Julius von Sachs suggested that plant form was attained through the action of specific "organ-forming" substances, such as "leaf-forming" substances, "root-forming" substances, and "flower-forming" substances. Early efforts to isolate and identify such substances were unsuccessful and Sachs' views were not strongly supported by other botanists of his time. A more widely accepted notion asserted that plant form resulted from the maintenance of specific levels of organic constituents, such as carbohydrates, soluble nitrogen, protein, or other substances. Support for this view came from studies on the chemical composition of plants at different stages of development when grown under various levels of inorganic nutrition, light, and temperature. During this period there was great interest in the isolation and identification of plant constituents. Such compounds as starch, sucrose, glucose, fructose, organic acids, amino acids, protein, and nucleic acids were found in plants and methods were developed for their analysis.

Later work indicates that the two points of view are not mutually exclusive. Instead of specific organ-forming substances, as suggested by Sachs, plants contain substances that trigger or initiate biochemical processes, which, in turn, ultimately lead to organ formation or to other aspects of growth. Several distinctly different groups of compounds are recognized as triggering substances: auxins, gibberellins, cytokinins, and phenolics. In addition, several specific compounds have been found—ethylene and abscisic acid. Structures of representative examples of these substances are shown in Figure 14-1. With the exception of ethylene, all have carbon ring structures of varying degrees of complexity. Indole-3-acetic acid and cytokinins contain nitrogen whereas the gibberellins, phenolics, and abscisic acid are composed of carbon,

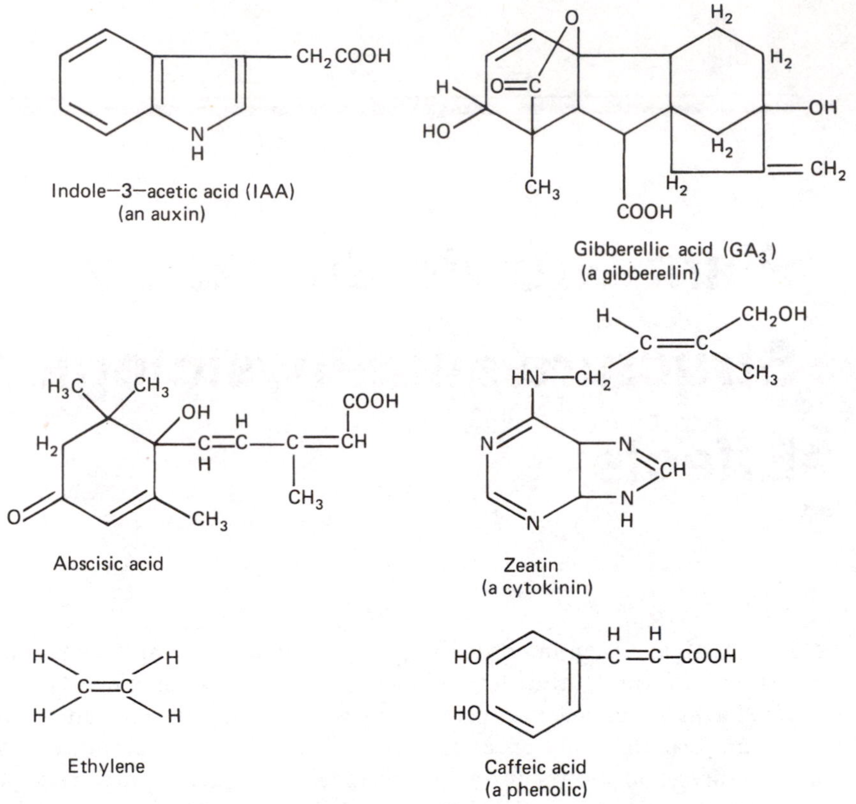

Figure 14-1   Some endogenous plant growth substances.

hydrogen, and oxygen. Probably still other substances with growth-regulating activity will be found in plants because extracts from plant tissues and organs have been isolated and found to modify growth and form. The chemical identity of the active principles in these extracts is as yet unknown.

The triggering substances, referred to as phytohormones, initiate biochemical reactions and changes in chemical composition within the plant. Accompanying the changes in chemical composition are changes in growth patterns that lead to the formation of roots, shoots, leaves, flowers, and other structural entities characteristic of the plant. Environmental factors, such as light and temperature, interact with the phytohormones and biochemical processes during growth and differentiation.

HORMONE CONCEPT

The term *hormone* was developed by animal physiologists to denote a specific organic substance, effective in low concentrations, which is synthesized by cells in one part of the organism and then transported to another part of the organism, where it produces a characteristic physiological response. Because animals have a well-organized circulatory system, much of the early work on hormones stressed their transportability. The circulatory system provided a means of collecting sufficient material for isolation and identification procedures and also enabled investigators to trace the sites of synthesis and specific targets of hormone action. Subsequent research has provided additional information, but the major elements of the animal hormone concept are illustrated in Figure 14-2(a).

Early plant physiologists were influenced by the concepts from animal physiology in their search for similar substances in plants. The properties and behavior of certain plant substances were thought to be sufficiently similar to an animal hormone to justify use of the term *plant hormone* or *phytohormone*. Recent studies indicate, however, that plant hormones do not fit the model of an animal hormone as just described. In plants all metabolically active cells appear capable of synthesizing hormones under appropriate conditions. The situation is different for animals, where well-defined groups of cells or tissues (glands) perform this function. Furthermore, although transport of plant hormones occurs in xylem and phloem tissue, in many instances, the hormone synthesized within a cell may modify metabolic processes

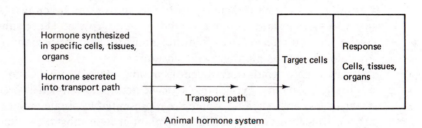

Animal hormone system

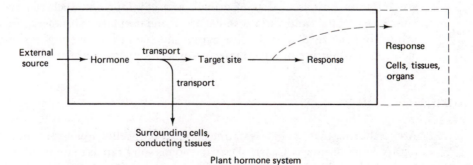

Plant hormone system

Figure 14-2 Models of animal and plant hormone systems. Both involve synthesis and transport of substances to target sites where a physiological response is elicited. In the animal model, the sites of synthesis and response are separate but connected by a transport path. In plants all elements—synthesis, transport, and response—may occur within a single cell.

within that cell or in surrounding cells. A model for plant hormone action is shown in Figure 14–2(b). It is a collapsed version of the animal hormone concept with the various elements occurring within a single cell. The important questions regarding either the plant or animal model are as follows. What initiates hormone synthesis? What is the pathway of synthesis? What are the hormone target sites? How is the response mediated (physical, biochemical)? One other complication arises when considering plants. Most plant physiological responses appear to involve the combined effects of several hormones rather than a single hormone. For this reason, it has been difficult to relate specific plant hormones with well-defined physiological responses.

The hormone concept developed by animal physiologists considered the active material to be organic in nature. There is no reason, however, for excluding inorganic ions from the definition. The general notion of a nonnutritive substance being active at low concentrations certainly can encompass inorganic ions as well as the more widely recognized organic molecules. The point to keep in mind is that hormones are substances that trigger changes in metabolic patterns that then lead to physiological responses. This is a broad view, but it removes some of the restraining effects of a narrowly defined concept on the development of new ideas of regulatory mechanisms in plants.

Many plant physiologists use the term *plant growth substance* rather than plant hormone, for it can include both the native (endogenous) and the synthetic (exogenous) substances found to modify plant growth. Those substances elaborated by the plant are referred to as phytohormones (or simply hormones) whereas the others are called synthetic plant growth substances.

Five major kinds of endogenous plant growth substances are present in plants—auxins, gibberellins, cytokinins, abscisic acid, and ethylene. With the exception of abscisic acid and ethylene, which are represented by single molecules in plants, there are multiple forms of the endogenous plant growth substances. Such a multiplicity of active molecules creates problems in attempting to understand the mechanisms of action of growth substances. Moreover, plants contain numerous other molecules that are active in various aspects of growth and development. Included are phenolics, vitamins, cyclitols, and various so-called secondary plant substances. This chapter is primarily concerned with the five major endogenous plant growth substances, but mention is also made of other molecules with biological activity.

AUXINS

About the time that Sachs was stating his ideas of specific organ-forming substances in plants, experiments by Charles Darwin and his son Francis in England started several lines of research that eventually confirmed many of Sachs' early ideas. Darwin was interested in plant movements, commonly called tropisms (movements in a non-predetermined manner), in response to such external stimuli as light (phototropism), gravity (geotropism), touch (thigmatropism), and chemicals (chemotropism). He also

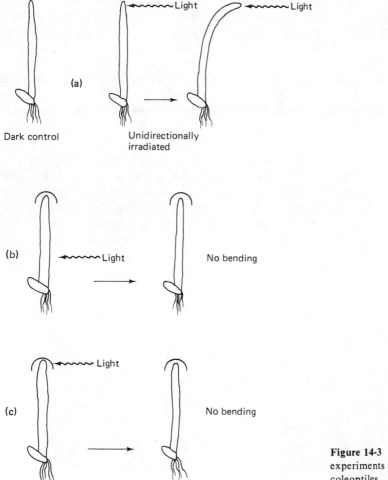

(a)

Dark control

Unidirectionally
irradiated

(b) Light

No bending

(c) Light

No bending

Figure 14-3 Some of Darwin's
experiments on phototropism in grass
coleoptiles.

investigated the phenomenon of circumnutation, the twining or nodding habit of
stems. The results of these studies were published in 1880 in a book entitled *The Power
of Movement in Plants*. In his studies on phototropism Darwin used the coleoptiles of
several grasses as test objects. When the seeds were germinated in darkness, the
coleoptiles grew straight [Figure 14-3(a), control]. If the tip of the coleoptile was
unilaterally irradiated, the coleoptile bent toward the light [Figure 14-3(a)]. Darwin
noted that if the base of the coleoptile [Figure 14-3(b)] rather than the tip was
irradiated, no bending occurred. Also, if the tip was covered with a light-tight cap
[Figure 14-3(c)], no bending occurred following unilateral illumination.

Darwin's observations were made in the 1870s, but it was not until the early 1900s that plant physiologists returned to the problem of phototropism. In 1907 Fitting showed that lateral incisions made below the tip of the coleoptile did not prevent the coleoptile from bending toward a light source. Boysen Jenson showed that if a coleoptile is decapitated and a layer of gelatin or agar is inserted between the tip and the coleoptile stump, the coleoptile continues to grow. If a water-impermeable substance, such as mica or lead foil, is inserted between the tip and coleoptile stump, however, no growth occurs. A. Paál demonstrated in 1918 that when an excised tip is replaced asymmetrically on one side of the coleoptile stump, growth is accelerated below the tip and bending occurs. Finally, in 1928 F. W. Went carried out a series of experiments demonstrating that coleoptile tips contain a substance capable of promoting the elongation of decapitated coleoptiles. Went used the coleoptiles of oat (*Avena sativa*) seedlings, but the coleoptiles of other grass seedlings (wheat, maize, etc.) have also been studied. He decapitated dark-grown coleoptiles [Figure 14-4(a)] and placed the tips in the dark on sheets of gelatin or agar [Figure 14-4(b)], where they remained for several hours. All subsequent manipulations were carried out in darkness. The coleoptile tips were removed from the agar sheets, after which the sheets were cut into small blocks [Figure 14-4(c)]. When such a block is placed on a decapitated coleoptile for several hours [Figure 14-4(d)], coleoptile growth is comparable to that of an intact coleoptile, demonstrating that the coleoptile tips contain a diffusible substance that is capable of stimulating the elongation of the basal sections of the coleoptiles. Went also showed that if a block of agar containing tip diffusate is placed

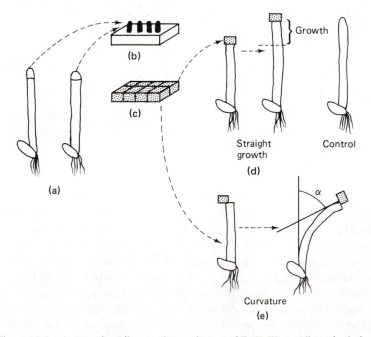

Figure 14-4 *Avena* coleoptile growth experiments of F. W. Went. All manipulations were carried out in darkness.

asymmetrically on a decapitated coleoptile stump for several hours [Figure 14-4(e)], the coleoptile bends as it grows, the degree of bending being proportional to the concentration of growth substance in the block. This observation is the basis of a widely used technique for measuring auxin activity—the Went *Avena* coleoptile test. Indoleacetic acid (IAA) has been isolated from coleoptile tips (1 μg isolated from a pooled sample of 10,000 coleoptile tips). Identifiable amounts of IAA have been isolated from only a few plants, but chromatographic evidence indicates that it is widely distributed in the plant kingdom.

Auxin is a general term used to denote substances that promote the elongation of coleoptile tissues, particularly when tested in the *Avena* coleoptile test or in several other bioassay techniques. Indoleacetic acid is an auxin that occurs naturally in plants.

Physiological Effects of Auxins

Indoleacetic acid and other auxins participate in a number of aspects of plant development, as shown in Figure 14-5. Some of the major effects are described briefly.

Cell enlargement Early studies on coleoptile growth showed that IAA and other auxins promote cell enlargement. Elongation of coleoptiles and stems occurs as a result of cell enlargement. An unequal distribution of IAA in stems and roots produces a pattern of differential cell enlargement, together with organ bending (geotropism, phototropism). Meristematic cells in callus tissue or in tissue cultures also enlarge under the influence of IAA.

Inhibition of lateral buds The development of axillary (lateral) buds is inhibited by IAA produced at the apical meristem and transported down the stem. If the source of auxin is removed by excising the apical meristem, the lateral buds are released from the inhibitory state and undergo development.

Leaf abscissioin Leaves separate from the stem following changes in the chemical and physical properties of cells in the abscission zone, a group of cells at the base of the petiole (Figure 14-5). The concentration of IAA in cells near or within the abscission zone appears to bear some relation to the abscission process.

Cambial activity Secondary growth of stems involves cell division in the cambium and the formation of xylem and phloem tissue. Auxins promote cell division within the cambial region.

Root growth As noted, cell enlargement is usually promoted by IAA. Over a range of concentrations the enlargement is proportional to the amount of IAA present. In roots, however, the usual effect of IAA is to inhibit cell enlargement, with only very low IAA concentrations promoting cell enlargement.

The effects of IAA on plant development just described present only a partial view of what actually occurs. The effects discussed are based on the assumption that

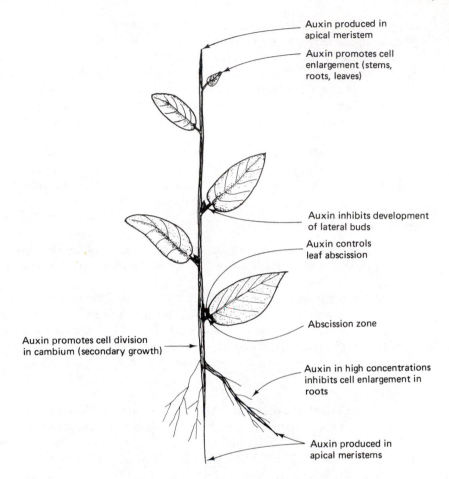

Figure 14-5   Some auxin effects in plants.

IAA is the sole phytohormone involved. It is now known that other plant hormones (gibberellins, cytokinins, abscisic acid, ethylene, and possibly others) participate with IAA in many of the physiological responses outlined here and shown schematically in Figure 14-5. Several of these interactions are described in later chapters.

Other Plant Indole Compounds

Plants contain a considerable number of indole compounds in addition to indoleacetic acid (Figure 14-6). As noted later, some of these indole compounds are probably intermediates in the biogenesis of IAA in plants. Others (methylene oxindole) are breakdown products of IAA catabolism. Such compounds as IAA-aspartic acid, IAA-*myo* inositol, and IAA-glucose appear to function as storage or transport forms of IAA. Most of these indole compounds do not elicit a growth response in plants

I Possible intermediates in IAA synthesis

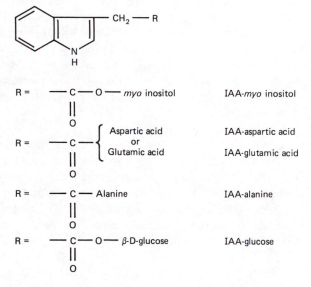

$R =$ $-CH_2C\overset{\overset{O}{\parallel}}{C}COOH$ In-pyruvic acid

$R =$ $-CH_2COOH$ In-ethanol

$R =$ $-CH_2\overset{\overset{O}{\parallel}}{C}H$ In-acetaldehyde

$R =$ $-CH_2CH\,NH_2COOH$ Tryptophan

$R =$ $-CH_2CH_2CH_2NH_2$ Tryptamine

II Product of IAA catabolism

Methylene oxindole

III Conjugated forms of IAA

$R =$ $-\overset{\overset{}{\underset{\underset{O}{\parallel}}{C}}}{C}-O-$ *myo* inositol   IAA-*myo* inositol

$R =$ $-\overset{}{\underset{\underset{O}{\parallel}}{C}}-\left\{ \begin{array}{l} \text{Aspartic acid} \\ \text{or} \\ \text{Glutamic acid} \end{array} \right.$ IAA-aspartic acid

                                                IAA-glutamic acid

$R =$ $-\overset{}{\underset{\underset{O}{\parallel}}{C}}-$ Alanine   IAA-alanine

$R =$ $-\overset{}{\underset{\underset{O}{\parallel}}{C}}-O-$ β-D-glucose   IAA-glucose

Figure 14-6   Some indole derivatives isolated and identified in plants.

unless they are converted to IAA by enzymic activity. Appropriate enzymes have been found in plants capable of converting the compounds to IAA. Because of the multiplicity of indole compounds and the presence of different enzymes for converting them to IAA or of destroying IAA, there are many opportunities for regulating the level of IAA at a particular active site.

Other Plant Auxins

As noted, the term auxin is used to designate substances that promote cell elongation, particularly in coleoptile tisse. Although indole-related auxins occur generally in plants, some plants have a nonindole auxin—phenylacetic acid.

$$\text{—CH}_2\text{COOH}$$

Tomato plants have been shown to be rich in phenylacetic acid, which appears to play a role similar to indoleacetic acid. With improvements in separation procedures and analytical techniques, other nonindole auxins will probably be found in other plants.

Synthetic Auxins

Soon after the recognition of the importance of IAA as a plant hormone, compounds similar in structure were synthesized and tested for biological activity. A few are shown in Figure 14-7. Among the first compounds studied were substituted indoles, such as indole-3-propionic acid (II) and indole-3-butyric acid (III). Both compounds are biologically active and commonly used as rooting hormones in horticultural work. Both have the same indole ring as IAA (I) and a terminal carboxyl group but differ in their side chains. If longer side chains are added to the indole ring, the compounds generally lack biological activity. Certain species of plants, however, possess enzymes capable of shortening the side chains and will convert the compounds to a biologically active molecule.

Compounds lacking the indole ring but retaining the acetic acid side chain present in IAA (I) are also biologically active. Naphthaleneacetic acid (IV) is such a compound and it is used as a rooting hormone for certain plants. Another biologically active synthetic auxin lacking the indole ring is 2,4-dichlorophenoxyacetic acid (V). This compound, known as 2,4-D, is a potent auxin and is used as a weed killer. It is probably the most widely used of the synthetic auxins in commercial crop production. The carbamate compound (VI) was developed for use as a fungicide but was also found to have auxin activity. Note that it lacks a ring structure but does possess an acetic acid side chain.

Hundreds of other molecules have been synthesized and tested for auxin activity. Only a few have been found to possess biological activity. What is necessary for a molecule to have auxin activity? By looking at the compounds depicted in Figure 14-7; we see that they are of approximately the same size and shape. If three-dimensional

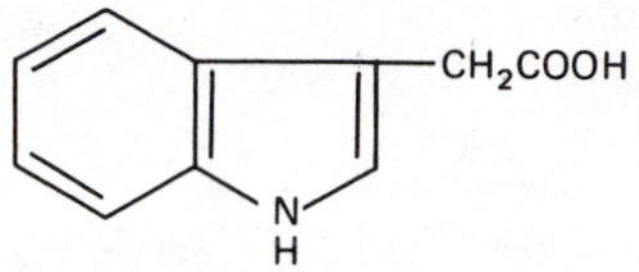

Indole-3-acetic acid
(I)

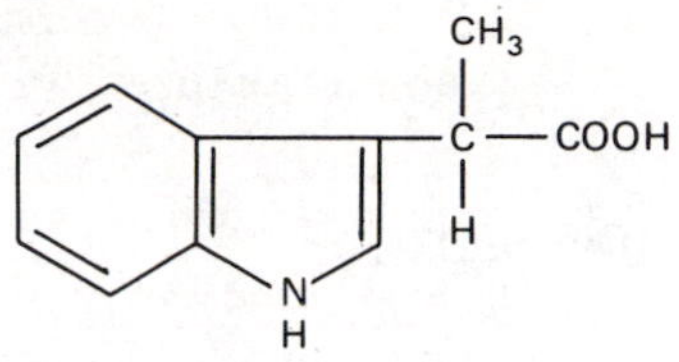

Indole-3-propionic acid
(II)

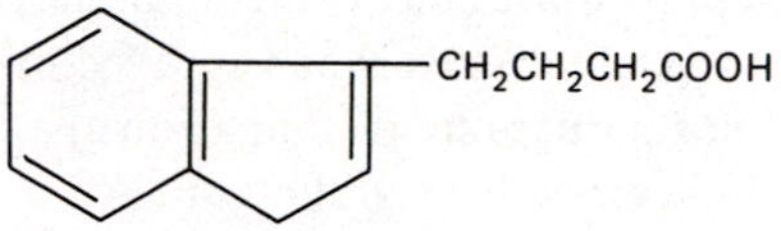

Indole-3-butyric acid
(III)

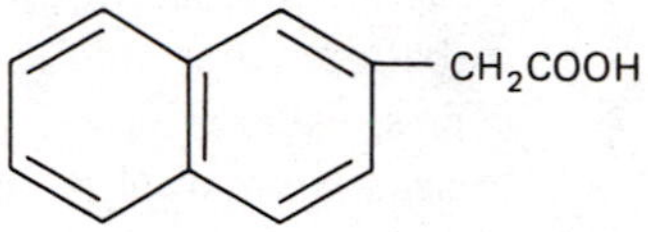

Naphthaleneacetic acid
(IV)

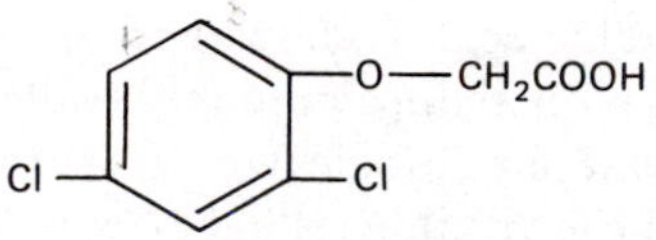

2,4-dichlorophenoxy-acetic acid
(V)

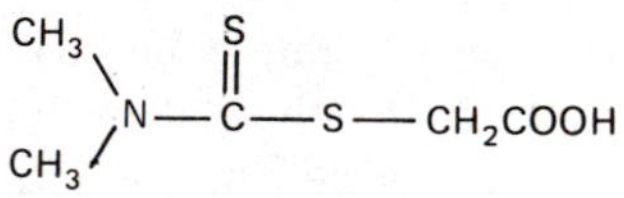

S-carboxymethyl-N,N-dimethyl dithio carbamate
(VI)

Figure 14-7 Structures of IAA and some synthetic auxins.

models are constructed, the size and shape relationship is shown even better. They seem to fill about the same space. Furthermore, their electronic structures are similar in that one portion is more electronegative than the other. This property may be important in orienting the molecule at some particular site (active site?) within the cell. In addition, other factors influence the activity of a synthetic auxin:

1. Ability of the substance to penetrate the cuticle or waxy epidermis.
2. Translocation within the plant.
3. Mode of inactivation within the plant (destruction, conjugation).
4. Interaction with other hormones.
5. Species of plant.
6. Stage of plant development.
7. External environment (temperature, radiation, moisture).

When all these factors are considered, it can be seen that the design and synthesis of a synthetic auxin is a very complex problem.

GIBBERELLINS

During the 1920s, when Went and other plant physiologists were elucidating the nature of IAA and its role as a plant growth substance, several Japanese workers were investigating a fungal disease of rice. The plant disorder was characterized in the seedling stage by elongated stems. It was determined that the fungus responsible for this disease is *Gibberella fujikuroi*. In 1926 E. Kurosawa demonstrated that when grown under sterile conditions in an appropriate culture medium, the fungus secretes a substance that induces symptoms of the rice disease when sprayed on healthy plants. The substance, called gibberellin A, also induces stem elongation of a number of other plants when applied as a spray. In 1938 crystalline gibberellin A was isolated by the Japanese workers from sterile culture filtrates of *Gibberella fujikuroi*.

The Japanese work attracted little outside attention until the conclusion of World War II, when several teams of scientists from Great Britain and the United States visited Japan and became aware of the gibberellin work. Following intensive studies in all three countries, it was found that gibberellin A is actually a mixture of at least six gibberellins referred to as GA_1, GA_2, GA_3, GA_4, GA_7, and GA_9. The structure of gibberellin A_3, also called gibberellic acid, is shown in Figure 14-1. The other gibberellins have similar structures. The gibberellin commonly available is GA_3, and it is the gibberellin most widely used in plant physiological research. In discussing the role of these substances in plant growth, the term gibberellin or GA will be used to designate the chemically known gibberellins (GA_1, GA_3, GA_7, etc.) whereas substances that behave biologically as gibberellins but whose structure is not known will be referred to as gibberellinlike compounds.

Gibberellins in Green Plants

Early studies of the physiological effects of GA on plant development were carried out with material isolated from culture filtrates of *Gibberella fujikuroi*. Industrial mycologists experienced in producing fungal metabolites, such as penicillin, streptomycin, and other antibiotics, developed procedures for turning out large quantities of gibberellins. Depending on the particular strain of *Gibberella* used, several different GAs are produced. In the late 1950s sizable quantities of GA_3 were released to plant scientists for experimental purposes and from these studies it became apparent that the gibberellins have a broad range of effects on plant growth and development.

With the development of new analytical procedures, many plant extracts were found to possess gibberellinlike activity. Subsequent studies showed that green plants contain gibberellins, some identical with those isolated from *Gibberella fujikuroi* and others with new chemical structures. At present, over 50 GAs are known and over 40 are known to occur in green plants. GA_1, GA_{3-8}, and GA_{17-20} occur rather commonly; the others are found only in certain plants. Thus gibberellins are not simply fungal

metabolites with interesting physiological effects on green plants; instead they are a
class of endogenous plant growth substances. They are present in a variety of organs
and tissues—roots, shoots, buds, leaves, floral apices, root nodules, fruits, and callus
tissues.

Physiological Effects of Gibberellins

Many plants respond to applications of GA by a marked increase in stem length.
Figure 14-8 shows the effects of weekly applications of gibberellic acid (GA_3) on the
growth of cabbage plants. The effect is primarily one of internode elongation, with the
internode cells increasing both in size and number. Brian and Hemming noted that GA
had a differential growth effect on normal and dwarf strains of plants. When dwarf
peas are sprayed with GA, the internodes elongate and the plants assume the
appearance of normal tall varieties. The tall varieties of peas do not respond to
gibberellins. Maize shows a similar response. There are about 20 known genetic dwarf
strains of maize, half of which respond to applications of exogenous GA by an increase
in plant height. It is not clear why all the dwarf maize strains do not respond to GA.
There is no reason to believe that the dwarf habit is solely dependent on GA
relationships and those that do not respond to GA may have a different biochemical
basis for their dwarf habit.

In addition to their effects on stem elongation, gibberellins increase the size of
leaves of a number of different plants when applied as a spray to intact plants.
Similarly, the size of flower blossoms is increased in camellias, geraniums, and a
number of other plants. The size of some fruits increases following a spray application
of GA; several grape varieties are routinely treated to increase fruit size. Gibberellins
also induce parthenocarpic fruit set in a number of plants. Such fruits are commonly
seedless.

Besides affecting organ size, gibberellins influence other plant responses. Many
plants require a period of low temperature (2 to 4° C) followed by long days before they
can be induced to flower. If the low temperature is not received, stem elongation
(referred to as bolting) is inhibited and a rosette of leaves is formed. When flowering is
induced, the stem elongates. Gibberellin can substitute for low temperature in a
number of these plants and induce flowering.

It has also been observed that the dormancy of some seeds and buds is broken if
they are treated with gibberellins. As will be shown later (Chapter 16), the level of
endogenous gibberellin increases in some seeds, leading eventually to germination.
The application of exogenous gibberellin promotes the germination of such seeds
where dormancy apparently is due to a low level of endogenous gibberellin. A similar
dormancy mechanism has been observed in some buds.

Many physiological responses to gibberellins involve both cell division and an
increase in cell size. In these respects, gibberellin effects are similar to those induced by
auxin treatment. One difference, however, is that gibberellins are more effective when
applied to intact plants, whereas the major auxin effects are noted on excised organs—
coleoptiles, internodes, roots and so forth.

Figure 14-8 Response of cabbage plant to gibberellic acid, 100 μg applied to stem apex weekly. (Courtesy S. H. Wittwer, Michigan State University).

Forms of Gibberellins in Plants

As stated previously, over 50 different GAs have been isolated and identified. Generally a particular plant species will possess only a few different gibberellins. Metzger and Zeevaart (1980) found six GAs in spinach shoots—GA_{17}, GA_{19}, GA_{20}, GA_{29}, GA_{44}, and GA_{53}. The possibility of other GAs being present was not ruled out, but they would occur in very low concentrations. The same array of GAs occur in immature seeds of *Vicia faba*. In maturing seeds of *Pisum sativum* cv Progress No. 9, Sponsel and MacMillan (1978) found seven GAs—GA_9, GA_{17}, GA_{20}, GA_{29}, GA_{38}, GA_{44}, and GA_{51}. As the seeds matured, the GAs first increased and then decreased in concentration and at maturity no free acidic GAs could be found. A GAlike structure, which they suggested might represent an end product in GA catabolism, was isolated and identified, however. The molecule had no GA activity.

In addition to free gibberellins, several forms of bound or conjugated forms of gibberellin have been identified in plant tissues. Two such forms are shown in Figure 14-9. In both, glucose is linked to GA—in one case through a hydroxyl group (GA_1 glucoside) and in the other through a carboxyl group (GA_4 glucosyl ester). It is not clear whether these conjugated GAs function as transport forms of GA or whether they represent storage forms. They might serve both functions.

Why plants should have such an array of gibberellins is not known. Probably a particular plant may only possess a few different GAs and they appear to change as the plant matures. The differences between the forms of GA are not great. In Figure 14-9 note that the difference between GA_1 and GA_4 is the presence of a hydroxyl group (—OH) at position 13 in GA_1 (see numbered positions in GA_4, Figure 14-9, top). The multiplicity of GAs may be an artifact related to the procedures necessary to extract

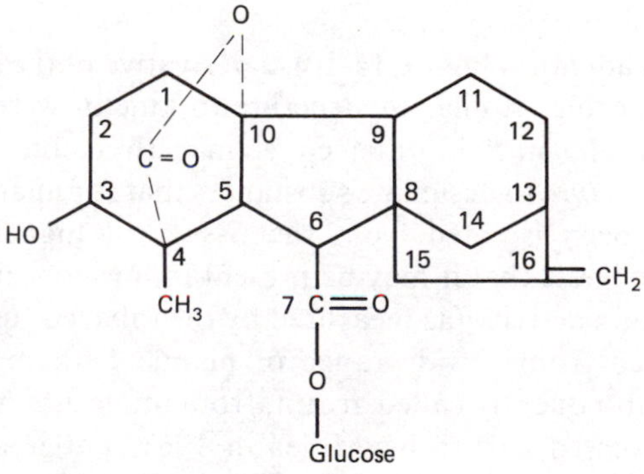

GA_4 glucosyl ester

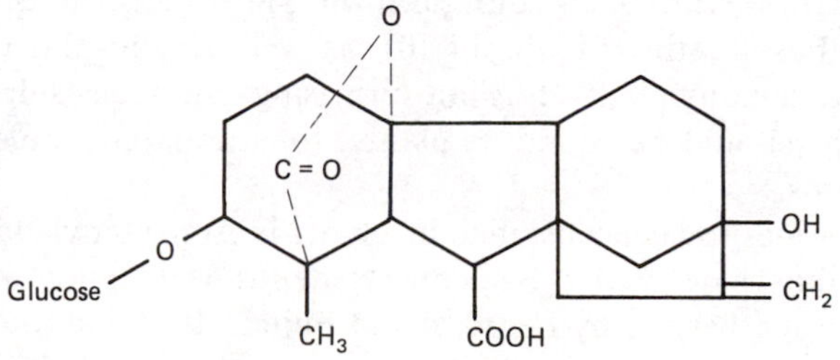

GA_1 glucoside

Figure 14-9 Gibberellin conjugates. In the GA_4 conjugate (top) glucose is esterified with a carboxyl group of GA. In the GA_1 conjugate (bottom) glucose forms a glycosidic bond through a hydroxyl group of GA.

and purify tissues in preparation for analysis. There may only be a few biologically active GAs. Further research is needed to ascertain the biologically active forms of GA in plant tissues and the relative activities of the free GAs as against the so-called bound or conjugated GAs.

CYTOKININS

During investigations of the nutritional requirements of callus tissue cultures F. Skoog and C. O. Miller discovered a substance that stimulated cell division. They used callus tissue derived from tobacco stem pith cultured on a basal medium containing inorganic ions, sucrose, vitamins, and glycine. The callus cultures grew slowly on the basal medium, but growth could be promoted by adding additional materials. Indole-3-acetic acid provided a brief growth stimulation (cell enlargement), but it was not maintained. Coconut milk and yeast extract provided the stimulus for rapid, sustained growth of the callus tissue when added with IAA to the basal medium. Other natural products were also examined for their ability to sustain growth in the callus cultures. Nucleic acids, especially RNA, were found to be exceptionally rich sources of growth-stimulatory materials. Further studies led to the isolation, identification, and chemical synthesis of the active fraction. The substance was called *kinetin* because of its promoting effect on cell division (cytokinesis).

Kinetin is N^6-(furfuryl) adenine (Figure 14-10), a derivative of the nitrogen base adenine. Other synthetic molecules similar in structure to kinetin were found to be effective in stimulating cell division in tobacco stem pith callus tissue. Plant physiologists use the term *cytokinin* to designate substances that stimulate cell division in plants. Kinetin has not been isolated from the tissues of higher plants, but chromatographic evidence suggests that it may be present in low concentrations.

Substances with cytokinin activity (as measured by the tobacco stem pith callus technique) have been isolated from a wide range of plants. Letham isolated and characterized a cytokinin, subsequently called zeatin, from immature maize kernels. This substance had been isolated earlier by Miller in 1961, but it had not been chemically identified. Zeatin (Figure 14-10) is a N^6-substituted adenine found in RNA hydrolysates from peas, spinach, wheat germ, cotton ovules, potato tubers, and other plants. The *t*RNA fraction of RNA is especially rich in zeatin. As shown in Figure 14-10, zeatin can assume either the trans or cis configuration. The trans form seems to be more prevalent. Also, ribosylzeatin (Figure 14-10), as well as phosphorylated ribosylzeatin, has been found in many plants. It is not surprising that nucleosides and nucleotides of the cytokinins should be found in plants, for the parent molecule, adenine, exists in similar forms.

Another cytokinin of widespread occurrence in plants is isopentenyladenine, which is found with the ribosyl derivative isopentenyladenosine (Figure 14-10). Isopentenyladenine had been synthesized by Leonard and Fujii in 1961 and found to have cytokinin activity (tobacco stem pith assay) before it was isolated and identified

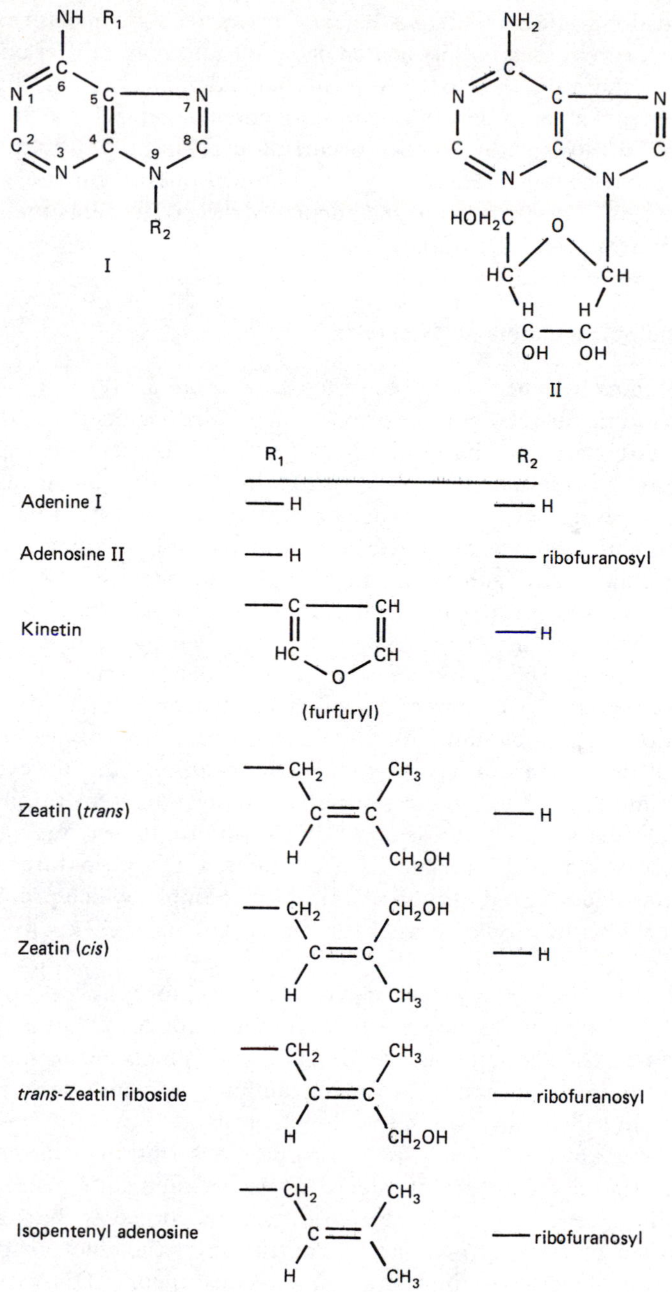

Figure 14-10 Structures of adenine, kinetin, and some native cytokinins.

from plants. It is a constituent of specific transfer RNAs (tRNAs) and has also been found along with isopentenyladenosine in the free state in plant extracts. From Figure 14-10 it can be seen that isopentenyladenine and zeatin differ only in the structure of the side chain attached to the N^6 position of adenine. There is a hydroxyl group on the isopentenyl group of zeatin whereas isopentenyladenine lacks the hydroxyl group. As far as is known, all naturally occurring cytokinins possess the basic structure of isopentenyladenine. There may be modifications of the isopentenyl side chain or addition of other groups to the adenine structure (such as the addition of ribose at position 9 in ribosylzeatin).

Physiological Effects of Cytokinins

Cytokinins have been found to influence a broad array of physiological processes in plants. A major activity is the promotion of cell division. It was this activity that led to their discovery and the promotion of cell division remains a prime requisite for a substance to be classified as a cytokinin. Cell division in many root meristems, however, is inhibited by an exogenous application of cytokinin. Both the stimulatory and inhibitory effects of cytokinins on cell division apparently require the presence of other plant growth substances, particularly auxins. Just how varying levels of auxins and cytokinins control the promotion or inhibition of cell division is not known.

Cytokinins also participate in the orderly development of embryos during seed development. The liquid endosperm of coconuts (coconut milk) has long been recognized as a rich source of growth substances necessary for embryonic development. Although coconut milk contains a number of active materials (see Chapter 16), one of the substances has been identified as an endogenous cytokinin. Here, again, cytokinin appears to interact with other endogenous growth substances.

Under some conditions cytokinins enhance the expansion of cells in leaf disks and cotyledons. These cells are considered to be mature and under ordinary circumstances do not expand. Auxins, for example, which promote the expansion of parenchymatous cells, have little influence on the expansion of leaf disks or cotyledons.

Cytokinins delay the breakdown of chlorophyll in detached leaves. The biochemical activities in detached leaves are characterized by catabolic or degradative processes. Similar activities are found in intact plants during the senescence of leaves, fruits, and other organs. Cytokinins inhibit these degradative reactions in detached leaves and slow down senescence in intact plants.

It is obvious from the preceding brief description of the action of cytokinins in plants that a wide array of effects has been noted. It is believed that these diverse physiological effects share some basic initial reaction with the subsequent events being modified by other growth factors, nutrients, or the environment. The similarity in structure of most cytokinins to adenine, a constituent of DNA and RNA, suggests that the basic effect of a cytokinin might be at the level of protein synthesis.

Synthetic Cytokinins

During studies on the isolation and identification of kinetin it was found that a number of substituted adenine compounds had cytokinin activity in the growth of tobacco stem pith callus. The structures of several of these molecules are shown in Figure 14-11. 6-Benzyladenine is similar in structure to kinetin (Figure 14-10) and is very active in promoting callus growth. The isomeric form, 1-benzyladenine (Figure 14-11), has limited activity. It is believed that 1-benzyladenine must first be converted to 6-benzyladenine by enzymic activity before it acquires cytokinin activity.

Skoog, Leonard, and associates have synthesized numerous derivatives of adenine and many are active as cytokinins in the growth of tobacco stem pith callus and other callus tissues. Those derivatives substituted in the 6 position (as 6-benzyladenine) are the most active cytokinins. Molecules substituted in other positions of the adenine nucleus (Figure 14-11) may be converted to the 6 isomer by enzymic activity and thereby achieve some cytokinin activity.

Benzimadazole (Figure 14-11) and adenine both have limited cytokinin activity. Benzimadazole is similar in structure to adenine and both lack the substituent groups at position 6 found in zeatin, isopentenyladenine, and other cytokinins. It is possible that both adenine and benzimadazole are not in themselves active but function only after being converted to a 6-substituted molecule. It is known, for example, that adenine is converted to isopentenyladenosine during the synthesis of *t*RNA.

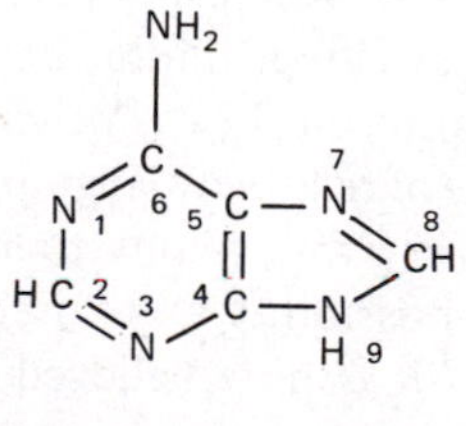
Adenine
(6-Aminopurine)

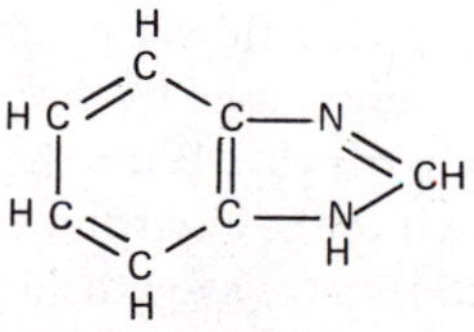
Benzimadazole

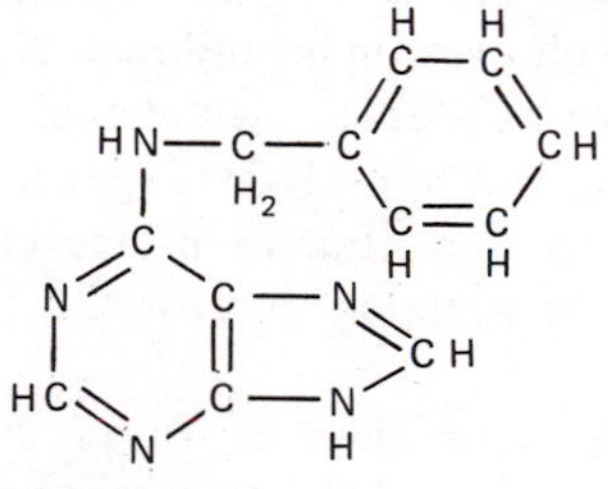
6-Benzyladenine
(6-Benzylaminopurine)

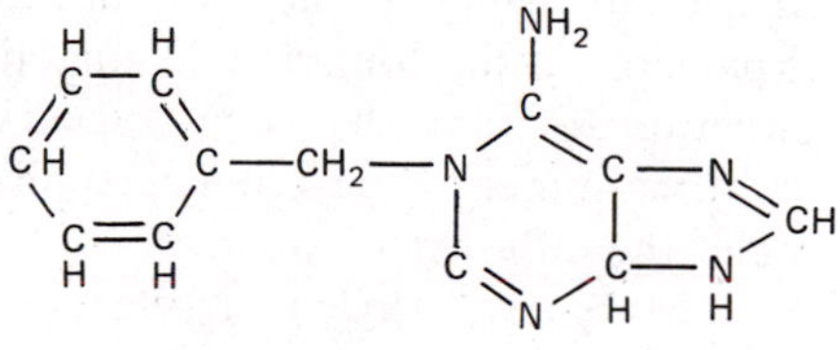
1-Benzyladenine

Figure 14-11 Some synthetic cytokinins.

ETHYLENE

The physiological effects of ethylene on plant growth have been recognized for 75 years. Ethylene is a volatile gas formed by the incomplete combustion of carbon-rich compounds, such as coal, petroleum, and natural gas. It is a component of smoke, automobile exhaust, and other industrial gases. Soon after the introduction of illuminating gas (produced from coal) for home and street lighting, evidence of plant damage was observed. Leaf abscission, abnormal curling of leaf blade and petiole, flower petal discoloration, stem swelling, inhibition of stem elongation, and inhibition of root growth were some of the symptoms of a disorder that was finally traced to the presence of ethylene.

Subsequent physiological studies led to the discovery that plants produce ethylene by metabolic processes during growth and development. Ripening fruits, in particular, synthesize quantities of ethylene that build up to rather high concentrations within intracellular spaces of fruit tissue. Ethylene is also produced in other tissues and organs, such as flowers, leaves, stems, roots, tubers, and seeds. The amount of ethylene normally present in tissues is quite small, usually less than 0.1 ppm, but it is probable that locally high concentrations may develop at certain times during growth and development.

Physiological Effects of Ethylene

As noted, ethylene is responsible for a number of growth responses in plants, including leaf bending (epinasty), leaf abscission, stem swelling, inhibition of stem and root growth, fruit ripening, and flower petal discoloration. It was originally believed that these growth effects were caused by the presence of relatively large amounts of ethylene in the external environment. With the discovery that plants themselves synthesize ethylene and secrete it to the environment, the possibility was raised that endogenous ethylene may function as a growth hormone. It is now believed that ethylene is a natural plant growth hormone and it has been shown to participate in many different physiological processes.

From Figure 14-12 it can be seen that ethylene modifies the growth of etiolated seedlings in several different ways. Stem elongation is inhibited. Normally the cells in such plants elongate under the influence of indole-3-acetic acid. Ethylene prevents expansion in the longitudinal direction and stimulates expansion in a transverse direction so that the stem appears swollen. Furthermore, ethylene induces a modification of the geotropic response in etiolated stem tissue. This process is also under the influence of auxin.

Moreover, ethylene accelerates the abscission of leaves, stems, flowers, and fruits. There is an interaction with auxins and the metabolic processes accompanying senescence. Senescence is discussed in Chapter 18.

Many practical aspects of fruit storage involve ethylene. Ethylene triggers ripening in many fruits and during fruit ripening relatively large quantities of ethylene

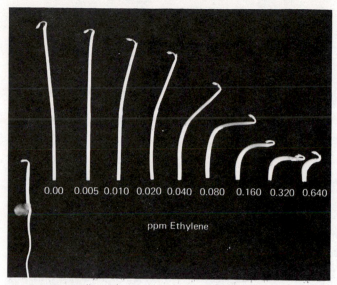

0.00   0.005   0.010   0.020   0.040   0.080   0.160   0.320   0.640

ppm Ethylene

Initial size (3 days)

Figure 14-12   The effect of various concentrations of ethylene on the growth of etiolated pea seedlings. Three-day-old etiolated plants were treated for 48 hours at different C_2H_4 concentrations in a flowthrough system that allowed C_2H_4 to pass continuously over the tissues. (Courtesy J. D. Goeschl, Texas A & M University).

are formed and released to the surrounding air. The ripening process can be delayed by removing ethylene from the vicinity of developing fruits or by maintaining a low oxygen level around the fruit. Such fruits as oranges, lemons, grapefruits, melons, apples, bananas, and avocados can be maintained for considerable periods of time in storage by manipulating the gas storage conditions. Fruits maintained under controlled storage conditions can be induced to ripen at any time by changing the levels of ethylene or oxygen.

ABSCISIC ACID

The problem of how plants retain or shed leaves, stems, flowers, and fruits is one of practical and theoretical interest. Frequently ripening fruits drop from the plant before they can be harvested. In other instances, it may be desirable to remove leaves or fruits before the normal abscission processes develop. Fruit trees often set more fruits than the tree can nourish and ripen and thinning procedures are needed. In harvesting cotton with a mechanical picker, it is desirable that the plants be defoliated before the ripe fruits (cotton bolls) are picked. These few examples show some of the practical interests in organ abscission. Of more general interest is the fact that the growth and form of a plant depend on the orderly shedding of organs.

Both IAA and ethylene are known to be involved in certain aspects of organ abscission and many investigators have studied the role of these phytohormones in abscission. In 1955 Osborne found that senescent leaves contained a diffusible

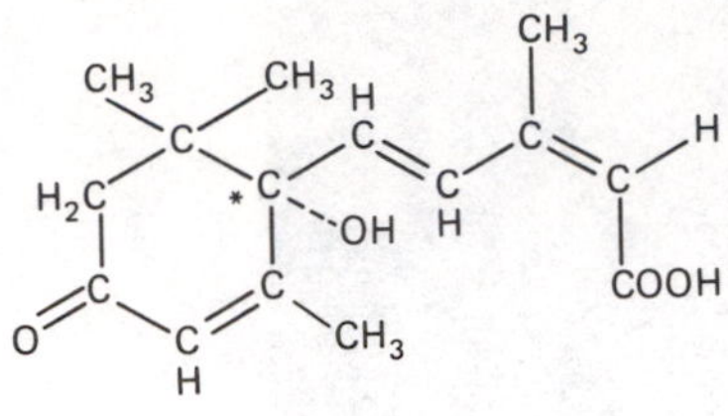

(+) — Abscisic acid, ABA

(Abscisin II, Dormin)

Figure 14-13 Structure of (S)-abscisic acid. The structure is viewed as follows. The ring structure lies in the plane of the page whereas the two groups around the asymmetric carbon atom (C^*) lie above and behind the plane of the page. The designation (S) indicates that the —OH group is above the plane of the page and the side chain lies behind the plane of the page.

abscission-accelerating substance with chemical properties quite different from IAA or any of the other known grown substances. Carns, Addicott, and coworkers isolated several abscission-accelerating substances from cotton plants, which they named abscisin I and abscisin II. Abscisin I was isolated from the burrs of mature cotton fruit whereas abscisin II was isolated from immature cotton fruits.

Additional information concerning abscission-accelerating substances and their role in plants came from several unexpected sources. Wareing and associates, during the course of studies on bud dormancy in woody plants, isolated a dormancy-inducing substance from the leaves of *Acer*, which they termed *dormin*. When dormin is applied to the leaves of many woody plants growing under vegetative conditions, bud dormancy is induced. Another group of investigators isolated a substance from lupins that was involved in the abscission of fruit pods. It was subsequently found that dormin and the lupin abscission factor are identical to abscisin II.

The chemical structure of abscisin II shown in Figure 14-13 was proposed by Ohkuma and colleagues in 1965. The structure was confirmed by Cornforth and associates, who also synthesized the material and showed dormin and abscisin II to be identical. It was decided in 1967 that abscisin II-dormin should be named abscisic acid (ABA). It is seen in Figure 14-13 that ABA has an asymmetric carbon atom, which means that two optical isomers are possible. The naturally occurring isomer of ABA has the configuration shown in Figure 14-13 and is referred to as (+)-abscisic acid. Abscisic acid prepared by chemical synthesis is a mixture of (+)-abscisic acid and (−)-abscisic acid and is designated as (±)-abscisic acid. Both of the isomers, (+)-ABA and (−)-ABA, and the mixture (±)-ABA are active in a number of biological tests. In some instances, however, the unnatural isomer (−)-ABA is not biologically active. Synthetic abscisic acid, (±)-ABA, is the usual form of the growth substance available for physiological studies, for the natural isomer, (+)-ABA, is difficult to isolate and purify in sufficient amounts for experimental purposes.

Physiological Effects of Abscisic Acid

The original studies on abscisic acid were directed toward leaf and fruit abscission and dormancy. With the realization that a specific substance, ABA, is involved in these physiological processes and that ABA occurs widely in plant tissues and organs, the role of ABA in plant growth and development has assumed new significance. ABA interacts with other plant growth substances, usually in an inhibitory manner. In many instances, the inhibitory effects of ABA can be overcome by adding more of some other plant growth substance. The promotive effect of IAA on *Avena* coleoptile curvature or on coleoptile straight growth, for instance, is prevented by ABA. If further amounts of IAA are provided, the inhibitory effect of ABA is overcome. In other growth responses inhibited by ABA—lettuce seed germination and growth of the aquatic plant *Lemna*—IAA does not reverse the inhibition. In these situations, growth substances other than IAA may be more directly involved.

Abscisic acid also interacts in an interesting fashion with gibberellins. The GA-induced synthesis of α-amylase and other hydrolytic enzymes in barley aleurone cells is inhibited by ABA. This inhibition can be reversed by increasing the amount of GA supplied.

Abscisic acid accumulates in many seeds during maturation and apparently contributes to seed dormancy. In seeds that require stratification (see Chapter 16) ABA decreases during stratification whereas GA increases. Also, ABA concentrations are high in dormant buds and ABA decreases as bud dormancy is broken.

A quite different role of ABA has been noted in plants growing under stress conditions—drought, flooding, mineral deficiency, injury. Under such conditions the leaves are found to contain high concentrations of ABA. In moisture stress leaves respond by loss of turgor (wilting), accompanied by stomatal closure. ABA accumulates quickly in leaves with the onset of moisture stress, and if the plants are watered, they regain normal turgor and ABA concentrations decrease. ABA appears to be synthesized de novo in the leaves during moisture stress and then destroyed or inactivated when the moisture stress is relieved.

It has been pointed out by Thimann and coworkers that ABA seems to be involved in senescence. The onset of senescence (see Chapter 18) is correlated with stomatal closure. The ABA content of aging leaves increases markedly as senescence is initiated. Indicators of senescence are loss of chlorophyll, decrease in rate of photosynthesis, changes in nucleic acid and protein catabolism. In this view, ABA can be considered a chemical signal initiating the senescence phase of plant development.

Compounds Related to Abscisic Acid

Several compounds similar in structure to ABA have been found in plants. Xanthoxin (Figure 14-14) is formed by photooxidation of carotenoids and has many of the inhibitory effects of ABA. In some plants it has been observed that xanthoxin is a precursor of ABA. The significance of this synthetic pathway is not understood at this time. Phaseic acid (Figure 14-14) is formed during the oxidative catabolism of ABA and lacks biological activity. It is possible that phaseic acid plays a role in the maintenance of physiological levels of ABA in tissues.

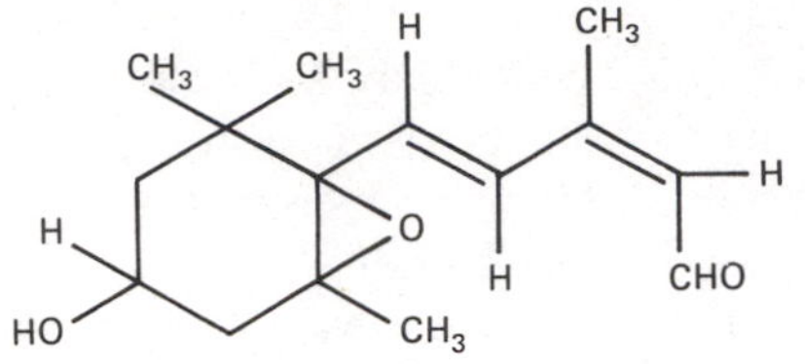

Xanthoxin

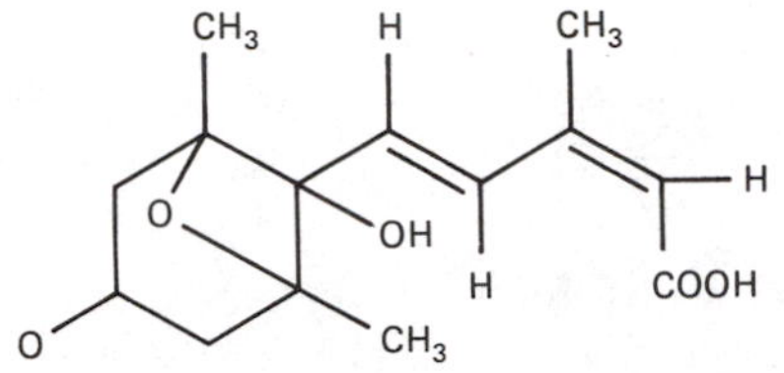

Phaseic acid

Figure 14-14 Some naturally occurring compounds with structures similar to ABA. Xanthoxin is formed by the breakdown of carotenoids. Phaseic acid is a product of the oxidative catabolism of ABA.

OTHER PLANT CONSTITUENTS WITH BIOLOGICAL ACTIVITY

In addition to auxins, gibberellins, cytokinins, abscisic acid, and ethylene, plants contain numerous other organic molecules. Many of these compounds have been shown to have growth regulating activity when tested on a variety of intact plants, organs, tissues, or isolated cells. Some have been shown to increase the yields of field crops and vegetables. Whether these compounds are required for orderly growth and developmental processes is not known, but the subject is an area of intense research interest at present. Only a few of the many interesting organic molecules found to have biological activity are described here.

Phenolics

Plants contain a large number of compounds collectively referred to as *phenolics*. They are derivatives of phenol and are quite diverse in their chemical structure (Figure 14-15), ranging from such compounds as catechol, caffeic acid, and aesculin to anthocyanidins and other complex polyphenolic compounds. Many polyphenols are brightly hued pigments (blue, red, yellow, orange) and are responsible for the coloration in flower petals, leaves, and other tissues. Phenolics frequently occur in plants coupled with sugar molecules (anthocyanidin + glucose = anthocyanin) in the form of glycosides.

The phenolics have been recognized as plant constituents for many years and numerous functions have been proposed. Some of the simple phenols are powerful fungicidal and bacteriocidal agents and it has been suggested that they protect the plant from invasion by fungi and bacteria. If a plant is wounded, it is frequently observed that phenolic compounds become concentrated around the damaged tissue and tend to "seal off" the area from the rest of the plant. Plants infected with fungi or

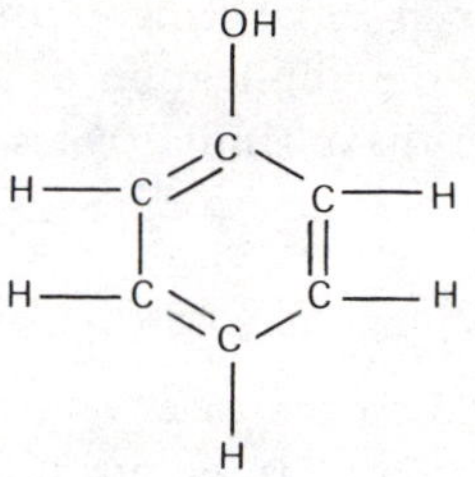

Phenol

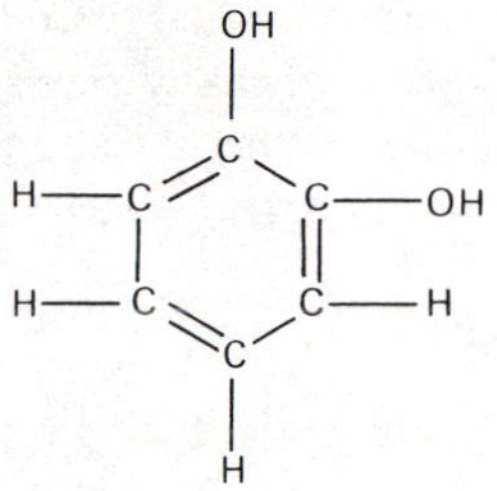

Catechol

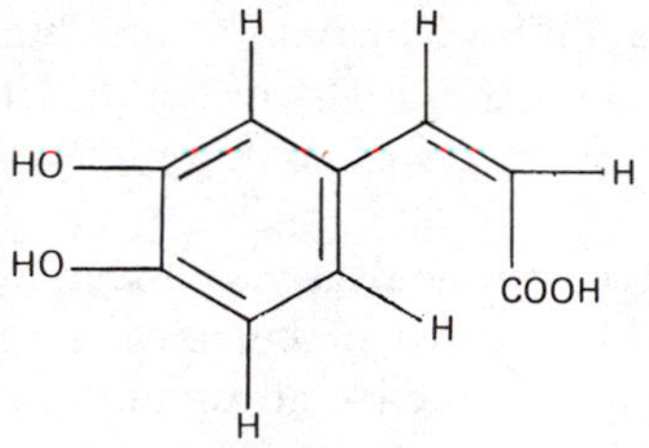

Caffeic acid

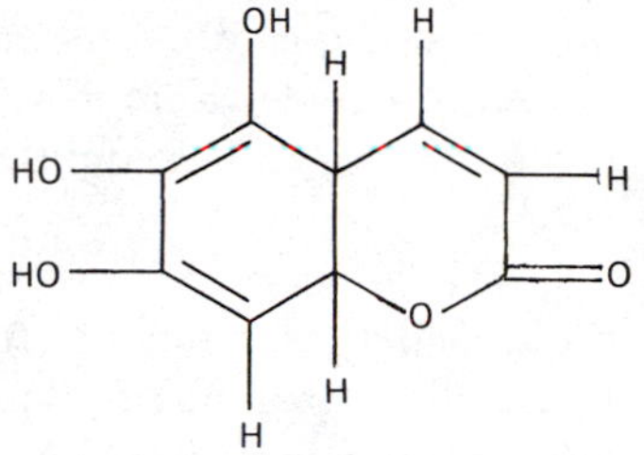

Aesculin

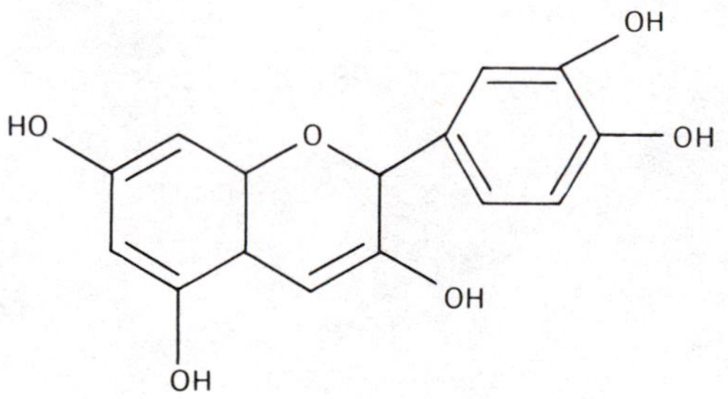

Cyanidin (an anthocyanidin)

Figure 14-15 Some phenolic compounds identified in plants.

bacteria often respond by producing phenolic compounds that tend to limit the extent of invasion of the pathogen. Large, complex phenolic molecules frequently are found in plant vacuoles or in the lumen of dead cells and it has been proposed that these insoluble phenolics are waste products of metabolic reactions. Heartwood in woody plants, for example, is often heavily impregnated with polymeric phenolics, such as lignin and tannin.

The most commonly observed effect of an exogenously applied phenolic is an inhibition of growth. Cell division and cell enlargement are inhibited and the germination of many seeds is prevented by the application of a variety of phenolic compounds. Whether endogenous phenolics have similar roles within the plant is a problem of current research interest.

Vitamins

Studies on human nutritional disorders led to the discovery and identification of essential dietary factors known as vitamins. Humans require about a dozen vitamins, which are ordinarily supplied in a diet containing plants or parts of plants (leaves, stems, petioles, seeds, tubers, roots, fruits). The vitamins are generally classified as water-soluble vitamins or fat-soluble vitamins. Included in the water-soluble category are vitamin C (ascorbic acid) and a group referred to as the B vitamins, which includes vitamin B_1 (thiamine), vitamin B_2 (riboflavin), vitamin B_6 (pyridoxine), folic acid, nicotinamide, pantothenic acid, vitamin B_{12} (cobalamin), and biotin. The fat-soluble vitamins include vitamin A (carotene), vitamin D, vitamin E, vitamin K, vitamin Q (ubiquinone), and vitamin F. The structures of some representative examples are shown in Figure 14-16.

The function of most vitamins in mammals is well understood. The B vitamins are essential components (prosthetic groups) of coenzymes necessary in cellular metabolism; thiamine pyrophosphate is the active group in the carboxylase enzyme, nicotinamide is a component of NAD and NADP, and pantothenic acid is a part of coenzyme A. Vitamin A is involved in the pigment system of the eye whereas vitamin K is one of a group of compounds known as quinones that function in the electron transport chain in photosynthesis. Because most function as cofactors in enzymic

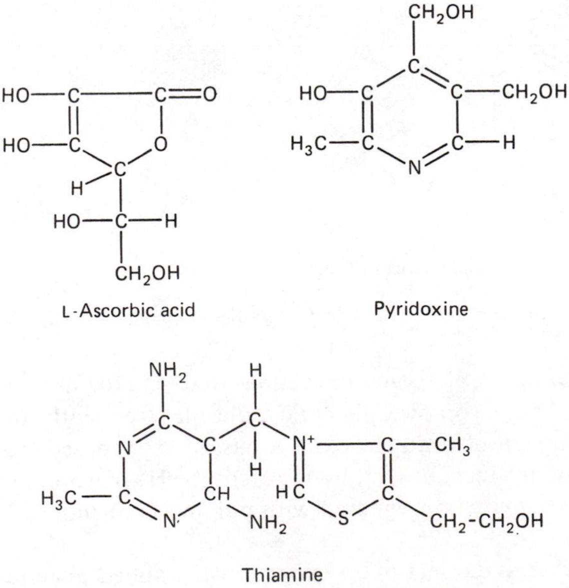

L-Ascorbic acid   Pyridoxine

Thiamine

Figure 14-16 Structures of several vitamins.

reactions, vitamins are usually found within the cell in minute amounts. In plants vitamins participate in the same sort of enzymic reactions as in mammalian systems. Plants are able to synthesize vitamins, however, whereas animals must depend on plants for their supply of vitamins.

The suggestion that vitamins might be considered plant growth substances came from studies of the growth of isolated root systems. Excised roots grow well in a sterile nutrient solution containing inorganic ions and sucrose, provided that several B vitamins are supplied. The roots of most plant species require thiamine, plus either nicotinamide or pyridoxine. Intact plants do not require an exogenous vitamin supply. The vitamins appear to be synthesized in leaves and are then transported to meristematic regions in the root and shoot tip. Why excised roots are unable to synthesize vitamins is not clear. Presumably root cells have the same genetic information as leaf cells, but for some reason the enzymic apparatus necessary to synthesize vitamins is inoperative. A similar situation is found in young embryos. Very young embryos cannot be grown on a synthetic medium unless it is fortified with several vitamins. If older embryos are examined, however, it is found that they do not need an exogenous supply of vitamins. Apparently at some critical age the embryo develops the capacity to synthesize the factors necessary for growth.

Cyclitols

In studies on the chemical composition of coconut milk (Chapter 16) Steward and associates found that the so-called neutral fraction contains rather substantial amounts of several cyclitols, *myo* inositol and *scyllo* inositol (Figure 14-17). By themselves, the inositols have no effect on the growth of carrot callus tissue, but in combination with the *active fraction*, they promote the growth of this tissue. It has also been found that inositol stimulates the growth of other kinds of callus tissue in the presence of auxin, kinetin, and vitamins. There is some question as to whether all callus tissues require inositol, but at least a few appear to have an absolute requirement.

The role of inositol in the growth of callus tissue is not known at this time. Recent findings, however, indicate the participation of inositol in several important metabolic reactions related to the growth of cells. Inositol is an intermediate in the conversion of glucose into glucuronic acid and galacturonic acid, precursors of constituents in primary cell walls. Inositol is also found in plasma membranes as a constituent of the lipid fraction and may play a role in maintaining membrane integrity. Moreover, inositol is found conjugated with indoleacetic acid and may be a storage form of IAA. The mixed calcium, potassium, and magnesium salts of phytic acid, the hexaphosphate ester of inositol, are major constituents in storage tissues of many seeds. During germination potassium, calcium, magnesium, and phosphate ions, as well as inositol, are made available for early seedling growth.

Brassinolide

A number of years ago John W. Mitchell of the U.S. Department of Agriculture, Beltsville, Maryland, initiated a program of examining plant pollens as sources of plant growth regulators. Extracts from the pollen of rape flowers (*Brassica napus* L.)

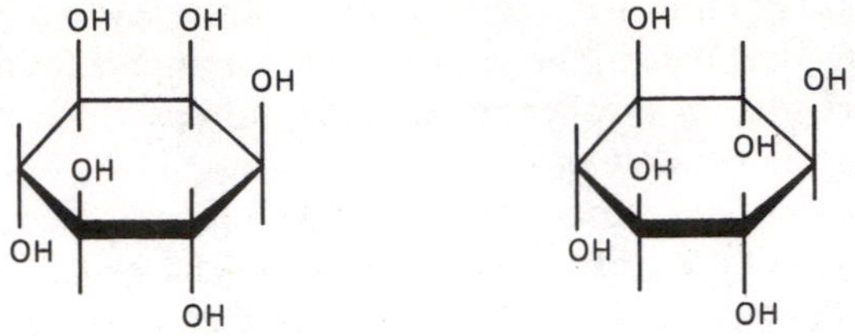
OH
OH
OH
OH
OH
OH
OH
OH
OH
OH
OH
OH

Myo-inositol *Scyllo*-inositol

(a)

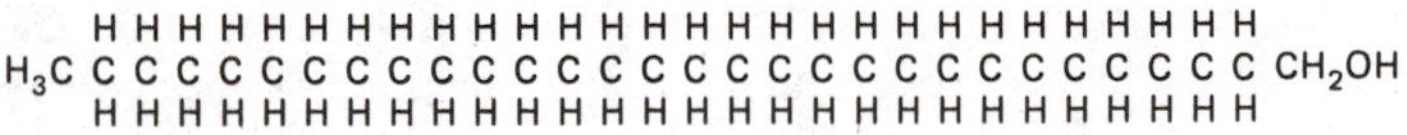
H H H H H H H H H H H H H H H H H H H H H H H H H H H H
H3C C C C C C C C C C C C C C C C C C C C C C C C C C C C C CH2OH
H H H H H H H H H H H H H H H H H H H H H H H H H H H H

Triacontanol (1-Hydroxytriacontane; $CH_3(CH_2)_{28}CH_2OH$)

(b)

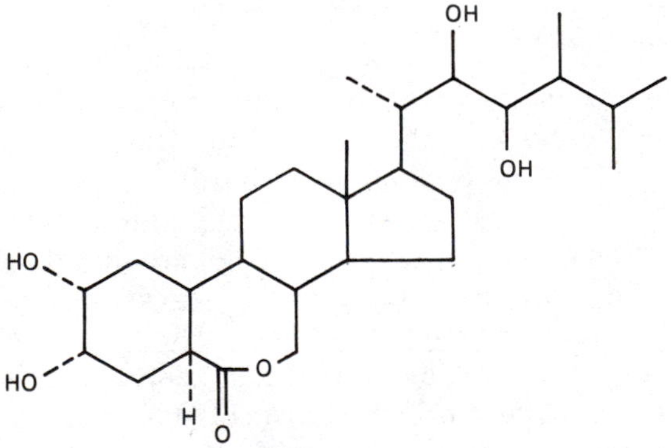
OH
OH
HO
HO
O
H
O

Brassinolide

(c)

Figure 14-17 Structures of some naturally occurring plant growth substances: (a) inositol, (b) triacontanol, (c) brassinolide.

were found to enhance the growth of bean seedlings in glasshouse experiments. Initially the active material was believed to be a lipid, but further purification showed it to be a steroid (Figure 14-17). Concentrations in the range of 1 ng per plant were effective in bean seedling bioassays. The substance, named brassinolide, enhances the yields of a number of plants—radish, potato, bean, lettuce—when low concentrations are sprayed on plants.

Brassinolide has been synthesized and the synthetic material is as active as the natural substance isolated from rape pollen. In addition, several analogs of brassinolide have been synthesized and shown to be biologically active. At present, brassinolide is too expensive for widespread use on field and horticultural crops, but further research may suggest ways of using the material to enhance crop yields.

Triacontanol

In 1977 S. K. Ries of Michigan State University found that a soil application of powdered alfalfa (*Medicago sativa* L.) leaves enhanced the growth and yield of such plants as soybean, corn, wheat, rice, tomato, and carrot. Ries and colleagues subsequently found that the active ingredient in alfalfa leaves was a long, straight-chain primary alcohol, 1-hydroxytriacontane (Figure 14-17). The trivial name, triacontanol, has been applied to the substance. Triacontanol had been known as a constituent of beeswax and the cuticle of many leaves.

Pure triacontanol, when applied as a spray on the leaves of many plants, also increases plant growth. Concentrations in the range of 10 mg per acre were effective. Studies now are under way to determine if triacontanol is practical for use in increasing growth and yield of field-grown crops.

Flowering Hormone

There is considerable evidence (discussed in Chapter 18) that flowering may be under the influence of a chemical stimulus that is transported within the plant. The Soviet plant physiologist M. Kh. Chailakhyan has called this substance *florigen*. Several investigators have obtained extracts from plants capable of stimulating flower induction, but the active material has not been characterized or identified. Such experiments have been difficult to confirm and the existence of an endogenous flowering hormone remains tentative. Auxins, gibberellins, cytokinins, and abscisic acid also influence flowering in some species of plants and it may be unnecessary to postulate the existence of a specific flowering hormone.

GROWTH-RETARDING CHEMICALS

Thousands of new organic compounds are synthesized each year by organic chemists. The usual fate of such compounds is that they are described in the literature and remain as chemical curiosities on the shelf in some laboratory. Their potential usefulness in biology is not realized unless they are screened against a number of different biological systems. The National Academy of Sciences and National Research Council established, in cooperation with U.S. Department of Agriculture, a screening program to test compounds submitted to them for plant growth activity. Screening programs were also established for anticancer, antiviral, and antibiotic activity. A number of interesting compounds were found, some of which have proved to be of value in agriculture and medicine. Today many of the agricultural chemical companies maintain their own screening facilities for testing new compounds for biological activity.

During the early years of the screening program several chemicals were found to reduce stem elongation. Subsequently it has been found that a number of different kinds of compounds act as growth retardants. Figure 14-18 shows that the chemical structures of these growth-retarding chemicals are quite diverse. Amo 1618 and Cycocel contain a quaternary ammonium group (nitrogen atom to which is attached four chemical groups) whereas Phosfon-D contains a phosphonium group (phosphorus atom to which are attached four chemical groups). This type of structure is similar to choline (Figure 14-18), an extremely important biological molecule involved in membrane structure and activity. It is not known, however, whether Amo 1618, Cycocel, and Phosfon-D antagonize or inhibit choline activity.

Maleic hydrazide is extensively used by tobacco growers in retarding the development of axillary buds following removal of the apical shoot.

Growth retardants have other biological effects besides retarding stem elongation. Leaves of treated plants are frequently darker green than in untreated plants. Flowering is also accelerated in some plants following treatment with a growth retardant. Greenhouse-grown azaleas, for instance, can be brought into flower much earlier when treated with a growth retardant, together with appropriate temperature regimes.

The reduction in plant height following the application of several growth retardants—Amo 1618, Phosfon-D, and Cycocel—can be overcome by treatment with a gibberellin. There is no obvious structural similarity between growth retardants and gibberellins and it is difficult to visualize how they might interact in a competitive manner at some growth site. It was found by Kende and associates that Amo 1618 and Cycocel inhibit the biosynthesis of gibberellin in the fungus *Gibberella fujikuroi* and further work with higher plants corroborates this discovery. It is believed that some of the dwarfing compounds inhibit steps in the pathways leading to the formation of gibberellins. The precise nature of the inhibitory effects of the growth retardants is not known, but there seem little doubt that their ultimate effect is on the biosynthesis of gibberellin.

No endogenous growth-retarding substances have as yet been recognized in plants, although quaternary ammonium compounds, such as choline, are commonly present. These compounds are active in metabolic processes, particularly in lipid metabolism and the formation of cell membranes. The quaternary ammonium compounds are labile and difficult to handle. Future work may reveal that endogenous quaternary ammonium compounds function as plant growth substances.

SECONDARY PLANT SUBSTANCES: ALLELOCHEMICS

Plants contain a bewildering array of constituents whose function is not understood. It is not certain that each of these substances has a physiological role in the life of the plant. Many, however, frequently referred to as *secondary plant substances*, have been

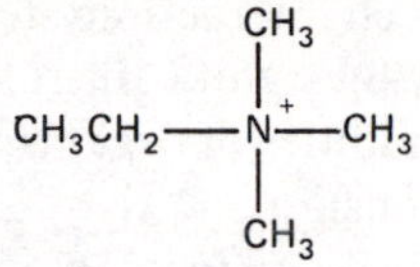

Choline

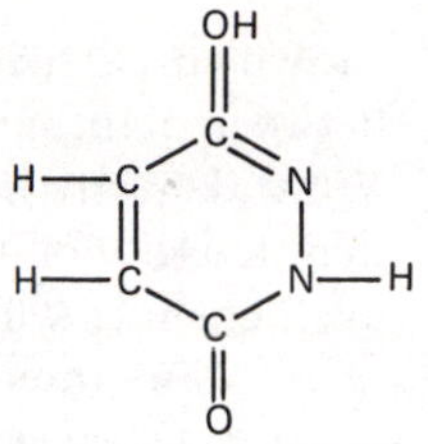

Maleic hydrazide

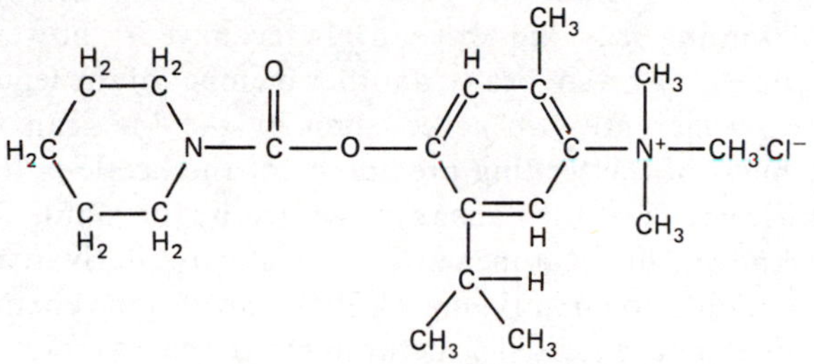

Amo 1618

(2-Isopropyl-4-dimethylamino-5-methylphenyl-
1-piperidine-carboxylate methyl chloride)

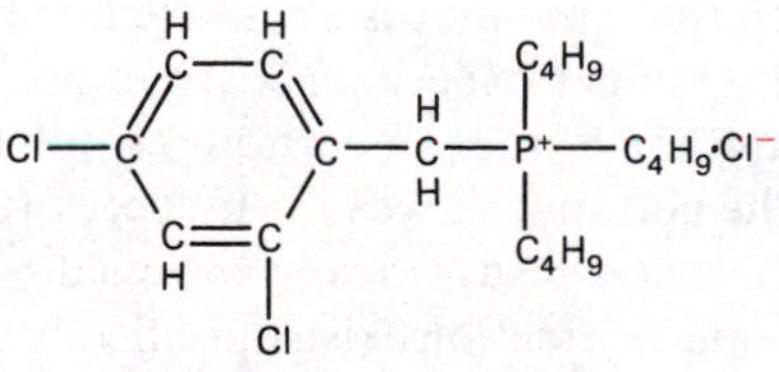

Phosfon-D (2,4-dichlorobenzyl-tributyl phosphonium chloride)

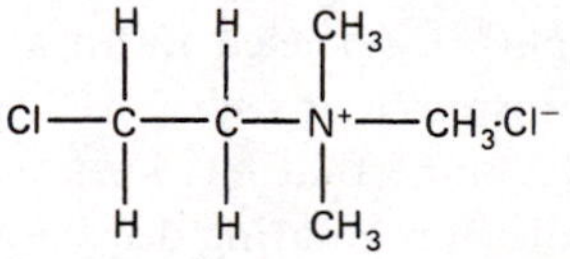

Cycocel [(2-chloroethyl)trimethyl ammonium chloride]

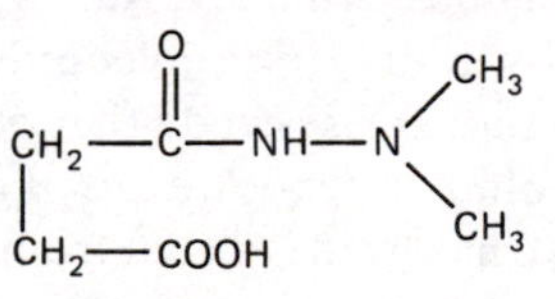

B-Nine (N-dimethyl amino succinamic acid)

Figure 14-18 Structures of several growth-retarding chemicals.

shown to participate in important interactions between species of plants as well as between plants and other organisms. Such interactions, termed *allelochemic* by R. H. Whittaker, are now recognized as playing significant roles in the establishment and maintenance of populations and communities of organisms. A few examples of plant allelochemics will illustrate these relationships.

Many insects obtain their food solely by feeding on plants. Insects presumably are not different from the higher animals in their nutritional requirements and must have in their diet carbohydrates, amino acids, vitamins, fats, and inorganic ions. The growing parts of practically any plant provides such materials, but insects do not feed indiscriminately. Rather they may feed on a single species or closely related species. Other insects may feed on a wider range of species within a family. The Mexican bean beetle, for example, generally feeds on plants within the genus *Phaseolus* (includes mung bean, kidney bean). In recent years, however, the beetle has become adapted to feeding on soybeans, another member of the legume family. Some leguminous species apparently are never eaten by the Mexican bean beetle. It is thought that the basis of the feeding preference of the beetle is the presence of an attractant, a specific chemical substance synthesized by the plant. The identity of the substance is not known, but it appears to be a glucose derivative of a terpenoid. Squalene, a triterpenoid shown in Figure 14-19, illustrates the chemical nature of the molecule. The terpenoids are related structurally to the carotenoids, gibberellins, and rubber.

The Mexican bean beetle is a major pest of soybeans and a great deal of time and money is spent in applying insecticides to control it. Attempts are under way to see if it is possible to develop a soybean plant resistant to the beetle. It might be done by breeding experiments in which the attractant is eliminated or by introducing a new chemical that repels the beetle. Similar experiments are in progress on other plant-eating insects—that is, corn and the European corn borer, tobacco and the tobacco hornworm, and cotton and the cotton-boll weavil. If successful, the current heavy applications of many insecticides may be drastically curtailed.

Another type of allelochemic interaction exists specifically between plants. Some plants release chemicals that inhibit the growth of other plants, a phenomenon referred to as allelopathy. The harmful substance may be released in a volatile form or excreted in a soluble form. The substance might be released in an active form or require conversion to an active form in the atmosphere or soil.

Naringenin (Figure 14-19) is a flavonine that has been isolated from dormant peach buds and is thought to play a role in regulating dormancy. Bud dormancy in many plants has been found to involve abscisic acid and it is not clear whether both naringenin and abscisic acid act in peach bud dormancy. Naringenin is a polyphenolic compound and strongly inhibits the growth of *Avena* coleoptiles.

It has been known for many years that grasses and other plants do not grow well in the immediate vicinity of black walnut trees (*Juglans nigra*). A polyphenolic naphthaquinone, juglone (Figure 14-19), has been isolated from roots, bark, and fruit hulls. If juglone is added to water cultures of grasses, tomato, or potato, the plants wilt and die. Juglone apparently diffuses from the roots or is leached from fallen fruits or

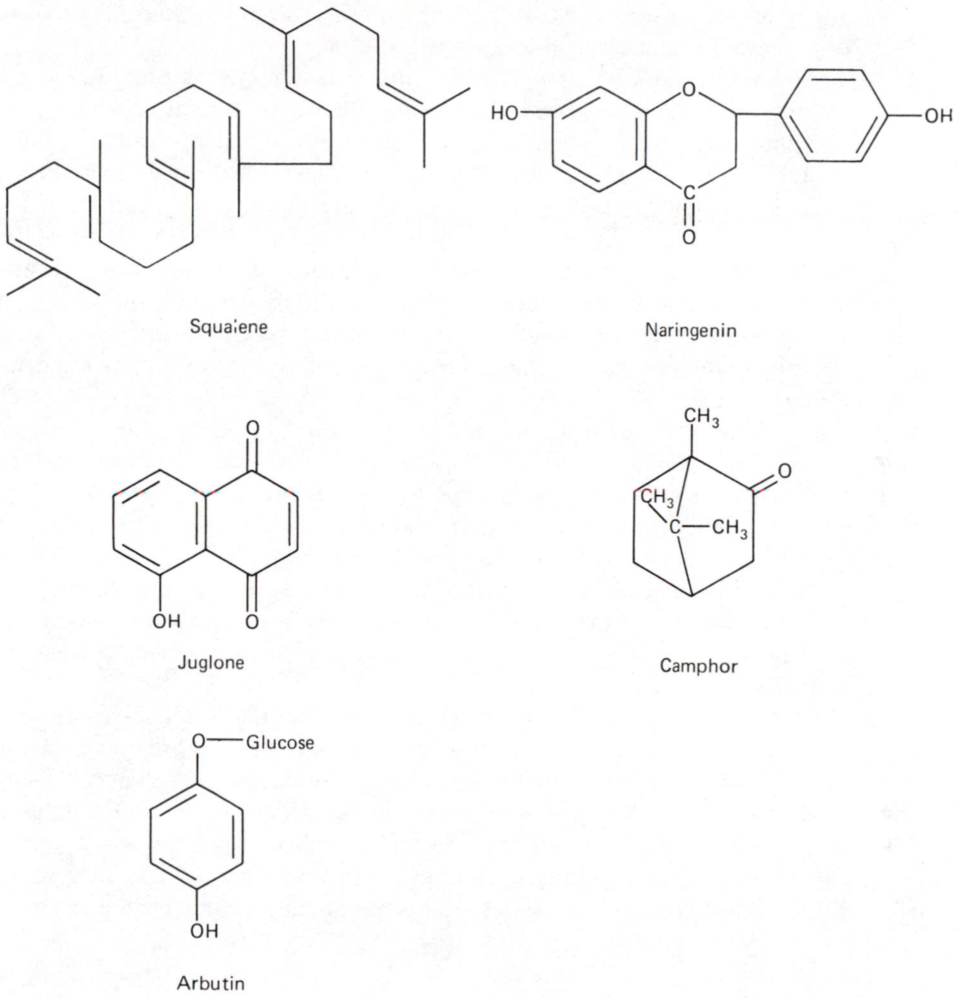

Figure 14-19   Structures of some plant constituents with growth-regulating activity.

from bark and leaves. Within the plant juglone is found as an inactive glucoside, but when it enters the soil it is converted into an active form.

Another interesting example of allelopathy is that found in the chaparral shrub vegetation of southern California. Several shrubs are dominant; and when they were left undisturbed, it was noted that low-growing herbaceous plants were absent. After fires had completely destroyed the shrub vegetation, many herbaceous plants developed from seeds dormant in the soil. A shrub vegetation also developed from seeds and underground roots. After 5 to 7 years the herbaceous plants again had disappeared and the shrub vegetation once more dominated the scene. C. H. Muller and associates discovered that the suppression of the herbaceous plants was caused by phytotoxins produced by the shrub vegetation.

The plant *Adenostoma fasciculatum*, a member of the rose family, is one of the widespread dominant species in the chaparral. Water extracts of the leafy branches of this plant were found to contain at least nine phenolic compounds, including arbutin (Figure 14-19). These phenolics (both as free phenolic acids or as glycosides) appear to be formed in the leaves and excreted on the leaf surfaces during periods of dry weather. They are then washed off by rain and become concentrated in the surface soil, where they inhibit seed germination and seedling growth of herbaceous annuals. The phytotoxins do not accumulate and remain for long periods in the soil but are readily leached out by rain. To maintain an inhibition of herb growth, the toxic materials from *Adenostoma* are constantly deposited in the soil following rain or fog drip from the foliage.

Another allelopathic interaction was found in the southern California grasslands. The grasses are mainly annuals that germinate, emerge, and reproduce in response to rainfall patterns. In some areas the shrub *Salvia leucophylla*, a member of the mint family, invades the grasslands and inhibits the germination and growth of the grasses. Volatile terpenes, such as cineole and camphor (Figure 14-19), are produced by the foliar branches of *Salvia* and these terpenes are adsorbed on the surface of dry soil particles. When the grass seeds germinate following a rain, the terpenes inhibit seedling development and prevent establishment of a grass cover around the base of the *Salvia* plants.

These few examples illustrate some of the allelochemic reactions of plants. Normal metabolites, such as sugars, amino acids, and organic acids, as well as inorganic ions, are lost from leaves and roots by leaching. They may have a profound influence on the growth of soil microorganisms, which, in turn, influence plant growth. There is great interest in the secondary plant substances as possible agents for use in the control of insects, fungi, nematodes, and other plant pests as replacements for DDT, chlorinated hydrocarbons, and other nondegradable chemicals.

REFERENCES

References for this chapter are given at the end of Chapter 15.

15

Plant Growth Substances: Biosynthesis, Analysis, Transport, and Mechanism of Action

BIOSYNTHESIS OF PLANT GROWTH SUBSTANCES

The maintenance of effective levels of plant growth substances at specific sites in the plant depends on synthesis, destruction, transport, inactivation (conjugate formation), and localization or compartmentation. Of these, biosynthetic processes are important because they provide for the formation of plant growth substances from molecules of intermediary metabolism. A simplified scheme of intermediary metabolism was outlined in Figure 3-26, Chapter 3, in which a few key metabolites were shown to be precursors of saccharides, lipids, amino acids, nucleic acids, and other primary plant constituents. These same key metabolites—acetyl-CoA, triose phosphate, glycolytic intermediates, TCA cycle intermediates—are also involved in pathways of synthesis of other plant constituents, including alkaloids, terpenes, rubber, phenolics, and plant growth substances. These pathways have not been as thoroughly investigated as those involving the primary plant constituents, but there is much current interest in this area of plant physiological research. Without going into detail, pathways for the biosynthesis of major groups of plant growth substances are outlined in Figure 15-1.

A major pathway of IAA synthesis involves the amino acid tryptophan. Several intermediates are recognized between tryptophan and IAA—indolepyruvic acid, tryptamine, indoleacetaldehyde (see Figure 14-6, for structures). Tryptophan itself is synthesized from phosphoenolpyruvate (PEP) and erythrose-4-phosphate. The pathway also provides intermediates in the synthesis of a wide range of phenolic compounds. Indole is an intermediate and IAA can be formed directly by a reaction between the amino acid serine and indole. The pathway from indole to tryptophan involves several intermediates. The relative importance of the different pathways probably depends on environmental conditions and species variability.

The usual fate of tryptophan is to be incorporated into cellular proteins. If tryptophan is to be made available for IAA synthesis, cellular proteins must undergo proteolysis, a process associated with senescence and death. The British plant biochemist A. R. Sheldrake has suggested that IAA is synthesized from tryptophan released during cell autolysis in the differentiation of xylem and phloem elements. When xylem and phloem are laid down, they are typical meristematic cells, but as they mature and become functional transport cells, the cellular contents are hydrolyzed and their breakdown products become available for metabolic reactions in surrounding cells. According to this view, the sites of IAA synthesis are not only in meristematic cells but also in regions where vascular differentiation occurs. Other products of cellular autolysis during vascular differentiation include nucleic acids (involved in cytokinin synthesis?), other amino acids (methionine in ethylene synthesis?), and intermediates of phenolic synthesis. There is no agreement among plant physiologists on the idea that products of cell autolysis are of major importance in plant growth substance synthesis. It illustrates, however, the complexity of attempting to pinpoint centers of synthesis.

Both gibberellins and abscisic acid are synthesized from acetate (Figure 15-1). The intermediate, isopentenyl pyrophosphate, appears to be a branch point in the biosynthetic pathways. There is some evidence that ABA can be formed from degradation products of carotenoids that are also synthesized via isopentenyl pyrophosphate. The anabolic pathway from isopentenyl pyrophosphate to ABA may predominate in actively growing tissue whereas the catabolic pathway from carotenoids may occur during aging and senescence. Isopentenyl pyrophosphate is also an intermediate in the synthesis of cytokinins.

Less is known about the synthesis of cytokinins in plants than the synthesis of other growth substances. Nucleotide synthesis in bacteria is rather well understood, but similar information for plants is lacking. The cytokinin isopentenyladenine Figure 14-10) apparently is formed by the addition of isopentenyl to adenine. Additional information is needed on cytokinin biosynthesis in plants.

The synthesis of ethylene in plants has been of interest for years. Because of its simple structure, it might be imagined that ethylene could be formed from metabolic intermediate in ethylene synthesis (Figure 15-2). The conversion of S-adenosylmethionine (SAM) to ACC is mediated by ACC synthase and the synthesis and activity of this enzyme may be modified by such factors as anaerobiosis, wounding. IAA, cytokinin, Ca^{2+}, and some exogenous inhibitors. Moreover, the conversion of

It has been found that 1-aminocyclopropane-1-carboxylic acid (ACC) is the key intermediate in ethylene synthesis (Figure 15-2). The conversion of S-adenosylmethionine (SAM) to ACC is mediated by ACC synthase and the synthesis and activity of this enzyme may be modified by such factors as anaerobiosis, wounding, IAA, cytokinin, Ca^{2+}, and some exogenous inhibitors. Moreover, the conversion of ACC to ethylene is enzymic and this enzyme is affected by a number of external and internal factors. Thus it is clear that the synthesis of ethylene is closely regulated and its formation occurs under rather well-defined conditions.

(a) Auxins

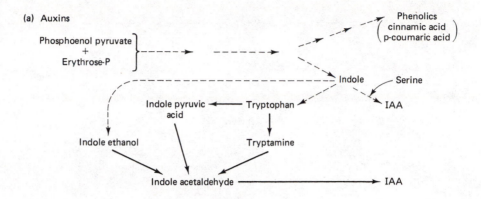

(b) Gibberellins, abscisic acid

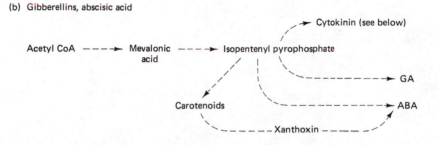

(c) Cytokinins

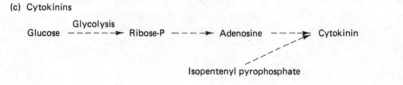

(d) Ethylene (see Figure 15-2)

Oxaloacetic $\text{---}\blacktriangleright$ Methionine $\text{--- }\blacktriangleright$ 1-Aminocyclopropane-1-carboxylic acid $\longrightarrow$ Ethylene
acid

Figure 15-1 Summary of proposed pathways of plant growth substance synthesis.

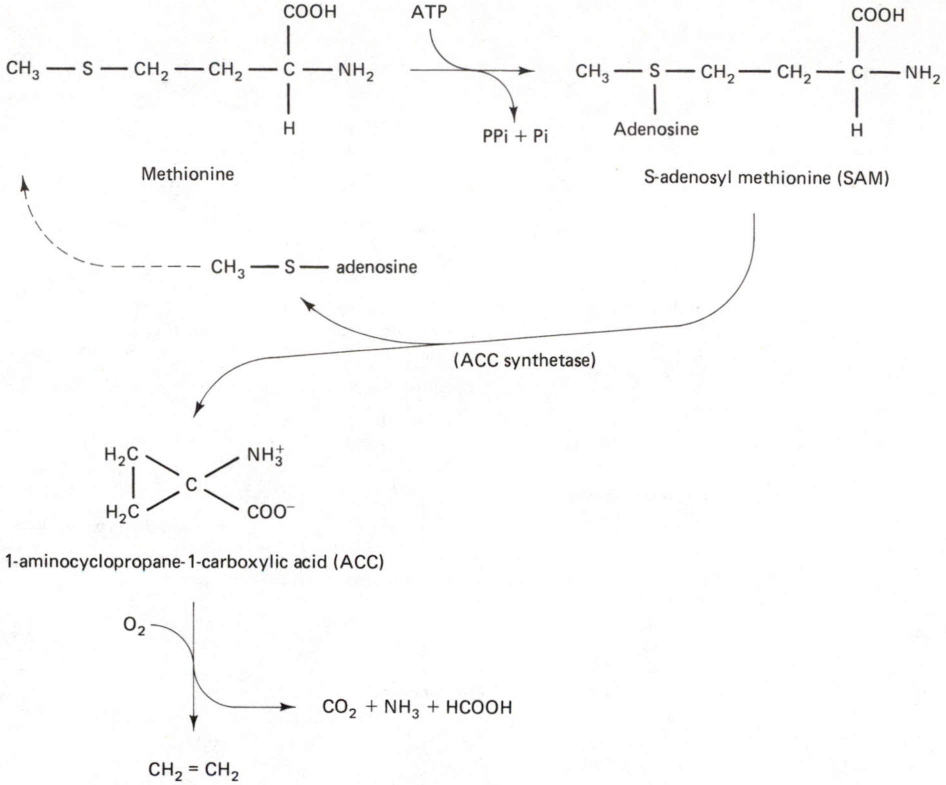

Figure 15-2 Some intermediates in the biosynthesis of ethylene. The rate-limiting step is the conversion of *S*-adenosyl-methionine (SAM) to 1-aminocyclopropane-1-carboxylic acid (ACC) by the enzyme ACC synthetase. The enzymic conversion of ACC to ethylene may also be involved in ethylene formation under certain environmental conditions (anaerobiosis, high temperature).

ANALYSIS OF PLANT GROWTH SUBSTANCES

From the material presented so far it is clear that plant growth substances are of widespread occurrence in plants. In some instances, there are multiple forms of plant growth substances and it is not known if all are of equal importance in modifying or regulating growth and development. Furthermore, there is a lack of information on the localization or compartmentation of the growth substances within the plant. Are they present in all parts of the plant? Do their concentrations vary with age of plant and with different growing conditions? Are they bound to membranes? Are they present in the cytosol or organelles? How much of a particular growth substance is present in a particular compartment? These are important questions that must be answered before a comprehensive view can be developed of the role of plant growth substances in growth and development.

Answers to most of these questions depend on being able to measure how much of a particular growth substance is present. In most instances, they occur in plants in very low concentrations (i.e., ABA is present in the nanogram per gram fresh weight range). The common chemical procedures used in analyzing plants for carbohydrates, proteins, amino acids, nucleic acids, lipids, and so forth ordinarily are not sensitive enough to measure plant growth substances. Two kinds of problems are encountered: The extraction of the growth substance from the plant or plant part and the actual measurement of the amount present in the extract.

Extraction Procedures

Plants contain numerous small soluble molecules, such as sugars, organic acids, amino acids, and inorganic ions. Because plant growth substances are present in much lower concentrations than these molecules, generally it is necessary to use procedures for eliminating them in order to minimize the possibility of their interference with growth substance measurements. Also, such procedures can assist in concentrating plant extracts, thereby facilitating eventual analytical techniques.

One method of handling this problem was used by Went in studies with IAA. As shown in Figure 14-4, Went placed coleoptile tips on agar slabs and found that a substance (later identified as IAA) diffused into the agar from the coleoptiles. Because this procedure did not involve crushing or macerating the tissue, little cellular damage was done and sugars and other cellular constituents did not move into the agar along with IAA. The IAA must have moved out of the intact cells, raising the question of how it was moved through cell membranes. Moreover, subsequent studies have shown that IAA conjugates (Figure 14-6) are present in coleoptiles and probably did not move with free IAA into the agar. Coleoptile tips also contain gibberellins and other plant growth substances. These complications, however, do not diminish the usefulness of the diffusion technique as a method of obtaining plant growth substances free of other cellular constituents. In addition, Went showed that many coleoptile tips could be applied to agar slabs, thereby concentrating the IAA in a small volume.

The more usual method of separating growth substances from plant tissue is to extract with an organic solvent. The tissue may be ground up in a homogenizer or cut into small pieces before extraction. The choice of solvent is important, as is the temperature at which the extraction is carried out. The main purpose of the extraction is to obtain as much as possible of the growth substance in question, uncontaminated by other plant constituents, as well as to prevent enzymic reactions that might alter the growth substance—destroy it or convert it into an inactive form. Hundreds of procedures have been devised; many are suitable for only a limited objective whereas others have broader use for more than one growth substance. The investigator must develop appropriate techniques for the growth substance being studied, the type of tissue being studied, and the measurement technique to be followed.

Once a satisfactory extraction procedure is achieved, the next step is to purify the extract so that only the substance being studied is present. This process generally involves some sort of chromatographic procedure—paper, column, liquid, gas. Figure

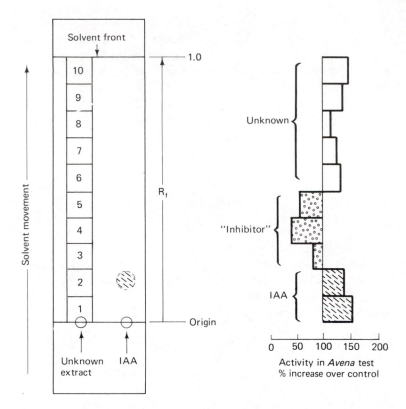

Figure 15-3   Chromatographic separation and bioassay of growth substances.

15-3 illustrates the use of paper chromatography. Following solvent extraction, an aliquot of the extract is applied to the paper along with known markers—for instance, IAA. The chromatogram is then developed with an appropriate solvent. After removal of the solvent by evaporation, the chromatogram is cut into sections that are then analyzed separately. In this example, the sections were eluted with water and the water extracts analyzed for IAA by an appropriate bioassay procedure (see next section).

It can be seen from Figure 15-3 that the plant extract contained growth-promoting and growth-inhibiting substances. The material in the first segments of the chromatogram probably is IAA, inasmuch as synthetic IAA migrates to the same relative position. The inhibitory materials in segments 3, 4, and 5 were not identified, but it is known that several phenolics (discussed in Chapter 14) migrate to this region. Segments 6, 7, 8, 9, and 10 contained substances stimulatory in the bioassay procedure, but their chemical identity is not known. The usual practice in chromatographic studies is to use several different solvents to develop the chromatograms. If the unknown and known substance follow identical migration patterns, the likelihood of correct identification is enhanced.

Other chromatographic procedures, such as silica gel plate chromatography, high-performance liquid chromatography (HPLC), column chromatography, and gas chromatography, have been developed for concentrating and separating plant growth substances extracted from plants. Some chromatographic processes are used solely for removing interfering molecules before the application of analytical procedures. In some chromatographic methods it is necessary to convert the growth substance to a derivative that can then be chromatographically separated. Such procedures tend to be time consuming and it is essential that known pure standard compounds be available for comparative purposes. The key in most of these techniques is to have a very specific method for identifying the molecule being studied. For some growth substances (ethylene, GAs, IAAs), such techniques are capable of measuring nanogram (10^{-9} g) and picogram (10^{-12} g) quantities of plant growth substances.

Quantification Techniques

As noted, what is needed in plant growth substance studies is a technique for measuring the amount of a substance, usually present in low concentration, with some degree of certainty that the substance being measured is not an artifact of the extraction and isolation procedure. Physical techniques used in organic chemistry and biochemistry—colorimetry, spectrophotometry, fluorometry, spectroscopy, mass spectrometry, and so forth—work well in many cases if care is taken to remove interfering compounds. Biological methods have been used in meeting many problems encountered in assaying for growth substances.

Bioassay procedures In the discussion of Figure 15-3 it was stated that the identity of the material present in different segments of the chromatogram was determined by a bioassay procedure. A bioassay utilizes the response of biological material to the compound being tested. If the appropriate biological material is selected, bioassays are capable of measuring very small concentrations of plant growth substances. Several representative bioassay methods are illustrated in Figure 15-4. The *Avena* coleoptile (other grass coleoptiles, such as wheat, may be used) straight growth test for IAA is a variant of the *Avena* coleoptile bending test developed by Went (Figure 14-4). The bending test involves a number of manipulations and strict adherence to environmental conditions of temperature and humidity. The straight growth test does not require special humidity control, for the coleoptile sections are floated on the test solutions. IAA concentrations in the range of 0.001 to 1 mg L^{-1} can be measured. A straight growth test was used to measure the growth substances and inhibitors separated by paper chromatography, as illustrated in Figure 15-3.

The barley endosperm bioassay for GA [Figure 15-4(b)] is one of a number of such tests. One of the specific roles of GA is the induction of α-amylase synthesis in the aleurone cells of germinating cereal grains. The barley endosperm test measures this activity by determining the amount of α-amylase released from the endosperm of barley. The α-amylase is measured by its ability to hydrolyze starch to soluble sugar. Either the disappearance of starch from a standard starch solution or the amount of soluble sugar formed may be determined.

Cytokinins are measured by their ability to increase the growth of callus tissue

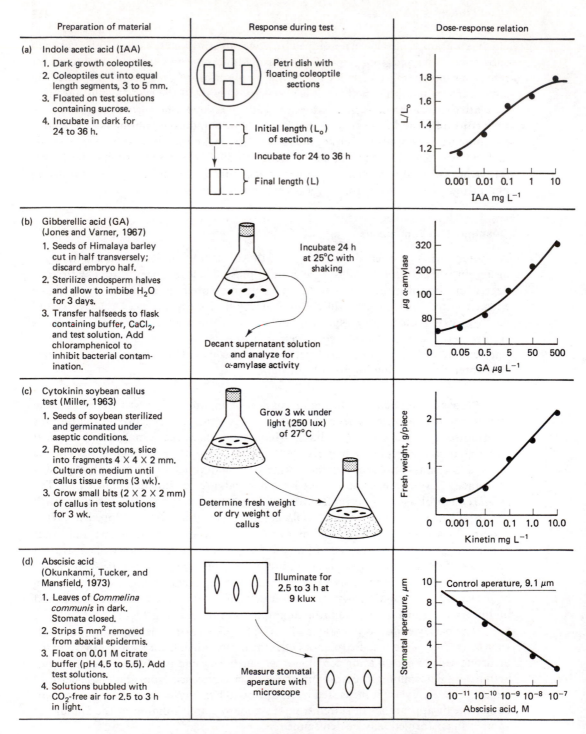

Figure 15-4 Some representative bioassays for plant growth substances. (a) *Avena* coleoptile straight growth test for auxins. (b) Barley endosperm test for gibberellins. (c) Soybean callus test for cytokinins. (d) Stomatal opening test for abscisic acid.

[Figure 15-4(c)]. The discovery of cytokinins was based on the use of tobacco pith callus tissue as a bioassay. Soybean callus tissue grows somewhat more rapidly than tobacco pith callus.

The inhibitory effects of abscisic acid on seed germination and coleoptile growth have been used in ABA bioassays. Other constituents of plants (phenolics), however, are also inhibitory in these tests; so such bioassays are unsatisfactory. The bioassay outlined in Figure 15-4(d) takes advantage of the ability of ABA to regulate stomatal closing. Interference from phenolics is minimal. Leaf epidermal strips from plants kept in the dark (stomata closed) are floated on test solutions and illuminated. In the absence of ABA the stomata will open, but stomatal opening is inhibited in the presence of ABA.

There is no good bioassay for ethylene. The so-called triple response of etiolated seedlings to ethylene was shown in Figure 14-12. Ethylene inhibits elongation, induces a bending response, and causes an increase in diameter of the stems. The "triple response" has been used as a bioassay for ethylene, but it requires a constant flowthrough of ethylene and is not practical for measuring small concentrations of ethylene. There is, however, a very good gas chromatographic technique for measuring ethylene and this procedure is generally used in studies of ethylene physiology.

Immunoassay procedures   In recent years plant physiologists have turned to another biological procedure for determining plant constituents, including plant growth substances. The technique, known as *immunoassay*, depends on the immunological properties of foreign substances introduced into the bloodstream of animals. The well-known examples of human allergies to foreign substances, such as plant pollen and animal hair, and the development of immunity to certain diseases, such as chicken pox, smallpox, and mumps, illustrate the phenomenon.

Without going into great detail (Figure 15-5), the technique involves the introduction of a test substance, an antigen, into the bloodstream of an animal (rabbits commonly used), where the natural defense mechanisms of the animal synthesize specific proteins known as antibodies. Antigens are usually proteinaceous in nature— proteins, glycoproteins, lipoproteins, toxins, bacteria, viruses, for example. Plant growth substances, such as abscisic acid and the cytokinins, are not antigenic, but they can be converted to antigenic molecules by coupling to a soluble protein, such as human serum albumin. If such a synthetic antigen is then injected into a rabbit, a specific antibody develops.

Antibodies are proteins known as immunoglobulins and are present in the serum fraction of blood. This is the fraction free of cellular components, such as red blood cells. To prepare this fraction, blood is removed from the animal 30 to 60 days after injection of an antigen and the cellular fraction is removed by centrifugation. The antibody (antiserum) fraction is stable and can be stored under refrigeration for several years. The serum fraction contains many globular proteins but only one, the antibody formed in response to the injected antigen, reacts when challenged by the antigenic molecule. A precipitate forms when the antibody and antigen are brought into contact with one another.

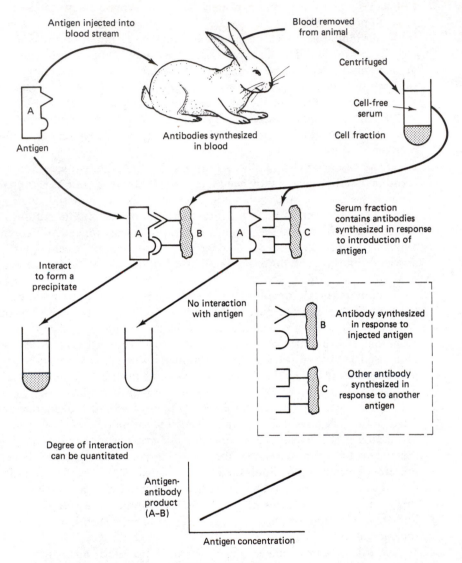

Antigen injected into
blood stream

Blood removed
from animal

Centrifuged

A

Antigen

Antibodies synthesized
in blood

Cell-free
serum

Cell fraction

A   B   A   C

Serum fraction
contains antibodies
synthesized in response
to introduction of
antigen

Interact
to form a
precipitate

No interaction
with antigen

B   Antibody synthesized
in response to
injected antigen

C   Other antibody
synthesized in
response to another
antigen

Degree of interaction
can be quantitated

Antigen-
antibody
product
(A–B)

Antigen concentration

Sensitivity of immunoassay can be increased by adding
tracer-labeled antigen with the test antigen at the
time of antigen-antibody interaction

Figure 15-5 Schematic representation of immunoassay procedure. See text for
details.

To perform an immunoassay, constant amounts of antiserum formed in response to the test compound (antigen) are treated with increasing amounts of the test compound and the amount of precipitate formed is measured. From this interaction a standard curve is prepared relating antigen concentration to the quantity of antigen–antibody precipitate. The quantity of precipitate can be determined by weighing or by radioactivity measurements if a radioactive antigen is used. Tritium (^{3}H) or iodine (^{125}I) have been used in such determinations. Radioactivity measurements enable the investigator to determine very low concentrations of the test substance. E. W. Weiler, for example, used a radioimmunoassay for determining abscisic acid (ABA) and found that concentrations in the range of 0.4×10^{-12} to 20×10^{-12} mol (0.1 to 5 ng) could be measured. Similar techniques have been applied to cytokinins by Weiler and by Leonard and colleagues.

In addition to the sensitivity of radioimmunoassay techniques, the procedure is of value because it discriminates between closely related compounds and can be applied to plant extracts without going through an extensive extraction and purification scheme. Also, Weiler has automated the procedure and indicates that it is possible to analyze several hundred samples per day. Radioimmunoassays appear to be a very promising technique for measuring plant growth substances, particularly when multiple forms of the molecule are present.

TRANSPORT OF PLANT GROWTH SUBSTANCES

The question of transport of growth substances in plants has been an area of much concern. Charles and Francis Darwin interpreted their work on phototropism (Chapter 14, Figure 14-3) of etiolated grass coleoptiles as indicating that some "influence" occurred between the site of light perception and the region of bending. When IAA was found to be involved, considerable importance was placed on the fact that IAA is transported through the coleoptile. Such an effect was compatible with the animal physiologists' idea of a hormone (see section on *Hormone Concept* in Chapter 14) and transportability became a significant element in the development of research on plant growth substances. It appears, however, that most plant meristematic cells are capable of synthesizing growth substances and transport (at least over long distances) is not of prime importance in defining a plant growth substance. Growth substance movement may occur within cells (between organelles), between cells by protoplasmic connections (symplasm), or in the apoplasm, or in xylem and phloem tissue.

One of the most thoroughly studied examples of growth substance transport is that of indoleacetic acid (IAA). The distribution of IAA in monocot and dicot seedlings is shown in Figure 15-6. In monocot seedlings IAA is found in highest concentrations in the coleoptile tip, with lesser amounts in other parts of the coleoptile and in the root. Such a distribution pattern will develop if IAA from the coleoptile tip is transported away from the tip to other parts of the seedling. During movement from the tip IAA is utilized in growth processes, immobilized by complexing with other organic molecules, or destroyed or inactivated by enzymic action.

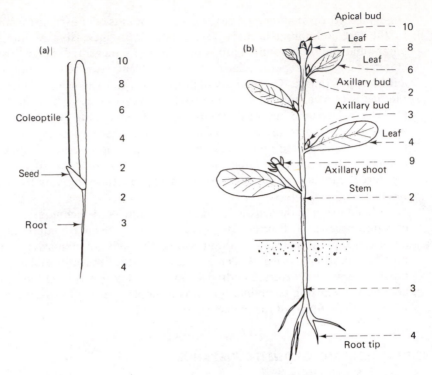

Figure 15-6   The relative concentrations of auxin in different parts of an etiolated monocot seedling (a) and a green dicot seedling (b). Relative units of auxin: 10 = high, 1 = low.

IAA at the tip of the coleoptile is formed from IAA conjugates (inositol, glucose, or amino acid, Figure 14-6) transported in coleoptile xylem elements from the embryonic tissues in the seed. During seed maturation IAA is synthesized by the developing embryo and stored in a conjugated form in endosperm tissue. At seed germination the IAA conjugates move to the coleoptile tip, where they are enzymically hydrolyzed to IAA, which is then transported away from the tip to establish the distribution pattern shown in Figure 15-6. Root meristems synthesize very little IAA and most of the IAA in roots is provided by transport from coleoptiles or stems.

The pattern of IAA distribution in dicot seedlings (Figure 15-6) is somewhat more complex but understandable if the IAA is synthesized in meristematic regions in the shoot, young leaves, and axillary shoots. Transport, immobilization, and destruction of IAA can establish different patterns of concentration within the plant.

An important feature of the IAA distribution pattern in plants is the transport of IAA away from the shoot apex. The movement of auxin from the apex toward the base of the plant (organ or tissue) is referred to as polar transport. Polar transport is not a matter of simple diffusion in response to a concentration gradient of IAA but involves the activity of living cells. Several features of polar transport of IAA in stem tissue are illustrated in Figure 15-7. Stem segments are removed from seedlings growing in a

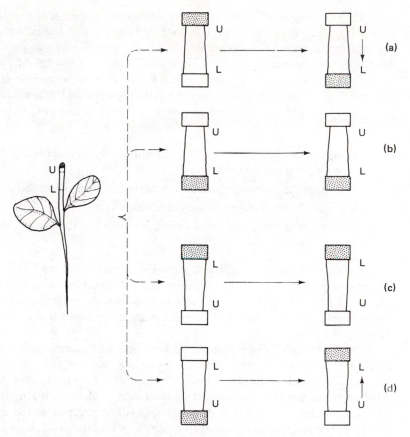

Figure 15-7 Polar transport of IAA in stem segments. Movement occurs from the morphological apex toward the base.

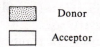

▨▨▨ Donor

☐ Acceptor

vertical position. An agar block containing IAA (donor block) is appressed to the upper (U) end of the segment and an agar block devoid of IAA (acceptor block) is appressed to the lower (L) end of the stem segment. After several hours the two agar blocks are assayed for IAA by an appropriate bioassay procedure. The IAA has moved through the stem segment from the morphological apex to the base [Figure 15-7(a)]. If the position of the donor and acceptor blocks is reversed, no IAA moves from the base of the segment toward the apex. If the stem segments are turned upside down [Figure 15-7(c) and (d)], IAA moves from the morphological tip of the segment toward the base [Figure 15-7(d)] rather than against the morphological gradient [Figure 15-7(c)]. Polar transport is negated if the stem segments are maintained under anaerobic conditions or treated with respiratory inhibitors.

Polar transport of IAA is strongly developed in monocot coleoptiles. In other plants, however, polar transport is less strongly developed and is found to depend on the age and metabolic status of the tissue, the position of the tissue in the plant, developmental status (vegetative or reproductive) of the plant, wounding, and environmental factors. The physiological basis for polar transport is not known at present. Polar transport is an attribute of living cells and it is reasonable that the phenomenon of polarity resides within each cell. Such polarity might be due to various factors:

1. An asymmetric distribution of subcellular organelles (mitochondria, sphero-somes, ribosomes, plastids).
2. Asymmetric membranes (plasma membrane, tonoplast, endoplasmic reticulum).
3. Asymmetric distribution of enzymes within the cell.
4. Differences in distribution of inorganic ions or other small molecules within the cell.

A temperature gradient might also establish a metabolic gradient through a rate-limiting enzyme reaction. Small differences in IAA concentration between the apical and basal parts of a cell are sufficient to establish the different degrees of polarity observed in plants, for the small differences are magnified when a number of cells are placed end to end.

Polar transport of cytokinins, gibberellins, and abscisic acid does not seem to occur. Instead these growth substances move in xylem and phloem tissue as well as in the symplasm and apoplasm. There is considerable evidence that gibberellins and cytokinins are not transported as free molecules. Gibberellin glucosides have been isolated from bleeding sap (xylem exudate) of sunflower, pea, grape, and other plants and it has been suggested that the glucosides may function as storage forms of GAs as well as transport forms. Similarly, conjugated forms of cytokinins and abscisic acid have been reported in xylem and phloem exudates. There seems little doubt that plant growth substances can be transported readily within plants. The precise pathways, however, are not known in every case.

The movement of ethylene in plants has concerned plant physiologists for some time. As a gas, ethylene readily diffuses throughout the plant and eventually makes its way out of the plant. Many cells, tissues, or organs may be in contact with concentrations of ethylene in the physiologically active range. Not all parts of the plant appear to respond to ethylene, however, and the concept of "target" sites has been advanced to explain why only certain cells or tissues respond specifically to ethylene. The target may depend on the particular stage of plant development or on certain environmental conditions, such as anaerobiosis, temperature, and metabolite concentration.[1]

---

[1]It might be noted that the "target" concept need not apply only to ethylene. If all meristematic cells are capable of synthesizing all plant growth substances, only specific cells might respond to appropriate internal and external factors that trigger synthesis, activation, or release from bound forms of the growth substance in question.

Elucidation of the pathway of ethylene biosynthesis (Figure 15-2) has provided additional insight into this matter. The key intermediate in ethylene synthesis is 1-aminocyclopropane-1-carboxylic acid (ACC) and its formation is catalyzed by the enzyme ACC synthase. If the enzyme is not present or is in an inactive form, ACC—and eventually ethylene—is not made. An ethylene target cell may be one in which sufficient ACC is present so as to provide for ethylene formation. Either the enzymic apparatus is active in forming ACC from methionine or ACC may be transported to the target cell in the xylem or phloem or by diffusion from neighbor cells. According to this concept, a single cell may be capable of synthesizing ethylene, which then exerts its physiological effect in that cell. Long-distance transport of ethylene is not essential, but ACC might be transported to target sites in the plant.

MECHANISMS OF ACTION OF PLANT GROWTH SUBSTANCES

The nature of the mechanisms involved in the regulation of growth and development is not completely understood today. Studies with microorganisms and, to a lesser extent, multicellular organisms indicate that several different kinds of regulatory systems are to be found. It is generally agreed that the basic information necessary for the production of proteins (enzymes) that regulate development is stored in DNA of the nucleus. Chloroplasts and mitochondria also contain DNA responsible for production of specific proteins characteristic of these organelles. The flow of information proceeds from DNA by way of RNA to the synthesis of a multiplicity of proteins having structural and enzymic properties. Growth and development depend on organized and regulated metabolic processes responsible for the synthesis of nucleotides, proteins, polysaccharides, lipids, porphyrins, growth regulators, and so forth.

Besides the regulation of protein synthesis, metabolic processes present a number of possibilities for regulatory mechanisms. Included are such factors as inorganic and organic cofactors for enzyme reactions, ATP/AMP–ADP ratios, ratios of NAD^+ to NADH or $NADP^+$ to NADPH, availability of carbon skeletons for biosynthetic processes, availability of reduced nitrogen, transport processes, membrane permeability, and the like. The subcellular organization of the cell involving the compartmentalization of metabolic reactions offers additional opportunities for the operation of regulatory systems. Some features of these control processes were discussed in Chapter 3.

The Role of Plant Growth Substances in Regulatory Processes

When indole-3-acetic acid was recognized as a natural plant hormone, numerous attempts were made to obtain information on its probable mode of action. Because IAA appears to be involved in a wide array of physiological processes, it was reasonable to assume that it might influence some basic molecular or cellular event in cell growth. As earlier chapters have shown, the growth of a cell encompasses biochemical and biophysical processes that lead to the transformation of simple molecules, such as carbon dioxide, water, sugars, amino acids, and inorganic ions, into proteins, nucleic acids, polysaccharides, and other large complex molecules. These large molecules are further elaborated into subcellular organelles, membranes, primary cell walls, and the like. Finally, groups of cells are organized into tissues and organs. From this complex array of cellular activities investigators tried to identify a so-called primary event or key metabolic process particularly amenable to modification through the action of IAA. Following the discovery of gibberellins, cytokinins, abscisic acid, ethylene, and other natural plant growth substances, similar attempts were made to link their action to molecular and cellular activities of cell growth.

Despite numerous attempts to localize an IAA-sensitive metabolite, no specific site of action was discovered. It was found that auxins influence the levels of starch, sugars, phosphorylated sugars, fatty acids, lipids, amino acids, and organic acids, but these changes in the chemical composition of auxin-treated plants seem to be secondary rather than primary effects.

It was not until the relationship between nucleic acids (DNA and RNA) and protein synthesis was elucidated that biologists were provided with a model system for studying the action of plant growth substances. Both animal and plant physiologists have developed schemes for the role of hormones in the regulation of growth and protein synthesis. Because enzymes are essential in metabolism, any process that might regulate the synthesis, activation, or degradation of enzymes is of major significance. A single enzyme is capable of catalyzing the transformation of a large number of substrate molecules into new products. If a growth substance can influence enzyme action, then a few molecules of a plant growth substance might be able to effect a very large chemical or physiological change. This is precisely the observed effect of many plant growth substances—a few molecules initiate a large physiological change.

Although much of current research on the mode of action of plant growth substances is concerned with their role in regulating protein synthesis, it should be noted that other physiological or biochemical processes might also be affected. Changes in rates of protoplasmic streaming or changes in cell turgor, for instance, occur almost immediately following the application of an exogenous plant growth substance. The initiation of protein synthesis, on the other hand, usually requires a much longer period of time following treatment with a plant growth substance. It is unlikely that plant growth substances act through a single physiological or biochemical event but rather at several levels of cellular activity, depending on internal and external factors. Cell in meristematic regions, for example, may react in a different manner to a growth substance than cells undergoing expansion and differentiation. Furthermore, external factors, such as temperature and radiant energy, may influence the pattern of reaction to a plant growth substance.

Some of the ways that interactions between plant growth substances and the plasma membrane might influence cell behavior are illustrated in Figure 15-8. The primary event is shown to be the binding of the growth substance to a site on the plasma membrane. Inasmuch as the membrane is composed of proteins, glycoproteins, and lipids, there are opportunities for different kinds of binding between the growth

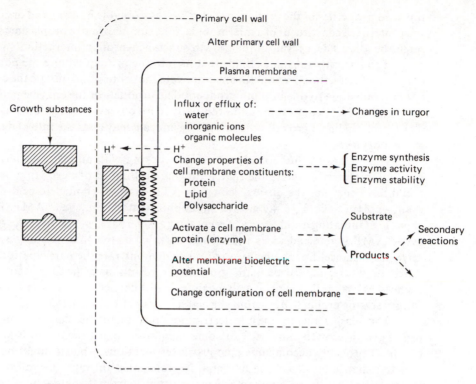

Figure 15-8   Some possible changes in cellular activities following attachment of a growth substance molecule to a specific site on the plasma membrane (or cell wall). The initial effects may be very rapid—change membrane potential, alter configuration of some membrane constituent, permit influx or efflux of solutes—whereas other effects may follow more slowly.

substance and membrane. Also, it is possible that the growth substance might bind to some constituent in the cell wall and thereby initiate a physiological or biochemical event. Furthermore, the growth substance might interact at other membrane surfaces within the cell—nucleus, chloroplast, dictyosome, tonoplast, ribosome, microtubules, and so forth. In the discussion that follows it is assumed that the growth substance binds to a protein (or glycoprotein) molecule in the plasma membrane.

Because of the growth substance binding to a protein within the membrane, the conformation of the protein might change so as to alter the transport properties of the membrane. Water, inorganic ions, or organic solutes may then enter or leave the cell, thereby altering the osmotic environment within the cell. With the change in osmotic environment, biochemical reactions are influenced and cell turgor is altered. Afterward a whole series of secondary processes might then occur that ultimately lead to a visible change in the growth form of the plant—bending, organ initiation, change in chemical composition, and the like.

The binding of the growth substance to a membrane protein may alter the physical properties of the protein in such a way as to enable it to expand or contract. If the protein is a constituent of a microtubule, for example, then protoplasmic streaming might be altered as a result of the growth substance–protein interaction.

If the growth substance binds to a membrane protein having the potential for enzymic activity, the act of binding may activate the protein and initiate the conversion of some substrate to one or more products. The products of the enzyme reaction may then lead to secondary events and eventually to an observable physiological process. The act of binding a growth substance to an enzyme may alter the rate of degradation of the enzyme.

The studies of Sutherland and others working with animal cells have been of particular interest. It was found that, following the binding of a growth substance to a membrane enzyme, the enzyme converts ATP to another nucleic acid derivative, cyclic-AMP (cyclic-3′, 5′-adenosine monophosphate). The cyclic-AMP, in turn, is capable of initiating a wide array of secondary and tertiary events. In such systems, the cyclic-AMP is referred to as the *second messenger*, the first messenger being the growth substance. Despite many attempts to demonstrate the participation of cyclic-AMP in the regulation of plant growth, the results have not been conclusive. If cyclic-AMP is applied to plant cells, changes in cellular activity are noted, but a system for generating cyclic-AMP within the cell has not yet been isolated.

The plasma membrane bioelectric potential might be changed following the binding of a growth substance to some membrane constituent. Such a change in potential probably accompanies the processes mentioned, but it might be imagined that the change in membrane potential is propagated through several or many cell membranes (or cells) to a site some distance removed from the initial binding reaction. Thus a biochemical or physiological process might be initiated at a site different from the point of the growth substance–membrane binding.

In the preceding processes it is seen that the primary action of the growth substance is binding with some constituent of the cell membrane. Such a binding may have an immediate effect—protein conformation, turgor change—or trigger some process that takes some time before a visible event is observed. As noted, one of the current areas of interest to plant physiologists studying the mode of action of plant growth substances is their effect on nucleic acid and protein synthesis. Some of these studies will now be discussed.

Plant Growth Substances and Protein Synthesis

The current ideas of the relationships between nucleic acids and protein synthesis were discussed in Chapter 3. There is a flow of information from DNA to several RNA molecules that subsequently are used in the synthesis of new proteins. A plant growth substance might induce or repress the synthesis of the various RNA molecules (rRNA, tRNA, mRNA) on the DNA template. Such a mode of regulation is said to operate at the transcriptional level. Or the growth substance might interact with one of the RNAs in such a manner as to facilitate or inhibit protein synthesis. If regulation is exerted here, it is said to operate at the translational level. Growth substance–RNA complexes have been isolated from plant tissues, but the RNAs have not been identified and the role of the complexes is not understood. Finally, a growth substance might be responsible for regulating the amount of DNA available for transcription. In the nucleus the DNA is coated with protein, forming a DNA-protein complex referred to as chromatin. Transcription occurs when a segment of the DNA is exposed by stripping away some of the protein coating. The plant growth substance might play a role in maintaining the integrity of the DNA-protein complex or in removing specific segments of protein from the DNA core.

Growth substances could also influence enzyme activity without being directly involved in some aspect of protein synthesis. A growth substance might directly interact with an enzyme and alter its tertiary structure in such a way as to modify the enzymic activity of the protein. Another possibility is that a growth substance might convert an inactive enzyme (a zymogen) into an active enzyme. Such effects have been noted in enzymes from mammals but have not been reported in plants. Finally, it might be imagined that a growth substance could enhance the activity of an enzyme by preventing its degradation. Some sort of growth substance–enzyme complex may be formed.

Clearly there are many possible sites of action where plant growth substances could interact with enzymes—synthesis, activation, degradation, conformation, and so forth. Most of the research on this problem by plant physiologists has dealt with various aspects of enzyme synthesis. Several examples are discussed to illustrate the nature of the problem and the techniques used.

THE MECHANISM OF ACTION OF INDOLE-3-ACETIC ACID

Many studies have shown that IAA interacts with one or more components of the biochemical systems involved in the synthesis of proteins. Yet it has not been possible to identify the precise step where IAA exerts an effect. As noted in previous sections, there are many points where regulation might occur in the interplay between DNA, RNA, and the synthesis of a specific protein. The fact that IAA appears to participate in several physiological processes suggests that more than one protein or enzyme may be involved in growth substance action. Because of the difficulties of resolving the multiplicity of IAA effects, investigators have turned to one of the first observed effects of IAA on plant growth—the extension growth of cells in coleoptiles and stems.

It was shown earlier that IAA promotes the elongation of cells in grass coleoptiles and in internodal stem tissue. The cells elongate, mainly in a vertical direction, by an increase in cell size and fresh weight. The fresh weight increase is due almost entirely to the uptake of water. The elongating cells are enclosed by a primary cell wall that must increase in area in order to accommodate the increase in cell volume as water enters (see Chapter 12).

A cell can expand in several possible ways. Some osmotically active substance, such as a sugar, might be introduced into the vacuole. Water will then enter the cell and the cell wall will stretch until sufficient wall pressure develops to stop further water uptake. The cell wall is then stabilized by the addition of new cell wall constituents. In this scheme, one of the roles of IAA might be to alter the osmotically active contents of the cell vacuole. It is unlikely that IAA itself is osmotically active, for only a few IAA molecules are necessary to induce cell elongation. Another possibility is that IAA influences the synthesis or activity of an enzyme in the cytoplasm or cell wall that hydrolyzes polysaccharides to yield osmotically active monosaccharides, such as glucose or fructose. Although such an effect has been observed in storage tissue, it does not seem important in coleoptile and stem tissue.

Another possible mechanism of cell expansion involves a loosening of the primary cell wall. With the decrease in wall pressure, water enters the vacuole and the cell volume increases until a new equilibrium is attained with the stretched cell wall. The cell wall is then stabilized around the enlarged cell. The role of IAA in this scheme of cell enlargement might be at some step in the process leading to the loosening of the cell wall. Much of the current research on the role of IAA is now centered on various aspects of the structure and function of the cell wall. Before considering some recent evidence on this problem, it will be useful to examine briefly several mechanical and chemical properties of plant cell walls.

Some Mechanical Properties of Plant Cell Walls

Plant cell walls behave like other organic polymers when placed under tension. Thus a rubber band stretches under tension and a polyethylene rod bends if a weight is hung from one end. If a polymer, such as rubber, is stretched under tension, the rubber instantly returns to its original shape when the tension is released. Such a reversible process is referred to as *elastic extension*. If the polymeric material does not return to its original shape on the removal of the applied tension or force, the process is referred to as *plastic* or *irreversible extension*. Also, plastic extension, unlike elastic extension, is time dependent in that more deformation occurs the longer the tension or force is applied.

The two types of extension, elastic and plastic, have been measured in coleoptile segments by several procedures. One technique involves securely fixing a coleoptile in a horizontal position and then attaching a weight to the tip (Figure 15-9). Over a period of 20 to 30 minutes the coleoptile will be displaced from the horizontal by some measurable distance, which we can call M. If the weight is removed, the coleoptile will move toward the horizontal but will not return to its original position. The distance that the coleoptile is displaced from the horizontal we can call N. The distance N is a measure of the plastic component (irreversible) of extension, whereas $M - N$ is a measure of the elastic or reversible component of extension. It should be noted that growth of the coleoptile is not involved. If the cells in the coleoptile are killed by immersing the coleoptile in boiling alcohol, plastic and elastic extension can still be measured.

If the coleoptiles are treated with a solution of IAA before being subjected to the manipulation described, the plastic, irreversible component of extension is increased.

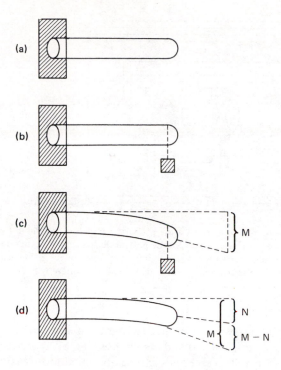

Figure 15-9 Demonstration of plastic and elastic components of extension in primary cell walls of coleoptiles. The coleoptile segment is clamped firmly in a horizontal position (a) and then a small weight is attached to the tip of the segment (b). After 20 to 30 minutes (c) the coleoptile has been displaced some distance (M) from the horizontal. If the weight is removed, the coleoptile will return to some intermediate position (N) between the position reached in (c) and the initial, horizontal position (b). (d) The distance (N) is a measure of the plastic component of extension whereas (M − N) is a measure of the elastic component. If the coleoptile is treated with IAA before the application of a weight, the plastic component is increased.

The elastic, reversible component of extension growth is not greatly altered. Such results are interpreted as an indication that IAA modifies the plasticity of the cell wall. Although the cell wall component susceptible to IAA is not known, recent studies of the chemical structure of cell walls provide a basis for considering several possible sites of action.

Chemical Structure of Primary Cell Walls

It was shown in Table 3-3 that the primary cell wall of sycamore is composed of cellulose, pectic polysaccharide, protein, and a polysaccharide fraction composed of xyloglucans, arabinogalactans, and rhamnogalacturonans. Most of the cell wall components are bound together by covalent linkages that are relatively stable. It has been proposed, however, that the cellulose microfibrils are not covalently attached to the other wall components but rather are bound by hydrogen bonds to one of the polysaccharide fractions, xyloglucan. Such a model of the primary cell wall is shown in Figure 15-10.

The cellulose microfibrils are embedded in a structured matrix composed of protein, pectic polysaccharides, and hemicellulose (polysaccharides). The matrix materials are tied together through covalent bonds—bonds that are formed through

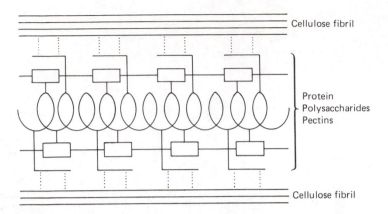

Cellulose fibril

Protein
Polysaccharides
Pectins

Cellulose fibril

Figure 15-10 Schematic representation of the primary cell wall of sycamore (see Figure 3-18, Chapter 3). The xyloglucans, pectic polysaccharides, polysaccharides, and protein are linked together with covalent bonds. Segments of the xyloglucan molecules are aligned with the cellulose fibrils and are held in position by hydrogen bonds (shown as dotted lines).

enzymic activity. One of the matrix components, a xyloglucan, is covalently bound to other matrix components, but it is aligned and held in position by hydrogen bonds to the cellulose microfibrils. The hydrogen bonds can be disrupted more easily than the covalent bonds holding the matrix constituents together. Thus the xyloglucan sheath surrounding the cellulose microfibrils might slip and allow stretching or bending of the cell wall.

The Role of IAA in Cell Wall Extension

If the structure shown in Figure 15-10 is a reasonable representation of the primary cell wall, what is the role of IAA in promoting cell extension? As noted, IAA appears to have a role in the loosening of cell wall components. In terms of the cell wall model illustrated in Figure 15-10, this suggests that IAA is involved in modifying the bonds holding together the matrix material and cellulose microfibrils.

Consider, first, the hydrogen bonds. Such bonds can be weakened in several ways. An increase in temperature, for instance, will loosen hydrogen bonds. A more probable event, however, is the influence of hydrogen ions (H^+, protons) on the integrity of hydrogen bonds. It has been known for some time that coleoptile segments, or dicot hypocotyl sections, elongate if immersed in acid solutions in the absence of IAA. From such studies have come what is known as the "acid growth" hypothesis of tissue elongation. Indole-3-acetic acid itself is not a strong enough acid to increase the hydrogen ion concentration in the cell wall. Studies indicate that a pH of around 4 must be established in the cell wall if tissue extension is to occur. What could the source of the protons in the cell wall be? As described in Chapter 11, both anions and cations (including H^+) are moved through the plasma membrane by active processes referred to as ion pumps. One such pump might transport protons (H^+) across the plasma membrane, giving rise to locally high concentrations of protons in the cell wall. The role of IAA might be the activation of the proton pump in the plasma membrane, followed by the secretion of the protons into the cell wall where wall loosening occurs.

Tissue extension involves more than loosening of components of the cell wall. After cell volume has increased and the cell has enlarged, the cell wall must be stabilized. This process seems to involve the formation of new cell wall material. A number of studies have shown that IAA modifies the activity of enzymes engaged in the synthesis of cell wall components (polysaccharides, glycoproteins, etc.). Thus to sustain tissue expansion, new cell wall material must be synthesized and organized into stable matrix components.

The role of IAA in cell wall extension appears twofold. First, IAA may activate a proton pump, either in the plasma membrane or cell wall, which maintains the cell wall in an acid condition of around pH 4. Under these conditions cell wall components are loosened. With the relaxation of the wall around the cell, wall pressure is reduced and water enters the cell, accompanied by an increase in cell volume. As the cell increases in volume, the cell wall matrix components are modified and new cell wall material is necessary to stabilize the wall. A second role of IAA may be in activating or modifying enzymes involved in synthesizing cell wall material.

THE MECHANISM OF ACTION OF GIBBERELLINS

One of the physiological effects of gibberellic acid is to increase the activity of hydrolytic enzymes during the germination of barley and rice seeds. Original interest in this problem came from observations of the chemical changes in barley grains that occur during malting. In this process, barley grains imbibe water and start to germinate, during which time starch in the endosperm is converted to sugar. The converted grain, now known as malt, is then used to support the growth of yeast, which transforms the sugar to alcohol. It was noted that some diffusible substance in the barley seeds is responsible for initiating the formation of an enzyme that converts starch to sugar. From those studies came the general picture of the changes occurring during the germination of barley seeds. Gibberellic acid is released from the embryo and diffuses to the aleurone cells, where hydrolytic enzymes (α-amylase, protease, ribonuclease, β-glucanase, phosphatase, and possibly others) are synthesized or activated. These enzymes then diffuse into the endosperm, where they catalyze the digestion of stored macromolecules and convert them to soluble sugars, amino acids, nucleosides, and so forth, which support growth of the embryonic plant during germination and seedling emergence (see Chapter 16 for details).

Because of the ease of measuring α-amylase activity, the barley seed provides a useful system for studying the role of gibberellic acid in enzyme synthesis. Barley seeds are cut at right angles to the long axis of the seed and the embryo half is discarded. The remaining half consists of an outer layer of aleurone cells (living cells) surrounding dead endosperm cells filled with starch (and other stored materials, such as proteins, lipids, and nucleic acids). When the half-seeds are incubated in a medium containing gibberellic acid, the aleurone cells synthesize α-amylase, which is excreted into the medium. The activity of the α-amylase in the medium, or what remains in the aleurone layer, is measured by determining its ability to hydrolyze a standard starch preparation. Either the disappearance of starch or the appearance of soluble sugar is measured. Rather than work with half-seeds, it is possible to excise the aleurone layer cells and study their responses to applied gibberellic acid.

The time course for the development of α-amylase activity in half-seeds, or isolated aleurone layers, in the presence of 10^{-6} M gibberellic acid is shown in Figure 15-11. There is about a 7-hour lag before α-amylase activity appears. After the initial lag period, α-amylase production increases until at 33 hours there is a drop in enzyme production. Differences in the amount of α-amylase formed in the half-seeds compared with the isolated aleurone cells are caused by a leakage of metabolites from the isolated aleurone cells. With appropriate manipulation of the incubation medium, isolated aleurone cells synthesize α-amylase as vigorously as intact seeds.

Another feature of the relationship between gibberellic acid and α-amylase

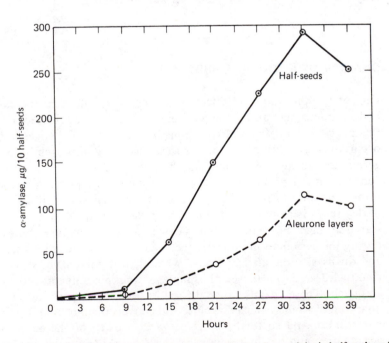

Figure 15-11   Time course for the development of α-amylase activity in half seeds and in isolated aleurone layers of barley. Gibberellic acid (10^{-6} M) added at 0 time. (After Varner and Chandra, 1964; Figure 1.)

production in aleurone layers of barley is shown in Figure 15-12. In this experiment the aleurone layers were incubated in GA for 7 hours (lag period in Figure 15-11), after which the GA was removed by four consecutive half-hour rinses. Gibberellic acid was then added to one set of aleurone layers and α-amylase production was measured at 15 and 23 hours. Another set of aleurone layers did not receive any GA and α-amylase production was determined at 15 and 23 hours. A third set of aleurone layers was incubated in the absence of GA for 6 hours, followed by the addition of GA during incubation for an additional 8 hours. The production of α-amylase drops off in the absence of GA, but on the addition of GA, α-amylase is again produced at rates comparable to those achieved in the presence of GA. These results indicate that GA must be present in order to maintain the production of α-amylase.

Experiments by Varner and associates have provided further information about the role of gibberellic acid in germinating barley seeds. Dry seeds contain no detectable amounts of α-amylase, but about 7 to 8 hours after the onset of water imbibition (the

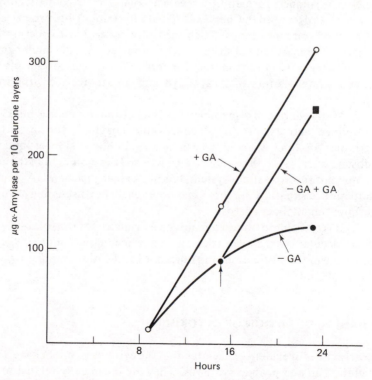

Figure 15-12 Effect of removing gibberellic acid at end of lag period. Isolated aleurone layers were incubated for 7 hours in $0.5\,\mu$M of GA. The GA was then removed by four consecutive 0.5-hour rinses. The aleurone layers were further incubated either in GA ($-\bigcirc-\bigcirc-$), without GA ($-\bullet-\bullet-$), or in the absence of GA for 6 hours, followed by the addition of GA ($-\blacksquare-\blacksquare-$). (After Chrispeels and Varner, 1967; Figure 3.)

lag phase in Figure 15-11) the enzyme appears, being formed by de novo protein synthesis. Despite this finding it has not been possible to delineate precisely the specific stage of protein synthesis where gibberellic acid might function during transcription when specific RNAs are formed on DNA templates. Of the RNAs transcribed, a messenger RNA (mRNA) containing the information for the synthesis of α-amylase might be involved. Or gibberellic acid might participate in the formation of some essential ribosomal RNA (rRNA). Still another possibility is that GA participates in the translation process when mRNA, rRNA, and transfer RNA (tRNA) interact to form the protein molecule α-amylase. With the evidence now at hand, the specific role of GA in protein synthesis cannot be identified.

The role of gibberellins in the induction of de novo synthesis of α-amylase and other hydrolytic enzymes during germination of barley and other monocot seeds is well documented. Comparable studies with dicot seeds indicate that gibberellins are also involved in mobilizing seed storage reserves during germination and seedling emergence. Studies with germinating castor bean seeds have shown that gibberellic acid (GA_3) induces formation of several enzymes involved in the conversion of lipids to sucrose. As described in Chapter 16 (Seed Physiology), lipid reserves in the endosperm of castor bean and other fat-rich seeds are converted to sucrose by gluconeogenic enzymes compartmented in the cytosol and several subcellular organelles—that is, glyoxysomes and mitochondria. The term *gluconeogenesis* means synthesis of sugar from substances other than carbohydrates. In fat-rich seeds the source of sucrose is lipid.

Many enzymes are present in the cytosol and organelles, but GA_3 appears mainly to induce activity of the gluconeogenic enzymes during early stages of seed germination. Such specificity ensures a rapid conversion of lipid to sucrose, which subsequently is used in supporting growth and development of the embryonic axis to a competent root and shoot system. It is not known just how GA_3 is involved in enzyme induction, a situation similar to that noted earlier in discussing α-amylase induction in germinating monocot seeds.

It is difficult to obtain unequivocal evidence for the role of a growth hormone at the molecular level. Nevertheless, with new techniques of organelle isolation, enzyme isolation and identification, and hormone localization, rapid progress will be made on this problem.

THE MECHANISM OF ACTION OF CYTOKININS

Several lines of evidence suggested that the cytokinins might have a role in nucleic acid metabolism and protein synthesis. They are chemically related to adenine, a purine base found in both DNA and RNA. Also, cytokinins have been isolated from meristematically active parts of the plant where vigorous nucleic acid and protein synthesis occurs. Moreover, cytokinins prevent the appearance of leaf yellowing during senescence. Ordinarily a senescing leaf turns yellow as chlorophyll is degraded, but cytokinins appear to activate a number of metabolic processes at the site of application and inhibit chlorophyll breakdown. With the change in metabolism at the site of cytokinin application, such molecules as amino acids, inorganic phosphate, sugars, and other solutes accumulate. Although these effects of cytokinin are consistent with the idea that cytokinins have some role in nucleic acid metabolism and protein synthesis, the specific role of cytokinins has yet to be discovered.

One possible role of cytokinins in plant metabolism is suggested by the discovery that some tRNAs possess cytokinin activity. Recall from Chapter 3 that protein synthesis involves several forms of RNA (mRNA, rRNA, and tRNA). The role of tRNA in protein synthesis is to supply specific amino acids to the ribosomes where the protein molecule is assembled (Figure 3-12). Because there are some 20 amino acids, there are at least 20 different tRNA molecules. Each tRNA molecule is composed of about 70 to 80 nucleotides (nitrogen base-sugar-phosphate). Some of these nucleotides are potent cytokinins.

It has been observed that cytokinins influence the activity of certain enzymes. In some situations, cytokinins substitute for gibberellins in inducing the de novo synthesis of α-amylase in germinating seeds. It was noted, however, in the previous section that the site of action of gibberellins in protein synthesis is not known.

In the case of the inhibition of leaf senescence by cytokinins, it is suggested that the cytokinins act on some enzymes responsible for the formation of certain amino acids. Also, cytokinins inhibit the breakdown of proteins during senescence.

From these various effects of cytokinins on metabolic processes it seems clear that they play a role in some phase of nucleic acid metabolism or protein metabolism. In such a role, the cytokinins are important in numerous phases of growth and development.

THE MECHANISM OF ACTION OF ABSCISIC ACID

Abscisic acid interacts with indole-3-acetic acid, cytokinins, and gibberellins in numerous physiological activities. In some of these interactions ABA has a promotive effect whereas in others it negates or inhibits the physiological action of another plant growth substance. The most reasonable explanation for such a multiplicity of effects is that ABA and the other plant growth substances exert their influence through some aspect of nucleic acid metabolism and protein synthesis. As has been pointed out, however, direct evidence has been difficult to obtain and no specific site of action of ABA has been discovered.

The effect of ABA on guard cells and stomatal activity is poorly understood. Apparently ABA alters some membrane property so that the membrane becomes "leaky" to K^+. With the loss of K^+ from the guard cells, solute concentration drops and water moves from the guard cells into surrounding cells. The loss in guard cell turgor leads to stomatal closure. When stress conditions are relieved, ABA disappears and K^+ is transported into the guard cells and the drop in solute concentration is reversed. The specific effect of ABA at the plasma membrane where K^+ "leaks" from the guard cells is unknown.

THE MECHANISM OF ACTION OF ETHYLENE

Many of the effects of ethylene on plants—organ abscission, fruit ripening, senescence, flower petal discoloration, epinasty—appear to be associated with alterations in the properties of cellular membranes. It is thought that ethylene binds to some component of the membrane thereby initiating secondary responses in a manner similar to those shown in Figure 15-8. However, there is little information on the sequence of events following membrane binding and subsequent biochemical or physiological responses.

USES OF PLANT GROWTH SUBSTANCES IN AGRICULTURE AND HORTICULTURE

The lack of a satisfactory model of the mechanism of action of plant growth substances has not prevented their use in agricultural and horticultural practices. From the early work on IAA came the synthesis of large numbers of molecules with auxin activity. Several—indole butyric acid and naphthaleneacetic acid (Figure 14-7)—were found useful as rooting hormones for woody and herbaceous plants and they are widely used today for this purpose. Other compounds, such as 2,4-D (Figure 14-7), were found to be potent inhibitors of plant growth and have been used as weed killers (herbicides). Because plants differ in their susceptibility to the synthetic auxins, it has been possible to develop compounds to kill monocots (grasses) growing in fields of dicots (soybeans) or to kill dicots (chickweed, pigweed) growing in grass (maize, wheat, turf). The use of synthetic auxins as herbicides accounts for about one-half the money spent worldwide on agricultural chemicals. The other half includes chemicals used as insecticides, fungicides, and growth regulators.

In 1980 it was estimated that about 8000 million dollars was spent on agrichemicals worldwide and it is estimated that this figure will increase to about 10,000 million dollars in 1984. Of this amount, the largest increase is estimated to occur in the use of growth regulators. Why this great interest in growth regulators? What aspects of plant growth and development are target areas for exploitation?

Interest in the use of growth regulators in crop production arises from the beliefs of plant physiologists and agriculturists that maximum levels of plant productivity have been reached by current technology. The large increases in crop yields achieved over the past 30 years have utilized the technologies of the green revolution—improved seeds, new varieties of plants, fertilizers, irrigation, pesticides, (herbicides, fungicides, insecticides, nematocides), and mechanization. Yields of many crops have now reached a plateau and additional increments of fertilizer, pesticides, water, and similiar items are no longer economical. Thus agriculturists and horticulturists are looking for ways to break present yield limits by chemical modification of the plant through the use of chemical growth regulators.

What aspects of plant growth and development are likely targets for modification? Nickell (1981) has listed some possibilities along with several of the chemicals now being used (Table 15-1).

TABLE 15-1   PLANT GROWTH REGULATORS PRESENTLY IN USE IN U.S. AGRICULTURE

Common name	Chemical name	Structure	Use
Chlormequat (Cycocel)	2-Chloroethyl trimethyl-ammonium chloride	$\left[\begin{array}{c} CH_3 \\ | \\ CH_3-N-C_2H_4Cl \\ | \\ CH_3 \end{array}\right]^{+} Cl^{-}$	Dwarfing agent in wheat and poinsettias.
Ethephon	2-Chloroethyl phosphonic acid	$ClCH_2CH_2PO_3H_2$	Induce flowering in pineapple. Sugarcane ripener.
Maleic hydrazide	1,2-Dihydro-3,6-pyridazine-dione	(structure: pyridazinedione ring)	Growth retardant. Sucker control on tobacco. Turf grass inhibitor.
Triacontanol	1-Hydroxy-triacontane	$CH_3(CH_2)_{28}CH_2OH$	General growth stimulant.
Gibberellins	(See Figure 14-1)		Enhance berry size of grapes.
Diaminozide	Succinic acid-2,2-dimethyl hydrazine	(structure with CH_3, O, $NNHCCH_2CH_2COOH$)	Enhance size and color of various fruits.
Glyphosine	N,N-Bis(phosphono-methyl)glycine	(structure: $HOOC\,CH_2N$ with $CH_2PO_3H_2$ branches)	Sugarcane ripener.

Nickell, 1981; Hardy, 1979

Some chemicals presently used in commercial agriculture and horticulture are structurally related to endogenous plant growth substances. Table 15-1 shows, however, that many biologically active compounds bear little resemblance to natural substances. In the absence of a molecular basis for growth substance activity, it is necessary simply to apply various organic molecules to plants and observe the results. Biological screening procedures can be developed for selecting compounds for growth stimulation, yield enhancement, abscission, dwarfing, and other aspects of plant growth development. Many agricultural chemical companies now combine basic research on studying mechanisms of growth substance activity and empirical screening of new and extraordinary organic molecules. In addition to screening for biological activity in plants, it is essential that the effect of the compound on mammals (including humans) and other organisms be evaluated. Most of the compounds shown in Table 15-1 are not natural products and it is almost impossible to predict what effect they may have on other living creatures. Already there are a number of examples of foreign chemicals having an adverse effect on birds, fish, small mammals, and other organisms. Federal laws now require that new chemicals pass through rigorous screening tests before they are approved for use. Such requirements slow down the production of new agrichemicals and increase the cost but are essential to ensure that they are not deleterious to humans.

REFERENCES: Chapters 14 and 15

ABELES, F. B. 1973. *Ethylene in Plant Biology*. Academic Press, New York.

ADAMS, D. O., and S. F. YANG. 1979. Ethylene biosynthesis: Identification of 1-aminocyclo-propane-1-carboxylic acid as an intermediate in the conversion of methionine to ethylene. *Proc. Nat. Acad. Sci. U.S.* 76:170–174.

ALBERSHEIM, P. 1975. The walls of growing plant cells. *Sci. Amer.* 232(4):81–95.

BANDURSKI, R. S. 1979. Chemistry and physiology of conjugates of indole-3-acetic acid. Pages 1–17 *in* N.B. Mandava, ed. *Plant Growth Substances*, ACS Symp. Series No. 111. Washington, D.C.

BURG, S. P., and ELLEN A. BURG. 1967. Auxin stimulated ethylene formation: Its relationship to auxin inhibited growth, root geotropism and other plant processes. Pages 1275–1294 *in* F. Wightman and G. Setterfield eds. *Biochemistry and Physiology of Plant Growth Substances*. The Runge Press Ltd. Ottawa.

CHRISPEELS, M. J., and J. E. VARNER. 1967. Hormonal control of enzyme synthesis: On the mode of action of gibberellic acid and abscisin in aleurone layers of barley. *Plant Physiol.* 42:1008–1016.

EPSTEIN, E., J. D. COHEN, and R.S. BANDURSKI. 1980. Concentration and metabolic turnover of indoles in germinating kernels of *Zea mays* L. *Plant Physiol.* 65:415–421.

GONZALES, E., and M. A. DELSOL. 1981. Induction of glyconeogenic enzymes by gibberellin A_3 in endosperm of castor bean seedlings. *Plant Physiol.* 67:550–554.

GROSS, D. 1975. Growth regulating substances of plant origin. *Phytochem.* 14:2105–2112.

HALL, R. H., L. CSONKA, H. DAVID, and B. McLENNON. 1967. Cytokinins in the soluble RNA of plant tissues. *Science* 156:69–71.

HANGARTER, R. P., M. D. PETERSON, and N. E. GOOD. 1980. Biological activities of indoleacetylamino acids and their use as auxins in tissue culture. *Plant Physiol.* 65:761–767.

HARDY, R. W. F. 1979. Chemical plant growth regulation in world agriculture. Pages 165–206 *in* T. K. Scott, ed. *Plant Regulation and World Agriculture*. Plenum Press, New York and London.

HILLMAN, J. R., ed. 1978. *Isolation of Plant Growth Substances*. Cambridge University Press, Cambridge.

JACOBS, W. P. 1979. *Plant Hormones and Plant Development.* Cambridge University Press, Cambridge.

JENNINGS, D. H., ed. 1977. *Integration of Activity in the Higher Plant. 30th Symp. Soc. Exper. Biology.* Cambridge University Press, Cambridge.

KEY, J. L. 1964. Ribonucleic acid and protein synthesis as essential processes for cell elongation. *Plant Physiol.* 39:365–370.

LABAVITCH, J. M., and P. M. RAY. 1974. Relationship between promotion of xyloglucan metabolism and induction of elongation by indoleacetic acid. *Plant Physiol.* 54:499–502.

LETHAM, D. S., P. B. GOODWIN, and T. J. V. HIGGINS, eds. 1978. *Phytohormones and Related Compounds—A Comprehensive Treatise.* Vol. 1, *The Biochemistry of Phytohormones and Related Compounds.* Vol. 2, *Phytohormones and the Development of Higher Plants.* Elsevier-North-Holland Biomedical Press, Amsterdam, Oxford, New York.

LÜRSSEN, K., K. NAUMANN, and R. SCHRÖDER. 1979. 1-aminocyclopropane-1-carboxylic acid—an intermediate of the ethylene biosynthesis in higher plants. *Z. Pflanzenphysiol.* 95:285–294.

MANDAVA, N. B., ed., 1979. *Plant Growth Substances.* ACS Symposium Series III. American Chemical Society, Washington, D.C.

METZGER, J. D., and J. A. D. ZEEVAART. 1980. Identification of six endogenous gibberellins in spinach shoots. *Plant Physiol.* 75:623–626.

MILLER, C. O. 1963. Kinetin and kinetinlike compounds. Pages 194–202 *in* H. F. Linskens and M. V. Tracey, eds. *Modern Methods of Plant Analysis*, vol. 6. Springer-Verlag, Berlin.

MITCHELL, J. W., N. MANDAVA, J. F. WORLEY, J. R. PLIMMER, and M. V. SMITH, 1970. Brassins—a new family of plant hormones from rape pollen. *Nature* 225:2065.

MOORE, T. C. 1979. *Biochemistry and Physiology of Plant Hormones.* Springer-Verlag, New York, Heidelberg, Berlin.

NICKELL, L. G. 1981. *Plant Growth Regulators, Agricultural Uses.* Springer-Verlag, New York, Heidelberg, Berlin.

OKUNKANMI, A. B., D. J. TUCKER, and T. A. MANSFIELD. 1973. An improved bioassay for abscisic acid and other antitranspirants. *New Phytol.* 72:277–282.

OSBORNE, DAPHNE J. 1977. Ethylene and target cells in the growth of plants. *Sci. Prog.* (Oxford) 64:53–65.

RAPPAPORT, L., and D. ADAMS. 1978. Gibberellins: synthesis, compartmentation and physiological process. *Phil. Trans. Royal Soc.* (*London*) B284:521–539.

RIES, S. K., V. WERT, C. C. SWEELEY, and R. A. LEAVITT. 1977. Triacontanol: a new naturally occurring plant growth regulator. *Science* 195:1339–1341.

SCOTT, T. K., ed. 1979. *Plant Regulation and World Agriculture.* NATO Advanced Study Institutes Series, A: Life Sciences. Plenum Publishing Corp., New York.

SEMBDNER, G., J. WEILAND, D. AURCH, and K. SCHREIBER. 1968. Isolation, structure and metabolism of a gibberellic glucoside. Pages 70–86 *in* F. Wightman and G. Setterfield, eds. *Biochemistry and Physiology of Plant Growth Substances.* The Runge Press Ltd., Ottawa.

SHELDRAKE, A.R. 1973. The production of hormones in higher plants. *Biol. Rev.* 48:509–559.

SHIBAOKA, H., M. FURUYA, M. KATSUMI, and A. TAKIMOTO, eds. 1978. *Controlling Factors in Plant Development.* Special Issue No. 1, *Botanical Magazine*, Tokyo.

Skoog, F. 1973. A survey of cytokinins and cytokinin antagonists with reference to nucleic acid and protein metabolism. *Biochem. Soc. Symp.* 38:195–215.

Skoog, F., ed., 1980. *Plant Growth Substances in 1979*. Springer-Verlag, Berlin, Heidelberg, New York.

Skoog, F., and C. O. Miller. 1957. Chemical regulation of growth and organ formation in plant tissues cultured in vitro. *Symp. Soc. Expt. Biol.* 11:118–131.

Skoog, F., and N. J. Leonard. 1968. Sources and structure: activity relationships of cytokinins. Pages 1–18 *in* F. Wightman and G. Setterfield, eds. *Biochemistry and Physiology of Plant Growth Substances*. The Runge Press Ltd. Ottawa.

Sponsel, V. M., and J. MacMillan. 1978. Metabolism of gibberellin A$_{29}$ in seeds of *Pisum sativum* cv Progress No. 9; use of ^{2}H and ^{3}H GA's and the identification of a new GA catabolite. *Planta* 144:69–78.

Steward, F.C., and H. Y. Mohan Ram. 1962. Determining factors in cell growth: Some implications for morphogenesis in plants. *Adv. Morphogenesis* 1:189–265.

Taylor, H. F., and T. A. Smith. 1967. Production of plant growth inhibitors from xanthophylls: a possible source of dormin. *Nature* 215:1513–1514.

Thimann, K. V. 1972. The natural plant hormones. Pages 3–332 *in* F. C. Steward, ed. *Plant Physiology, a Treatise*, vol. 6 B. Academic Press, New York.

Thimann, K. V. 1977. *Hormone Action in the Whole Life of Plants*. University of Massachusetts Press, Amherst.

Trewavas, A. J. 1976. Plant growth substances. Pages 249–326 *in* J. A. Bryant, ed. *Molecular Aspects of Gene Expression in Plants*. Academic Press, London, New York.

Valent, B.S., and P. Albersheim, 1964. The structure of plant cell walls. V. On the binding of xyloglucan to cellulose fibers. *Plant Physiol.* 54:105–108.

Vanderhoef, L. N., and R. R. Dute. 1981. Auxin-regulated wall loosening and sustained growth in elongation. *Plant Physiol.* 67:146–149.

Varner, J. E., and G. Ram Chandra. 1964. Hormonal control of enzyme synthesis in barley endosperm. *Proc. Nat. Acad. Sci. U.S.* 52:100–106.

Vold, B. S., and N. J. Leonard. 1981. Production and characterization of antibodies and establishment of a radioimmunoassay for ribosylzeatin. *Plant Physiol.* 67:401–403.

Wareing, P. F., and I. D. J. Phillips. 1978. *The Control of Growth and Differentiation in Plants*, 2nd ed. Pergamon Press, Elmsford, N.Y.

Weiler, E. W. 1979. Radioimmunoassay for the determination of free and conjugated abscisic acid. *Planta* 144:255–263.

Weiler, E. W. 1980. Radioimmunoassays for *trans*-zeatin and related cytokinins. *Planta* 149:155–162.

Went, F. W., and K. V. Thimann. 1937. *Phytohormones*. The Macmillan Company, New York.

16

The Physiology of Seeds

Seeds[1] have been of interest to mankind for a long time. The early interests were largely practical, for seeds are a major source of food in most parts of the world. Consequently, information on seeds concerns their nutritive value, chemical composition, changes in composition during storage, ability to store, retention of viability, and so forth. Farmers and horticulturists are interested in the factors related to seed germination because a large part of conventional agriculture is engaged in growing plants for their seeds. Plant physiologists have also used seeds to study the influence of temperature, moisture, oxygen, light, and other factors on germination and seedling emergence.

Much of the success of modern agriculture as practiced in the United States and western Europe depends on the availability of high-quality seeds with good genetic potential and proven performance in germination, emergence, and vigorous vegetative growth. As pressure has developed for increased food production throughout the world, the use of improved and superior seeds has been stressed. Similarly, the needs for other plant products, such as oils, fibers, and industrial chemicals, have created additional demands for new seed varieties.

Although the importance of seed quality is well documented, a scientific base for seed quality has not been established. Studies on this problem are in progress in a number of laboratories, but the application of modern tools of biochemistry and molecular biology to seed physiology is just beginning. It is the purpose of this chapter to discuss some of the recent work on seed physiology, particularly as it might apply to seed quality.

---

[1]Botanically a seed consists of an embryo, variable amounts of endosperm (sometimes none), and the seed coat, or testa. In this discussion the term *seed* will be used to denote a unit of dispersal and will include true seeds and certain fruits, such as caryopses, achenes, and nuts.

SEED DEVELOPMENT

The embryo, or embryonic plant, is the beginning of a new generation. A reserve of stored food, either as cotyledons attached to the embryonic axis or as endosperm tissue, functions as an initial source of nourishment for the embryonic plant until it attains an independent autotrophic existence. The seed coat serves to protect the embryo against adverse environmental conditions and, in some cases, is adapted as a means of seed dispersal. The morphological events that occur in the flower of angiosperms preceding seed development are illustrated in Figure 16-1. It should be noted that there is considerable variation among members of the plant kingdom in the details of seed development, but generally the events shown in Figure 16-1 illustrate the basic features of the process.

The early developmental stages of the embryo sac are nourished by the cells of the surrounding ovulary tissue. These cells are originally rich in starch, lipids, and proteins, which are subsequently hydrolyzed to form soluble sugars, amino acids, organic acids, and other metabolically active materials. The ovule is also connected to the main transport system of the plant by a vascular strand through which water, ions, and other solutes are supplied to the developing seed.

The diploid nucleus of the zygote receives one chromosome complement from the female parent (egg) and one from the male parent (sperm). Thus the zygote contains all the genetic information necessary for development of the zygote into a mature plant. It is obvious, however, that the cells in a seedling or mature plant are not alike and that these differences arise very early in the life of the plant. The first division of the zygote forms two cells, which are generally different in size. Further divisions of these cells lead to embryonic development and the differentiation of tissues and organ systems characteristic of the plant.

The formation of diverse cell types and their organization into tissues and organs encompass the field of differentiation, which is one aspect of the overall process of development. How much influence does the internal environment of the embryo sac or the external environment of the parent plant have on the differentiation and develop-ment of the embryo? If there are any effects, do they also influence the subsequent growth of the seed into a mature plant? It is not possible at present to supply complete answers to these questions, but there is ample evidence that the embryo and its future development into a mature plant may be significantly modified by events that occur while the embryo is still attached to the parent plant.

ENDOSPERM DEVELOPMENT AND COMPOSITION

Endosperm tissue is composed of cells with three chromosome sets ($3n$), two from the maternal and one from the paternal parent. This is the situation encountered in many plants, but in the gymnosperms, such as pine and hemlock, the functional equivalent of the angiosperm endosperm has a different chromosome complement and is derived from the female gametophyte, which is composed of haploid ($1n$) cells. Regardless of its origin and chromosome number, the endosperm serves a very special function in

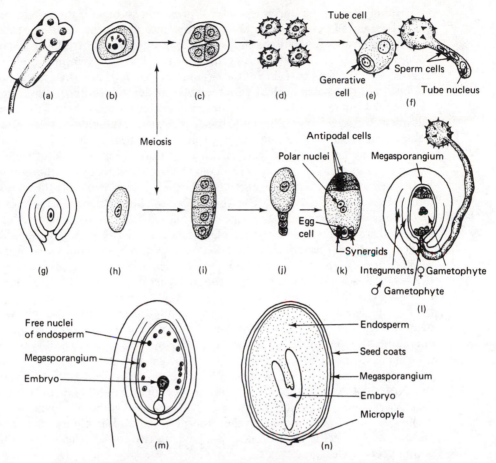

Figure 16-1 Summary of the reproductive processes in angiosperms. The megaspore mother cell within the ovule undergoes a meiotic division, giving rise to four megaspores, each containing a haploid ($1n$) chromosome set. Usually one of the megaspores survives to give rise to an embryo sac whereas the others abort. The nucleus within the embryo sac undergoes three successive divisions to form 8 nuclei: an egg nucleus, 2 synergid nuclei, 3 antipodal nuclei, and 2 polar nuclei. In the anther tissue the microspore mother cell undergoes a meiotic division to form four haploid microspores that develop into pollen grains. The pollen grains contain two cells, a tube cell and a generative cell. At pollination the pollen grains are deposited on the stigma of the ovule, where they germinate and develop pollen tubes that grow through the stylar tissue and into the embryo sac. During the growth of the pollen tube the generative cell undergoes a division to form two sperm nuclei. After penetration of the embryo sac, one of the sperm nuclei unites with the egg nucleus to form the zygote whereas the other fuses with the two polar nuclei in the embryo sac to form a triploid ($3n$) nucleus that undergoes division to give rise to the endosperm. Further development of the zygote leads to formation of the embryo. These developments are accompanied by the formation of the endosperm and by changes in certain ovulary tissues that lead to the development of the seed coat. (After Bold, 1961; Figure 9-6.)

nourishing the embryo during early seed formation and maturation and later during seed germination before the embryo develops into an independent plant.

In angiosperms where the endosperm is commonly in the triploid ($3n$) condition, the development of the endosperm generally precedes the development of the zygote. That is, even though the fusion of the egg and sperm nuclei to form the zygote ($2n$) and the fusion of the sperm nucleus with the polar nuclei to form the endosperm nucleus may occur simultaneously, the $3n$ endosperm nucleus usually divides to form numerous nuclei before the zygote begins to divide. Frequently the endosperm develops in the free nuclear condition without forming cell wall material so that a liquid endosperm containing many free nuclei results. Coconut milk is an example of such a liquid endosperm, but maize, peanut, walnut, and horse chestnut also contain similar liquid endosperms during the early stages of seed development. Probably the endosperm of many other plants passes through a similar free-cell liquid stage. During later stages of endosperm development cell walls form and a cellular, or solid, endosperm is produced. In many dicotyledonous plants the endosperm is absorbed by the cotyledons of the developing embryo. The food reserves of the cotyledons serve as a nutrient source during germination. In other plants, particularly monocots (maize, wheat, etc.), the solid endosperm persists and becomes a part of the seed, where it functions to nourish the developing embryo during seed germination.

The special role of the endosperm in supporting embryo growth was clearly demonstrated by the work of Blakeslee and van Overbeek in 1941, who found that coconut milk supported the growth of cultured immature *Datura* embryos. Since that time many workers have used coconut milk as a source of nutrients for culturing embryos and other tissues and organs. Studies by Steward and associates have shown that coconut milk and other liquid endosperms contain a number of different chemical substances, some of which are active by themselves, some of which interact with nonactive compounds present in the endosperm, and some compounds that apparently have no specific growth effect but have nutritive roles (sugars, sugar alcohols, amino acids, amides, etc.).

Carbohydrates constitute up to 85% of the dry weight of coconut milk, with glucose as the major sugar. The bulk of the carbohydrate fraction is sorbitol, a six-carbon alcohol. Two other sugar alcohols, or cyclitols, are also present in substantial amount—*myo* inositol and *scyllo* inositol. The carbohydrate components appear to serve primarily as sources of carbon for the synthesis of new protoplasm and as metabolites of respiration. The function of the inositols is less certain. In certain microorganisms they are required for normal growth and are classified as vitamins. Inositol is also an intermediate in the formation of certain cell wall components (galacturonic acid). Possibly some aspect of inositol metabolism is involved in the regulation of the free-nuclei stage of endosperm development and subsequent wall formation.

During the maturation phase of seed development inositol is transformed into phytin, the mixed potassium, magnesium, calcium salt of inositol hexaphosphate. In members of the grass family (maize, wheat) phytin is stored largely in aleurone tissue, where it is subsequently mobilized during germination. Phytin thus represents a rich store of inorganic ions that are redistributed to the developing embryonic axis.

The endosperm also contains a number of nitrogenous compounds, such as proteins, free amino acids, amides, and nucleic acids. These materials serve to maintain the nitrogen nutrition of the embryo and associated structures of the developing seed. At least some of the carbohydrates and nitrogenous components of coconut milk have no specific growth-promoting properties for supporting the growth of isolated embryos. They can be replaced by one of several sugars (sucrose, glucose, mannose, maltose) and by a mixture of amino acids, such as are present in casein hydrolysate. In some instances, tissues will grow on casein hydrolysate but will not grow in the presence of coconut milk. It is not certain that the amino acids are required, for it has been possible to grow isolated embryos of some plants in a medium in which the nitrogen is present in the form of calcium and potassium nitrate. In such cases, it is likely that the cells contain, or are capable of forming, the nitrate reductase enzyme system (see Chapter 9).

The so-called active fraction of coconut milk and other liquid endosperms, as well as the solid endosperm present in seeds of members of the grass family (maize, rice, etc.), contain several kinds of growth substances. Indoleacetic acid, gibberellins, and cytokinins have been identified and several unknown substances have been shown to be present by bioassay procedures. In maize, Bandurski and colleagues have demonstrated that indoleacetic acid is largely esterified with *myo* inositol as IAA-*myo* inositol or IAA-*myo* inositol glycosides. Very little free IAA is present. During germination IAA conjugates are transported to the embryonic axis, where they supply IAA and *myo* inositol for growth and development (see page 500).

It is clear that the endosperm contains a complex mixture of compounds, some of which are known to play an important role in the regulation of plant growth. These substances act alone or in combination with other compounds to support the development and growth of the embryo. Their source is not definitely known, but it is possible that they are partly elaborated by the endosperm from the cellular remains of the synergid and antipodal cells within the embryo sac as well as from materials drawn from the ovulary tissues and the mother plant.

THE PHYSICAL ENVIRONMENT OF THE DEVELOPING SEED

During development of the seed the embryo is subjected to the influences of a number of chemical and physical factors. In the previous section it was shown that there is considerable information concerning the chemical substances in the embryo sac that may influence embryo development. Relatively little is known, however, of the nature of the physical factors that might modify development of the embryo. As noted, the early stages of embryonic development frequently occur in a liquid endosperm. It has been found that when very young embryos are removed from the embryo sac, the success of their further development often depends on their being nurtured in a liquid medium. Older embryos do not require a liquid medium and usually develop more quickly on a semisolid substrate, such as is provided by agar. It has been suggested that the initiation of a root primordium in an embryo may depend on its being in a liquid environment whereas the shoot primordium is initiated after the embryo becomes attached and oriented in a fixed position.

Frequently the embryo sac develops a hydrostatic pressure as a result of the development of a liquid endosperm. This situation can be demonstrated by puncturing an embryo sac with a fine hollow needle; the liquid endosperm will squirt out. The interior of the embryo sac is also likely to be low in oxygen, but there is no evidence that severe anaerobic conditions develop or that respiration is inhibited.

Early stages of zygote formation and embryo growth occur within the flower and developing fruit. During these periods of development the plant may be subjected to environmental conditions that alter the normal patterns of seed development. Temperature and radiant energy fluctuations and moisture stress during flowering have been found to modify fruit and seed development. Excessive rainfall during seed development may modify subsequent seed performance by leaching organic materials out of the seed.

From these examples it can be seen that the physical environment of the embryo within the developing seed is rather specialized and may have considerable influence on embryonic development and subsequent growth of the embryo and young plant. Furthermore, the external environment surrounding the mother plant and its developing embryo may also have a marked effect on the developmental pattern of the embryo. Organic nutrients, such as carbohydrates, fatty acids, amino acids, nucleotides, and growth substances, are elaborated in the mother plant and translocated to the developing fruit and seeds. Any disturbance in the orderly synthesis of these compounds or in their movement to the fruit and seeds will modify the pattern of seed development. There are numerous reports of aberrant seed germination or seedling growth that cannot be explained on the basis of treatment effects on mature seeds. In a number of cases, it has been shown that the difficulties encountered in germination, emergence, or subsequent growth are due to altered developmental patterns during seed maturation. Some of these developmental patterns are described in the next section.

MORPHOLOGICAL AND BIOCHEMICAL CHANGES ACCOMPANYING SEED DEVELOPMENT

The morphological aspects of embryo development (embryogenesis) following pollination, fertilization, and zygote formation have been described for many plant species. The events shown in Figure 16-1 illustrate the general pattern of embryogenesis. Less information is available concerning the biochemistry of embryogenesis. Studies have been made of a number of important crop plants, such as maize, rice, pea, cotton, tomato, and cucumber, as well as a few woody plants (apple, pine). Biochemical changes accompanying embryogenesis and seed development are characterized by vigorous anabolic processes, resulting in the formation of new cells, tissues, and organs rich in proteins, nucleic acids, carbohydrates, and fats.

The general pattern of biochemical events accompanying seed development is shown in Figure 16-2, where seed dry weight is plotted against days from flower opening (anthesis). The early stages of seed development, Phase I, Figure 16-2, involve pollination, fertilization, and zygote formation, processes that contribute very little to

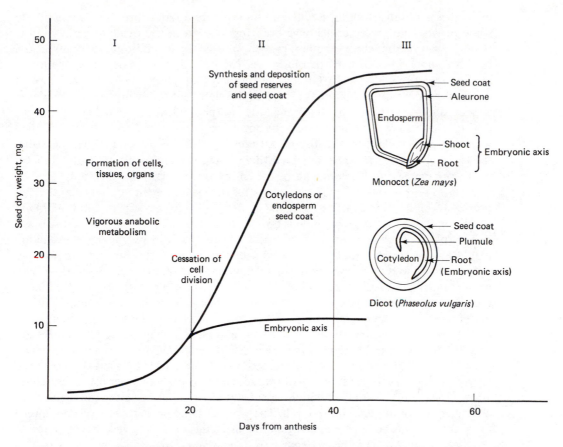

Figure 16-2 Diagrammatic representation of changes in dry weight during seed maturation. Phase I—development of embryonic axis (shoot, root). Phase II—synthesis of storage reserves in endosperm or cotyledons. Phase III—seed dessication and seed coat formation. [*Note*: Seed dry weight and the time of seed development (days from anthesis) represent relative values. Some seeds may be smaller or larger and the times of development may range between 15 days and 100 days.]

dry weight formation but must involve intense metabolic activity. Details of the biochemical changes during this period have not been studied. Much more is known of the biochemistry of embryogenesis, initiated by cell division of the zygote. The embryo increases in dry weight as new cells are formed and cellular constituents synthesized. This is a period of intense metabolic activity with a high demand for low molecular weight precursors, such as sucrose, amino acids, fatty acids, nucleosides, organic acids, water, and inorganic ions. The bulk of these materials is supplied by the parent plant through vascular connections, but some also comes from the dissolution of cellular material in the ovule and embryo sac [Figure 16-1 (k), (l), and (m)]. Phase I comes to an end when the embryonic plant is fully differentiated and cell division ceases.

Full-term embryos, excised and nourished by a suitable array of organic molecules and inorganic ions, will continue to develop and form mature plants. In some plants younger embryos have been removed from the ovule and grown. The young embryos generally require growth substances in addition to organic and inorganic nutrients. Also, in other instances, the excised embryos require exposures to low temperatures (2 to 10°C) for several days or weeks before normal plant development occurs. As far as future plant development following seed germination and seedling emergence is concerned, the morphological and biochemical events during Phase I are very important. The developing seed is in direct vascular contact with the parent plant. If environmental factors, such as low or high temperature, reduced light, moisture stress, or mineral deficiency, alter the metabolism of the parent plant, the pattern of development during embryogenesis may be altered.

If, as noted, excised embryos are capable of developing into mature plants, what prevents embryos within the ovule from developing? They do in some plants, a condition known as *vivipary*, but it is not the usual course of events. If it were to occur in crop plants, such as maize, cotton, and soybeans, early or precocious germination would have a detrimental effect on seed production and yield. Studies by Dure and colleagues on seed development in cotton have shown that precocious germination is prevented by the action of an inhibitor from ovule tissue. The inhibitor, abscisic acid, may move into the ovule through vascular connections from the parent plant or it may be synthesized in the ovule. Whatever the source, abscisic acid prevents premature or precocious germination. Later in seed development, when vascular connections between the ovule and parent plant are broken by desiccation, low seed water content prevents premature germination. It is not known if abscisic acid plays a similar role in preventing premature germination during seed development in other plants. Seeds are known to contain a variety of inhibitory substances, as described later, but their role in dormancy is not understood too well. The work of Dure provides an excellent model for studying embryogenesis and regulatory mechanisms during seed development.

As stated earlier, Phase I comes to an end when the embryonic plant is fully formed and cell division ceases. Seed dry weight continues to increase rapidly during Phase II, however, because of the synthesis and deposition of seed storage materials—starch, protein, fats, phytin, etc.—in endosperm or cotyledonary tissues. In monocots, such as maize and rice, the endosperm cells lose their nuclear material and fill up with starch and phytin. A specialized tissue, the aleurone layer, forms to the outside of the endosperm, next to the developing seed coat. The aleurone layer may be several cells thick and is composed of dense living cells filled with protein. In dicots, such as pea and bean, the contents of endosperm cells are absorbed by the cotyledons, a part of the embryonic axis (Figure 16-2). The cotyledonary cells retain their nuclear contents (remain alive) and become packed with starch, protein, lipids, and phytin.

Phase II is the period of maximum seed dry weight increase. The storage materials are synthesized from small precursor molecules from the parent plant. The synthesis and deposition of storage molecules in developing seeds constitute a major sink for carbohydrate and nitrogenous components made by the parent plant. Sucrose, the major product of leaf photosynthesis, supplies carbon skeletons for starch and fats. Moreover, sucrose is a source of carbon skeletons for nitrogenous constituents—amino acids, amides, nucleotides. During seed filling the demand for carbonaceous and nitrogenous molecules is high and may not be met by current CO_2, NO_3^-, or N_2 assimilation. In such cases, reserve materials in the parent plant may be mobilized and transported to developing seeds. To obtain maximum seed yields, especially in food plants such as maize, peas, soybeans, and beans, it is essential that the leaves and other assimilatory organs of the parent plant be kept active as long as possible. In soybeans it has been observed that leaf nitrogenous compounds have been hydrolyzed and transported to developing seeds under conditions when the roots cannot supply enough nitrogenous material to support seed filling. The loss of leaf nitrogen leads to premature leaf senescence and the loss of photosynthetic surfaces for carbon assimilation.

Cell division ceases in the embryonic axis when the embryo has reached full term (end of Phase I). It has been noted, however, that in many dicots nuclear division continues in cotyledonary cells. The process, referred to as *endoreduplication*, leads to polyploid cells with high DNA contents. It is not clear how DNA acts in this case. It may simply be hydrolyzed during germination to supply nucleotides necessary for new cell formation. On the other hand extra DNA may have a specific informational role during seed germination. Another possibility is that during seed dormancy some DNA is lost by degradation. The polyploid condition may ensure sufficient active DNA molecules for germination to occur. It is an interesting problem and a number of scientists are studying it.

Phase II (Figure 16-2) comes to an end as the seed begins to lose water. The synthesis of storage molecules involves the elimination of water molecules, but there appears to be an accelerated process of water loss, possibly through an alteration of membrane structure. Vascular connections between the developing seed and parent plant are broken so that no water or solutes can move into the seed. Moisture content during seed filling (deposition of storage material) may be in the range of 50 to 60%, but after the desiccation process is under way, water content drops to 10 to 15% at maturity. Water loss is not uniform in all parts of the seed. The embryonic axis, composed of nonvacuolated parenchyma cells, contains relatively little free water, but the structural components are hydrated. Cells in endosperm and cotyledonary tissues, however, contain low amounts of water. Also, the tissues surrounding the seed that develop into seed coats undergo desiccation and sclerification, forming a hard protective structure.

During Phase III (Figure 16-2) the desiccation process continues, attaining moisture levels of between 5 and 15% (total seed). With desiccation, the subcellular organelles in cotyledonary cells seem to lose their structural integrity. In addition, organized ribosomes (polysomes) essential for protein synthesis break up into single ribosomes. The entire picture is one of very low metabolic activity, and if seed moisture remains low, further development of the embryonic axis into a mature plant does not occur. The seed is said to be dormant. In such a condition, seeds may remain in a quiescent state for considerable periods of time. As described later, this dormant condition can be terminated by supplying water to the dry seeds.

To obtain seeds of high vitality and vigor, it is important that seed moisture levels be brought to at least 10 to 12% during maturation. In some instances, it is necessary to dry seeds by artificial heat if they are harvested early. Several hybrid maize varieties, for instance, have been developed for regions with short growing seasons. If the growing season is terminated early by frost and cold weather, the normal seed desiccation process may not bring the seed moisture levels sufficiently low for safe storage. The grain is harvested and then put through a slow drying process to attain moisture levels of around 10% or lower. This process is costly and adds to the expense of producing the crop. If the grain is not dried, it may spoil in storage by fermentive respiratory processes or by fungal and bacterial growth. In the dry state seeds can withstand environmental conditions generally unfavorable to growth: Low temperature, drought, excessive water, fire, and toxic materials in the soil.

Commercial seed producers, especially flower seed growers, follow rather stringent procedures of seed drying, storage, and packaging so as to provide the home gardener with a quality product—seeds that will readily germinate to produce a vigorous plant. Agronomic seeds, such as maize, cotton, soybeans, peanuts, cereals (wheat, oats, rice), and tobacco, also benefit from adequate harvesting, drying, and storing procedures, but the sheer bulk of the seed to be handled prevents most seed producers from maintaining close control of their seed-handling processes. Consequently, the germination and emergence of many agronomic seeds are quite variable. With the present emphasis on increasing food production, the planting of high-quality seed is as vital to increasing plant yields as irrigation, fertilizers, pest control, and improved cultural conditions.

In a seed with low moisture content, achieved either during natural ripening and maturation processes or by artificial means, the embryo is at a low level of activity, whether this activity is measured in terms of metabolic processes (e.g., respiratory loss of CO_2) or cellular processes (e.g., cell division). The embryos remain alive and capable of resuming growth when the appropriate conditions are satisfied. Seeds with such embryos are said to be viable. The length of time seeds retain their viability is quite variable and much research has been carried out in an attempt to delineate the factors responsible for maintaining seed viability. The problem is of considerable economic importance.

SEED GERMINATION

Longevity of Seeds

The length of time that embryos retain their viability, often referred to as their lifespan, varies enormously. Periods ranging from a few days to several thousand years have been reported. In general, it can be said that seed viability is governed by external environmental factors to which the seed is exposed following maturation. Many seeds can be stored for months or even years under conditions of low moisture, low oxygen levels, and low temperatures. For other seeds, dry storage at ordinary levels of oxygen and temperature is necessary to maintain seed viability. Still other seeds appear to require storage under cool, moist conditions.

The longest lifespans reported are for seeds that have been discovered accidentally in different natural situations. Seeds of the Indian lotus (*Nelumbo nucifera*) were found to be viable after being buried for about 1000 years in peat moss under a dry lake bed. Storage conditions were characterized by low oxygen levels. Several years ago seeds of a lupine (*Lupinus arcticus*) were found to have retained their viability for about 10,000 years. They had been buried in dry, frozen soil in the Arctic in the burrows of a small mammal. The ages of these seeds were estimated by geological and radiological (^{14}C-dating) methods.

Several controlled experiments on the longevity of seeds have been carried out in various parts of the world. One of the best known was started in 1879 by Professor W. J. Beal at Michigan State University. He prepared 20 bottles containing 50 seeds each of 20 species of weedy plants common in the Midwest. The seeds were mixed with sandy soil and the 20 bottles were buried at a depth of 45 cm with mouths tilted downward so that water would not collect in them. The bottles, however, were open in the soil so that the soil within the bottles obtained some moisture. It had been planned to remove a bottle every 5 years and study the germination behavior of the 20 species. Since 1920 a bottle has been removed at 10 year intervals to extend the time of the experiment.

In 1960 three species germinated: *Oenothera biennis*, *Rumex crispus*, and *Verbascum blattaria*. In 1970 only one species retained its viability: *Verbascum blattaria*, commonly known as moth mullein. Only preliminary results are available from the bottle dug up in 1980, the hundredth year of the experiment. One seedling, probably *Verbascum blattaria*, has appeared (R. S. Bandurski, personal communication).

Seeds of many plants, particularly those from subtropical and tropical regions, retain their viability for short periods—frequently less than a week. Most of the observations have been made on seeds collected from native plants. Frequently no special precautions were taken to store the seeds and under conditions of high temperatures and high humidity seed viability quickly disappears. There is no reason to expect that such seeds cannot retain their viability if stored under proper conditions.

Because of the economic importance of seeds, many countries have established seed laboratories to study various aspects of seed physiology. The U. S. Department of Agriculture maintains such a laboratory at Fort Collins, Colorado. Here are maintained in storage seeds of major agricultural and horticultural crop plants from around the world. Seeds of many native plants and wild relatives of agronomic plants are also in storage. Such a collection (seed bank) is a valuable asset to plant breeders interested in obtaining germ plasm that might be useful in improving crop plants. Another important activity of the seed laboratory is research on methods of improving seed storage so as to maintain seed viability.

Germination

As noted, the block to embryonic growth in many seeds is overcome by placing the seeds in an appropriate environment. The major environmental conditions necessary are access to water and air; a suitable range of temperatures; freedom from high concentrations of inorganic salts, poisons, and inhibitors; and, for some seeds, exposure to light. When dry seeds are placed in water or exposed to humid atmospheres, the first measurable process associated with germination is an increase in water content of the seed and its associated components (seed coat, endosperm or cotyledons, embryonic axis). The initial entrance of water to the seed occurs by imbibition.

Much of the recent work in seed physiology has been directed toward the determination of the sequence of events initiated when seeds imbibe water. It has been helpful in these studies to divide the overall process into several substages of development. The term *germination* is used to designate those processes beginning with the imbibition of water by a dry seed and ending when a portion of the embryo penetrates the seed coat. Ordinarily it is the radicle or root that first penetrates the seed coat, but in some plants the plumule or shoot emerges first. The *emergence* phase of growth begins when the embryo penetrates the seed coat and ends when the shoot system is able to sustain the growth of the plant by photosynthesis. During emergence a root and shoot system develops. The arbitrary separation of these two phases, germination and emergence, is for convenience and has no special physiological significance. In commercial seed laboratories where seeds are tested and certified for germinability, the usual germination criterion is seedling emergence and the establishment of a vigorous plant.

The period of the plant life cycle encompassing germination and emergence may be completed within a few days or may be spread out over a period of months. Regardless of the time required, environmental factors, such as temperature, light, and moisture, interact with the physiological and biochemical process accompanying germination and seedling emergence.

Environment and Germination

The effects of the environment on seed germination are quite complex because of interactions and internal factors that modify germination patterns. A few of the major environmental factors will be described.

Temperature   Seeds have been extensively used for studying the effect of temperature on physiological processes. Dry seeds are frequently able to withstand a broad range of temperatures, but after the germination process has been set in motion by the imbibition of water, most seeds appear to tolerate a much narrower range of temperatures. Many early studies led to the concept of minimal, maximal, and optimal temperatures for seed germination. The minimal and maximal temperatures are those temperatures that just permit germination whereas the optimal temperature is considered the one that permits the highest percentage of germination in the shortest period of time. Such studies established that plants vary widely in their response to temperature, as shown in Table 16-1. Among the cereals, wheat, barley, and rye can tolerate temperatures of 3 to 5°C whereas maize and rice require temperatures above 8 to 12°C for germination. Cantaloupe requires even higher temperatures for germination and can tolerate temperatures between 45 to 50°C. *Lepidium draba*, a common weed, can germinate when the temperature is just above freezing (0.5 to 3°C).

TABLE 16-1 TEMPERATURE RANGES FOR THE GERMINATION OF SEEDS

Seeds	Temperatures (°C)		
	Minimum	Optimum	Maximum
Triticum sativum (wheat)	3–5	15–31	30–43
Hordeum sativum (barley)	3–5	19–27	30–40
Secale cereale (rye)	3–5	25–31	30–40
Zea mays (maize)	8–10	32–35	40–44
Oryza sativa (rice)	10–12	30–37	40–42
Cucumis melo (canteloupe)	10–19	30–40	45–50
Lepidium draba (whitetop)	0.5–3	20–35	35–40

After Mayer and Poljakoff-Mayber, 1975.

The data on minimal, optimal, and maximal temperatures in Table 16-1 were obtained under conditions of constant temperature. Under natural growing conditions seeds are not ordinarily subjected to uniform temperatures but rather to alternations of high and low temperatures. The temperatures fluctuate on a diurnal cycle and on a seasonal cycle. In addition, the temperature requirements of certain seeds depend on their age or physical condition. Freshly harvested seeds frequently require a very narrow range of temperatures for germination, but as the seeds age, the temperature requirements become less exacting and eventually germination proceeds over a broad range of temperatures. Also, there is a strong interaction between temperature and light on seed germination.

The data shown in Table 16-1 do not provide a clue as to why seeds of some plants are able to germinate at low temperatures whereas others require high ones. Inasmuch as a number of different kinds of processes occur during germination, it is to be expected that one or more of these processes might be especially temperature sensitive and thereby set the pace for germination. The germination of lima bean seeds is known to be inhibited by brief periods of chilling and studies have shown that low temperatures during the imbibitional phase have a marked influence on later stages of development. A temperature of 5°C during the first hour of imbibition not only leads to an immediate depression in the rate of respiration of the embryonic axis but also to the death of the embryo within 5 days. Chilling for periods as short as 30 minutes during imbibition are sufficient to inhibit respiration.

It appears that a brief period of low temperature affects membranes within the chilled cells, for the damaged regions were observed to leak nucleotidelike organic materials into the external medium. It has been shown by other workers that one result of chilling damage to fruits is the induction of leakiness in the membranes of mitochondria and other subcellular organelles. Techniques have been developed for following changes in the conformation of cellular membranes through a range of temperatures. Membranes from chilling-sensitive plants display a change in membrane conformation at temperatures of 10 to 14°C. Membranes from chilling-tolerant plants, on the other hand, do not display a change in membrane structure in a range of temperatures between 1 and 25°C. Chilling effects have been observed in many plants in the temperature range between 10 and 14°C, but species differ and chilling temperatures around 7 to 10°C have been observed in some plants. The point, however, is that in chilling-sensitive plants there is an alteration of membrane conformation at some characteristic low temperature.

Gaseous environment   Respiratory processes in seeds are stimulated soon after they imbibe water. Cells in the embryonic axis of dry seeds contain reduced amounts of water, but after water addition the cells absorb water by imbibition and increase in volume. In many seeds this increase in embryonic volume is sufficient to break the seed coat and facilitate organ (radicle or plumule) emergence. With breaking of the seed coat, gas exchange can occur, thus providing an aerobic environment for further metabolism. The seed coats of some seeds are extremely hard and physically prevent enlargement of the embryonic axis, although water has entered by imbibition. The role of the seed coat in germination is discussed more fully in a later section.

The influence of carbon dioxide, carbon monoxide, nitrogen, and other gases on germination can be understood in terms of their effects on metabolic processes. Because of the interest in space travel, some research has been carried out on the effects on germination of gaseous mixtures that might be encountered in outer space or on planetary surfaces. Published results are scanty, but the effects are consistent with those noted concerning the need for oxygen during cell division.

Poisons and inhibitors   Many different kinds of compounds are known to affect seed germination. In some cases, the effects are quite specific—for example, the action of hydrogen cyanide. Low concentrations of hydrogen cyanide, or more specifically cyanide (CN^-), will poison and kill growing embryos. It is unlikely that cyanide will be found frequently in the immediate environment of a germinating seed. Some fruits, however, contain amygdalin, a glucoside composed of sugar and cyanide. If the fruit tissue is damaged, enzyme activity may hydrolyze the glucoside and release cyanide in concentrations great enough to kill the growing embryo. In other instances, the effect of a compound on seed germination may result from its influence on the moisture status of the immediate seed environment. Thus high salt concentrations (fertilizers or other inorganic salts) in contact with the seed may prevent the seed (by an osmotic effect) from obtaining enough water to initiate germination; or if the radicle does manage to protrude through the seed coat, the embryonic tissue may become dehydrated and killed. Many soils contain high salt concentrations, either from irrigation waters or from soil minerals, which drastically inhibit seed germination.

Extracts or *leachates* from fruits, leaves, twigs, and roots have also been found to inhibit seed germination. The seeds of tomato, for instance, will not germinate as long as they are enclosed within the fruit, but if they are removed and thoroughly washed free of fruit tissue, they will germinate when moistened. Leachates from leaves have been found to inhibit the germination of seeds other than those from the plant itself. Extracts from roots and stems also prevent seeds from germinating. Various types of complex organic molecules have been isolated and identified from these extracts— alkaloids, essential oils, amino acids, coumarin, mustard oils, to name a few. These compounds also inhibit the growth of a mature plant under appropriate conditions and, in most instances, it is not known what specific effect the plant extract has on germination and growth. There is evidence in some cases that the effect is osmotic, similar to the effect of a high inorganic salt concentration, but in many other situations the organic leachates influence growth in concentrations too dilute to be ascribed to an osmotic effect. In a few cases, the plant leachate has been identified as abscisic acid, a natural plant growth substance (Chapter 14). Recall that abscisic acid has been shown to be involved in maintaining the dormancy of embryos during seed development. Its presence in leaves and other plant parts suggests that abscisic acid may play a wider role as an inhibitor of plant growth and development.

Many substances foreign to the natural seed environment also influence germination. Some organic herbicides block seed germination and have been effectively used to prevent the growth of certain weeds. Industrial wastes and pollutants often prevent seed germination. They may be quite diverse—acids, alkalis, salts of metals (mercury, lead, silver, aluminum, etc.), phenolics, fluorides, and so forth. The germinating seed is very sensitive to foreign materials and under most conditions it is essential that the seed environment be kept free of harmful chemicals. Yet there are some examples of plants whose seeds and seedlings can tolerate chemicals that ordinarily kill other plants. The tolerant plants are able to live successfully under conditions that prevent the competition of other less tolerant species. The ecological implications of such results are not thoroughly understood, particularly in areas where pollution is a major problem.

MORPHOLOGICAL AND BIOCHEMICAL CHANGES
ACCOMPANYING SEED GERMINATION

Although the exact sequence of events in seed germination varies among different plant species, the basic processes are similar: water imbibition, cell expansion, hydrolysis of food reserves in endosperm or cotyledons, transport of soluble metabolites to the embryo, and synthesis of cellular constituents in the embryo, accompanied by cell division. These processes are outlined in Figure 16-3, where seed dry weight is plotted against time of beginning of water imbibition.

The seed coat takes up water by imbibition and the enclosed embryonic axis is gradually hydrated. In many seeds the embryonic root region (radicle) of the axis takes up water more quickly than the rest of the axis and emerges first through the seed coat. In seeds of most crop plants radicle emergence (Phase I, Figure 16-3) occurs within 24 to 36 hours after the onset of water imbibition.

Even though water may be readily available for entering a dry seed, cells in the embryonic axis may not hydrate uniformly. Cells in the radicle may hydrate before those in the hypocotyl, thereby setting up localized stress areas. If chilling or freezing temperatures are encountered, it is possible that cell or tissue damage may occur and diminish germination and/or seedling emergence. Embryonic axis damage of this kind can be reduced by exposing dry seeds to limiting amounts of water (presoaking), which enables all parts of the seed to become hydrated uniformly before planting.

As the seed imbibes water, all the cells in the embryo, cotyledons, and endosperm become hydrated, resulting in cell expansion and size increase. The complete

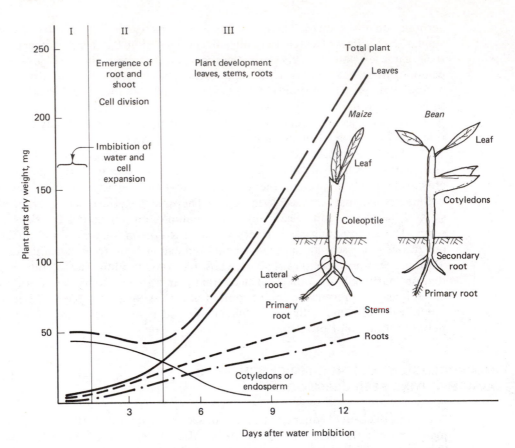

Figure 16-3 Diagrammatic representation of changes in dry weight during seed germination. Phase I—imbibition of water, cell expansion. Phase II—emergence of root and shoot, onset of active metabolism in embryonic axis and enzymic hydrolysis of reserves in endosperm or cotyledons. Phase III—autotrophic growth with development of root system and leafy shoot system.

hydration process may take 40 to 60 hours, depending on the temperature and availability of water. These events occur early in Phase II, Figure 16-3. No changes in dry weight take place during the first 24 to 36 hours following the onset of water imbibition. Hydration, however, enables the cells in the embryonic axis and cotyledons to attain full turgor, accompanied by reorganization of the subcellular organelles and cellular membranes. Respiratory activities are initiated and some dry weight loss occurs. The food reserves in the embryo are not adequate to sustain cell division and new tissue and organ formation. Food reserves in the endosperm or cotyledons are mobilized to provide substrates for continued growth of the embryonic axis.

The major food reserves in seeds are starch, fat, protein, nucleic acid, and phytin. As shown in Figure 16-4, these reserves, when mobilized and metabolized, provide the

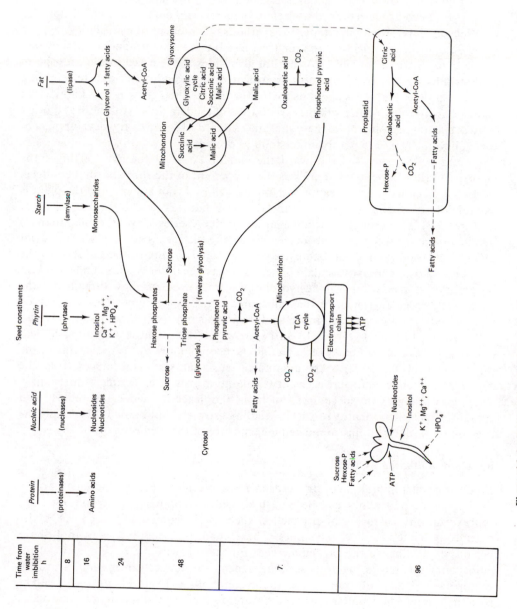

Figure 16-4 Some metabolic events during seed germination. The dashed line (– – –) to the right between glyoxysome and proplastid indicates that citric acid formed in the glyoxylic acid cycle may be converted to fatty acids in proplastids. (See text for details.) (After B. J. Rapp and D. D. Randall, 1980; I. J. Mettler and H. Beevers, 1980; and M. Nishimura and H. Beevers, 1981.)

embryonic axis with amino acids, nucleotides and nucleosides, inositol, sucrose, fatty acids, and some inorganic ions. Also shown in Figure 16-4 are several metabolic pathways involved in the conversion of storage fats to sucrose, an important pathway in fat-rich seeds. These pathways are discussed more fully later on.

In monocots cells in the hydrated embryo secrete gibberellic acid, which moves to the aleurone layer, where it initiates the synthesis of a number of hydrolytic enzymes: α-amylase, protease, nuclease, and phytase. Bandurski and colleagues have shown in maize that IAA-inositol and IAA-inositol glycosides are released in the endosperm following hydration. IAA-inositol is translocated to the embryonic axis (coleoptile), where it supplies both IAA and inositol for further seedling growth. It is known that endosperm and cotyledonary tissues are sources of gibberellins, cytokinins, and other growth substances. The role of these substances in orderly growth and development of the embryonic plant are poorly understood at present.

Figure 16-3 shows that toward the end of Phase II the dry weight of the endosperm (in monocots) decreases as the dry weight of the root and shoot systems increases. Eventually green leaves develop, as well as a root system, and the plant is capable of an independent mode of metabolism; that is, it is autotrophic. Phase II (Figure 16-3) ends as the plant becomes autotrophic. A series of events similar to those just described for seeds of monocots occurs during the germination of seeds of dicot plants. The food reserves of dicots are largely stored in cotyledons attached to the embryonic axis. There are vascular connections between the cotyledons and the rest of the embryo; so at least some of the hydrolyzed storage products move directly into the developing vascular system of the embryonic plant. A portion of the hydrolyzed reserve material, however, will move by diffusion, as in germinating monocot seeds.

During Phase III, Figure 16-3, the growth of the plant is supported by leaf photosynthesis and the uptake of water and inorganic solutes by the roots. During the early stages of Phase III there are indications that some substances from the cotyledons, or endosperm, are necessary to maintain growth of the emerging seedling. They may be growth substances, such as indoleacetic acid, gibberellins, and cytokinins, but their identity or role in seedling growth is not known. Eventually the developing seedling attains complete independence of the seed reserves.

Initiation of Enzyme Activity

Following water imbibition by dry seeds, macromolecules—protein, polysaccharides, nucleic acids, phospholipids—become hydrated. Accompanying hydration is the reorganization of subcellular organelles, such as ribosomes, mitochondria, endoplasmic reticulum, and glyoxysomes. These changes occur over a 6- to 12-hour period, depending on the particular seed. Further growth and development of the embryonic axis into a vigorous seedling depend on new cell formation, a process requiring ATP and substrates for anabolic metabolism. From Figure 16-4 it can be seen that about 8 hours after the onset of water imbibition the activity of some major hydrolytic enzymes can be detected. These enzymes catalyze the breakdown of inert storage macromolecules into small molecules or ions that can be transported to active metabolic centers within the seed. Some of these small molecules (or ions) may be absorbed directly by the embryonic axis whereas others are further metabolized before entering the metabolism of the embryo.

Enzymic activity by the hydrolytic enzymes may occur in several ways. Either the enzymes are present in an inactive form, having been synthesized during the maturation phase of seed development, or the enzymes must be synthesized de novo. There is evidence in support of both processes. Several proteinases, nucleases, lipases, and phytase apparently are synthesized during seed maturation but are present in the dry seed in an inactive, dehydrated form. Following water imbibition, the enzymes (proteins) assume an active conformation and are capable of catalytic activity. α-Amylase, on the other hand, is not present in dry seeds but is synthesized de novo after water imbibition. The role of gibberellic acid in initiating α-amylase synthesis was described in Chapter 15.

In general, most of the enzymes involved in the metabolic reactions outlined in Figure 16-4 appear to be synthesized de novo after water imbibition. Apparently during seed maturation components of the protein-synthesizing apparatus are degraded or destroyed and must be reconstituted after water imbibition by dry seeds.

Dure and colleagues found in cotton seeds that two enzymes involved in germination are synthesized de novo following water imbibition. One, a carboxypeptidase, catalyzes the breakdown of protein to amino acids whereas the other, isocitric acid lyase, is a key enzyme in the glyoxylic acid cycle (Figure 16-5). An

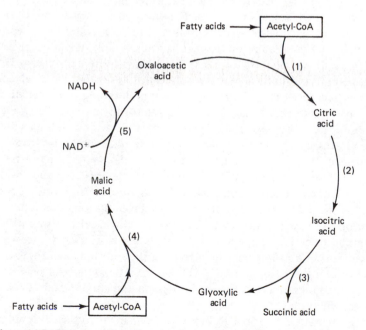

Figure 16-5 The glyoxylic acid cycle. *Cis*-aconitic acid (the intermediate between citric and isocitric acids) is not shown here. The enzymes are (1) citric acid synthase (also referred to as the condensing enzyme); (2) aconitase; (3) isocitric acid lyase; (4) malic acid synthase; (5) malic acid dehydrogenase.

interesting feature of the synthesis of these two enzymes is the observation that the m RNA responsible for their coding is present in dry seeds. During seed maturation m RNAs are synthesized and then stored in the dry seed. On water imbibition m RNAs and other elements of the protein-synthesizing system are reconstituted and the translation process proceeds with protein synthesis. Dure refers to such enzymes as "germination enzymes" and suggests that there may be groups or packages of them processed to the m RNA stage, ready for the translation stage. Such an arrangement would provide for rapid hydrolysis of food reserves and the initiation of other metabolic processes observed during germination.

Metabolism in Germinating Fat-Rich Seeds

In seeds containing starch-filled endosperm cells, during germination the starch is hydrolyzed to glucose and further metabolized by glycolysis, the TCA cycle, and the electron transport chain (Figure 16-4). During these metabolic events ATP and carbon skeletons are provided for anabolic reactions essential in supporting the developing embryonic plant until it attains an autotrophic condition. There are, however, many seeds in which the stored food in the endosperm or cotyledons is predominately fat. These fats also provide ATP and carbon skeletons for the developing embryonic plant during germination, but different metabolic pathways are involved, as shown on the right-hand side of Figure 16-4.

In the fat-rich castor bean seed, which has been extensively studied, fat reserves are contained largely in endosperm tissue, where they account for more than 60% of the dry weight of the seed. During the first several days of germination these fat reserves are used as respiratory substrates by conversion to sucrose and fatty acids (Figure 16-4). Sucrose is transported out of the endosperm into the developing shoot–root system, where it is respired to CO_2 and water by glycolysis and the TCA cycle. The developing embryonic plant is undergoing active cell division and relatively large amounts of fatty acids are required for formation of new membranes—plasma membrane, tonoplast, chloroplast, mitochondrion, endoplasmic reticulum, Golgi apparatus, and so forth. Some of the fatty acids may arise from acetyl-CoA formed during glycolysis (Figure 16-4), but additional fatty acids may arise more directly from the fat reserves via the glyoxylic acid cycle (Figures 16-4 and 16-5).

The conversion of fat to sucrose in endosperm tissue of the castor bean seed begins with the hydrolysis of fats to glycerol and long-chain fatty acids through the action of lipase (Figure 16-4). Glycerol is converted to triose phosphate and further metabolized. The long-chain fatty acids are converted to acetyl-CoA by a process known as *beta oxidation*, which involves the addition of Coenzyme A, consumption of ATP, and the reduction of NAD^+ to NADH. Details of beta oxidation have been described for numerous tissues, both plant and animal, and can be found in textbooks of biochemistry.

Acetyl-CoA formed by beta oxidation in fat-rich seeds is not metabolized via the TCA cycle (described fully in Chapter 4) but rather by the glyoxylic acid cycle, as outlined in Figure 16-5. In castor bean seeds fats are stored in subcellular organelles known as glyoxysomes (Chapter 2) and the enzymes involved in fat hydrolysis, beta oxidation, and the glyoxylic cycle are all contained in the organelle. Thus acetyl-CoA formed in the glyoxysomes is sequestered in a compartment in which it undergoes a different metabolic fate than acetyl-CoA formed during glycolysis.

Further conversion of acetyl-CoA to sucrose and fatty acids occurs in four metabolic phases (Figure 16-5):

1. Formation of citric acid, succinic acid, and malic acid in glyoxysomes via the glyoxylic acid cycle.
2. Metabolism of succinic acid to malic acid in mitochondria.
3. Conversion of malic acid to sucrose in the cytosol by a reversal of glycolysis (Figure 16-6).
4. Conversion of citric acid (produced in glyoxysomes) to fatty acids in proplastids.

Details of the glyoxylic acid cycle and accompanying metabolic processes have been elucidated in recent years through the research of the American plant physiologist H. Beevers and his colleagues.

The glyoxylic acid cycle uses certain members of the TCA cycle but not others. Unlike the TCA cycle, however, the glyoxylic acid cycle is a synthetic rather than degradative process. As shown in Figure 16-5, the major metabolites are citric acid, succinic acid, glyoxylic acid, malic acid, and oxaloacetic acid. Acetic acid (as acetyl-CoA) enters the cycle by condensing with oxaloacetic acid to form citric acid, which is transformed to isocitric acid. But isocitric acid does not undergo oxidative decarboxylation as in the TCA cycle (Figure 4-6). Instead isocitric acid is cleaved to succinic acid and glyoxylic acid in a reaction catalyzed by isocitric acid lyase.

Glyoxylic acid then condenses with another molecule of acetyl-CoA to form malic acid in the presence of the enzyme malic acid synthetase. Finally, the cyclic sequence of reactions (Figure 16-5) is completed with the transformation of malic acid to oxaloacetic acid, which then couples with acetyl-CoA to continue the glyoxylic acid cycle. In summary, for one turn of the glyoxylic acid cycle, two molecules of acetyl-CoA are consumed and one molecule of succinic acid is produced.

Intermediates of the glyoxylic acid cycle (citric acid and malic acid) and the product, succinic acid, are transported to the cytosol or other organelles where further metabolic conversions occur. Succinic acid is transformed to malic acid (via TCA cycle), which is then transported out of the mitochondrion to the cytosol (Figure 16-4). Malic acid from the glyoxysome is also transported to the cytosol, where it is converted to sucrose by reverse glycolysis via phosphoenolpyruvic acid and triose phosphate, as outlined in Figure 16-6. A feature to note about the reactions of the glyoxylic acid cycle, as well as the conversion of succinic acid to malic acid in the mitochondrion, is that acetyl-CoA is converted to malic acid with no loss of CO_2. The metabolism of acetyl-CoA via the TCA cycle, on the other hand, involves the loss of CO_2 (Figure 16-4).

The metabolic reactions for converting malic acid to sucrose are shown in Figure 16-6. These reactions occur in the cytosol and involve a reversal of the glycolytic pathway. The metabolic reactions require substantial amounts of ATP, which is

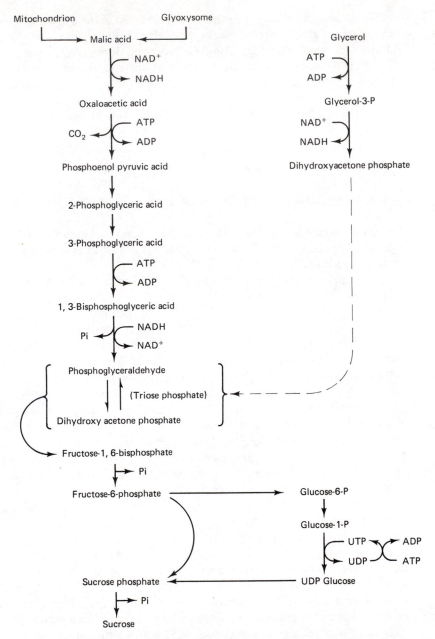

Figure 16-6 The conversion of malic acid to sucrose by reverse glycolysis (gluconeogenesis). Also shown (upper right) is the pathway for the conversion of glycerol (from fat hydrolysis) to sucrose.

generated by the electron transport chain in mitochondria. Also shown in Figure 16-6 are reactions concerned with the conversion of glycerol (from fat hydrolysis) to sucrose. The conversion of fatty acids and glycerol to sucrose is referred to as *gluconeogenesis* (i.e., the formation of glucose units from substances other than carbohydrates).

The metabolic reactions outlined in Figures 16-4, 16-5, and 16-6 illustrate enzyme compartmentation and substrate metabolism in converting carbon skeletons of fat reserves into sucrose or fatty acids. Carbon is conserved and diverted to metabolites used in synthesizing new cells, tissues, and organs of the developing shoot–root axis. Many of the enzymes catalyzing these reactions are not present in dry seeds but must be synthesized de novo after water imbibition. Clearly the early stages of seed germination require numerous metabolic substrates supplied by seed reserves and involve considerable metabolic activity in forming new cellular material. The reactions illustrated in Figures 16-5 and 16-6 have emphasized the roles of fat and starch. Less detailed information is available concerning other seed constituents— phytin, nucleic acid, and protein—as well as various growth substances, such as abscisic acid, gibberellic acid, cytokinin, and auxin. All, however, play a role in seed germination.

DORMANCY

The seeds of most agricultural and horticultural plants usually germinate promptly if given access to moisture and air, if provided with a suitable range of temperatures and, in some instances, if exposed to a proper sequence of light and dark. There is, however, a more numerous group of plants whose seeds do not readily germinate even though they are placed under favorable conditions of moisture, air, temperature, and light. Germination may be delayed for days, weeks, or even months. The seeds of such plants are said to be in a dormant condition. The phenomenon of dormancy is not unique to seeds but is also found in other plant organs, such as the buds of woody and herbaceous plants as well as the buds of tubers, rhizomes, and bulbs. Other organisms—bacteria, fungi, insects, and animals—also undergo periods of reduced growth activity and the term *dormancy* is used to denote such a condition. The common feature in all these examples is reduced growth; in seeds growth of the embryo is blocked whereas in buds growth of the apical meristem is inhibited.

It is unlikely that the same basic mechanisms are involved in establishing the dormant condition in all the organisms mentioned, for they differ enormously in their growth habits, nutrition, temperature and water regulation mechanism, and metabolism. The dormant condition, however, appears to serve a common purpose in the various organisms in that it enables them to endure periods of unfavorable environmental conditions, thereby providing a mechanism for survival. In seeds an additional feature of the dormant condition is the provision for dispersal afforded by various modifications of the seed covering. Dispersal is accomplished while the embryo is quiescent and growth is initiated when the block to embryonic growth is overcome. Thus continuity of the species is achieved.

Recognizing that the term *dormancy* has been used to describe a number of quite different physiological processes, we shall use the term to mean the blocking of embryo growth. All parts of the seed—seed coverings, endosperm, and the embryo itself—may be responsible for blocking growth of the embryo. In most cases, the mechanism of dormancy is quite complex, involving interactions between external environmental factors, internal factors, and time.

Seed dormancy usually develops as the result of the action of two kinds of processes: one within the embryo and one in the seed coat and other tissues external to the embryo. There are, of course, interactions between the embryo and other parts of the seed.

Dormancy Due to the Condition of the Embryo

Two kinds of embryo dormancy have been noted. In a few cases, the embryo is rudimentary and poorly developed at seed maturity. It is necessary for such embryos to continue their development during the dormant period before they can germinate successfully. In the second type of embryo dormancy the embryos appear to be fully developed but unable to resume growth promptly when provided with a suitable environment. This phenomenon is noticed especially in the seeds of Rosaceous plants (apple, pear, hawthorne, cherry, blackberry, etc.) and conifers. The seeds can be induced to germinate if stored moist but well aerated under low-temperature conditions, a treatment referred to as stratification. Sometimes the process is called *afterripening* and it is often said that the embryos of seeds that require stratification are immature. The embryos, however, appear fully developed and there is little evidence to indicate that they must undergo a period of development before they can be induced to germinate. After the embryos imbibe water, the processes that must occur in the embryo before the root and shoot begin to grow involve a sequence of events that require a relatively long time. It is possible to remove and culture embryos from seeds that require afterripening. If placed at room temperature (20° C and above), the embryos germinate readily, but seedling development is often abnormal. Cultured embryos kept at low temperatures (2 to 5° C) grow more slowly, but subsequent seedling development is normal. It appears that the early germination processes of these embryos require the formation (or accumulation) of some promoting substance or the degradation (or diminution) of an inhibitory molecule. In some seeds the gibberellins (Chapter 14) appear to function as germination promoters while abscisic acid acts as an inhibitor. A proper balance between these two kinds of regulatory substances is achieved during the afterripening process.

Dormancy Due to Seed Coats

Much of the physiological work on seed dormancy has been directed toward an examination of the role of the seed coat. The seed coat (or testa) of most seeds is composed of several layers of cells derived from the integumentary tissues of the ovule. In addition, some seeds have additional coat layers derived from the endosperm or fruit tissues. From a chemical standpoint, seed coats consist of a complex mixture of polysaccharides, hemicellulose, fats, waxes, and proteins. During seed ripening the chemical components of the seed coat become dehydrated and form a hard, tough protective layer around the embryo. The seed coats have a strong influence on the resumption of growth of the embryo. Several different kinds of coat effects have been noted: seed coats that are impermeable to water or gases, seed coats that offer mechanical resistance to the growth of the embryo, and seed coats that contain inhibiting or promoting substances.

Water impermeability   Seeds of many plants are almost impermeable to water. This is particularly true of members of the legume family. If the seed coat is cracked or scarified so that water can gain entrance, the seeds usually germinate promptly. Under natural conditions in the soil these seed coats are acted on by fungi and bacteria. These organisms hydrolyze the polysaccharides and other coat components, thereby softening them so that water can penetrate to the embryo. It may take several weeks or even months for the seed coats to be degraded by biological activity.

Gas impermeability   The coats of some seeds, while permeable to water, appear to be impermeable to dissolved gases, such as oxygen and carbon dioxide. Because early respiratory activity is characteristic of the germination of many seeds, if oxygen is prevented from reaching the embryo, prompt germination may not be able to take place. Respiration also involves the release of CO_2 and some seed coats, while permeable to oxygen, may be impermeable to CO_2. The CO_2 accumulated in the vicinity of the embryo inhibits further germination processes. If the seed coats are broken or scarified, prompt germination can generally occur.

Mechanical resistance   Some plants have seeds whose seed coats are permeable to water and dissolved solutes, but the coats have such mechanical strength that they cannot be broken by the growing embryo. If the seed coats remain moist, they eventually weaken and break. If the coat softens and allows some embryo swelling and dries again, however, further growth of the embryo may be prevented. Recent work indicates that the mechanical resistance of the seed coats of many seeds may be the primary factor contributing to dormancy. If the coats of these seeds are fractured or removed, prompt germination can occur. During the germination of some seeds enzymes that hydrolyze the seed coat are secreted, thereby weakening it so that the growing embryo can continue its growth.

Seed coat treatments   Hard seed coats are softened under natural conditions in the soil by alternating temperatures, by drying and wetting, and by the biological activity of soil flora and fauna. Depending on the seed coat and the vigor of soil activities, it may take considerable periods of time for the seed coats to become softened to the point where germination can proceed. This time period may be important in carrying the seed over an unfavorable growing period when seedling growth might be harmed. If seeds are of agronomic importance, however, they are usually planted under favorable growing conditions, and it is essential that prompt germination be obtained. Various mechanical and chemical treatments have been used to ensure prompt germination, including mechanical scarification by cutting or chipping the seed coat, moist storage at high temperatures, use of organic solvents to remove waxy or fatty seed coat components, and treatment with acids to hydrolyze some of the seed coat components. Seeds are extremely variable in their responses to such treatments and great care must be taken to ensure that the treatments to not damage the embryo and thus reduce subsequent seedling growth.

Chemicals in Seeds

A large number of different types of chemicals affecting plant growth have been isolated from seeds as well as from other parts of the plant—fruits, leaves, stems, and roots (see Chapter 2 and 15). Some of the compounds inhibit plant growth whereas others promote growth. The list includes organic acids, phenolics, tannins, alkaloids, unsaturated lactones, mustard oil, ammonia-releasing substances, cyanide-releasing substances, gibberellins and other growth substances, indoles, and numerous unidentified substances. Not only do many of these compounds inhibit seed germination but they also inhibit the growth of seedlings and older plants (see the discussion of allelochemics in Chapter 15).

Because of their widespread occurrence in plants, it has been difficult to define precisely the role of these compounds in seed dormancy. In a few instances, it has been shown that an inhibitory chemical substance present in the seed covering blocks growth of the embryo. If the substance is leached out, the block disappears and embryonic growth is resumed. In other seeds the inhibitory chemicals seldom act alone in blocking growth of the embryo but interact with one or more of the other dormancy factors just described.

The germination of some dormant seeds can be promoted by the application of growth substances, such as gibberellins and cytokinins (Chapter 15). Also, the germination of nondormant seeds can be inhibited by exogenous abscisic acid, another plant growth substance. Because these three growth substances also occur in seeds and other plant parts, it has been suggested that dormancy and germination are controlled by interactions between these growth-promoting and growth-inhibiting substances.

The Role of Light in Seed Dormancy

The effect of radiant energy on seed germination is quite diverse and many patterns of germination behavior have been observed. The responses of seeds to sunlight (white light) fall into three categories.

1. Seeds are induced to germinate by exposure to a single irradiation (positive photoblastic seeds). Depending on the intensity of the radiation source, the single exposure may be as brief as a few seconds or as long as several hours.
2. Seeds are prevented from germinating by exposure to light (negative photoblastic seeds). Such seeds require total darkness for optimal germination.
3. Seeds germinate in either light or dark (nonphotoblastic seeds).

The primary effect of light on seed germination is mediated by phytochrome (Chapter 5), a pigment composed of a chromophore molecule and a protein. Phytochrome within the seed can be modified by such factors as temperature, moisture, hydrogen ion concentration, age of seed, seed-ripening conditions, and the quantity and quality of irradiation. These factors acting through phytochrome are responsible for the different pattern of germination-light interaction mentioned.

Phytochrome does not exist solely in the P_r or P_{fr} form following red or far-red irradiation. Rather some ratio of P_r to P_{fr} will be found because the absorption spectra of the two forms overlap (Figure 5-12). Phytochrome irradiated with red light (660 nm) will be 81% P_{fr} and 19% P_r. Far-red light (730 nm) establishes phytochrome at about 98% P_r and 2% P_{fr}. With white light, for example, sunlight, the ratio of P_r to P_{fr} at ground level will vary considerably at different times of day because of the angle of the sun above the horizon, the extent of cloud cover, and the leaf canopy overhead. A seed under a layer of soil is exposed to variable irradiation environments and the two forms of phytochrome can be established in different proportions.

Several other features of phytochrome should be mentioned before the influence of light on seed dormancy and germination is discussed. Phytochrome in the P_{fr} form slowly reverts to P_r in the dark. When phytochrome is synthesized, it is believed that the P_r form is made. Finally, P_{fr} is the active form and it is destroyed, or inactivated, if it is not stabilized by interacting with a molecule, referred to as [X], to form $[P_{fr} \cdot X]$. All these features of phytochrome can be visualized as follows:

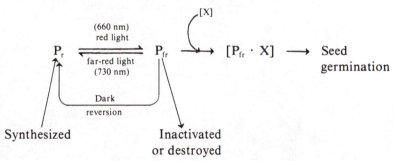

With the preceding scheme in mind, how can the three classes of light requirement for seed germination be understood? Keep in mind that $[P_{fr} \cdot X]$ or an appropriate ratio of $P_{fr} \cdot X / P_r$ must be present to trigger germination. Positive photoblastic seeds (lettuce var. Grand Rapids, *Chenopodium rubrum*, *Pinus taeda*, and many others) require a single exposure to white light (containing light at 660 nm) and such light is adequate to establish a satisfactory level of $P_{fr} \cdot X / P_r$ to initiate germination. If the appropriate phytochrome ratio is not achieved, the seeds do not germinate but remain dormant.

Nonphotoblastic seeds (seeds of many cultivated plants, such as tomato, cucumber, some lettuce varieties) are not sensitive to light and germinate in either light or darkness. It is believed that an appropriate ratio of $P_{fr} \cdot X$ to P_r is present in these seeds and that light is not needed to convert P_r to P_{fr}. The mechanism for establishing the appropriate levels of phytochrome is not clear, but possibly it is established during seed maturation and ripening. Shropshire, Klein, and associates found that phytochrome could be photoconverted during seed ripening. If the seed continues to lose moisture during the ripening process, the phytochrome is "set" at some particular P_r to P_{fr} ratio. When the dry seed then imbibes water, the phytochrome is also hydrated and capable of triggering germination if the appropriate phytochrome ratio is present. The process is illustrated in Figure 16-7. If this phenomenon is a common occurrence during seed ripening, it is not difficult to imagine that a particular seed lot may be quite variable with respect to the light requirement for germination. The individual seeds ripen at different times under variable light regimes and each seed may have a different ratio of P_r to P_{fr} at maturity.

Finally, what is the condition in negative photoblastic seeds? These seeds (e.g., *Nemophila insignis* and *Phacelia tanaietifolia*) will not germinate in white light but

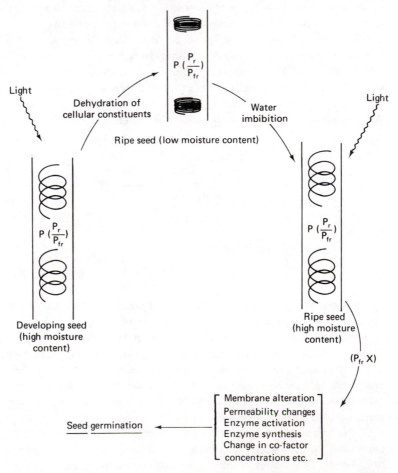

Figure 16-7 Hypothetical scheme for the development of functional phytochrome in seeds.

require relatively long periods of darkness. Again, it can be assumed that in some way an appropriate level of $[P_{fr} \cdot X]$ is achieved so as to initiate germination. It is probable that the level of phytochrome necessary to initiate is low compared to the levels needed in other seeds. Also, it might be essential that a low level of $[P_{fr} \cdot X]$ be maintained for a considerable period of time after water imbibition of the seeds. Another possibility is that the amount of $[X]$ is maintained by metabolic activity and darkness may allow an appropriate level of $[X]$ to be achieved to permit germination.

Although clearly phytochrome participates in seed dormancy and germination, the mechanism of action is not thoroughly understood. During seed development on the parent plant phytochrome is synthesized. It is not known whether phytochrome is localized at a few places in the embryonic plant or at many sites. It is believed, however, that phytochrome is localized within membranes. During seed development phytochrome is photoconvertible and various levels of P_r to P_{fr} can be established by different light regimes. During the desiccation phase of development (Figure 16-2) the various organelles and their membrane surfaces may be dehydrated with phytochrome immobilized at some particular ratio of P_r to P_{fr}. Following water imbibition, the membrane surfaces and phytochrome become hydrated and organized so as to establish the potential for metabolic activity. Phytochrome now initiates some physiological or biochemical events that culminate in seed germination.

From the preceding discussion it can be seen that many factors, both external and internal, may interact with phytochrome in establishing seed dormancy and initiating germination. Seeds have a life history and at various points during seed development and maturation the phytochrome system may be altered. Light, temperature, moisture, and nutritional conditions acting on the parent plant may influence subsequent seed performance during storage, germination, seedling emergence, and vegetative and reproductive growth. Moreover, the same environmental factors may act on the mature seed or during germination. It is little wonder that so many different patterns of seed germination behavior have been observed. The possibilities seem unlimited.

SEED QUALITY

Throughout this chapter reference has been made to the importance of seeds in food production and other agricultural and horticultural activities. An essential feature of modern intensive agriculture is the use of seed of improved varieties of crops. The seeds of these new varieties are of high quality and are produced under optimal cultural conditions and are handled so as to maintain their quality when planted. In the United States, where a very efficient agricultural system operates, the planting of high-quality seed is also of major importance. Through the use of fertilizer, irrigation, pest control, and soil management, crop growth has been brought to the point where the quality of the seed planted may be a limiting factor in crop production.

What is meant by *seed quality*? Sometimes the term *seed vigor* is used. The following attributes appear to be important. Seeds should maintain their quality under proper storage conditions. They should germinate rapidly and uniformly when planted. Seed emergence should be prompt and early vegetative growth should be vigorous. Yields of grain, fruit, roots, leaves, or other harvested parts of the plant should be of superior quality. If these are important characteristics of seed quality, can a seed be evaluated as to its probable performance before it is planted?

Seed testing laboratories, staffed with trained seed technololgists, are maintained in all states in the United States and national standards are also established by the U.S. Department of Agriculture. Similarly, foreign countries have seed testing laboratories. As far as germinability is concerned, seed analysts can provide an accurate estimate of the percent germination of a particular seed lot and can determine if weed seed or other foreign matter is present. Also, damaged seeds can be identified and experienced seed analysts can judge the probable performance of a seed lot under optimal field conditions. Field performance, however, is seldom reported by seed testing laboratories.

No single test has been developed that can measure field performance or yield potential. It is possible that a series of tests might identify different aspects of seed quality and provide a profile of seed factors related to the establishment of a vigorous field stand. The following seed factors are possible components of such a profile: Seed coat (thickness, hardness, presence or absence of cracks, presence or absence of inhibitors or promoters), embryonic axis (stage of maturity, hydration, ultrastructure, ability to develop "root thrust," presence or absence of inhibitors or promoters), stored reserves in cotyledons or endosperm, seed size, seed density, age, and conditions of seed storage. Some of these characteristics can be measured; further research will show if they can be used to develop a predictive value of seed vigor.

This discussion on seed quality has focused on agricultural and horticultural seeds. Seed quality is also an attribute of the seeds of wild, or native, plants. The seeds of most of these plants possess dormancy mechanisms that prevent their immediate germination and emergence, but when the block to embryonic growth is overcome, the appearance of a vigorous plant is essential to the maintenance and survival of the species.

REFERENCES

BEEVERS, H. 1975. Organelles from castor bean seedlings: biochemical roles in gluconeogenesis and phospholipid biosynthesis. Pages 287–299 in T. Galliard and E. I. Mercer, eds. *Recent Advances in the Chemistry and Biochemistry of Lipids*. Phytochemical Society Symp. Series, No. 12. Academic Press, London, New York.

BEWLEY, C. C., and M. BLACK. 1978. *Physiology and Biochemistry of Seeds in Relation to Germination*. Vol. 1, *Development, Germination and Growth*. Springer-Verlag, Berlin, Heidelberg, New York.

BOLD, H. C. 1961. *The Plant Kingdom*, 2nd ed. Prentice-Hall, Englewood Cliffs, N.J.

BREIDENBACH, R. W., N. L. WADE, and J. M. LYONS. 1974. Effect of chilling temperatures on the activities of glyoxysomal and mitochondrial enzymes from castor bean seedlings. *Plant Physiol*. 54:324–327.

CHEAH, K. S. E., and D. J. OSBORNE. 1978. DNA lesions occur with loss of viability in embryos of ageing rye seed. *Nature* 272:593–599.

CHING, T. M. 1973. Biochemical aspects of seed vigor. *Seed Sci. & Technol.* 1:73–88.

DURE, L. S. III. 1975. Seed formation. *Ann. Rev. Plant Physiol.* 26:259–278.

EASTWOOD, D. and D. L. LAIDMAN. 1971. The mobilization of macronutrients in the germinating wheat grain. *Phytochemistry* 10:1275–1284.

EASTWOOD, D., and D. L. LAIDMAN. 1971. The hormonal control of inorganic ion release from wheat aleurone tissue. *Phytochemistry* 10:1459–1467.

EPSTEIN, E., J. D. COHEN, and R. S. BANDURSKI. 1980. Concentration and metabolic turnover of indoles in germinating kernels of *Zea mays* L. *Plant Physiol.* 65:415–421.

EVANARI, M. 1965. Physiology of seed dormancy, after ripening and germination. *Proc. Internatl. Seed Test. Assoc.* 30:49–71.

FLINN, A. M., and J. S. PATE. 1968. Biochemical and physiological changes during maturation of the field pea (*Pisum arvense* L.). *Ann. Bot.* 32:479–495.

HAYES, R. G., and W. H. KLEIN. 1974. Spectral quality influence of light during development of *Arabidopsis thaliana* plants in regulating seed germination. *Plant Cell Physiol.* 14:643–653.

HENDRICKS, S. B., and R. B. TAYLORSON. 1979. Dependence of thermal responses of seeds on membrane transitions. *Proc. Natl. Acad. Sci. U.S.* 76:778–781.

HEYDECKER, W., ed. 1973. *Seed Ecology*. Pennsylvania State University Press, University Park.

IHLE, J. N., and L. S. DURE, III. 1972. The developmental biochemistry of cottonseed embryogenesis and germination. III. Regulation of the biosynthesis of enzymes utilized in germination. *J. Biol. Chem.* 247:5048–5055.

KIGEL, J., M. OFIR, and D. KOLLER. 1977. Control of the germination responses of *Amaranthus retroflexus* L. seeds by their parental photothermal environment. *J. Expt. Bot.* 28:1125–1136.

KIVILAAN, A., and R. S. BANDURSKI. 1973. The ninety-year period for Dr. Beal's seed viability experiment. *Am. J. Bot.* 60:140–145.

KOZLOWSKI, T. T., ed. 1972. *Seed Biology*, 3 vols. Academic Press, New York.

LIST, A., JR., and F. C. STEWARD. 1965. The nucellus, embryo sac, endosperm, and embryo of *Aesculus* and their interdependence during growth. *Ann. Bot.* 29:1–14.

MARCUS, H., and J. FEELEY. 1964. Activation of protein synthesis in the imbibition phase of seed germination. *Proc. Nat. Acad. Sci. U.S.* 51:1075–1079.

MARRÉ, E. 1967. Ribosome and enzyme changes during maturation and germination of the castor bean seed. *Current Topics Devel. Biol.* 2:75–105.

MAYER, A. M., and A. POLJAKOFF-MAYBER. 1975. *The Germination of Seeds*, 2nd ed. Pergamon Press, Elmsford, N. Y., The Macmillan Company, New York.

METTLER, I. J., and H. BEEVERS. 1980. Oxidation of NADH in glyoxysomes by a malate-aspartate shuttle. *Plant Physiol.* 66:555–560.

MOHAPATRA, N., E. W. SMITH, R. C. FITES, and G. R. NOGGLE. 1970. Chilling temperature depression of isocitratase activity from cotyledons of germinating cotton. *Biochem. Biophys. Res. Comm.* 40:1253–1258.

NISHIMURA, M., and H. BEEVERS. 1981. Isoenzymes of sugar phosphate metabolism in endosperm of germinating castor beans. *Plant Physiol.* 67:1255–1258.

RAISON, J. K., and E. A. CHAPMAN. 1976. Membrane phase changes in chilling-sensitive *Vigna radiata* and their significance in growth. *Aust. J. Plant Physiol.* 3:291–299.

RAPP, B. J., and D. D. RANDALL. 1980. Pyruvate dehydrogenase complex from germinating castor bean endosperm. *Plant Physiol.* 65:314–318.

ROBERTS, E. H., ed. 1972. *The Viability of Seeds.* Chapman & Hall, London.

ROLLINS, P. 1972. Phytochrome control of seed germination. Pages 229–254 *in* K. Mitrakos and W. Shropshire, Jr., eds. *Phytochrome.* Academic Press, New York.

SHROPSHIRE, W., JR. 1973. Photoinduced parental control of seed germination and the spectral quality of solar radiation. *Solar Energy* 15:99–105.

TAYLORSON, R. B., and S. B. HENDRICKS. 1977. Dormancy in seeds. *Ann. Rev. Plant Physiol.* 28:331–354.

TOOLE, V. K. 1973. Effects of light, temperature and their interactions on the germination of seeds. *Seed Sci. & Technol.* 1:339–396.

VARNER, J. E., and G. RAM CHANDRA. 1964. Hormonal control of enzyme synthesis in barley endosperm. *Proc. Natl. Acad. Sci. U.S.* 52:100–106.

WALBOT, VIRGINIA. 1971. RNA metabolism during embryo development and germination of *Phaseolus vulgaris. Devel. Biol.* 26:369–379.

WALKER, K. A. 1974. Changes in phytic acid and phytase during early development of *Phaseolus vulgaris* L. *Planta* 116:91–98.

WARDLAW, C. W. 1955. *Embryogenesis in Plants.* Methuen & Co. Ltd., London.

17

Vegetative Plant Growth

As noted in the previous chapter on seed physiology, the formation of new cells during germination and seedling emergence occurs at the expense of metabolites and growth substances from the seed food reserves. As these reserves are exhausted, the heterotrophic mode of metabolism declines and further growth depends on the establishment of an autotrophic mode of metabolism in which water, inorganic ions, oxygen, and carbon dioxide are converted to new cellular material by photosynthetic and other anabolic processes. The transition from heterotrophic to autotrophic growth takes place with the establishment of a root and green shoot system and marks the onset of a new phase of plant growth. The duration of this new growth phase is variable, but in annual plants it is terminated by the initiation of flowering. Plant growth between the periods of seedling emergence and flower initiation is referred to as vegetative growth.

GROWTH: TERMINOLOGY AND DEFINITIONS

Before examining some characteristics of vegetative growth, let us consider briefly some terminology related to growth. All of us are familiar with the term *growth* and generally equate growth with an increase in size. Growth may be evaluated by measurements of mass, length or height, surface area, or volume. Growth is usually accompanied by changes in form or shape. During plant growth, for example, the fertilized egg (zygote) changes from a single cell to a multicellular plant body composed of roots, leaves, and stems.

Many biologists restrict the term *growth* to an irreversible increase in size. The irreversible condition rules out the use of growth to describe the increase in size of a block of dry wood composed of dead cells on being wetted (imbibition and swelling). Growth is thus restricted to living cells and is accomplished by metabolic processes involving the synthesis of macromolecules, such as nucleic acids, proteins, lipids, and polysaccharides, at the expense of metabolic energy. Growth at the cellular level is also accompanied by the organization of macromolecules into assemblages of membranes, plastids, mitochondria, ribosomes, and other organelles. Cells do not indefinitely increase in size but divide, giving rise to daughter cells. An important process during cell division is the synthesis and replication of nuclear DNA in the chromosomes, which is then passed on to the daughter cells. Cell division provides for the transmission of the genetic information established in the zygote to all cells of the plant body. We shall use the term growth in the broad sense as discussed to denote an increase in size by cell enlargement and cell division, together with the synthesis of new cellular material and the organization of subcellular organelles.

The metabolic activities of a growing cell (or tissue or organ) are not uniformly distributed but are localized in centers of activity. Considering the zygote again, centers of activity are established in the fertilized egg that result in an unequal distribution of organelles and macromolecules. When the zygote divides, initiating processes that in time lead to a mature plant, the daughter cells differ in size and content. The genetic information encoded as DNA in the chromosomes is equal in each daughter cell, but the information is used in different ways. The term *differentiation* is used to denote the processes involved in the establishment of localized differences in biochemical and metabolic activity and in structural organization that result in new patterns of growth.

A central problem of modern biology is to understand the factors responsible for initiating cell differentiation. With each meristematic cell possessing the sample genetic information, why should one cell become a sieve tube cell in a phloem tissue whereas another ends up as a vessel element in xylem tissue? Or why should a particular region of the shoot apex give rise to leaves at one time and to flowers later? The establishment of localized areas of differential activity can be visualized as occurring by different pathways. At the molecular level, it can be imagined that a cell destined to become a sieve tube cell, for example, possesses a distinctive cluster of proteins (both enzymic and structural) that differ from those in a cell destined to become a vessel element. Specific chemical compounds, such as growth substances, nucleotides, and ions, might be responsible for the differential activation and synthesis of the distinctive protein clusters. These compounds may arise within the cell or they may come from some neighboring cells or exogenous source. Environmental factors, such as temperature and light, also influence differentiation, presumably by localized effects on enzyme activity or on the processes leading to protein synthesis. Although many mechanisms have been proposed to explain differentiation, direct experimental evidence is needed.

Growth and differentiation over time lead to the formation of an organism with metabolic and structural complexity. The term *development* is used to encompass the activities resulting from growth and differentiation. During development there unfolds an orderly progression of transformations that culminate in a plant body of

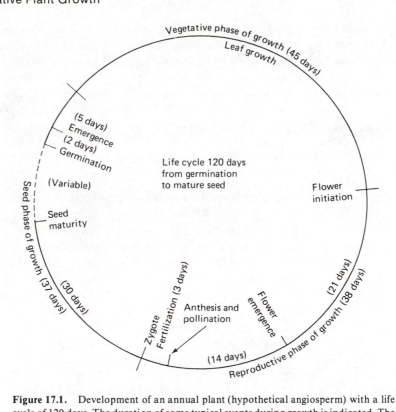

Figure 17.1. Development of an annual plant (hypothetical angiosperm) with a life cycle of 120 days. The duration of some typical events during growth is indicated. The block in germination, whether due to deficiency of moisture or to some dormancy factor, is of variable duration.

characteristic form and chemical composition. In the course of development certain crucial events occur that are necessary for subsequent growth and differentiation. Consider Figure 17-1, an outline of the life cycle of a hypothetical annual plant. From seed germination to seed maturity a number of different events are indicated: Germination, seedling emergence, leaf growth, flower initiation, flower emergence, anthesis, pollination, fertilization, and seed development. Under a particular set of environmental conditions a certain time is necessary for the completion of these stages of development—for instance, 2 days for germination and 5 days for emergence, with a total elapsed time of 120 days for the complete life cycle.

Under a different set of environmental conditions the life cycle may be shortened or lengthened, but no developmental stage can be omitted if a mature seed is to be formed. Most stages are discernible to the eye, but some—flower initiation, pollination, and fertilization, for example—can be detected only by microscopic examination or by waiting until the subsequent event occurs. Environmental conditions may be such as to block a particular stage of development completely. In this case, the life cycle is not completed. Lack of moisture or low temperature, for example, may prevent seedling emergence. Or flower initiation may not occur because of unfavorable photoperiods (see Chapter 18). Or pollination may be prevented because of a heavy rainfall or absence of a suitable insect pollinator.

The terms *growth*, *differentiation*, and *development* thus encompass the events related to the progressive unfolding of the plant's genetic information. The plant receives various internal and external stimuli that interact with the genetic information to cause changes in metabolic activity and structural organization. A certain degree of flexibility is allowed, but each species of plant shows optimal development under rather restricted environmental conditions. In the following pages we shall consider the factors that influence the vegetative phase of growth.

METHODS OF MEASURING GROWTH AND DIFFERENTIATION

Growth can be measured in many different ways. For some kinds of experiments, a plant height measurement may suffice, but ordinarily more information is desirable. Individual leaf size (length, width, area); plant fresh weight and dry weight partitioned among organs, such as roots, stems, leaves, and fruits; cell numbers in tissues and organs; and concentrations of specific chemical constituents (nucleic acids, soluble nitrogen, protein nitrogen, lipids, carbohydrates) in tissues and organs are examples of growth data that have been reported. The degree of complexity (a measure of differentiation) of a plant can be ascertained by noting the time of appearance of leaves, flowers, flower opening (anthesis), and so forth. Differentiation, however, starts at the subcellular and cellular levels and changes in internal complexity occur long before any external changes are evident. Microscopic examination is necessary to detect these internal changes accompanying differentiation. Studies with light microscopy, electron microscopy, and cytochemical and histochemical procedures have provided useful information.

SOME QUANTITATIVE ASPECTS OF GROWTH OF ANNUAL PLANTS

Growth, as measured by height and dry weight, of barley (*Hordeum vulgare*) is shown in Figure 17-2. In this study seeds were planted in the field and growth measurements were begun 15 days later when the first leaf emerged; at this time a well-developed root system is also present. Growth (as measured by height) proceeds rapidly after three or four leaves are produced and continues at this pace until day 90, when height increase declines. Also shown in Figure 17-2 are some developmental stages during growth. The period of rapid growth is characterized by the production of leaves. The decline in plant height increase accompanies internal events associated with flowering. Flower initiation occurs about 30 days before flower emergence, but plant height increase continues because the leaves formed before flower initiation continue to grow. After flower initiation, no new leaves are formed.

Dry weight measurements follow a course similar to plant height except that the period of rapid increase occurs somewhat later. The plant dry weight declines after day 100. If the weight of the developing grain is included, however, dry weight continues to

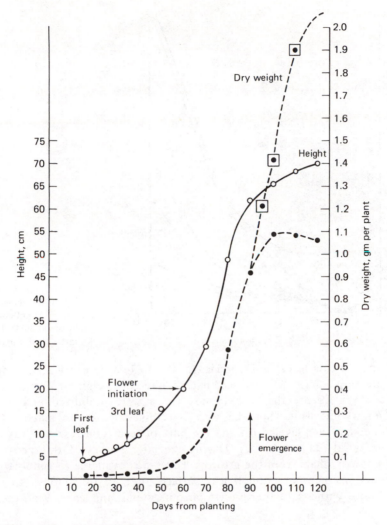

Figure 17-2. Growth curve of field-grown Hannchen barley. ○—○ Plant height;
● --- ● Dry weight per plant (shoot without grain); ▣ --- ▣ Dry weight per plant
(shoot plus grain). (After Pope, 1932; Figures 6 and 7.)

increase for an additional time during seed development and maturation. Strictly
speaking, the changes occurring at this period (90 to 120 days) are those that
accompany seed development in the next plant generation.

The data in Figure 17-2 do not give any indication of the individual contributions
of roots, stems, and leaves to vegetative growth. Such information during the first 24
days of growth of wheat (*Triticum aestivum*) is shown in Figure 17-3. The plants were
grown under continuous illumination (10 klux at plant level) in a growth room at 20° C
and supplied with a complete mineral nutrient solution. The plants were separated into

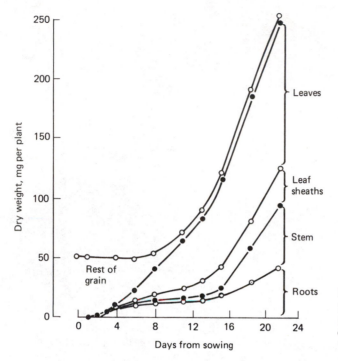

Figure 17-3. Dry weights of principal plant parts during early seedling growth. Constant environmental conditions at 20° C, 10 klux (After Williams, 1960; Figure 3.)

roots, stems, leaf sheaths, and leaves at 1- to 3-day intervals and the dry weights of each fraction determined. The seed had an initial dry weight of 50 mg. During germination and emergence the seed food reserves were mobilized and transported to the embryonic plant, where they were first converted into root material (days 3 to 6). Root increase then leveled off and the bulk of seed food reserves was diverted into leaf material (days 6 to 12). During the later phases of this period (days 6 to 12) photosynthate from the growing leaves was available to support growth. With the establishment of enough leaf surface after 12 to 14 days to support vigorous photosynthesis, further growth of roots, stems, and leaves was accelerated.

Growth Curves

The growth curve of barley (Figure 17-2) is characterized by an early period when the increase in size is quite slow. Then comes a period of rapid increase in size (height and dry weight), followed by another period of decreased growth. In terms of plant physiological processes, the three periods of growth may be visualized as follows. Early growth of the plant is limited by the amount of food reserves in the seed. When the emerged seedling develops an adequate root system and enough leaf surface to support vigorous photosynthesis and anabolism, a period of rapid increase in size is possible. High metabolic rates are not maintained indefinitely and eventually processes are set in motion that lead to a cessation of growth. The factors responsible for the decrease in growth are not thoroughly understood (see senescence in Chapter 18), but several have been proposed. They include competition for essential metabolites, growth substances, water, light, or the accumulation of inhibitors, toxic substances, or waste materials.

Growth curves of the shape and form shown in Figure 17-2 have been obtained for many plants or plant parts as well as other organisms (unicellular, multicellular, populations). The barley growth curve is a composite of the growth of individual tissues and organs of the plant body. Although only the early growth of leaves, stems, and roots of wheat plants is shown in Figure 17-3, it can be seen that the growth curves of these separate organs follow a pattern that, if carried to completion, would be similar to that of the intact plant.

Numerous attempts have been made to gain insight into growth processes by a mathematical analysis of the growth curve. The curve is sigmoid, or S shaped, and lends itself to mathematical manipulation. If, for example, the growth data for barley (Figure 17-2) are plotted on semilogarithmic paper, the curve shown in Figure 17-4 is obtained. The same curve is obtained if the logarithm of plant height is plotted against time. The three periods of growth noted are clearly seen. The first period is referred to as the *lag phase* of growth. In Figure 17-4 it extends between days 15 and 20 and it has been found that a similar slow phase of growth extends to water imbibition by the seed and the initiation of germination. Between days 20 and 90 the period of rapid increase in height is referred to as the *log phase* of growth. The period of reduced size increase following the log phase is referred to as the *senescence phase* of growth.

The log phase of growth is frequently called the phase of exponential growth. Characteristics of exponential growth can be illustrated as follows. Consider a meristematic cell that divides to form 2 cells and these divide to give a total of 4 cells and so forth, giving the series 2, 4, 8, 16, 32, 64, etc. If the time interval between successive divisions is approximately equal, a plot of time versus the series (2, 4, 8, 16, 32, 64, etc.) gives a smooth curve similar in shape to that shown in Figure 17-2. If the logarithms of the numbers in the series are plotted against time, a straight line is obtained, as shown in Figure 17-4. The relationship between time and number of cells formed is given as $N = 2^t$, where N is the number of individuals and t the time interval. Note that t is an exponent of 2—thus the use of the term *exponential growth*.

The growth curve shown in Figure 17-2 is a plot of the cumulative increase in height during the growing period. It is of interest to know the daily or weekly increments of growth contributing to the final plant height. These increments—size increase per unit interval of time—are referred to as the relative rate of growth and can be obtained as illustrated in Figure 17-5. As shown for A and B, a rate measurement is obtained by placing a straightedge tangent to the growth curve and connecting two points on the tangent by a 90° triangle. In the examples shown the lower legs of the triangle are taken as equal to 10 days; the other leg is then a measure of the size increase during the 10-day period. From Figure 17-5 it is seen that the growth rate at A is 0.17 cm per day whereas at B the growth rate is 0.75 cm per day. By taking shorter time increments, the instantaneous rate of growth (also called relative rate of growth) at any point on the growth curve can be calculated. It is best done mathematically rather than graphically.

If, instead of plotting accumulated size against time (Figure 17-2), the relative

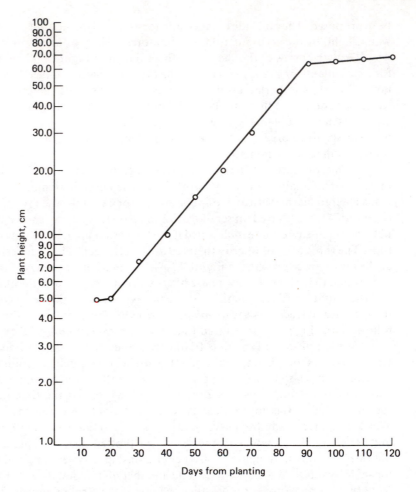

Figure 17-4. Growth data for barley from Figure 17-2 is plotted on semilogarithmic paper. Note three periods of growth as shown by three straight lines.

rate of growth is plotted against time, curves of the form shown in Figure 17-6 are obtained. Here the growth rates change continuously during growth. The growth rates of barley (Figures 17-2 and 17-5) follow a time course similar to that shown for curve A in Figure 17-6, with the maximum rate of growth being maintained only for a few days. In other plants the maximum rate of growth may be maintained over considerable periods of time, as illustrated by curve B in Figure 17-6.

The English plant physiologist V. H. Blackman (1919) suggested that the growth of plants can be represented by equation

$$W_1 = W_0 \cdot e^{rt} \tag{17-1}$$

where W_1 is the final size (weight, height, etc.) after time t, W_0 is the initial size at the beginning of the time period, r is the rate at which plant substance is laid down during

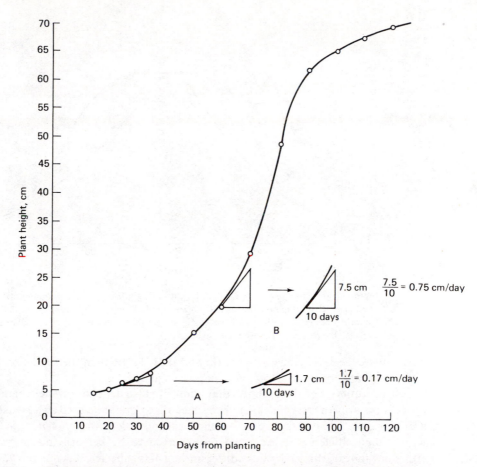

Figure 17-5.   Graphic method of determining the rate of growth. The growth curve of field-grown barley (from Figure 17-2) is shown. A straight line tangent to the growth curve is drawn touching the curve at 25 days (A). A triangle is then constructed as illustrated and the growth rate calculated from the units of the coordinates of the graph. The growth rate on the 25th day (A) is 0.17 cm day^{-1}; on the 60th day (B) the growth rate is 0.75 cm day^{-1}.

time t, and e is the base of natural logarithms. It should be noted that r is the relative growth rate as discussed earlier. Blackman pointed out that Eq. (17-1) also describes the way in which money placed at compound interest increases with time; the term *compound interest law* is used to describe such phenomena. Banks usually apply compound interest quarterly or annually so that the increase in amount occurs as a jump. With plants or other biological systems, compound interest is applied continuously and size increase follows a smooth curve.

From Eq. (17-1) the final size of an organism (W_1) is seen to depend on the initial size (W_0). It has been shown that seedling growth from seeds of different size follow such an expectation, with the larger seeds giving a larger plant. This is true, however,

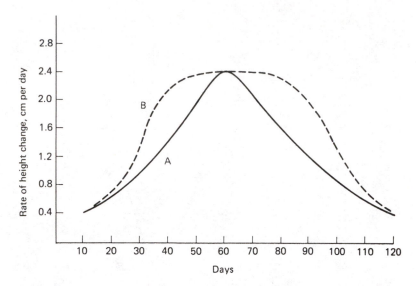

Figure 17-6. Rate of change of plant height during growth. Curve A represents a plant in which a rapid rate of growth is obtained for a brief period of time (at day 60). Curve B represents a plant in which a rapid rate of growth is maintained over a considerable period of time (days 40 to 80).

only during the early stages of growth; and if t is large, then the final sizes of plants from small and/or large seeds frequently are equal.

In addition, Eq. (17-1) shows that plant size also depends on the magnitude of r, the relative growth rate. Blackman suggested that r might be used as a measure of the ability of a plant to produce new plant material and called r the *efficiency index*. Plants with a high efficiency index could be expected to outperform plants with a low efficiency index. Although r does differ among plant species, it is not constant during the life of a plant. Furthermore, all parts of a plant are not equally involved in synthesizing new plant substance; some material goes to storage or is catabolized. The usefulness of the efficiency index as a predictive factor in evaluating performance or yield is limited.

Growth Analysis

Equation (17-1) can be put in the logarithmic form $\log_e W_1/W_0 = rt$. Several other English plant physiologists (Gregory in 1917 and Briggs, Kidd, and West in 1920) showed that the logarithmic expression of Eq. (17-1) can be written

$$r = \frac{\log_e W_2 - \log_e W_1}{t_2 - t_1} \tag{17-2}$$

where W_2 and W_1 are the plant dry weights at times t_2 and t_1, respectively. Rather than use r, the efficiency index, in this expression, the relationship between size increase and time is termed the relative growth rate. If logarithms to base 10 instead of base e are used, the relative growth rate (RGR) is

$$RGR = \frac{2.303 \,(\log_{10} W_2 - \log_{10} W_1)}{t_2 - t_1} \tag{17-3}$$

Briggs, Kidd, West, and others went on to point out that the relative growth rate has two components: The rate of increase of dry weight per unit time per unit area of leaf surface, termed the net assimilation rate (NAR); and the ratio of leaf area to dry weight, termed the leaf area ratio (LAR). These two components are expressed as follows:

$$NAR = \frac{(W_2 - W_1) \, 2.303 \,(\log_{10} A_2 - \log_{10} A_1)}{(t_2 - t_1)(A_2 - A_1)} \tag{17-4}$$

$$LAR = \frac{(A_2 - A_1) \, 2.303 \,(\log_{10} W_2 - \log_{10} W_1)}{2.303 \,(\log_{10} A_2 - \log_{10} A_1)(W_2 - W_1)} \tag{17-5}$$

where W_1, A_1, W_2, A_2 represent dry weights and leaf areas at time intervals t_1 and t_2. That RGR in Eq. (17-3) is equal to the product of NAR and LAR can be determined by multiplying together the right-hand terms of Eqs. (17-4) and (17-5).

The net assimilation rate is a measure of the amount of photosynthetic product going into plant material; that is, it is an estimate of net photosynthesis (carbon assimilated by photosynthesis minus the carbon lost by respiration). The NAR can be determined by measuring plant dry weight (roots plus shoots) and leaf area periodically during growth and is commonly reported as grams of dry weight per square centimeter of leaf surface per week (g cm^{-2} week^{-1}). The leaf area ratio is a measure of the proportion of the plant that is engaged in photosynthetic processes. The LAR is reported as the area of leaf surface in square centimeters per gram of dry weight (cm^2 g^{-1}). The relative growth rate is reported as the grams of dry weight increase per gram of dry weight present per day (g g^{-1} day^{-1}). Other units of area (dm^2, m^2) might be used as well as other time periods.

An application of the growth analysis technique to a study of plant growth is illustrated by Figure 17-7. Seedlings of sunflower (*Helianthus annuus* cv. Jupiter) and maize (*Zea mays* cv. Standfast) were raised in growth cabinets under 30 klux of fluorescent light (16-hour photoperiod) at five different temperatures. Measurements of dry weight and leaf area were obtained at two times—immediately after the third foliage leaf or pair of leaves appeared and just before the sixth leaf or pair of leaves appeared. It is seen [Figure 17-7(a)] that the RGR follows a similar pattern in both sunflower and maize, increasing to a maximum at 28° C and then decreasing at a higher temperature (34° C). Maize failed to germinate at 10° C.

The contributions of net assimilation rate (NAR) and leaf area ratio (LAR) to the relative growth rate (RGR) of the two species at different temperatures are shown in Figure 17-7(b) and (c). The differences in RGR between sunflower and maize are shown to be largely due to differences in NAR between the two species, for there is little difference in their LARs except at the highest temperature. Temperature appears to influence both components of plant growth—the efficiency of the photosynthetic apparatus as measured by net assimilation rate and the size of the photosynthetic

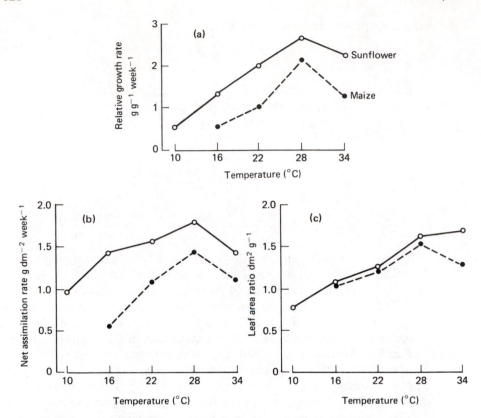

Figure 17-7. Effects of temperature on the relative growth rate (a), net assimilation rate (b), and leaf area ratio (c) of sunflower ($\bigcirc$—$\bigcirc$) and maize ($\bullet$---$\bullet$) seedlings grown in 16-hour photoperiods under 30 klux of fluorescent light in growth chambers. (After Warren Wilson, 1966; Figure 1.)

apparatus as measured by leaf area ratio. The two species of plants, however, react in a different manner to temperature.

Growth analysis is a useful tool in studying the complex interactions between plant growth and the environment. The measurements do not require elaborate equipment or extensive laboratory facilities, just an oven and a balance for dry weight determinations and a method of measuring leaf area. Throughout the balance of the chapter, applications of growth analysis to problems of vegetative growth will be shown.

INTERNAL AND EXTERNAL FACTORS AFFECTING VEGETATIVE GROWTH

The pattern of development during vegetative growth is influenced by various internal and external factors: Genetical, nutritional, environmental, and hormonal. The interplay among these various factors and their effects on growth and differentiation are complex. In many instances, we have only a rough idea of how a particular factor—for example, temperature—influences development. Advances in techniques of measuring chemical and physical processes and the development of facilities for handling large quantities of data (computers), however, now make it possible to study these problems.

Genetic Factors

Biologists have long recognized that units of information are present in the chromosomes within the nucleus. These units, or genes, consist of specific sequences of nucleotides in DNA molecules (see Chapter 3). The genetic complement of a plant is acquired at the time of zygote formation from male and female gametes. During subsequent cell divisions the genetic information is passed on to the daughter cells, thereby providing each cell with a copy of the original genetic complement. The copying is not absolutely perfect and small modifications may occur in the nucleotide sequence in DNA from chemical changes in individual nitrogen bases. As a rule, these modifications do not greatly alter the patterns of growth and development. At times, however, the change in the DNA sequences is sufficiently great to be lethal. In such a case, it is possible that some critical enzyme can no longer be synthesized because the proper sequence of DNA does not occur. Occasionally a major change in DNA sequences may occur, giving rise to an essentially new plant. Such changes, known as mutations, may develop because of some environmental shock (high or low temperature) to DNA at a critical period during the cell cycle. An example of such a mutation is the "seedless" orange which was found on a normal orange variety. The "seedless" plant is propagated vegetatively (by rooted cuttings) to maintain the plant in cultivation.

Our cultivated crop plants, such as cereal grains, legumes, root crops, flowers, and sugarcane, have been developed from wild progenitors over hundreds of years. Plants were selected from wild populations on the basis of increased seed size, root or tuber size, ease of harvesting, increased fiber length (cotton, hemp), flower petal color or shape, increased content of oil, sugar, or protein, or of flavor, aroma, or volatile oil, and so forth. Over the years many selections have been maintained as improvements have been made and there may be several thousand recognized breeding lines of a particular crop. They are maintained as a valuable genetic resource or gene pool for future selection and breeding experiments. Also, in many instances, the original wild ancestors are still in existence and they, too, make up a valuable genetic resource. Wild plants generally grow as weeds in their original habitats and tend to be lost as the land is developed for agricultural or economic purposes. International centers are now being established for preserving these genetic resources.

Several examples illustrate the importance of maintaining gene pools of our major crop plants. Susceptibility to fungi, bacteria, or insects is a major problem in agricultural production. Resistance to certain diseases is present in wild progenitors or in certain breeding collections. By crossing the resistant plants with plants having good yield characteristics, plant breeders have been able to develop disease-resistant lines of many major crops, particularly in cereal grains. In the future, plant breeders will seek to develop plants able to grow in saline habitats, or under limited moisture regimes, or in cool temperatures, or with increased protein content. As large a gene pool as possible must be available to assist them in these endeavors.

Another example of plant breeding for enhanced productivity is the production of dwarf varieties of some cereal grains—wheat, rice, sorghum. When normal (tall) varieties of rice are heavily fertilized to increase grain production, the plants respond by increases in stem length as well as grain size. This combination of a long stem with heavy grain head gave rise to plants likely to fall over (lodging). Such plants are difficult to harvest and so crop yield suffered. Another variety of rice was known to have short, thick stems; when these plants were heavily fertilized, the stems remained short and upright, with increased grain yield. The dwarf plants are native to regions of cooler climates (Japan, northern China, Korea) whereas the tall plants are native to warm, tropical regions (Philippines, India, Indonesia). By crossing the two types, scientists at the International Rice Institute in the Philippines were able to develop selections that combined the dwarf stem characteristics with the yield characteristics of the tropical varieties. Such selections could be heavily fertilized for increased grain production. Problems of insect and fungus infection arose in the new varieties, but additional breeding experiments yielded improvements in this area. It is not easy to select for a single characteristic and a large gene pool is essential for such research programs. Research similar to that described for rice has been carried out with wheat and later with sorghum. The new dwarf varieties respond to fertilization by increased yield but require additional inputs, such as pesticides for controlling insects and fungi, ample water, and the use of farm machinery on large hectarages for maximum benefit. These features are characteristic of intensive agriculture.

Genetic engineering   Much is said today about the prospects for improving plant productivity by using the techniques developed by molecular biologists. With bacteria, it is possible to transfer specific genetic information from one strain to another, to stabilize the new genetic information in the host cell, to have replication of the host-cell genome containing the introduced genetic information, and to have the host cell produce a new product (protein) based on the new genetic information. Of particular interest has been research showing that the nitrogen-fixing ability (the so-called *nif* gene) of one bacterium can be transferred to another bacterial species unable to fix atmospheric nitrogen (lacking the *nif* gene). The introduced *nif* gene (actually a cluster of genes essential for N_2 assimilation) then functions in the host, where atmospheric N_2 can be fixed. These results have led to research on the possibility of introducing *nif* genes from N_2-fixing bacteria into green plants. *Rhizobium*, the bacterium in root nodules of legumes, possesses *nif* genes and fixes N_2, which is then supplied to the plant. Current research, however, is directed toward the introduction of *nif* genes into the genomes of such plants as wheat, maize, and rice, thus enabling these plants to assimilate atmospheric N_2 and bypassing the need for nitrogen fertilizer. Nitrogen is the limiting factor in most plant production and nitrogen fertilizer is very costly, for its production is based on energy and hydrogen from natural gas or petroleum. The so-called *nif* gene is a cluster of about 15 genes whose products (enzymes) are involved in N_2 assimilation (see Chapter 9). It is unlikely that the entire *nif* cluster can easily be introduced into the genome of higher green plants. The possible long-term benefits of success in this research, however, indicate that the problem should be pursued.

Similar research is also being carried out on the identification of genes involved in a variety of plant processes—ability to grow in saline habitats, ability to grow under stress conditions of low water potential or adverse temperatures, ability to take up extra amounts of inorganic ions from the soil, production of useful chemical constituents (hydrocarbons, oils, etc.). If such gene clusters can be found and transferred to other plants, it might be possible to enhance plant productivity.

Hormonal Factors

In Chapter 16 it was pointed out that during germination and seedling emergence the embryonic plant depends on nutrients and growth substances from the endosperm and/or cotyledons to sustain growth until an autotrophic condition is attained. The importance of such nutrients as glucose (from starch), amino acids (from protein), inositol, K^+, Ca^{2+}, Mg^{2+}, PO_4^{3-} (from phytin), glycerol and fatty acids (from fats) in the synthesis of new cells is obvious. These materials, supplied by the seed, must be made available quickly after water imbition to ensure a successful vegetative plant. Just as important as the inorganic and organic nutrients in seedling emergence are the growth substances present in seeds—such substances as auxins, gibberellins, cytokinins, and abscisic acid.

When the embryonic axis starts to enlarge, it is important that the root and shoot be properly oriented so that the roots grow downward into the soil and the shoots grow upward into the air. This downward orientation of roots and upward orientation of shoots is a response to gravity and is called a *geotropic* response.

Geotropism   Several processes are involved in geotropism—the perception of the gravitational stimulus and the response of the plant to the stimulus. The perception of the stimulus seems similar in both shoots and roots and is discussed in a later section. Following perception of gravity, shoots and roots respond differently. In shoots the apical meristem is the source of indoleacetic acid (IAA). By polar transport (an active process), IAA moves from the shoot tip to cells below the meristem, where cell enlargement (elongation) occurs. If the shoot is oriented vertically, parallel to the gravitational field, the distribution of IAA from the stem tip is equal in all the cells in the stem and the shoot grows upward (Figure 17-8). If, however, the plant or stem is displaced from the vertical (shown as a 90° displacement in Figure 17-8) the stem will, after a certain period of time, bend upward. Cells in the lower half of the stem elongate more than those in the upper half, thereby producing unequal growth and an upward bending of the shoot. Cholodny in 1924 and Went in 1926 independently suggested that the geotropic response of horizontally oriented shoots is due to differential growth resulting from an unequal distribution of auxin (IAA) in the upper and lower halves of

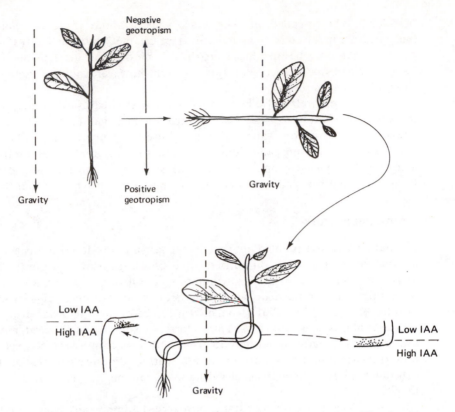

Figure 17-8. Growth responses of a horizontally oriented plant. The shoot is negatively geotropic and the root is positively geotropic.

the shoot. The proposal became known as the Cholodny–Went theory of geotropism, later expanded to include the bending reponse of plants to light (phototropism). The Cholodny–Went theory of geotropism postulated that IAA is displaced from the upper half of the shoot to the lower half when the shoot is displaced from a vertical orientation.

      Experimental data shown in Figure 17-9 indicate an unequal distribution of IAA in coleoptile tips following horizontal orientation. Coleoptile tips from dark-grown maize seedlings were oriented as illustrated (all manipulations and diffusion times were also carried out in darkness). The amounts of IAA found in the agar receiving blocks were measured by the *Avena* coleoptile test (Figure 14-4) and given as relative IAA units. With a vertical orientation, 40 units of IAA diffuse out of the coleoptile tip into the agar collecting block. This vertical orientation serves as the control for the other orientation.

      If the tip receives a horizontal orientation [Figure 17-9(b)], 40 units of IAA diffuse into the agar collecting block, an amount identical to that found with a vertical orientation [Figure 17-9(a)]. If a horizontally oriented coleoptile tip is split almost to the tip and the upper and lower halves separated by a glass cover slip (impermeable to

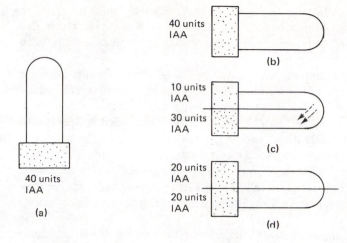

40 units
IAA

(a)

40 units
IAA

(b)

10 units
IAA

30 units
IAA

(c)

20 units
IAA

20 units
IAA

(d)

Figure 17-9. The lateral displacement of IAA in horizontally oriented coleoptile tips. (After Gillespie and Briggs, 1961.)

diffusible substances), more IAA diffuses out of the lower half of the split coleoptile than out of the upper half [Figure 17-9(c)]. That the differences in IAA orientation between the upper and lower halves of the coleoptile tip are due to lateral displacement of IAA is shown in Figure 17-9(d), where the two coleoptile halves are completely separated by a glass cover slip. Under these conditions similar quantities of IAA move into the two collecting blocks. Experiments similar to those described for coleoptiles from monocots have been carried out with apical stem tips of dicots. Essentially similar results have been observed and the Cholodny–Went hypothesis for geotropic responses in stems appears applicable to both monocots and dicots. In some dicot stem tips gibberellins and cytokinins are also present and show lateral displacement under geotropic stimulation. It is not known, however, if gibberellins and cytokinins play a general role in geotropism.

As shown in Figure 17-8, roots also respond to gravity and the Cholodny–Went theory was proposed to account for the downward orientation of roots displaced from the vertical (positive geotropism). As with shoots, the unequal growth of the upper and lower sections of horizontal roots was believed due to different amounts of IAA in the two sections, the differences in IAA concentration being caused by lateral transport or displacement, as shown in Figure 17-9 for shoots. The Cholodny–Went theory also suggested that root tips are centers of IAA synthesis or distribution and that IAA from the tip is transported from the tip toward the base of the plant (basipetal transport). The limited amount of research on roots does not support these ideas. Root tips contain little or no IAA and any IAA present has been transported from the base of the plant (shoots) toward the root tips (acropetal transport).

Research on the response of roots to geotropic stimulation is consistent with the notion that an inhibitor is synthesized in the root tip—more specifically, the root cap. The inhibitor is transported away from the tip and, under geotropic stimulation, will move laterally so as to attain a higher concentration on the lower half of the root. The inhibitor, identified as abscisic acid (ABA), inhibits elongation of cells in the lower half of the roots with the result that roots grow downward. In effect, the Cholodny–Went theory appears correct after all. It does not involve IAA, however, but ABA. Similar results have been found with both monocot and dicot roots. Roots also synthesize cytokinins and gibberellins, but their role in geotropic responses has not been established.

There is general agreement among plant physiologists that plants perceive gravity through a change in the distribution of some subcellular component in cells known as *statocysts*. In root–shoot meristems, all the cells appear to contain a displaceable subcellular element and would qualify for being termed statocysts. In other parts of the plant—that is, nodes of some grasses, which respond to gravity—only certain specialized cells are known as statocysts. The movable subcellular component is known as a *statolith*. It is generally agreed that specialized starch grains, known as amyloplasts, function as the subcellular organelles responsible for perceiving a geotropic stimulation. Amyloplasts are saclike bodies containing a number of starch grains.

The suggested role of starch statoliths in gravity perception is shown schematically in Figure 17-10. In a cell oriented parallel to the field of gravity [Figure 17-10(a)] amyloplasts are on the lower surface of the cell. It can be regarded as a stable configuration and the reactions leading to cell enlargement are symmetrical in all cells. If the cell is displaced [Figure 17-10(b) and (c)] amyloplasts fall through the cytoplasm to the new lower surface of the cell, attaining a new configuration [Figure 17-10(d)]. Just how the new amyloplast arrangement initiates reactions leading to cell enlargement is not known. It can be imagined that amyloplasts cover over reactive sites on the inner surface of the plasma membrane or cover up plasmodesmata that connect adjacent cells. These processes might alter the polarity of the cell and facilitate the transport of ions or growth substances. Amyloplasts need not actually come in contact with the new lower cell surface. The movement of amyloplasts through the cytoplasm might alter the configuration of subcellular organelles (endoplasmic reticulum, microtubules, microfilaments), thereby initiating some biophysical or biochemical events. The responses coupling gravity perception to growth are not well understood.

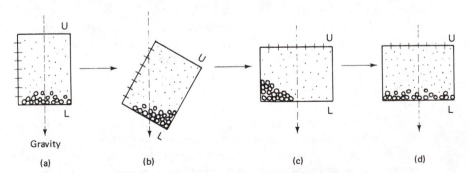

Figure 17-10. Perception of geotropic stimulus according to statolith theory. The statoliths are displaced when the cell receives a new orientation relative to the normal gravitational force. The change in statolith distribution is believed to bring about a change in IAA concentration within the cell. U, upper part of cell; L, lower part of cell.

Phototropism When the embryonic axis emerges from the seed coat and elongates so as to bring the stem tip into the light, a new orientation mechanism comes into play. Besides responding to gravity, the shoot now also responds to light. Cholodny and Went postulated that the response of shoots to light—phototropism—resulted from differential or asymmetric growth of shoot cells under the influence of an asymmetric distribution of IAA. Under a uniform light field IAA distributed from the apical meristem region is transported symmetrically down the stem in a polar (basipetal) fashion. All shoot cells receive identical amounts of IAA and cell elongation is symmetrical and the stems grow upright. If unilateral illumination is presented to the shoot tip, the transport of IAA from the tip is displaced away from the illuminated side and cells on the darkened side of the shoot receive more IAA. With the increase in IAA, cell elongation is greater and the shoot bends toward the source of illumination (see Figure 14-3).

Experiments shown in Figure 17-11 illustrate the relationship between illumination and the distribution of IAA in maize coleoptiles. Experimental conditions similar to those described for the geotropic experiments were followed. In all experiments the

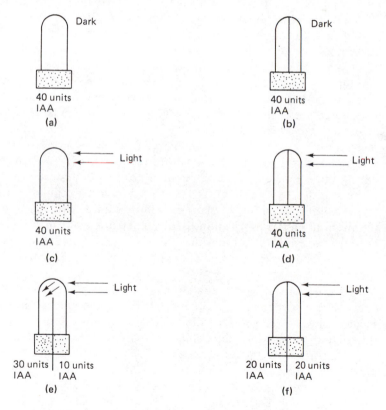

Figure 17-11. The lateral distribution of IAA in unilaterally illuminated maize coleoptile tips. (After Briggs et al., 1957; Figure 1.)

diffusible auxin was collected in agar blocks, which were then assayed by the *Avena* coleoptile curvature bioassay. The units shown below each block represent the relative amounts of IAA present. The same amount of diffusible IAA is collected from intact coleoptiles whether they remain in darkness or are illuminated [Figure 17-11(a) and (c)]. If coleoptiles are split but the half sections remain in contact (no impermeable barrier), similar results are obtained [Figure 17-11(b) and (d)]. If the two halves are separated by a thin glass cover slip, the same amounts of IAA are collected from the illuminated and darkened halves [Figure 17-11(f)]. When the coleoptile is partially separated at the base by a thin piece of glass, more auxin diffuses from the darkened half than the illuminated half [Figure 17-11(e)]. The results support the idea that unilateral illumination induces a lateral transport of IAA from the illuminated coleoptile tip to the darkened half of the coleoptile.

Although it seems clear that phototropism in coleoptile tissue is mediated by an asymmetrical distribution of IAA brought about by unilateral illumination, it is less certain how dicot stems react. Unilateral illumination has been shown to mediate lateral transport of other growth substances in addition to IAA. Gibberellins and cytokinins seem involved, but no satisfactory explanation for their role is available. IAA induces cell enlargement and an asymmetric distribution of IAA will produce an asymmetric elongation of cells, followed by bending. Just how the other growth substances interact with IAA is not known.

When the young seedling achieves the autotrophic condition with a well-developed root system and a shoot system capable of carrying out photosynthesis, further growth is rapid, provided that environmental conditions are appropriate. It was shown, for example, in Figure 17-2 that the independent existence of barley was reached about 15 days after planting. Before then the plant axis depended on food reserves and growth substances mobilized from endosperm tissue. With leaf development, photosynthesis is possible and further growth (dry weight or height) is rapid during the log, or exponential, phase of growth. In addition to the biosynthesis of carbohydrates, fats, proteins, DNA, RNA, and other cellular constituents, the plant also synthesizes a variety of plant growth substances. A schematic representation of the distribution of these substances in a dicot plant is shown in Figure 17-12. It should be kept in mind that the values shown are relative and can be changed by environmental factors and age of the plant and that they will vary in different plant species.

The idea that growth regulation occurs through interactions of several growth substances is particularly well documented in studies of vegetative growth. Development during this stage of the life cycle depends on cell division, enlargement, and differentiation. Auxins, gibberellins, cytokinins, abscisic acid, and ethylene all have variable effects on these processes. Consider, for example, cell division. Auxins enhance cell division in cambial growth while inhibiting the growth of lateral buds. Gibberellins also influence cambial growth and promote cell division in subapical regions of shoot meristems. Cytokinins promote cell division in callus tissue and promote growth of lateral buds. Abscisic acid and phenolics inhibit cell division, but the inhibition can be overcome by exogenous auxin, gibberellins, and cytokinins.

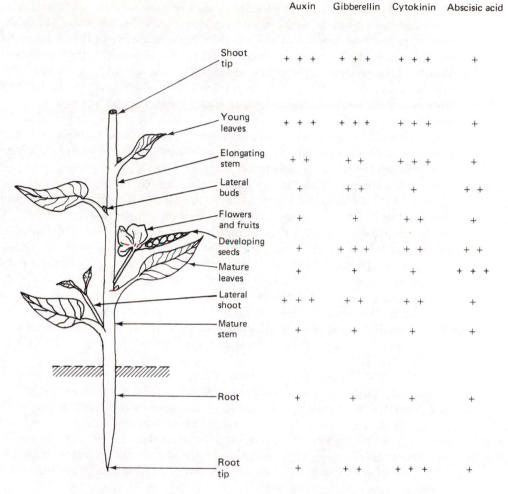

	Auxin	Gibberellin	Cytokinin	Abscisic acid
Shoot tip	+ + +	+ + +	+ + +	+
Young leaves	+ + +	+ + +	+ + +	+
Elongating stem	+ +	+ +	+ + +	+
Lateral buds	+	+ +	+	+ +
Flowers and fruits	+	+	+ +	+
Developing seeds	+	+ + +	+ +	+ +
Mature leaves	+	+	+	+ + +
Lateral shoot	+ + +	+ +	+ +	+
Mature stem	+	+	+	+
Root	+	+	+	+
Root tip	+	+ +	+ + +	+

Figure 17-12. Relative concentrations of some plant growth substances in various parts of the plant. +++ high concentration, ++ medium concentration, + low concentration.

Ethylene seems to have little effect on cell division, but it has a strong effect on cell enlargement. Several other effects of plant growth substances on growth and development are considered next.

Apical dominance The growth habit displayed by herbaceous and woody plants shown in Figure 17-12 (ignore the presence of flowers and fruits) has been explained on the basis of the influence of the stem apex on growth and development of other parts of the plant. Some features of this phenomenon, known as *apical dominance*, are as follows. Lateral (or axillary) buds are present in the leaf axils and the development of these buds is correlated with their distance from the apical meristem. Lateral buds close to the shoot apex remain dormant whereas those some distance from the apical meristem develop into shoots. Thimann and Skoog showed that when the apical meristem of *Vicia faba* (a close relative of the common garden bean) is removed, a bushy plant develops as the result of growth of the lateral buds into shoots. If the severed meristem is replaced by a source of IAA (in an agar block or lanolin paste), development of the lateral buds is again inhibited. (Note that IAA has both promotive and inhibitory effects on plant growth, depending on its concentration and the particular tissue in which IAA acts.) The practice of removing the apical bud is commonly used by gardeners and horticulturists to promote bushiness and enhance the production of floral buds. Similarly, woody shrubs and trees are frequently pruned to remove terminal shoots and change the growth habit of the plant.

The influence of the shoot meristem on axillary bud development can be understood if it is assumed that IAA is the growth substance involved and that the distribution of IAA in plants follows the pattern shown in Figure 17-12. Apical meristems are rich in IAA, either through biosynthesis or as a site of deposition from other parts of the plant. Young leaves are also rich in IAA, probably because of biosynthesis. The IAA from apical meristems and young leaves is transported down the stems, where it inhibits development of axillary buds. Older leaves do not synthesize IAA; also, lateral buds in the axils of these older leaves are more distant from the apical meristem. Under these conditions lateral buds in older leaf axils are not strongly inhibited and so develop into shoots. If the apical meristem is excised, nearby lateral buds can develop into shoots because lowered concentrations of IAA are present at the lateral bud.

Figure 17-12 shows that the apical meristem and other tissues and organs contain additional growth substances besides IAA. Recent studies have shown that gibberellins and cytokinins are probably involved in apical dominance. Thimann and coworkers found that axillary buds on intact plants can be stimulated to enlarge and elongate slightly if treated with an exogenous supply of a cytokinin. Full development of buds into shoots takes place if the cytokinin-stimulated buds are treated with IAA. It was also noted that endogenous levels of IAA and cytokinins in axillary buds on intact plants are low.

Based on these observations, a modified version of the role of growth substances in apical dominance might be as follows. Axillary buds lack direct vascular connections with the phloem and xylem elements of the stem and are out of the mainstream of organic solute (sugars, amino acids, etc.) transport. These buds may not develop because they lack metabolites and growth substances, such as IAA, cytokinins, and gibberellins, which are transported to the apical bud by xylem transport. Another possibility is that IAA in the stem apex enhances metabolic activity in meristematic cells, which is accompanied by a vigorous demand on organic solutes moving in the vascular tissue. This combination of a vigorous metabolic sink at the apical meristem, enhanced by a high IAA concentration, and axillary buds isolated from the main transport stream effectively inhibits the development of axillary buds.

As for the possible role of cytokinin in apical dominance, in Figure 17-12 apical meristems and roots have high concentrations of cytokinins whereas lateral buds contain low concentrations. The roots are considered a main center of cytokinin biosynthesis and it is transported in the xylem stream to other parts of the plant. In an intact plant the apical meristem, with its high IAA activity, serves as an effective sink for cytokinins from the roots. The lateral buds, on the other hand, are maintained with low levels of both IAA and cytokinin because there are no direct vascular connections between the lateral buds and the main stem. When the apical meristem is removed, or if an exogenous cytokinin is applied to lateral buds on intact plants, the cytokinin level of the lateral bud is increased. Perhaps the increase in cytokinin level at the base of the lateral bud stimulates cell division and possibly completes the vascular connections between the axillary buds and the main transport path. Axillary buds may then be supplied with IAA, cytokinins, and other organic solutes, thereby establishing a new center of metabolic activity, accompanied by shoot development.

It is unlikely that cytokinin and IAA are the only growth substances involved in apical dominance. Abscisic acid and gibberellins are present in axillary buds and they may also interact with IAA and cytokinins. Moreover, exogenous ethylene influences lateral bud development in many plants. It is not possible at this time to give a comprehensive explanation of apical dominance based on actions of all these growth substances. It can be pointed out that it is not essential to postulate that growth substances are involved in apical dominance. The observed effects might be achieved if some organic or inorganic solute (or solutes) is required in critical concentrations in the apical meristem. Such an explanation, known as the nutritional hypothesis, has considerable supporting data. It might be imagined that a growth substance, such as IAA or cytokinin, may create an active metabolic sink and that the movement of sugars or amino acids or other solutes to and from these sinks establishes apical dominance.

Plant growth substances at the cellular level   The preceding material on the influence of plant hormones on vegetative growth considered effects on tissues and organs. It is probable, however, that growth substances act at the cellular or subcellular (molecular) level. Several examples of such actions are discussed in the following pages.

Botanists have long been interested in whether plant cells are *totipotent*—that is, capable of recapitulating the embryonic stages of development with the eventual formation of a mature plant. The question is a practical one because of the possibility of improving techniques for increasing plant populations by vegetative reproduction. Clearly not all cells in a plant are totipotent, for many (tracheids, fibers, sclereids, vessel elements, etc.) lack cytoplasm and a nucleus. Each living (somatic) cell, however, has the same genetic complement as the zygote from which the plant developed; it also has the necessary genetic information for directing the orderly development of the cell into a mature plant.

Tissue culture studies have provided much information on the question of cell totipotency. Cultures of tissues from pith, cortex, cambium, leaf parenchyma, and storage parenchyma grow readily under aseptic conditions in a medium composed of inorganic ions, sucrose, vitamins, and an organic supplement (yeast extract, malt extract, coconut milk). In the absence of an organic supplement very little growth occurs. Growth on the enriched medium is by cell enlargement and division and forms a large mass of undifferentiated cells known as callus tissue.

Steward and associates have studied extensively the growth of cells isolated from callus tissue cultures. They removed small segments of tissue (explants) from the secondary phloem of domestic carrot roots that grew rapidly in an apparatus that maintained the explants in constant agitation so that they were intermittently bathed in nutrient solution and exposed to the atmosphere. During growth clumps of cells slough off and continue to enlarge and undergo division. Many of these cell masses resemble various stages of development of embryos derived from zygotes. Under appropriate nutritional and environmental conditions these cell aggregations, or embryoids, continue to develop, forming roots and shoots and eventually mature carrot plants (Figure 17-13). Steward considers the secondary phloem cells of the carrot root to be totipotent, but totipotency is not realized unless the cells are freed from their neighbors and placed in an environment resembling that found in an embryo sac. Coconut milk, a liquid endosperm, supplies such an environment. As described in Chapter 16, coconut milk and other liquid endosperms are complex mixtures of carbohydrates, cyclitols, amino acids, auxins, gibberellins, cytokinins, phenolics, and other substances.

Many investigators have observed that embryoids develop spontaneously in tissue cultures in the absence of coconut milk. Halperin and Wetherall made the interesting observation that explants of somatic cells from Queen Anne's lace, the wild carrot (*Daucus carota*), readily form embryoids on a medium containing inorganic ions, sucrose, vitamins, and 2,4-dichlorophenoxyacetic acid, a synthetic auxin. Explants removed from different parts of the plant form a callus tissue that spontaneously develops embryoids. The embryoids, if suitably nurtured, develop into plantlets and later into mature plants. Steward found that if explants were removed from embryonic tissue of Queen Anne's lace, as many as 100,000 embryoids developed from a single explant following callus formation.

Evidence to date indicates that somatic cells within the plant body are totipotent. Individual cells from some tissues, such as those from the embryo in the developing seed of Queen Anne's lace, readily undergo embryogenesis. Other cells from this same plant form embryoids less readily because of local deficiencies of other as yet unrecognized factors. Some plant families (Umbelliferae, which includes the carrot and Queen Anne's lace) form embryoids more readily than others. The reasons for the differences in ease of embryoid formation are unknown. Deficiencies in specific growth substances, lack of balance between inorganic ions, lack of some organic cofactor, or deficiency of adequate sugar or reduced nitrogen might contribute to the difficulties. Further studies on these problems will resolve many of the present uncertainties and give a clearer understanding of cell totipotency.

Practical application of embryoid formation   Many important horticultural plants (fruit trees, berries, etc.) are propagated by asexual methods, such as stem or leaf cuttings. If such plants can be propagated by the embryoid technique described, it will

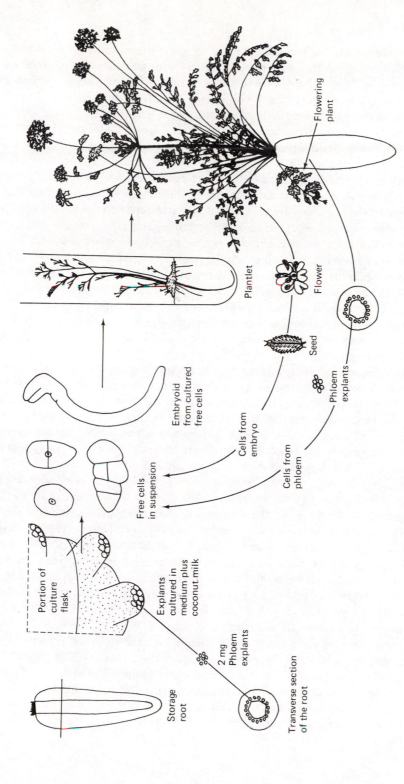

Figure 17-13. Diagrammatic representation of the cycle of growth of the carrot plant; successive cycles of growth are linked through free cells derived from phloem or the embryo. (After Steward et al., 1964; Figure 2.)

Storage root

Transverse section of the root

2 mg Phloem explants

Explants cultured in medium plus coconut milk

Portion of culture flask

Free cells in suspension

Cells from embryo

Cells from phloem

Phloem explants

Embryoid from cultured free cells

Plantlet

Seed

Flower

Flowering plant

help considerably in maintaining a "bank" of germ plasm of herbaceous and woody plants. Callus tissue cultures can be maintained for long periods on a minimal medium. When embryoids are needed for plantlet production, the callus cultures can be transferred to a medium enriched with the appropriate growth factors.

The embryoid technique is being successfully applied to orchid cultivation. Orchid seeds are extremely minute and are generally grown under aseptic conditions on an agar medium containing inorganic ions, sucrose, and vitamins. In many instances, it takes 4 to 7 years to obtain a flowering plant from the time of seed germination. Callus tissue cultures from meristematic tissues can be nurtured so as to form embryoids that can then be grown into plants. Of even greater interest, however, is the fact that many highly prized show orchids are of hybrid origin and either do not produce viable seed or fail to "breed true." Such plants must be propagated by vegetative means. Tissue explants taken from shoot meristems readily proliferate when cultured in a liquid medium to form embryoids, which can be further subdivided and cultivated to form still more embryoids. When transferred from a liquid medium to the surface of an agar medium, the embryoids develop into plantlets and eventually into mature plants. At present, only a few of the orchid hybrids have been propagated by this procedure. Yet hundreds of other hybrids now in commercial production could be similarly handled; it is a question of determining the particular chemical and physical requirements of each species or genus.

Another cell culture technique that holds great promise in the improvement of crop and horticultural plants involves the production and propagation of haploid plants. Such plants possess half the normal number of chromosomes found in the cells of growing vegetative plants. If the chromosome number is then doubled to form the normal diploid condition, in a single generation a completely homozygous, true-breeding plant can be produced. Plant breeders generally must grow many generations of plants in order to approach the homozygous condition.

The technique depends on the production of haploid plants from pollen grains. The pollen grains are the products of a meiotic division of the microspore mother cell (see Figure 16-1, Chapter 16) and thus contain half the chromosome complement of the parent plant. Two Japanese workers, Nakata and Tanaka (1968), developed procedures for producing haploid plants, using the tobacco plant. Anthers were removed from flower buds and cultured under aseptic conditions on medium containing minerals, vitamins, and sugar. Each pollen grain within an anther (about 30,000 to 40,000 pollen grains in a single anther) contains a haploid ($1N$) nucleus that further divides to form two haploid nuclei. At this particular stage of pollen development the anthers are removed for culturing.

Apparently only one of the two pollen nuclei, the so-called tube or vegetative nucleus, divides to form an embryoid and eventually a plantlet. The plantlets can be transferred to a soil medium and grown to mature haploid plants. These plants are sterile and not useful in plant-breeding programs. If the plantlets, however, are manipulated in such a way that the chromosome complement is doubled, the resulting diploid plants are fertile. Under some conditions chromosome doubling is spontaneous, but ordinarily the plantlets are treated with colchicine, an alkaloid shown to be effective in doubling chromosome number.

The production of haploid plants from pollen grains has been successful with tobacco and several of its close relatives. The technique seems viable for other species of plants. Many plant scientists are working on the problem.

Another development in cell culture should be mentioned. This is the practice of removing cell walls from cultured cells to form what are known as *protoplasts*. The protoplasts can be cultured and have been observed to fuse with other protoplasts and subsequently undergo embryogenesis to give plantlets and eventually form plants. If haploid cells are used, diploid plants may be produced by such a procedure. There also is the possibility of producing hybrid plants by the fusion of protoplasts from different species of plants. Such a hybrid has been produced between two different species of tobacco by Carlson, Smith, and Dearing (1972).

Another possible use of cultured protoplasts is to attempt to introduce nucleic acid from different sources and alter the genetic makeup of the cell. Similarly, it might be possible to introduce subcellular organelles, such as chloroplasts or mitochondria into protoplasts.

Growth substances and the growth of tissues and organs   Apical shoot meristems of many plants can be aseptically removed and cultured on a culture medium of known composition. Callus tissue may develop at the cut surfaces, frequently followed by the development of adventitious roots. Eventually an entire plant develops from the meristem culture. This is basically the practice followed by horticulturists in the asexual propagation of plants.

Similarly, apical root meristems can be excised and grown in nutrient culture. Root cultures initiate buds less frequently. Still, the roots of certain plants readily initiate vegetative buds. Probably the inability of most isolated root cultures to initiate buds is due to our lack of knowledge about the factors controlling the development of root cells. It might be mentioned that many other aspects of root physiology are poorly understood, a situation largely due to technical difficulties in studying roots.

Nutritional requirements of shoot meristems are satisfied by a medium containing inorganic ions and sucrose. Sucrose serves as an energy source and supplies carbon skeletons for the synthesis of new cell constituents. If the stem cultures are illuminated, leaves frequently develop and photosynthetic processes supply the necessary carbohydrates. Isolated roots also require a basal medium of inorganic ions and sucrose; in addition, several B vitamins must be supplied. Some root cultures need only thiamin (vitamin B_1) whereas the roots of other plants also require the vitamins nicotinic acid and pyridoxine. In the intact plant these vitamins are synthesized in the shoots and then translocated to the roots. Either root cells are unable to synthesize enough vitamins to maintain vigorous growth or the synthetic mechanism is repressed in some fashion, for it is presumed that the root cells contain the same genetic information as the shoot cells.

Studies of the growth of organ systems in culture suggest that there are different levels of organization within the plant, each level having a rather specific set of requirements for maintenance and development. Studies with embryos and free cell cultures support such a conclusion. An idea of the nature of the specific growth substance requirements at different levels of organization has come from investigations of the induction of organ development in tissue cultures. Skoog and coworkers while working with callus tissue cultures derived from tobacco stem pith, observed that the callus could be maintained on a minimal medium containing inorganic ions and sucrose. Varying degrees of organ initiation in the callus tissue are observed, depending on the relative concentrations of IAA and kinetin (in the presence of adenine) added to the basal medium (Figure 17-14). In the absence of kinetin, low concentrations of IAA induce cell enlargement whereas higher concentrations of IAA induce root initiation. Very high concentrations of IAA, in the absence of kinetin, induce both cell expansion and cell division without organ initiation. With concentrations of kinetin between 0.2 and 1.0 mg per liter, some bud initiation occurs in the absence of IAA. There is a profuse initiation of buds in the presence of 1.0 mg per liter of kinetin and 0.005 mg per liter of IAA. High concentrations of both IAA and kinetin promote cell division and cell expansion, but no organ initiation is observed.

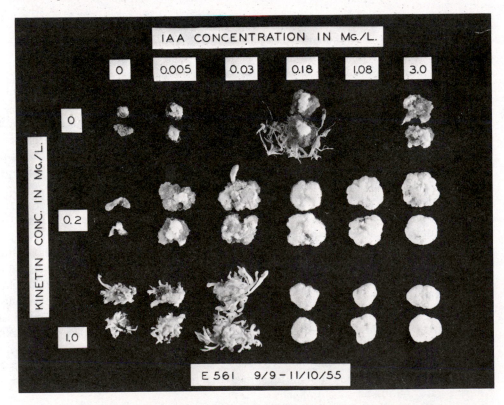

Figure 17-14. Effects of indoleacetic acid (IAA) and kinetin on organ formation in plant tissues cultured in vitro. (After Skoog and Miller, 1957; Plate 4.)

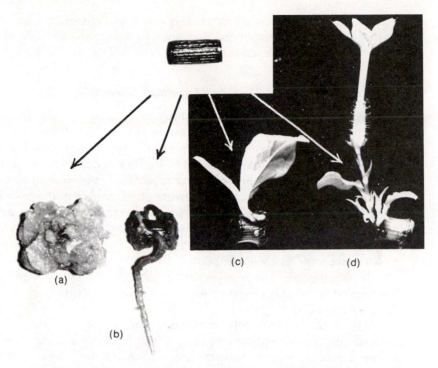

Figure 17-15. The in vitro induction of callus (a), roots (b), leaves (c), or flowers (d) in excised stems of *Plumbago indica*. (After Nitsch and Nitsch, 1967; Figure 1.)

Clearly organ initiation in tobacco pith callus cultures depends on a balance between the levels of IAA and kinetin present in the growth medium.

Further evidence on the role of plant growth substances in organ initiation is provided by studies of Nitsch and Nitsch. They found that internode segments from vegetative stems of *Plumbago indica* can be induced to form a callus, or roots, or vegetative shoots, or inflorescences (Figure 17-15), depending on the cultural conditions. The stem segments were cultured on a basal medium of inorganic ions, glycine, sugar, and vitamins, plus varying concentrations and combinations of growth substances and other organic compounds. A broad range of effects was noted, but the following generalizations give an idea of the complex interactions involved in organ initiation and development.

Gibberellins and auxins inhibit bud formation, but occasionally very low concentrations of IAA enhance bud initiation in the presence of cytokinin and adenine. External environmental conditions also modify the pattern of shoot development. *Plumbago indica* is a short-day plant. When stem segments are cultured under short-day conditions, more buds develop than under long-day conditions.

Short-day conditions are also necessary to obtain the initiation of flowers [Figure 17-15(d)]. High concentrations of sucrose, as well as cytokinin, adenine, and several other purines and pyrimidines, are essential for flower initiation. Abscisic acid stimulates flower bud production whereas auxins and gibberellins are inhibitory. The significance of these observations in relation to the general question of reproductive physiology is discussed in Chapter 18.

Effects of Environmental Factors on Vegetative Growth

Plant growth and development are subject to the action of physical and biological factors in the environment. The physical components include air, water, and soil and there is a ready flow between these components (Figure 17-16). The plant, anchored in position by its roots, is both influenced by the flow of energy between the air and soil and is also responsible for providing a pathway for the movement of energy between the compartments. Changes or movements of substances in the environment result primarily from radiation from the sun. Some of the incident radiant energy is reflected from the plant to the environment. Energy is also lost from the plant as heat or used to evaporate water lost by transpiration. A small fraction of the solar energy is conserved as chemical energy by photosynthesis.

In addition, the soil receives radiant energy from the sun and this energy is dissipated by processes similar to those noted for the plant. Moreover, the soil and plant exchange energy.

The flow of energy between the sun and air, water, and soil produces the elements of the environment collectively referred to as climate. These elements— radiation, precipitation, air temperature, soil temperature, moisture (relative humidity), cloudiness, CO_2 concentration, air pressure, wind, and so forth—can be measured by various physical methods and constitute the science of meteorology.

On a daily or short-term basis, meteorological data are used to describe local weather conditions. A statistical assemblage of meteorological data collected over a long period of time may be used to describe the climate of specific areas of the earth's surface. Plants interact closely with climate, and plant formations—grasslands, tundra, tropical forests, coniferous forests, and so forth—are integral elements of specific types of climates. Much of the meteorological data is collected on instruments located at least 2 m above the surface of the ground and at exposed sites free of vegetation. It is recognized, however, that climatic conditions at the surface of the ground and near vegetation are quite different from those at some distance above the ground. The term macroclimate is used to describe the assemblage of meteorological observations in the atmosphere and the term microclimate to describe surface climates. Within recent years meteorologists and biologists have pooled their efforts in studies of the microclimate, producing active research programs in biometeorology and agrometeorology.

Because of the many meteorological elements that make up the climate and of the ever-changing state of the plant as it proceeds through germination, seedling emergence, vegetative growth, flower production, fruiting, and senescence, it is difficult to determine the specific effect of a single environmental factor, such as temperature, on any particular phase of growth and development. Some workers feel that such problems can best be studied under field conditions whereas others have preferred

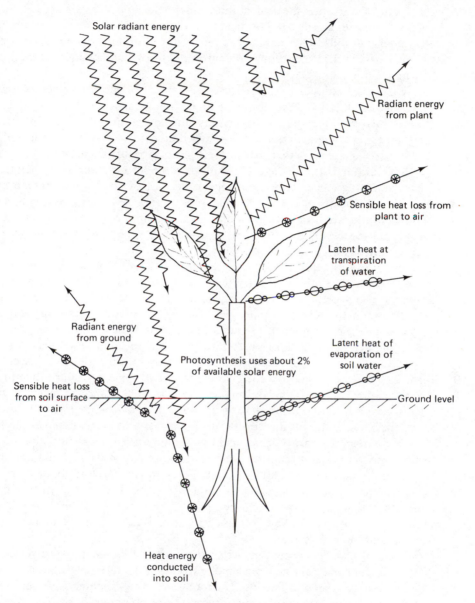

Figure 17-16. The energy balance of the plant and its environment. (After "The Plant" and Its Physical Environment," Anon., 1966 p. 2.)

controlled environment facilities (growth chambers, phytotrons, air-conditioned greenhouses). Because crop production ultimately depends on understanding the intact plant as it grows in the field, studies under field conditions are needed, but much useful information can also be obtained by studying the growth of organs, tissues, and cells under controlled environments. Studies of cell organelles and isolated enzyme systems also furnish information that contributes to an understanding of the growth of the whole plant.

Various techniques have been used to protect plants from uncertain weather conditions: Site location on south-facing slopes (northern hemisphere), wind screens of tall trees, an overstory of plants to provide shade, temporary covering of plants during cold weather. These practices are still followed in many parts of the world and enable agriculturists and gardeners to grow plants in unfavorable situations.

Several hundred years ago glasshouses came into use and such facilities offer rather complete control of the environment. Because of the thermal characteristics of glass, however, temperatures tend to rise in a glasshouse on a sunny day and during summer months it is difficult to grow plants unless some provision is made for artificial cooling. During the 1920s commercial air-conditioning equipment was developed and several plant research centers constructed air-conditioned greenhouses. The Boyce Thompson Institute for Plant Research in Yonkers, New York, and the Clark Greenhouses at the California Institute of Technology in Pasadena pioneered much of the early research on the use of air-conditioned glasshouses for studying plant growth.

Although the air-conditioned glasshouse provides a considerable degree of control over the environment, the next logical step is the construction of a unit for controlling all environmental aspects. Ideally such an installation should control light, temperature, moisture, nutrition, gaseous environment, wind, diseases, insects, and so forth. Early efforts to achieve such controlled environments were confined to small boxes or cabinets that handled only a few plants. In 1949, at the California Institute of Technology, the first large installation of controlled environment growth rooms was put in operation under the direction of F. W. Went. This unit, known as the Earhart Plant Research Laboratory, has served as a model for the development of similar installations throughout the world. Such units are expensive to build and operate; consequently, their planning and efficient operation involve close cooperation among engineers, meteorologists, statisticians, and biologists. The term phytotron has been coined to designate such an integrated plant research installation. Today there are phytotrons in Australia, New Zealand, France, Russia, The Netherlands, Japan, Canada, and the United States.

A number of "packaged" growth rooms are now commercially available. They are of different sizes, varying from small portable units to large "walk-in" rooms that can handle large numbers of plants. The rooms have self-contained refrigeration units that can supply a wide range of temperatures on a programmed basis. The light sources can provide as high as 40 to 80 klux at plant height and some special units may go higher. Figure 17-17 illustrates two of the individual plant growth rooms now available.

Figure 17-17. Cabinets and growth rooms for controlled environment studies. [Courtesy Sherer-Gillet Co., Marshall, Michigan (lower) and U.S. Department of Agriculture (upper).]

Radiant energy Light generally refers to the portion of the electromagnetic spectrum between 400 and 700 nm visible to the human eye. Plants, however, respond to a somewhat broader span of radiation—between 300 and 800 nm—and it is this region of the electromagnetic spectrum that is considered here (see Chapter 5). The term *light* will be used in this broader sense.

Light influences plant growth and development in different ways, depending on its intensity, duration, and spectral quality. They are, of course, interrelated, but it will be useful to consider them separately.

Intensity. Plants can accomplish some growth in the absence of light at the expense of stored food reserves. This situation is especially well demonstrated in the case of germinating seeds. Figure 17-18 shows the growth habit of pea seedlings grown in darkness for 10 to 12 days and in light for 21 to 28 days. The dark-grown plant is said to be etiolated and is characterized by a tall and spindly stem with long internodes,

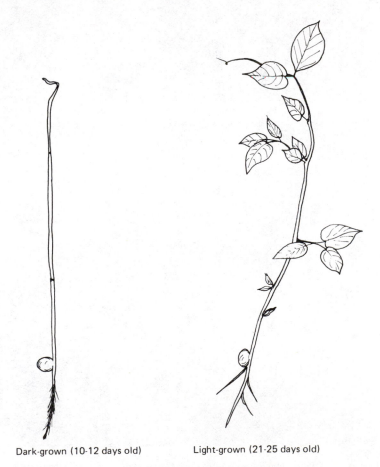

Dark-grown (10-12 days old)   Light-grown (21-25 days old)

Figure 17-18. Appearance of dark-grown (etiolated) and light-grown seedlings of *Pisum sativum.*

minute leaves that have failed to expand, and a lack of chlorophyll. The plant is pale yellow in color. The light-grown plant, on the other hand, has shorter internodes, expanded leaves, and a green color. The etiolated plant can survive in the dark for about 2 weeks, after which all food reserves in the cotyledons are exhausted. Small-seeded plants survive for shorter periods; large-seeded plants may live for 3 to 4 weeks in the dark. Unless light is provided, however, the plants perish.

Numerous studies have been carried out to determine the differences between etiolated and light-grown plants. No anatomical differences are apparent with respect to the kinds of tissue formed under the two conditions; instead light accelerates the phases of growth associated with the production of leaf primordia, the development of leaves, the early stages of cell division and cell elongation in the internodes, and cell differentiation in the internodes. Light also inhibits certain aspects of internode elongation, slowing down cell division and elongation. Thus light promotes some processes and inhibits others. The form and the size of the plant depend on a balance between these two kinds of processes.

Besides etiolation, light has other formative effects on plants. It has been recognized for many years that plants growing in the shade are different from those growing in full sun. Similarly, on a single plant, leaves may differ in appearance and activity, depending on whether they are on the outside of the leaf canopy (sun leaves) or within the interior of the canopy (shade leaves). Sun leaves are usually thicker than shade leaves, often composed of several layers of palisade parenchyma cells rather than a single layer as in shade. Sun leaves are also frequently smaller in area and have shorter petioles than shade leaves.

The physiological bases for the differences between sun and shade leaves are poorly understood. Certainly the marked differences in the morphological features of sun and shade leaves, or of sun and shade plants, suggest that cell differentiation has been altered and different patterns of metabolism have been established. In studies with sun and shade ecotypes of *Solidago virgaurea*, Björkman found that the rates of photosynthesis are much greater at high light intensities in the sun ecotypes than in the shade ecotypes. Furthermore, when the latter are exposed to high light intensities, some part of the photosynthetic system appears to be damaged. Accompanying the high rates of photosynthesis in leaves of sun ecotypes is a greater concentration of ribulose bisphosphate carboxylase, the enzyme responsible for the entrance of carbon dioxide into the photosynthetic carbon reduction pathway of CO_2 assimilation. From these studies it is apparent that sun and shade ecotypes do differ in their metabolic properties. How these metabolic differences are related to anatomical differences in cell size and shape is less clear. Also, it is not known if some specific pigment, such as phytochrome, is responsible for the initiation of these different photoprocesses.

One additional aspect of the influence of light intensity on plant growth should be mentioned—the effect on photosynthesis. The rate of photosynthesis increases with an increase in light intensity. There are, however, genetic differences between plants with respect to their response to light intensity. Moreover, the final size and form of a plant depend on how the photosynthate is distributed or apportioned to the roots, stems, and leaves. Light also has an effect on respiratory processes and plants with a high rate of photorespiration will have less photosynthate available for the formation of roots, stems, and leaves. While light intensity, through its influence on the assimilation of CO_2, is a major factor in vegetative growth (of annual plants), other factors alter the overall growth process.

Duration of light (day length). Almost 100 years ago it was noted that the length of the daily photoperiod influences the course of growth and development in plants. In 1920 Garner and Allard, plant physiologists with the U.S. Department of Agriculture, Beltsville, Maryland, showed that plants can be separated into several classes, depending on their response to day length. The most striking effect of day length, or photoperiod, is the shift from vegetative growth to reproductive growth. Plants that flower following an exposure to short days are termed *short-day plants* whereas those that flower after an exposure to long days are called *long-day plants*. Still another group of plants appears to flower regardless of the length of the photoperiod; such plants are termed *day-neutral plants*. A vast amount of work has been carried out on photoperiodism since the early observations of Garner and Allard. Their basic findings have been expanded and clarified and today there is a firm scientific base for many of the observed effects of day length on plant growth and development.

Although the effects of day length are most closely associated with the transition from vegetative to reproductive development, day length also influences vegetative growth. If a plant, for example, is grown under photoperiods that do not induce flower initiation (see Chapter 18), the length of the period of vegetative growth is extended. Figure 17-19 shows the influence of day length on vegetative growth of *Sesame indicum* var. Early Russian. The plants were grown in a greenhouse under natural illumination but with different dark periods so as to give photoperiods of 5 hours of light/19 hours of dark (5L/19D), 7 hours of light/17 hours of dark (7L/17 D), and 13 hours of light/11 hours of dark (13 L/11 D). At 2-day intervals stem length (from cotyledons to tip of stem) and leaf number were determined. Stem length is seen to increase as the duration of the light period is increased. Also, the number of leaves increases with the length of the photoperiod. Plants 61 days old had 6, 16, and 25 leaves under day lengths of 5 L/19 D, 7 L/17 D, and 13 L/11 D, respectively.

Sesame (var. Early Russian) is a short-day plant and floral initiation occurs in day lengths shorter than 10 hours. At photoperiods in excess of 10 hours, however, floral initiation is delayed and vegetative growth continues. With photoperiods of less than 10 hours duration, vegetative growth is limited by the amount of photosynthate formed. Other growth characteristics, such as time of appearance of first flower, number of flowers produced, and number of fruits set, are also influenced by day length (see Chapter 18).

Wavelength. The wavelength of light has a profound effect on plant development. Plants contain a number of pigments (discussed more fully in Chapter 5) that selectively absorb specific wavelengths of radiation, thereby initiating such photoprocesses as photosynthesis, phototropism, and photomorphogenesis. Photosynthetic processes are coupled closely to the pigments chlorophyll *a* and *b*, but radiant energy absorbed by carotenes and other pigments is also utilized in photosynthesis.

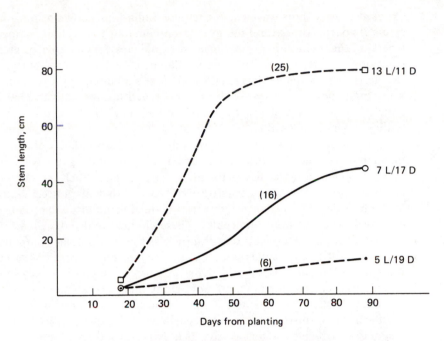

Figure 17-19. Effect of daylength on stem length of *Sesame indicum* var. Early Russian. Plants were grown in glasshouse under natural illumination but placed in the dark for different periods. 5 L/19 D = 5 hours of natural illumination and 19 hours in the dark. Numbers in parenthesis on day 61 indicate the number of leaves on the plants at that date. (After Smilde, 1960; Figure 6.)

Blue light (400 to 510 nm) is responsible for triggering growth reactions that culminate in the bending of leaves and stems. These growth reactions result in the orientation of leaves and stems in particular spatial patterns, depending on the direction and intensity of the radiation source. At low levels of irradiation leaves may become oriented so as to be perpendicular to the radiation source (sun), thereby presenting a greater expanse of leaf photosynthetic surface. With high levels of irradiation, the leaves of many plants are oriented parallel to the energy source. Such a growth reaction has the effect of minimizing the harmful effects of intense radiation levels.

Most photomorphogenetic effects are due to radiant energy absorbed by the pigment phytochrome. Phytochrome-mediated responses are initiated by low irradiances in contrast to the high irradiances utilized in photosynthesis. These low-irradiance reactions cover a broad range of physiological responses. The striking differences in growth form between light-grown and dark-grown seedlings (Figure 17-18) can be prevented by exposing dark-grown plants to brief exposures of red light (660 nm). Many different aspects of plant development depend on phytochrome-mediated processes and are discussed in Chapter 5.

Ionizing radiation. In addition to visible and ultraviolet radiations, plants are exposed to very short wavelength energetic radiations known as ionizing radiations. Some are part of the natural energy environment to which all living things are exposed whereas other radiations are manmade. Cosmic rays, for example, originate in nuclear reactions in outer space. They are similar in nature to x rays but are much more energetic. X rays and gamma rays, which have similar energy levels, are produced by x-ray machines and nuclear (radioisotope) reactions, respectively. Particles, such as neutrons, electrons, alpha particles, are products of nuclear reactors, accelerators, and radioisotope disintegration.

These radiations are known as ionizing radiations because in their passage through matter they produce ionizations (positively and negatively charged particles) that interact with molecules in the immediate vicinity. Such interactions frequently result in the production of reactive free radicals that disrupt cellular constituents. Water molecules, for instance, interact with ionizing radiation to produce positively or negatively charged water molecules. These charged molecules break down spontaneously to form free radicals that readily interact with cellular components.

In addition to cosmic rays, the natural environment contains several sources of ionizing radiation to which all living organisms are continuously exposed. These include radioactive carbon and potassium and several isotopes of radium and uranium. Radioactive carbon (^{14}C) is produced in the atmosphere by the action of cosmic rays on nitrogen gas; nitrogen-14 is transmuted to carbon-14. When ^{14}C disintegrates, it gives off an electron (beta particle), a form of ionizing radiation, and reverts to nitrogen-14.

It is interesting to note that the occurrence of ^{14}C in the atmosphere serves as a basis for determining the age of biological material. If it is assumed that the amount of ^{14}C in the atmosphere is constant and has remained constant for the last 100,000 years or so, it is seen that every living organism will contain a certain amount of ^{14}C. Plants assimilate ^{14}CO$_2$ from the atmosphere and incorporate it into carbohydrates, proteins, lipids, and other organic molecules. The tissues of animals contain ^{14}C because of eating plants and other animals and CO$_2$ assimilation. Both plants and animals carry out respiration, during which organic matter is degraded into carbon dioxide. So there is a turnover of ^{14}C in living organisms and the tissues of plants and animals reach an equilibrium with the ^{14}C of the atmosphere.

Consider a living tree that carries out photosynthesis and incorporates ^{14}C into organic matter. If the tree should die, it will no longer assimilate any new ^{14}C and what is present (cellulose, starch, lipid, protein, etc.) will start to disappear by radioactive decay. The disintegration rate of ^{14}C is known (half-life of about 5700 years), and by comparing the radioactivity of the dead tree with that of a living organism, the approximate age of the dead sample can be determined. Samples as old as 50,000 to 60,000 years can be dated by this technique.

The pollution of the environment by radioactive material from nuclear reactions—atom bomb tests, wastes from nuclear reactors, radioactive isotopes, etc.— is of great concern to biologists. The testing of atom bombs in the upper atmosphere deposits considerable quantities of radioactive isotopes in the biological environment. Many are short-lived, but others are long lived and accumulate in living material, where they remain a source of ionizing radiation. The widespread use of atomic energy in power plants and in industry, as well as the use of radioactive isotopes in chemical and biological research, has created a problem of the safe disposal of radioactive wastes. Peaceful uses of atomic energy will probably expand in the future and a great deal of research is needed to develop safe methods of handling and disposing of these materials.

Temperature Because of the energy exchange between the sun and the earth, air and soil temperatures fluctuate rather widely during a 24-hour period. Figure 17-20 shows air temperatures at several distances above the soil surface and soil temperatures at several depths on a cloudless day compared to a cloudy day. On the clear day both

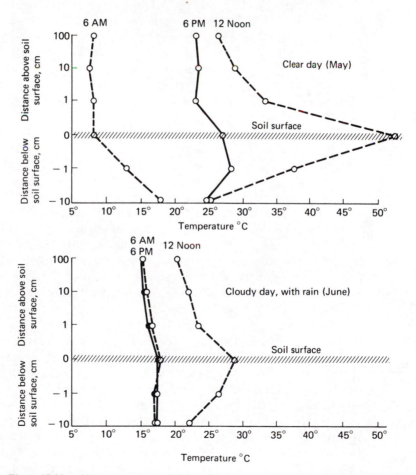

Figure 17.20. Air temperatures at heights of 1, 10, and 100 cm above the soil surface; at the soil surface; and at depths of 1 and 10 cm below the soil surface on a clear day (no clouds) and on a cloudy day (6% sun with afternoon shower) in Connecticut. (After Waggoner et al., 1959; Figures 4 and 5.)

air and soil temperatures increase during the day, reaching a maximum around noon, and then decrease in late afternoon. The greatest diurnal temperature differential exists at the soil surface, where the temperature may go from $10°C$ at 6 A.M. to over $50°C$ at noon. At distances of 10 to 100 cm above the soil surface the diurnal fluctuation may be as great as 15 to $20°C$. Soil temperatures are somewhat more uniform, but even at a depth of 10 cm the temperature may show a daily fluctuation of 5 to $10°C$. During the night the soil radiates heat into the atmosphere.

Similar diurnal fluctuations in soil and air temperatures occur on cloudy days (Figure 17-20), but the temperature differentials are not as great. Both air and soil temperatures do not drop as low during the night under cloudy skies as under clear skies because the clouds prevent loss of radiation from the soil.

The effect of air temperature on leaf temperature and several plant processes is shown in Figure 17-21. Two environmental situations are compared: A cool, cloudless day with a maximum day temperature of $25°C$ and a minimum night temperature of $12°C$ [Figure 17-21(a)] and a warm, cloudless day with a maximum day temperature of $40°C$ and a minimum night temperature of $20°C$ [Figure 17-21(b)]. Under both conditions a sunlit leaf is warmer than the surrounding air during the day and cooler than the ambient air during the night. A shaded leaf is always cooler than surrounding air.

Plant activity, which may be measured in terms of growth rate, protoplasmic streaming, enzyme activity, or some other vital process, is extremely sensitive to temperature. On cool days [Figure 17-21(c)] these processes in sunlit leaves increase as the leaf temperature increases until around noon there is a slight decrease in vital activity as a result of the elevated leaf temperature. Plant activity in the shaded leaf also increases with leaf temperature, but no midday decrease is observed, for the temperature of the leaf does not get too high. On a warm day [Figure 17-21(d)] the leaf temperatures of both sun and shade leaves at midday reach levels that are detrimental to metabolic processes and decreases in plant activity occur. The high midday temperatures are accompanied by elevated transpirational losses of water that result in loss of turgor and closing of stomata. As the leaf temperatures drop in the late afternoon, plant activity again increases.

Photosynthetic activity is also markedly affected by leaf temperature. During a cool day [Figure 17-21(e)] there are considerable differences between the activities of sun and shade leaves. Both follow the diurnal pattern of air temperature, but imposed on this is the effect of light on photosynthesis. The shade leaf does not receive as much radiation as the sun leaf and the limiting factor is light, not temperature. Respiration rates are also lower in the shade leaves and the net amount of photosynthate going into plant material is usually greater at the lower leaf temperatures. On a warm day [Figure 17-21(f)] the high leaf temperatures are deleterious to photosynthetic activity. Not only are metabolic processes reduced at the high leaf temperatures, but moisture stress (increase in transpiration) and loss of turgor result in stomatal closure, which decreases the supply of CO_2 to the chloroplasts. The interaction of these various effects causes a sharp reduction in photosynthetic activity of leaves exposed to full sunlight. The

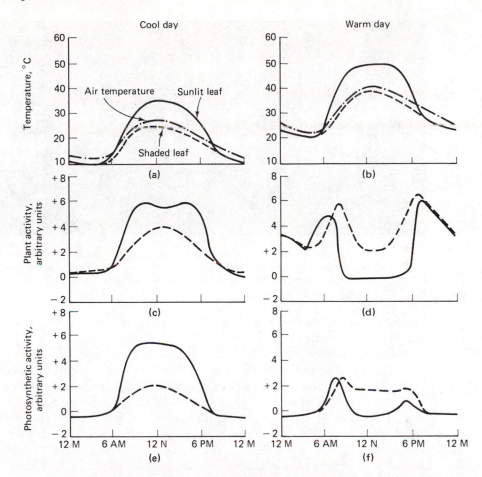

Figure 17-21. The influence of daily variations of air temperature on leaf temperature (a) and (b), plant activity (c) and (d), and photosynthetic activity (e) and (f). Conditions for cool and warm days: — sun leaves; - - - shade leaves; - · - air temperature. (After Gates, 1965, p. 82.)

temperature of shade leaves usually is slightly less than air temperature and the photosynthetic activity is not depressed during the midday rise in temperature.

The effects of temperature on plant growth described so far have concerned air temperature and shoot growth. While there is generally a few degrees' differential between air and soil temperatures (Figure 17-20), relatively few studies have investigated the influence of root temperature on plant growth. Results from one such study are shown in Figures 17-22 and 17-23. Following seed germination and growth in soil for 4 days at 25° C (soil temperature and air temperature), soil temperatures were adjusted so that the roots were exposed to temperatures ranging from 12 to 35°C in 1°C increments. Shoots were maintained at 25°C with an illumination (16-hour photoperiod) of 15 klux. The seedlings were harvested when 23 days old and their

Figure 17-22. The influence of root temperature on the growth of corn (*Zea mays*). The root temperatures were maintained at the indicated temperatures while the shoots were kept at a uniform 25° C temperature. (After Walker, 1969; Figure 5.) Photograph furnished through the courtesy of J. M. Walker (U.S.D.A. photo).

general appearance is shown in Figure 17-22. The plants increased in size as the root temperatures increased from 12 to 26° C. Root temperatures above 26° C resulted in a decrease in plant size.

The distribution of plant dry weight between shoots and roots is shown in Figure 17-23. Both shoot and root dry weight increased as root temperatures increased from 12 to 26° C and decreased between 26 and 35° C. The shoot–root ratio generally increased (Figure 17-23) as the root temperatures increased, but there is a depression in the curve at about 26° C. At root temperatures above 26° C there appears to be a greater decrease in root dry weight than shoot dry weight, as demonstrated by the increase in the shoot–root ratio. At the highest root temperatures (34 and 35° C) both root and shoot dry weight sharply decrease.

It is clear from this study that small changes in root temperature are responsible for rather striking changes in plant growth. Although organic constituents, such as carbohydrates and nitrogenous compounds, were not determined, the concentrations of a number of inorganic elements in the shoots were measured. In most instances, the uptake of inorganic ions from the soil increased as root temperatures increased, but there are differences in the patterns of uptake for all elements.

Thermoperiodicity. During the course of early studies of plants grown in controlled-environment rooms, Went found that tomatoes grown under uniform day and night temperatures did not grow as well as those grown under a fluctuating high day temperature (phototemperature) and low night temperature (nyctotemperature). Went referred to this phenomenon as *thermoperiodicity*. Some results of experiments with tomato plants are shown in Figure 17-24. Under nonfluctuating day and night temperatures the rate of stem elongation increases as the temperature increases from 5 to 25° C. If the phototemperature is maintained constant at 26° C (16-hour photo-

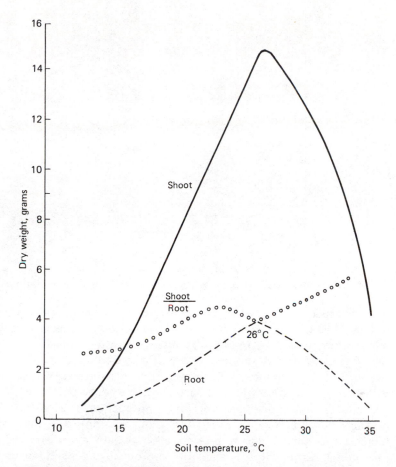

Figure 17-23. Shoot (—), root (---) dry weights, and shoot–root ratio (O O O) of 23-day-old maize seedlings grown at soil temperatures between 12 and 35° C. (After Walker, 1969; Figure 9.)

period) and the nyctotemperature (8-hour nyctoperiod) is maintained at a lower temperature, the optimal rate of stem elongation is achieved when the nyctotemperature is about 20° C. Other combinations of photo- and nyctotemperatures gave comparable results, provided that the temperature differential between day and night is about 6 to 10° C, the night temperatures being lower than the day temperatures. Studies with many other plants revealed similar responses to those described for the tomato plant.

Went noted that sugar accumulated in tomato stems during a low nyctotemperature and suggested that the rate of sugar translocation is sensitive to temperature, being slow at high temperatures and fast at low temperatures. Many studies, however, have shown that sugars accumulate in plants maintained at low temperatures and that the effect of low temperature is on a metabolic process rather

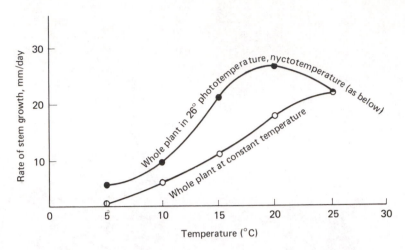

Figure 17-24. Growth rates of tomato plants (stem growth, mm day^{-1}) kept at 26° C constant temperatures ($\bigcirc$ — $\bigcirc$), and kept at 26° C phototemperature (16 hours) and different nyctotemperatures ($\bullet$ — $\bullet$) for 8 hours (After Went, 1944; Figure 17.)

than on sugar translocation. One possible explanation of thermoperiodicity is as follows. The rate of CO_2 assimilation increases with temperature; when the photo-temperature is high, the amount of photosynthate available for synthetic purposes is high. Synthetic processes are also increased at high temperatures and the photosynthate is diverted to leaf production. Respiratory rates also increase with temperature and dry weight loss is increased. Because plants differ with respect to photorespiration, the amount of photosynthate available for the production of new plant substance depends on the balance between photosynthesis and respiration. If the nycto- and photo-temperatures remain high, respiratory losses remain high during the dark period and not enough carbohydrate is available to produce new plant substance. If the night temperature is lowered, however, the rate of respiratory loss is decreased and the loss of dry weight is reduced. Low temperatures decrease not only the loss of carbon through respiratory processes but also the hydrolysis of sucrose or other translocation forms of sugar with the result that the soluble sugar content of the tissue remains high. It is also noted that low temperature influences the apportionment of carbohydrates into roots, stems, and leaves.

It shoud be noted, however, that not all plant species respond favorably to a combination of high phototemperatures and low nyctotemperatures. For many plants, the maximum rates of growth are achieved at uniformly high photo- and nycto-temperatures. The balance between photosynthesis, photorespiration, dark respiration, and anabolic processes must be different, but the precise nature of their interrelationships is not understood. Furthermore, temperature influences cell division, cell expansion, differentiation, and other morphogenetic responses. The interactions between temperature, light intensity, and light duration result in changes in rates of metabolic reactions and photomorphogenesis. Studies on these problems are difficult because of the complex environmental interactions, but improved facilities for controlling environmental factors (phytotrons, controlled-environment chambers, etc.) are changing the situation.

Thermoperiodicity in the natural environment. Plants are exposed to fluctuating day and night temperatures under natural conditions. Day temperatures are usually higher than night temperatures. The differential may be considerable during the spring and fall but of lesser magnitude during the summer. Near the equator the temperature differential between day and night is small (except at high elevations) compared to that found in the temperate zone. Thus plants in tropical regions have evolved and developed under a rather restricted range of phototemperatures and photoperiods. A tropical plant, such as sugarcane or coconut palm, cannot flourish outside a rather narrow set of environmental conditions. Similarly, a temperate zone plant, such as wheat, does poorly in the tropics. Using plant breeding, it is possible to widen the range of environmental conditions under which a plant can grow. Maize, for example, originated in the tropical highlands, but through breeding and hybridization varieties now grow as far north as southern Canada.

It is possible to determine the effects of photo- and nyctotemperatures on the growth of native vegetation or of horticultural and agronomic crops without the use of controlled environment facilities. Went devised a technique of evaluating photo- and nyctotemperatures by using weather data available from the records of the U.S. Weather Bureau at various localities. These stations report the monthly mean maximum and minimum temperatures. From these data Went derived what he called the *effective day temperature* and the *effective night temperature*,[1] which correspond to the photo- and nyctotemperatures.

The effective day and night temperatures at different times of the year in Pasadena, California, and Homestead, Florida, are shown in Figure 17-25. In January at Pasadena the effective day temperature is 14° C and the effective night temperature is 9° C whereas at Homestead they are 21 and 14° C, respectively. As the season progresses, both the effective day and night temperatures increase until in September and October they start to decline. At Pasadena the effective day temperatures are lower during the first part of the year than from August to December. The situation is reversed at Homestead, where the effective day temperatures are higher during the first part of the year. It should be kept in mind that these are mean monthly temperatures and that there may be individual days when the temperature conditions are quite different.

Also included in Figure 17-25 is the temperature requirement for the optimal growth and flowering of English daisy as determined in a phytotron. English daisies flower best when the phototemperature varies between 14 and 16° C and the nyctotemperature varies between 8 and 10° C. It can be seen that these conditions are

---

[1] Effective day and night temperatures (monthly basis) are calculated as follows:

Effective day temperature = monthly mean maximum — (mo. mean max. − mo. mean min. × 0.25)

Effective night temperature = monthly mean minimum + (mo. mean max. − mo. mean min. × 0.25)

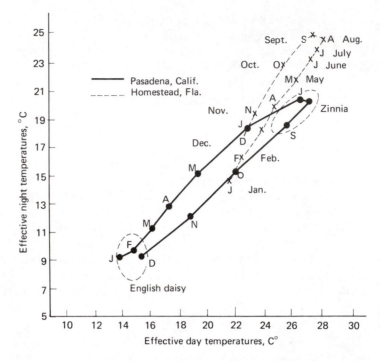

Figure 17-25. Monthly procession of effective day and night temperatures at Pasadena, California (●—●), and Homestead, Florida (×---×). See text for explanation of effective day and night temperatures.

obtained at Pasadena during December, January, and February. If the plants are to flower, they should be started from seed during the fall when days are warm and a vigorous vegetative growth is obtained. Then in December, January, and February, when days and nights are cool, the plants will flower. Conditions at Homestead are not suitable for English daisy because day and night temperatures are always too warm.

Moisture   The role of water in the physiology of plants is discussed in Chapters 12 and 13. The intention here is to consider how plant growth and development, particularly vegetative growth, are influenced by problems of water balance. The plant maintains a balance between water received from rainfall and irrigation and water used in photosynthesis and other chemical processes and between water stored in the plant and water lost by transpiration (Figure 17-26). The plant is part of a three-component system: the soil–plant–atmosphere continuum. In the soil–plant–atmosphere continuum the form and structure of the various tissues and organs of the plant, as well as their physiological activity, are important in determining the moisture status of the plant. Numerous features of the plant contribute to its ability to achieve some degree of stability with respect to internal water content. The loss of water by evaporation is controlled by the stomatal apparatus (Chapter 13) and other structural modifications,

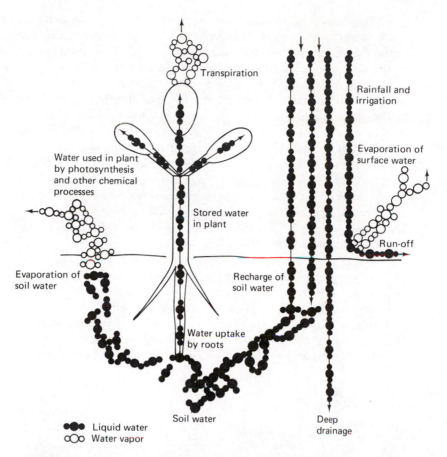

Transpiration

Rainfall and
irrigation

Evaporation of
surface water

Water used in plant
by photosynthesis
and other chemical
processes

Stored water
in plant

Run-off

Evaporation of
soil water

Recharge of
soil water

Water uptake
by roots

●●● Liquid water
○○○ Water vapor

Soil water

Deep
drainage

Figure 17-26. The water balance. To a close approximation, there is a balance
between the amount of water received in the plant environment and the sum of the
amounts lost (transpiration and evaporation) and stored (plant and soil). [After "The
Plant and Its Physical Environment." Anon., 1966 (p. 3).]

such as protective layers of cutin or suberin, epidermal hairs, thickened cell walls,
mucilaginous cellular constituents, reduced leaf area, and so forth. Entrance of water
into the plant occurs primarily through the roots; and the size of the root system, its
ramifications throughout the soil, the number of root hairs, and similar items govern
the amount of water that moves into the plant. Between the points of contact of the
plant with the soil and the atmosphere are large numbers of cells that must be
maintained at appropriate moisture levels. These cells differ in form and metabolic
activity; they are at different distances from the source of water or from the
evaporating surfaces; they are of different ages and at different stages of development.
It is obvious that water stress will affect them and their activities in many different
ways. Relatively little is known at this time, however, about the influence of internal
water stress on the cellular and subcellular activities of plants.

The amount of water in a soil that is available to plants depends on such factors as the nature and size distribution of the soil minerals, organic matter, content of soluble salts, compaction, and water supply. Depending on the relative importance of these factors, soils are capable of supplying variable amounts of water to plants. A sandy soil, for example, supplies a smaller amount of water to plants than a soil composed of clay particles or rich in organic matter. Soil scientists usually recognize two levels of soil moisture as being important in supporting plant growth. The upper level of soil moisture available to plants is referred to as *field capacity*. Field capacity is defined as the water content of a moist soil at the point at which capillary water movement is negligible. In sandy soils field capacity may be 5% (5% of the dry weight) whereas in clay soils field capacity may be as high as 45%. The lower level of available soil water is referred to as the *permanent wilting percentage*. This is the amount of water remaining in a soil when plants growing in it have reached a condition of permanent wilting, a condition under which the wilted plant will not recover its turgidity unless water is added to the soil.

The growth response of plants to soil moisture stress is shown in Figure 17-27(a). In this experiment tomato plants were grown from seed in containers for about 30 days, at which time the third pair of leaves was about one-third developed. During the growing period the soil, a clay loam, was maintained at field capacity (22% moisture) by regularly adding water to the containers. Water was then withheld from one group of plants while the control plants remained at field capacity. The plants were periodically sampled and the relative growth rate determined. During the experimental

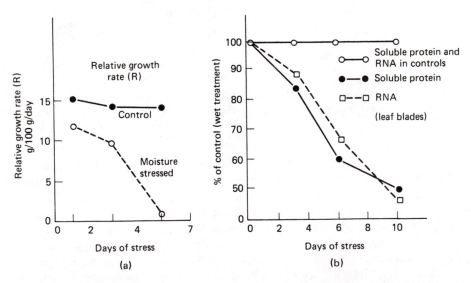

Figure 17-27. (a) Relative growth rates of tomato plants during a single cycle of moisture stress under greenhouse conditions. (After Gates and Bonner, 1959; Figure 2.) (b) Soluble protein and RNA in leaf blades of sugar beet plants during a single cycle of moisture stress under greenhouse conditions. (After Shah and Loomis, 1965; Figure 2.)

period of 7 days the soil moisture level dropped from field capacity to slightly above the permanent wilting percentage. It can be seen that the relative growth rate (measured as grams of dry matter produced per 100 g of plant material per day) of the control plants (at field capacity) decreased slightly during the experimental period whereas in the moisture-stressed plants the relative growth rate dropped to zero.

An idea of the nature of the chemical changes that occur during drought can be gained from Figure 17-27(b). Sugar beet plants were grown under ample moisture conditions in the greenhouse. At time 0, water was withheld from one group of plants while another group continued to receive moisture. It can be seen that the soluble protein and RNA concentrations immediately decreased as water was withheld. After 10 days of water stress, the soluble protein and RNA levels are about 50% lower than in the unstressed plants. It is frequently said that brief periods of moisture stress do not have a deleterious effect on plant growth. Figure 17-27(a) shows, however, that growth is slowed down almost immediately as moisture stress develops. If water is added quickly, the plant usually recovers, but its growth rate may never fully return to that of an unstressed plant.

Physiological processes, such as photosynthesis, respiration, uptake of nutrients, translocation, synthesis of cellular constituents, and hydrolysis of macromolecules, have all been observed to respond to moisture stress. The usual effect is that the rates of these processes are inhibited by a water deficiency. In some instances, however, the changes produced by water stress are agronomically useful. In the growth of sugarcane, for example, moisture stress during the later stages of growth increases the percentage of sucrose in the stem tissue. Frequently the oil content or concentration of secondary plant substances, such as alkaloids and flavor and aroma constituents, is enhanced by moderate moisture stress.

Most workers in this field have interpreted the effects of moisture stress on plant metabolism as though water were primarily a medium in which various physiological and biochemical processes occur. If water is deficient, the concentrations of reactants increase, thereby modifying their subsequent reactions. There is increasing evidence, however, that the water molecule, because of its configuration and tendency to form crystalline structures (see Chapter 12), may play a more significant role in metabolic reactions. If water molecules bind to protein molecules (enzymes), for instance, a small change in water concentration may alter the configuration of the enzyme and thereby alter its enzymic activities. If this is true, then water acts at the molecular level of activity, and new techniques and methods must be devised to study these effects.

Nutrition The roles of inorganic ions in the growth of plants are discussed in Chapters 8 and 9. Plants require 16 elements for normal growth and development and, in most instances, the specific function of each element is understood (Chapter 9). It has been emphasized in earlier sections that the vegetative phase of growth is characterized by the synthesis and organization of new plant material into an expanding root and shoot system. Photosynthetic processes supply the growing plant with carbon skeletons that are incorporated into amino acids, protein, phospholipids, nucleic acids, carbohydrates, porphyrins, and other cytoplasmic and structural constituents. These metabolic processes require adequate supplies of inorganic elements.

During vegetative growth the plant is exposed to a constantly changing environment. There are daily and seasonal fluctuations in day length, light intensity, light quality, air and soil temperature, air and soil moisture, and so forth, and all these factors influence plant growth by modifying metabolic processes. Despite the long-standing recognition of the roles of various inorganic ions on plant growth, relatively little is known of the effects of environmental factors on their uptake and metabolism within the plant. Such studies are difficult to carry out in the field, but the increasing availability of controlled-environment facilities will aid in studying these important physiological problems.

Results from a greenhouse study on the influence of day length and inorganic nutrition on the growth of mint (*Mentha piperita*) are shown in Figure 17-28. In this study rooted mint cuttings were grown in a complete nutrient solution for 9 to 10 days and then transferred to solutions deficient in a particular inorganic ion. Growth in the different solutions continued for an additional 5 to 6 weeks, whereupon the plants were harvested and separated into roots, leaves, and stems and the component parts analyzed for their soluble, insoluble (protein), and total nitrogen contents.

The plants were grown in a heated greenhouse during winter months when natural day lengths were between 6 and 8 hours (short days). Long days were obtained by extending the natural day length to 18 hours by overhead illumination from fluorescent lamps (about 3 klux). *Mentha piperita* is a long-day plant. Under the short days the plant remains vegetative and forms stolons but no inflorescences whereas under long days reproductive structures are formed.

Figure 17-28 shows that day length has a profound influence on the nitrogen composition of the leaves, stems, and roots of the mint plant. The overall aspects of the polygonal diagrams for long- and short-day plants are quite different. There is little difference between long- and short-day plants when grown on a complete nutrient solution with respect to the distribution of nitrogen between leaves, stems, and roots. Plants grown under nutrient-deficient conditions, however, show quite different patterns of total, soluble, and insoluble nitrogen in the roots, leaves, and stems.

Leaves of potassium-deficient (−K) plants grown under short days, for instance, contain about a third of the total nitrogen found in long-day plants. The stems and roots of −K short-day plants also contain less total nitrogen than −K long-day plants. Also note the striking influence of day length on the amounts of soluble and insoluble (protein) nitrogen in the leaves, stems, and roots of sulfur-deficient (−S) plants. Under long days more than half the nitrogen is in the soluble fraction, whereas under short days most of the nitrogen is in the insoluble form.

Pollutants   Environmental pollution has been recognized as a problem for many years. Only within the last few years, however, has it become of national concern, largely because of the adverse effects of pollution on humans. Smog, polluted waters,

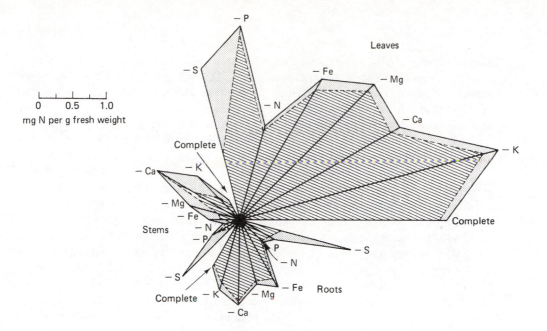

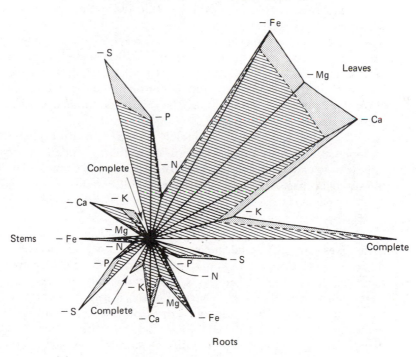

(a) Grown under long days at 70°F in the greenhouse

(b) Grown under short days at 70°F in the greenhouse

Figure 17-28. Total nitrogen (—), soluble nitrogen (- - -), and insoluble nitrogen (total minus soluble) in the leaves, stems and roots of *Mentha piperita* grown under long days (a) and short days (b) in the greenhouse at 70° F. (After Crane and Steward, 1962.)

discarded bottles, cans and papers, automobile exhausts, soot, and pesticides are but a few of the materials that foul our environment. Air pollution, in particular, has been recognized for well over 100 years as influencing plant growth. Ethylene, sulfur dioxide, and fluorides have been found to damage a wide variety of plants. More recently, several new classes of air pollutants have been recognized: Hydrocarbons from motor vehicle exhausts and photochemical products resulting from the interaction of hydrocarbons with nitrogen oxides in the presence of sunlight. Ozone and peroxyacetyl nitrate are two of the very active compounds that are formed in this way. Automobile exhaust fumes also contain large amounts of lead and boron, which are added to gasoline as antiknock compounds, and these, too, may have a detrimental effect on plant growth. Finally, the widespread use of organic pesticides (herbicides, fungicides, insecticides) has added additional foreign materials to the environment, many of which are harmful to plants. There are few places in the United States where uncontaminated plants can be grown. Even controlled-environment facilities must provide specially filtered and chemically purified air in several localities.

Air pollution will probably remain a problem insofar as plant growth is concerned, for the sources of air pollutants are increasing as new industries and highways are built and the urban population grows. It has already become almost impossible to grow certain plants in heavily polluted areas. Citrus fruits and gladioli, for example, are extremely sensitive to fluoride and are no longer grown in regions where the atmospheric fluoride concentrations are high. Tobacco grows very poorly where ozone and photochemical products reach high atmospheric levels. Cotton is very sensitive to ethylene and white pine is killed by high levels of ozone and other atmospheric oxidants. It has been easy to recognize plant damage that results from high levels of air pollutants, but little is known of the effects on plants of long-term, or chronic, exposures to low levels of atmospheric pollutants. Some authorities suggest that already these effects reduce crop yields 5 to 15%.

As noted, gladioli are extremely sensitive to atmospheric fluorides. Some varieties, however, are more susceptible than others. Similarly, occasional white pines appear to tolerate high concentrations of atmospheric ozone and other oxidants. The physiological basis for resistance to air pollution damage is not now known, but it might be possible to select and develop plants resistant to air pollution, as is now done in developing insect- and fungi-resistant plants.

One final aspect of the air pollution problem should be mentioned. Plants themselves emit volatile substances of a complex organic nature. The blue haze that hangs over the Smoky Mountains and other densely vegetated areas is due in part to these plant volatiles. Chemical studies show that terpenes and photochemical decomposition products of terpenes, as well as methane, are present in plant volatiles. These substances influence the growth of other plants and are part of the natural environment in which we live. Recent studies on the dynamics of ecosystems have demonstrated that plant volatiles (allelochemicals, see Chapter 14) play an important role in controlling plant growth and development.

REFERENCES

ANON. 1966. The plant and its physical environment. The Pye Laboratory. CSIRO, Division of Plant Industry, Canberra City, Australia.

BARRS, H. D. 1971. Cyclic variations in stomatal aperture, transpiration, and leaf water potential under constant environmental conditions. *Ann. Rev. Plant Physiol.* 22:223–236.

BJÖRKMAN, O. 1968. Further studies on differentiation of photosynthetic properties in sun and shade ecotypes of *Solidago virgaurea*. *Physiol. Plant.* 21:84–99.

BLACKMAN, G. E. 1961. Responses to environmental factors by plants in the vegetative phase. Pages 525–556 *in* M. X. Zarrow, ed. *Growth in Living Systems*. Basic Books, Publishers, New York.

BLACKMAN, V. H. 1919. The compound interest law and plant growth. *Ann. Bot.* 88:353–360.

BRIGGS, W. R., R. D. TOCHER, and J. F. WILSON. 1957. Phototropic auxin redistribution in corn coleoptiles. *Science* 126:210–212.

CARLSON, P. S., H. H. SMITH, and R. D. DEARING. 1972. Parasexual interspecific plant hybridization. *Proc. Nat. Acad. Sci. U.S.* 69:2292–2294.

CHALEFF, R. S. 1981. *Genetics of Higher Plants: Applications of Cell Culture*. Cambridge University Press, Cambridge.

CRANE, F. A., and F. C. STEWARD. 1961. Growth and nutrition of *Mentha piperita*. *Cornell Univ. Agr. Exp. Sta. Memoir 379*. Ithaca, New York.

DUGGER, W. M., and I. P. TING. 1970. Air pollution oxidants—their effects on metabolic processes in plants. *Ann. Rev. Plant Physiol.* 21:215–234.

EVANS, L. T., ed. 1963. *Environmental Control of Plant Growth*, Proceedings of a symposium, Canberra, Australia, 1962. Academic Press, New York.

EVANS, L. T., ed. 1975. *Crop Physiology, Some Case Histories*. Cambridge University Press, Cambridge.

FRIEND, D. J. C. 1965. The effects of light and temperature on the growth of cereals. Pages 181–199 *in* F. L. Milthorpe and J. D. Ivins, eds. *The Growth of Cereals and Grasses*. Butterworth & Co. Ltd., London.

GATES, C. T., and J. BONNER. 1959. The response of the young tomato plant to a brief period of water shortage. IV. Effects of water stress on the ribonucleic acid metabolism of tomato leaves. *Plant Physiol.* 34:49–55.

GATES, D. M. 1965. Heat transfer in plants. *Sci. Amer.* 213(12):76–84.

GILLESPIE, BARBARA, and W. R. BRIGGS. 1961. Mediation of geotropic response by lateral transport of auxin. *Plant Physiol.* 36:364–368.

HALPERIN, W., and D. F. WETHERALL. 1965. Ontogeny of adventive embryos of wild carrot. *Science* 147:756–758.

HOLLAENDER. A., et al., eds. 1977. *Genetic Engineering for Nitrogen Fixation*. Basic Life Sciences, vol. 9., Plenum Publishing Corp., New York.

KOZLOWSKI, T. T., ed. 1968. *Water Deficits and Plant Growth*, vols. I, II, III, IV, V, VI. Academic Press, New York.

KRAMER, P. J. 1969. *Plant and Soil-water Relationships: a Modern Synthesis.* McGraw-Hill Book Company, New York.

LANDSBERG, J. J., and C. V. CUTTING, eds. 1977. *Environmental Effects on Crop Physiology.* Academic Press, New York, London.

LEOPOLD, A. C., and P. E. KRIEDEMANN. 1975. *Plant Growth and Development*, 2nd ed. McGraw-Hill Book Company, New York.

LEVITT, J. 1980. *Responses of Plants to Environmental Stresses,* 2nd ed. Vol. 1, *Chilling, Freezing, and High Temperature Stresses.* Academic Press, New York.

LYONS, J. M., D. GRAHAM, and J. K. RAISON, eds. 1979. *Low Temperature Stress on Crop Plants. The Role of the Membranes.* Academic Press, New York, London.

MILTHORPE, F. L., ed. 1956. *The Growth of Leaves,* Proceedings of the 3rd Easter School in Agricultural Science, University of Nottingham, 1956. Butterworth & Co. Ltd., London.

NAKATA, K., and M. TANAKA. 1968. Differentiation of embryoids from developing germ cells of tobacco. *Japan. J. Genet.* 43:65–71.

NITSCH, C., and J. P. NITSCH. 1967. The induction of flowering *in vitro* in stem segments of *Plumbago indica* L. I. The production of vegetative buds. *Planta (Berl.)* 72:355–370.

NITSCH, J. P., and C. NITSCH. 1969. Haploid plants from pollen grains. *Science* 163:85–87.

POPE, M. N. 1932. The growth curve of barley. *J. Agr. Res.* 44:323–341.

RICHARDS, F. J. 1969. The quantitative analysis of growth. Pages 3–76 *in* F. C. Steward, ed. *Plant Physiology, a Treatise,* vol. 5A. Academic Press, New York.

RUBENSTEIN, I., B. GEGENBACH, R. L. PHILLIPS, and C. E. GREEN, eds. 1980. *Genetic Improvement of Crops: Emergent Techniques.* University of Minnesota Press, Minneapolis.

SHAH, C. B., and R. S. LOOMIS. 1965. Ribonucleic acid and protein metabolism in sugar beet during drought. *Physiol. Plant.* 18:240–254.

SKOOG, F., and C. O. MILLER. 1957. Chemical regulation of growth and organ formation in plant tissues cultured *in vitro. Symp. Soc. Expt. Biol.* 11:118–131.

SLATYER, R. D. 1967. *Plant-water Relationships.* Academic Press, New York.

SMILDE, K. W. 1960. The influence of some environmental factors on growth and development of *Sesame indicum. Meded. Landbouwhogeschool, Wageningen* 60:1–70.

SMITH, H., ed. 1976. *Light and Plant Development.* Proceedings of the 22nd Easter School in Agricultural Science, University of Nottingham, 1975. Butterworth & Co. Ltd., London.

STABA, E. J., ed. 1980. *Plant Tissue Culture as a Source of Biochemicals.* CRC Press, Boca Raton, Florida.

STEWARD, F. C., M. O. MAPES, A. E. KENT, and R. O. HOLSTON. 1964. Growth and development of cultured plant cells. *Science* 143:20–27.

THIMANN, K. V., and F. SKOOG. 1934. On the inhibition of bud development and other functions of growth substances in *Vicia faba. Proc. Roy. Soc. (London)* B14:317–339.

THORNLEY, J. H. M. 1976. *Mathematical Models in Plant Physiology.* Academic Press, London.

TIBBITTS, T. W., and T. T. KOZLOWSKI, eds. 1979. *Controlled Environment Guidelines for Plant Research.* Academic Press, New York.

WAGGONER, P. E., A. B. PARK, and W. E. REIFSNYDER. 1959. The climate of shade. *Conn. Agr. Expt. Sta. Bull. 626.* New Haven, Conn.

WALKER, J. M. 1969. One-degree increments in soil temperature affect maize seedling behavior. *Soil Sci. Soc. Amr. Proc.* 33:729–736.

WARREN WILSON, J. 1966. Effect of temperature on net assimilation rate. *Ann. Bot.* 20:753–761.

WATSON, D. J. 1952. The physiological basis of variation in yield. *Adv. Agron.* 4:101–145.

WENT, F. W. 1944. Plant growth under controlled conditions. III. Correlation between various physiological processes and growth in the tomato plant. *Am. J. Bot.* 31:597–618.

WENT, F. W. 1957. *The Experimental Control of Plant Growth.* The Ronald Press Company, New York.

WENT, F. W., and L. O. SHEPS. 1969. Environmental factors in regulation of growth and development: ecological factors. Pages 299–406 in F. C. Steward, ed. *Plant Physiology, a Treatise,* vol. 5A. Academic Press, New York.

WHITTINGTON, W. J., ed. 1969. *Root Growth.* Proceedings of the 15th Easter School in Agricultural Science. University of Nottingham, 1968. Plenum Publishing Corp., New York.

WILLIAMS, R. F. 1946. The physiology of plant growth with special reference to the concepts of net assimilation rate. *Ann. Bot.* 10:41–72.

18

Reproductive Growth

Some of the most colorful and spectacular aspects of plant growth are associated with the development of flowers and fruits. The change from vegetative to reproductive growth marks a major change in the life cycle of most plants. This is particularly true in annuals, where floral initiation signals the end of leaf production and the beginning of changes that culminate in seed production and death of the plant. In perennials floral initiation is not accompanied by such drastic changes in growth habit, for vegetative and reproductive growth go on simultaneously.

In the study of any complex physiological process it is often convenient to identify partial processes that might be separately studied. In photosynthesis research, for example, the identification of "light" and "dark" reactions provided new insight to the problem and enabled investigators to concentrate their efforts on the separate processes. Similarly, the separation of seed germination into the partial processes of water imbibition, radicle emergence, and shoot emergence provides a framework for the study of overall process. Reproductive growth is certainly a complex process and physiologists have recognized a number of partial processes that have been intensively studied.

PRINCIPAL EVENTS OF REPRODUCTIVE GROWTH

Ordinarily in thinking of reproductive growth, flower formation and fruit development come to mind. These events are obvious to the naked eye. Each of these processes, however, is the culmination of a number of other events, many of which are microscopic or submicroscopic. The following enumeration gives an idea of the complexity of plant reproduction:

570

1. Initiation of flower primordia
2. Bud development
3. Development and maturation of floral parts (sepal, petal, pistil, stamen, nectary, etc.)
4. Development of embryo sac with its enclosed nuclei (egg, synergid, antipodal, polar)
5. Development of pollen grains within the anthers
6. Anthesis (opening of flower)
7. Pollination
8. Growth of the pollen tube from the stigma through the style and into the ovule
9. Formation of two sperm nuclei from the generative nucleus in the pollen tube
10. Fertilization, involving formation of the zygote from fusion of the egg and sperm nuclei and formation of the primary endosperm nucleus from fusion of a sperm nucleus and two polar nuclei
11. Development of the embryo from the zygote
12. Development of the endosperm from the primary endosperm nucleus
13. Development of the seed from the ovule
14. Development of the fruit from tissues that support the ovule
15. Fruit ripening

Clearly reproductive growth is complex and encompasses a variety of anatomical, morphological, physiological, and biochemical processes. Such events as flower initiation, pollination, fertilization, embryo development, and fruit ripening have been intensively studied in a number of economically important plants. In the following discussion major emphasis is on only a few of the events listed, primarily because these events have been more thoroughly studied than some of the others.

INITIATION OF FLOWER PRIMORDIA

The initiation of flower primordia is a major event in the life cycle of a plant in that it involves a shift in the pattern of growth and development from vegetative to reproductive processes. The significance of flower initiation has been recognized by botanists for many years. In 1918 G. Klebs suggested that during the life cycle of a plant it passes through several phases of development. Before floral primordia can be initiated, the plant must complete a period of vegetative growth or attain some minimal leaf number. When this condition is attained, the plant is said to be ripe to flower. Ripeness-to-flower is not recognizable by any external characteristics, but it can be determined empirically by subjecting plants of varying age (from seedling emergence) to environmental conditions known to induce flowering. In most plants ripeness-to-flower is attained after the plant has produced several leaves. In some cereal grasses a minimum of seven leaves must be developed before the plant is ripe to flower. On the other hand, a plant like *Pharbitis nil*, the Japanese morning glory, is ripe

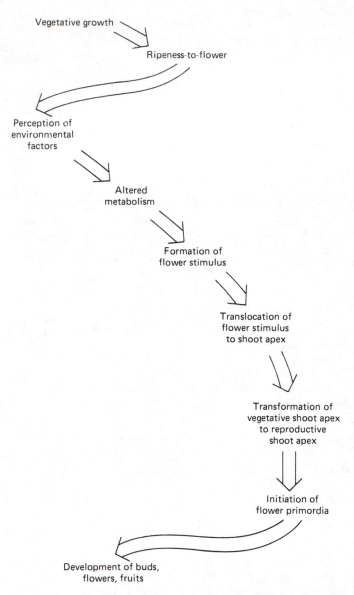

Figure 18-1 Some processes involved in the initiation of flower primordia.

to flower within a day after the cotyledons have emerged. Cotyledons presumably contain enough stored food to support subsequent reproductive development. Most common cultivated plants and weedy annuals, however, attain the ripe-to-flower condition 2 to 3 weeks after seedling emergence when a few leaves have fully developed.

Attainment of the ripe-to-flower condition does not automatically lead to the initiation of flower primordia. Certain environmental conditions must follow. These same environmental conditions, if presented to a plant that is not ripe to flower, elicit no flowering response. The importance of temperature and the promotion of reproductive development, a phenomenon referred to as vernalization, was described by G. Gassner in 1918. About the same time W. W. Garner and H. A. Allard (1920), two plant physiologists with the U.S. Department of Agriculture, found that day length, or the duration of light and dark periods within a 24-hour cycle, also influenced the initiation of flowering.

Since the early work of Klebs, Gassner, and Garner and Allard, plant physiologists have carried out detailed studies of the anatomical, chemical, and biochemical processes that accompany the shift from vegetative to reproductive growth. Although much is still not known about floral initiation, several different kinds of processes are thought to occur, as shown in Figure 18-1. The advantage of such a scheme is that in enables us, through the choice of appropriate plant material and experimental techniques, to isolate individual reactions for detailed study. The transport of the flower stimulus or the transformation of the shoot apex can be studied, for example. In certain plants it has been possible to determine the length of time necessary to complete some of the partial reactions.

THE ENVIRONMENT AND FLOWER INITIATION

After the plant attains the ripe-to-flower condition, further progress toward flower initiation depends on the environment. Both temperature and light are involved. In some instances, critical levels of either light or temperature seem necessary, but generally a proper balance between the two is needed.

Temperature Effects: Vernalization

Agronomists, horticulturists, and gardeners long recognized a relationship between temperature and flowering, but it remained for Gassner to point out that temperature was especially critical during early stages of seed germination. Furthermore, the presence or absence of appropriate temperatures during this critical stage determines the success or failure of flowering and fruiting. Gassner's studies were carried out on rye (*Secale cereale*, cv. Petkus). In Petkus rye, as in a number of other cereals grown in temperate regions, there are two varieties, winter and spring. Winter Petkus rye is sown in the field in the fall of the year, where it germinates and sends up a few leaves. Further growth is checked for several months by low winter temperatures. In many instances, the plants are covered by snow. When the temperature moderates in the spring, the plants resume growth, flower, and produce a grain crop during the summer. Spring Petkus rye, on the other hand, is sown in the spring, at which time it germinates and immediately proceeds through vegetative and reproductive growth to give a grain crop that is harvested in the summer.

If winter rye is planted in the spring, the seeds germinate and produce vegetative plants. The period of vegetative development is extended, however, and flower initiation does not occur until late in the growing season. Subsequent grain development and ripening may be prevented by drought or by low temperatures during the fall and a grain crop is not realized. Gassner found that the low-temperature requirement of winter Petkus rye that is satisfied by growing plants in the field during winter months can be accomplished by subjecting imbibed seeds to chilling temperatures (2 to 5° C) for 5 to 6 weeks. If the chilled seeds are then planted in the spring, they grow and flower on the same schedule as the ordinary spring variety of Petkus rye.

Winter varieties of rye and wheat are ordinarily grown in climatic regions with mild winters (the Midwest in the United States or southern Europe). They cannot survive the severe cold of winters in more northern latitudes (the Northern Plains in the United States and Canada or northern Europe and Russia). Spring varieties are grown in regions with severe winters. Winter varieties of rye and wheat possess milling and baking properties that enhance their value, and farmers were interested in extending the production of such varieties.

The practical significance of Gassner's observation on the chilling requirements of winter Petkus rye did not go unnoticed by agricultural workers. Chilled, imbibed seeds of winter rye or wheat, when planted in the spring, grow and produce a grain crop during the summer in much the same way as a spring variety. It is thus possible to grow the winter variety under environmental conditions that ordinarily do not allow it to produce ripe seed. Because winter varieties behave as spring varieties following a chilling treatment, the winter variety is said to be *vernalized* (vernal referring to spring). During the 1930s and 1940s the vernalization procedure was applied on a large scale to cereal crops, particularly wheat and rye, in northern Europe, where the chilled winter varieties were then grown as conventional spring varieties. In recent years, however, plant breeders and geneticists have developed cereal varieties that combine hardiness and quality (milling and baking properties) so that it is no longer necessary to use vernalization treatments.

Quantitative Studies of Vernalization

During the early practical uses of the vernalization technique extravagant claims were made by some workers concerning the treated plants. It was claimed, for example, that vernalization alters the genetic makeup in such a way that the progenies of vernalized winter varieties are permanently transformed into spring varieties. To understand these and other claims, the English plant physiologist F. G. Gregory, in collaboration with O. N. Purvis, initiated a careful study of vernalization procedures in the late 1930s. This work has continued, with the aid of numerous colleagues, until the present. Initial studies centered on developing a quantitative procedure for evaluating the progress of reproductive growth from flower initiation through fruit ripening. Without such a procedure it is impossible to compare the effects of different environmental conditions or experimental treatments on reproductive growth. Gregory and Purvis developed a procedure based on the external appearance of the shoot apex at various stages of development. Subsequent workers have used similar procedures in studying reproductive growth in many different plants.

The main results of the quantitative studies of Gregory and associates on the vernalization of cereals are as follows. Low temperature is most effective during early stages of germination, when the embryo is undergoing rapid cell division. It was found that the embryo can be vernalized as early as 5 days after fertilization while still attached to the mother plant. Embryos at this early stage do not contain more than a few hundred cells, but low temperature induces a change in the cells that is later expressed when the mature embryo (seed) germinates. The low temperatures required to produce the change in the embryo need only persist for a few days.

The fact that embryos can be vernalized at an early stage of development while still attached to the mother plant suggests some interesting questions relative to seed investigations. Unless the previous history of the seed is known from the time of fertilization, as well as the environmental conditions under which it developed to maturity, the later effects of light and temperature on vegetative and reproductive growth will be difficult to interpret. Such information is generally not available for commercially available seeds or for seeds collected in the wild.

Gregory and associates conceived of the vernalization process as being composed of several partial reactions. Cells within the shoot apex perceive low temperature, whereupon metabolic processes are initiated that trigger the synthesis of a flower stimulus. The flower stimulus then transforms localized areas within the shoot apex to flower primordia. Further developmental changes lead to flowers and fruits.

The German plant physiologist G. Melchers suggested that low temperature induces the formation of a growth substance, vernalin, which is responsible for initiating the synthesis of the flower stimulus. Such a growth substance has not been isolated from plants, but there is indirect evidence in support of such an idea. Major support comes from grafting experiments in which it is observed that a stimulus induced by low temperature is transported across a graft union to a nonvernalized plant that subsequently flowers without exposure to chilling temperatures. Many substances are known to move across a graft union, however, and until the chemical identity of vernalin is known, its role in vernalization remains in doubt.

Embryos, or apical meristems, are not the only tissues capable of being vernalized. It was noted that leaf and stem cuttings can be vernalized while forming new roots. The common feature of these cuttings and the embryos in seeds during vernalization is the presence of actively dividing apical meristems, which has been interpreted as indicating that the fundamental effect of low temperature during vernalization is on the nuclear apparatus and some aspect of nucleic acid metabolism. It has not been possible to demonstrate whether the effect of temperature is on DNA or RNA metabolism or perhaps on protein synthesis.

Although low temperature initiates the formation of flower primordia in many plants, it is not a requirement for all plants. Furthermore, in plants with a low-temperature requirement further progress toward flowering often depends on an exposure to an appropriate day length. In some plants a series of exposures to short days may replace the low-temperature requirement, but flowering does not occur unless the plants subsequently receive long days. Long days must also follow low temperature if flowering is to occur.

In 1957 A. Lang found that the low-temperature requirements of some long-day plants can be satisfied by treating the apical meristem with a gibberellin (GA). The effect of GA appears to be different from the flowering stimulus but is involved in some way in the formation or activation of the flowering stimulus. Gibberellin promotes cell division under some conditions; perhaps this is its role in the vernalization process. Yet not all plants that require vernalization and long days respond to GA. Further work is needed to clarify the role of GA in vernalization.

Day Length Effects: Photoperiodism

The crucial significance of light in flowering was not recognized until the work of Garner and Allard during 1915–1920. Garner and Allard's work came from observations on the flowering pattern of tobacco (*Nicotiana tabacum*) growing at the U.S. Department of Agriculture Station, Beltsville, Maryland. One widely grown commercial strain, Maryland Narrowleaf, flowered during the summer after a spring planting. Occasionally there appeared in fields of this strain a mutant plant that grew much taller and that produced very large leaves. These mutant plants, given the name Maryland Mammoth, had obvious commercial value for tobacco production and there was great interest in obtaining seed for further propagation. The Maryland Mammoth plants, however, did not flower during the summer and early fall and plants were killed by cold weather. If lifted from the field before being killed by frost and transplanted into a greenhouse, plants eventually flowered and set seed. From such seed it was found that Maryland Mammoth plants bred true and the strain was adapted to commercial production. It was necessary, however, to transplant field-grown plants into the greenhouse and grow them during the winter to produce seeds.

Further studies of this flowering pattern of Maryland Mammoth tobacco turned up the following facts. If seeds were germinated during winter in the greenhouse and the seedlings allowed to grow during early spring, small plants initiated flowers and produced an excellent seed crop. If the seeds were germinated in spring, however, and then allowed to continue their growth during summer the plants grew vegetatively to great heights but did not flower.

With the foregoing observations in mind, Garner and Allard initiated a series of experiments to study systematically the flowering behavior of Maryland Mammoth tobacco as well as several other plants. They constructed dark chambers that could be placed around field-grown plants and also built growth chambers in the greenhouse, where the length of light period could be varied. They recognized that temperature might also be a factor and took precautions to prevent high temperatures in their closed dark chambers. They published their results in 1920 and suggested that the duration of the light and dark period (within a 24-hour cycle) is a significant factor in the pattern of plant growth, particularly with respect to flowering. They referred to the photoperiods that promoted early flowering of Maryland Mammoth tobacco as short days and referred to Maryland Mammoth tobacco as a *short-day* plant. They found a number of other short-day plants, including several varieties of soybean (*Glycine max*), cosmos (*Cosmos bipinnata*), and ragweed (*Ambrosia artemisifolia*). Such plants as spinach (*Spinacia oleracea*), radish (*Raphanus sativus*), and lettuce (*Lactuca sativa* cv. Black Seeded Simpson) remained vegetative when grown under short days but flowered when exposed to long days. Garner and Allard called such plants *long-day* plants. They also noted several plants whose flowering pattern is not altered by either short or long days. When they reached the ripe-to-flower stage, they continued to flower without regard to photoperiod. Such plants were named *day-neutral* plants.

The Natural History of Photoperiodism

To help understand the observations made by Garner and Allard, it will be useful to consider the natural day length conditions encountered at Beltsville, Maryland and the seasonal changes during the course of the year. Beltsville, Maryland, lies at a latitude of about 39° N and the seasonal change in day length is shown in Figure 18-2. On March 21 and September 21 (the spring and fall equinoxes) the light and dark periods are of approximately equal length (12 hours), but at all other times of the year the days and

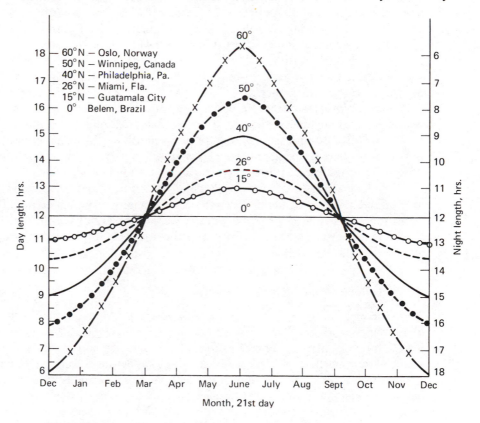

Figure 18-2   Day length cycle at various latitudes north of the equator. The curves depict the annual changes in day length. Washington, D.C., is at 39° N.

nights are of unequal length. From December 21 the days increase in length until they reach maximum length of 15 hours on June 21; thereafter the days decrease in length until December 21, when the days are 9 hours long.

With this pattern of seasonal day length change, the behavior of field-grown Maryland Mammoth tobacco, a short-day plant, is understandable. Tobacco seeds are germinated in seed beds during late February and early March. Seedlings are transplanted from these beds into the field during April and May, when favorable soil and air temperatures prevail. When the plants are placed in the field, the day lengths are about 14 hours and they increase to 15 hours in length on June 21, after which they decrease during the summer and fall. Garner and Allard found that Maryland Mammoth tobacco did not flower unless the plants received a photoperiod shorter than 12 hours. Figure 18-2 shows that such photoperiods were not obtained until after September 21. Even though flowering is initiated by a 12-hour photoperiod, it will take an additional several weeks for visible flower primordia to appear and an even longer time for flowers and seeds to develop. Thus even though flowering may be initiated in field-grown Maryland Mammoth tobacco around September 21, the onset of frost and cold weather kills the plants before flowering can occur and seeds be produced.

The day length data shown in Figure 18-2 are useful in interpreting the flowering patterns of plants growing at different localities. At the equator, for instance, days and nights are about 12 hours in length throughout the year. Many tropical plants are short-day plants and flower continuously, provided adequate moisture is available and temperatures are satisfactory.

As we move north from the equator, the flowering pattern is more seasonal and restricted to certain times of the year. Summer flowering annuals growing in the vicinity of Boston or Chicago usually require day lengths of 15 to 16 hours to initiate reproductive development. If these plants are moved south, they remain vegetative, producing leaves but no flowers because long days are not available.

Under greenhouse conditions plants can be artificially darkened and grown under any desired day length and their growth rather precisely manipulated as to the time of flowering. Thus chrysanthemums, poinsettias, and other horticultural plants can be brought into flower at any time of year by growing the plants under an appropriate day length schedule.

Classification of Photoperiodic Responses

Since the original work of Garner and Allard, the photoperiodic behavior of many plants has been studied. Table 18-1 shows that their original classification of short-day plants, long-day plants, and day-neutral plants has been considerably expanded. In addition to plants that specifically require long days before flowering is initiated, there are plants that flower on any day length but whose flowering is speeded up by long days. A similar situation is observed with short-day plants; there are those with a specific requirement for short days and those that are speeded up by short days.

Some plants, referred to as short–long-day plants in Table 18-1, require a number of short days, followed by long days, before flowering is induced. From the

TABLE 18-1 FLOWERING BEHAVIOR OF SOME SELECTED PLANTS

Long-Day Plants

Plants specifically requiring long days

Beta vulgaris	Sugar beet
Hordeum vulgare	Winter barley
Hyoscyamus niger	Black henbane
Lolium temulentum	
Spinacia oleracea	Spinach

Plants promoted by long days

Lactuca sativa	Lettuce
Poa pratensis	Bluegrass
Ricinus communis	Castor bean

Short-Day Plants

Plants specifically requiring short days

Ambrosia artemisifolia	Ragweed
Chenopodium rubrum	Red goosefoot
Chrysanthemum morifolium	Chrysanthemum
Nicotiana tabacum cv. *Maryland Mammoth*	Tobacco
Pharbitis nil	Japanese morning glory
Xanthium strumarium	Cocklebur

Plants promoted by short days

Cosmos bipinnatus	Cosmos
Gossypium hirsutum	Cotton

Day-Neutral Plants

Cucumis sativus	Cucumber
Impatiens balsamina	Balsam
Zea mays	Maize, corn

Long–Short-Day Plants

Bryophyllum daigremontianum	Bryophyllum
Cestrum nocturnum	Night-blooming jasmine

Short–Long-Day Plants

Secale cereale cv. Petkus	Winter rye
Iberis durandii	Candy tuft

After Naylor (1961) and Salisbury (1963)

seasonal shift in photoperiod shown in Figure 18-2 such plants flower in the late spring or early summer after the shift from short to long days. Still other plants flower after exposure to a sequence of long days, followed by short days (long–short-day plants in Table 18-1). Such plants will flower in the late summer or early fall.

Critical Day Length

Each plant in Table 18-1 flowers in response to a particular set of photoperiodic conditions. Consider Maryland Mammoth tobacco and *Xanthium strumarium*. Both are short-day plants, but Maryland Mammoth tobacco is induced to flower when the photoperiod is shorter than 12 hours (12 L/12 D) whereas *Xanthium* is induced to flower when the photoperiod is shorter than 15.5 hours (15.5 L/8.5 D). The photoperiod required to induce flowering is referred to as the *critical day length*. The critical day lengths for Maryland Mammoth tobacco and *Xanthium* are 12 and 15.5 hours, respectively. A short-day plant is one that flowers on photoperiods shorter than the critical day length and this critical day length must be determined experimentally for each species of plant.

Long-day plants, on the other hand, are induced to flower on photoperiods longer than the critical day length. For example, the critical day length for *Hyoscyamus niger* is 11 hours (11 L/13 D) and it is induced to flower on photoperiods longer than 11 hours.

Suppose that *Xanthium* and *Hyoscyamus* are exposed to a photoperiod of 14 hours of light and 10 hours of darkness (14 L/10 D). It can be seen that flowering will be induced in both plants. *Xanthium*, a short-day plant, will flower because the 14 L/10 D photoperiod is shorter than the critical day length of 15.5 hours. *Hyoscyamus*, a long-day plant, will flower because the 14 L/10 D is longer than the critical day length of 11 hours.

Photoperiodic Responses of Other Organisms

With the realization that the flowering pattern of plants is determined by day length, it was discovered that other organisms similarly respond to different photoperiodic regimes (Table 18-2). Long nights and short days promote reproductive activities in *Ulothrix*, an alga, and *Salvinia*, a fern, whereas short nights and long days promote vegetative growth. From Figure 18-2 it may be inferred that both plants are vegetative during summer months.

Both invertebrate and vertebrate animals also display periods of reproductive activity in response to day length. Red spider mites and pulmonate snails, for example, engage in reproductive activities under conditions of short nights and long days. Similarly, the reproductive activity of a number of vertebrates is controlled by day length.

Although likely that the basic mechanisms responsible for these photoperiodic responses differ in plants and animals, there is probably a fundamental similarity in that a time-measuring system (biological clock) is involved. More will be said of the biological clock later on.

TABLE 18-2   SOME REACTIONS OF PLANTS AND ANIMALS TO PHOTOPERIOD

Kinds of plants and animals	Seasonal responses	
	Long nights and short days	Short nights and long days
Algae (Ulothrix)	Zoospore production	Vegetative growth
Ferns (Salvinia)	Sporocarp production	Vegetative growth
Seed Plants		
Wheat	Vegetative growth	Flower formation
Chrysanthemum	Flower formation	Vegetative growth
Maple	Dormancy of buds	Vegetative growth
Irish potato	Tuber formation	Vegetative growth
Onion	Vegetative growth	Bulb formation
Lower Animals		
Red spider mites	Laying of dormant eggs	Laying of nondormant eggs
Pulmonate snails	Nonreproductive growth	Laying of eggs
Vertebrates		
Brook trout	Laying of eggs and changing of color	Nonreproductive growth
Junco	Nonreproductive growth and migration (southward)	Mating and migrating (northward)
Pheasant	Nonreproductive growth	Laying of eggs
Sheep	Mating	Bearing of young (6-month gestation)
Horse	Nonreproductive growth and winter coat formation	Mating and bearing of young (11-month gestation)

After Hendricks (1956) and Naylor (1961).

Quantitative Studies of Photoperiodism

Most of the studies on photoperiodism have been descriptive in nature but have been useful in demonstrating the diversity of the flowering responses. The flowering of many plants is now manipulated by shortening or extending the photoperiod and such plants as chrysanthemum and poinsettia can be brought into flower at any time of the year. Similarly, plant breeders manipulate the flowering behavior of such crops as rice, wheat, maize, sugarcane, and other agronomic plants so as to bring parent plants into flower simultaneously and thus expedite controlled pollination.

Despite the importance of the descriptive studies of photoperiodism, they have not provided an understanding of the mechanism of flower initiation and the subsequent development of flowers and fruits. It is difficult to ascertain the precise time at which the pattern of growth changes from vegetative to reproductive processes. By the time that flower primordia are macroscopically visible, molecular, subcellular, and cellular changes have progressed to a considerable degree. Then, too, the flowering response to photoperiod is quite diverse, as demonstrated by the different photoperiodic classes (Table 18-1). The majority of plants must be exposed to a number of consecutive photoperiods of appropriate duration before flower primordia are initiated. A few plants, however, behave quite differently and it is these plants that have provided the bulk of the current information on the quantitative aspects of photoperiodism.

Consider, for example, the behavior of *Xanthium strumarium*, a short-day plant with a critical day length of 15.5 hours. On a photoperiod of 16 hours of light/8 hours of darkness, *Xanthium* produces leaves and remains vegetative indefinitely. If it receives an exposure to a single photoperiod of 15 hours of light/9 hours of darkness and is then returned to the 16 hours of light/8 hours of darkness regime, microscopic evidence of flower initiation is visible within 2 to 3 days. The single short-day photoperiod is referred to as an *inductive* photoperiod. Another short-day plant, *Pharbitis nil*, also requires a single inductive photoperiod for the initiation of flower primordia. In addition, seedlings of *Pharbitis nil* may be induced to flower by exposing the cotyledons to an appropriate photoperiod rather than waiting until fully formed leaves are produced, as with *Xanthium*. The appearance of vegetative and reproductive plants of *Xanthium strumarium* and *Pharbitis nil* is shown in Figure 18-3.

Until recently no suitable long-day plants were available for study. A strain of black henbane (*Hyoscyamus niger*) has been used, but a number of inductive photocycles are required to initiate flower primordia. *Lolium temulentum*, a member of the grass family, has been shown to flower following a single inductive photocycle and is now widely used in photoperiodism studies. Vegetative and reproductive plants of *Hyoscyamus niger* and *Lolium temulentum* are shown in Figure 18-3.

Because most experimental work on photoperiodism has been carried out with *Xanthium* and *Pharbitis nil*, the discussion that follows is mainly concerned with short-day plants. It is believed, however, that many basic phenomena of photoperiodism are common to both short- and long-day plants. Inasmuch as flowering in *Xanthium*, *Pharbitis nil*, and *Lolium temulentum* is initiated by a single inductive photoperiod while the majority of plants require numerous inductive photocycles, some physiologists believe that our present knowledge of photoperiodism is at best fragmentary.

Site of Perception of the Photoperiod Stimulus

The photoperiodic stimulus is perceived by leaves or, in *Pharbitis nil*, by the fully expanded cotyledons. In either case, the organ of perception is leaflike. This situation is in contrast to the low-temperature stimulus (vernalization) perceived by the apical meristems. An experiment demonstrating the role of leaves in photoperiodism is shown in Figure 18-4. *Xanthium* is grown under vegetative conditions of 16 hours of

V Hyoscyamus R

R Xanthium V

Figure 18-3 Vegetative (V) and reproductive (R) plants of *Xanthium* and *Pharbitis* (short-day) and *Hyoscyamus* and *Lolium* (long-day). (Courtesy U.S.D.A. photo laboratory.)

R Pharbitis V

V Lolium R

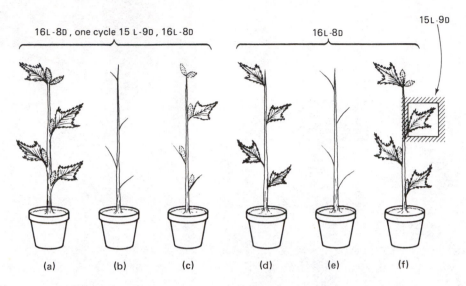

Figure 18-4 Responses of *Xanthium* to photoperiod. Plants were grown under a noninductive photoperiod (16 L-8 D) until treatment. Plants (a), (b), and (c) were given one inductive photoperiod (15 L-9 D) after the leaves were removed from plant (b) and all but one leaf from plant (c). Plants (a) and (c) flowered. Plants (d), (e), and (f) remained under noninductive photoperiods, plant (d) with all leaves and plant (e) with no leaves. Neither (d) nor (e) flowered. One leaf of plant (f) was given an inductive photoperiod while the balance of the plant remained under noninductive conditions. Plant (f) flowered. (After Naylor, 1952, p. 3.)

light and 8 hours of darkness (16 L/8 D) until four or five fully expanded leaves develop. One group of plants remains intact, another group is defoliated, and the third group is defoliated except for one fully expanded leaf. A set of these plants [Figure 18-4(a), (b), and (c)] is then given an inductive photocycle of 15 L/9 D and returned to 16 L/8 D for the remainder of the experiment, during which time flowering and fruiting may occur. It is seen that fruits (cockleburs) are produced on the plants with leaves or with a single leaf but not on defoliated plants.

Another set of plants similarly prepared did not receive an inductive short-day photoperiod (15 L/9 D) but remained under noninductive conditions. Flowering did not occur either in the presence or absence of leaves [Figure 18-4 (d) and (e)]. If a single leaf of a plant growing under a noninductive photoperiod is exposed to an inductive photocycle, however, flowering is induced in the entire plant [Figure 18-4(f)].

Importance of Light and Darkness in the Photoperiodic Cycle

In discussing the induction of flowering in plants, the duration of the photoperiod has been stressed. Clearly, however, in each 24-hour cycle there are periods of light and darkness. It is reasonable, therefore, to ask whether it is the light period or the dark period that is important in initiating flower primordia. Experiments with the short-day plant *Xanthium* in which the light and dark periods were interrupted showed that interruptions of the light period had little effect on flowering as long as there was a period of at least 2 hours of high-intensity light (sunlight or several klux of artifical light) immediately preceding the long dark period, but if the long dark period is interrupted by even a brief flash of light, the dark period is negated and flower initiation is prevented. The light interruption of the dark period is referred to as the *night break* and requires very low levels of illumination (only a few lux are sufficient). Thus the initiation of flower primordia in short-day plants is promoted by un-interrupted long dark periods.

Flower initiation in long-day plants, on the other hand, is inhibited by uninterrupted long dark periods. If the long dark period is interrupted (night break) by a brief flash of dim light, long-day plants initiate flower primordia. In fact, many long-day plants flower under continuous light and do not require any dark period.

Commercial flower growers take advantage of the night break phenomenon to regulate the time of flowering of certain plants. Chrysanthemums, for example, are short-day plants with a critical day length of about 13 hours. When grown out of doors in southern Florida during the late fall and winter, the natural day lengths (Figure 18-2) are shorter than critical day length and flowering is initiated. By interrupting the middle of the night with an hour or so of low-intensity incandescent illumination, the plants are prevented from flowering and remain in a vegetative condition. They can be brought into flower at any time by omitting the night break. In this way, the growers are able to extend the flowering period over several months. Similar practices are followed in growing horticultural plants in greenhouses. Both temperature and photoperiod can be regulated and plants can be brought into flower at any time.

The Effect of Light Quality on the Night Break

The specific effect of light during the night break on the flowering of plants is related to the wavelength of light used. To study this problem, *Xanthium* plants were grown under a light regime that prevents flower initiation (16 L / 8 D) until four or five mature leaves developed. All leaves were then removed except the most recently matured leaf [procedure similar to that shown in Figure 18-4, treatment (c)]. The plants were then exposed to a single inductive photocycle (15 L/9 D). Plants treated in this manner will flower unless, as shown, the dark period of the inductive photocycle (15 L/9 D) is interrupted by a brief flash of light. To determine the specific effect of light quality on the night break, the single leaves of the *Xanthium* plants were irradicated with light of different wavelengths. Of the different wavelengths used, it was found by Borthwick et al. in 1952 that red light (660 nm) and far-red light (730 nm) has the greatest influence on flower initation.

The effects of red and far-red light are illustrated in Figure 18-5. The relative lengths of the light and dark periods are those necessary to induce flowering in *Xanthium* (critical day lengths of 15.5 hours), but other photocycles are equally effective if the critical day length is achieved. *Xanthium* flowers with a single inductive photocycle (9 L/ 15 D), as shown in treatment (a). If the long dark period is interrupted with either white light [treatment (b)] or red light [treatment (c)], flowering is inhibited and the plants remain vegetative. Far-red light is without effect [treatment (d)]; the

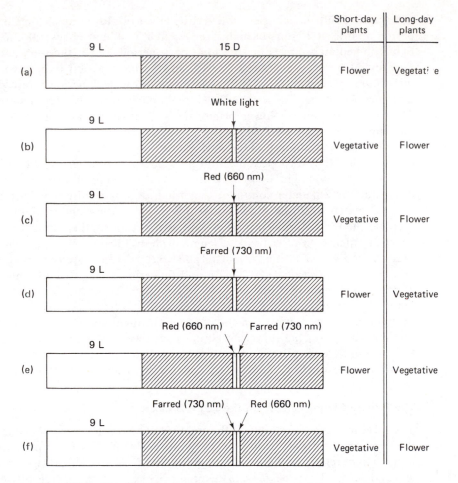

Figure 18-5 The effect of red and far-red light during a night break on the flowering of short- and long-day plants. Light during the major light period is either sunlight or high intensity radiation from incandescent lamp. (See text for an interpretation of the results.) Note the opposite effects of light treatments on short- and long-day plants.

plants behave as though exposed to continuous darkness [treatment (a)]. When a flash of far-red light immediately follows a flash of red light [treatment (e)], the inhibitory effect of red light [treatment (c)] is negated and flower initiation occurs. When red light follows far-red light [treatment (f)], the plants remain vegetative as after a night break of white light [treatment (b)] or red light [treatment (c)].

The effect of red and far-red light on flower initiation in long-day plants is also illustrated in Figure 18-5. Recall that long-day plants set flowers when the photocycle is longer than the critical day lengths and that flowering is inhibited by long, uninterrupted dark periods. Under a photoperiod of 9 L/15 D a long-day plant, such as *Hyoscyamus* (critical day length of 11 hours), remains vegetative [treatment (a)]. If the long dark period is interrupted by white light or red light [treatments (b) and (e)], flowering is initiated. Far-red light [treatment (d)] is without effect and the plants remain vegetative. However, red and far-red light interact [treatments (e) and (f)] with the flowering pattern depending on the wavelength last perceived.

The reversible action of red and far-red light on flower initiation established that phytochrome is involved, the same pigment involved in seed germination (Chapter 15) and other photomorphogenetic responses (Chapter 5).

The Role of Phytochrome in Flower Initiation

The precise role of phytochrome in flower initiation is not known. Phytochrome is not the flowering stimulus, but it is involved in triggering the formation (synthesis or activation) of the flowering stimulus. The level of phytochrome P_{fr}, the physiologically active form, depends on a number of factors (see Chapter 5).

Short-day plants   The effect of phytochrome on the flowering of short-day plants might be imagined as follows. At the end of the light period the concentration of P_{fr} is high and the ratio of P_{fr} to P_r is such that the formation of the flowering stimulus is prevented. During the long dark period P_{fr} reverts to P_r or is destroyed and the ratio of P_{fr} to P_r finally drops to a level in which metabolic processes are triggered that lead to the formation of the flowering stimulus. If the dark period is interrupted by red light, P_r is converted to P_{fr} and the P_{fr} to P_r level is such that the formation of the flower stimulus is prevented.

Long-day plants   The possible role of phytochrome in long-day plants might be as follows. Long-day plants require a high ratio of P_{fr} to P_r for the formation of the flowering stimulus. Such a high ratio of P_{fr} to P_r is attained at the end of a long day. If the night is too long, P_{fr} reverts to P_r or is destroyed and the flowering stimulus is prevented from being formed. When the night is interrupted by a flash of red light, P_r is converted to P_{fr}, thereby raising the ratio of P_{fr} to P_r to a level that allows the flower stimulus to form.

The preceding schemes for the participation of phytochrome in the flowering of short- and long-day plants help to explain the night break effects, but they do not provide information concerning the role of phytochrome in initiating the formation of the flowering stimulus. Another factor involved in flowering concerns the phasing of the light and dark cycles with the internal rhythms of the plant. This problem is discussed in the section on time measurement.

THE FLOWERING STIMULUS: FLORIGEN

There is ample evidence for the existence of a substance, the flowering stimulus, that is produced in leaves and translocated to apical and lateral meristems where flower formation is initiated. Experiments such as those outlined in Figure 18-4, treatment (c) and (f), are interpreted as indicating that the flowering stimulus is formed in a single leaf and translocated to flower-forming centers. Experiments with grafted or twin-shooted stems are also consistent with the interpretation that plants produce a mobile flower stimulus. The Russian plant physiologist M. Kh. Chailakhyan proposed in 1937 that the flowering stimulus, or flowering hormone, be called florigen.

Evidence that the flowering stimulus is similar in long- and short-day plants is provided by grafting experiments. Maryland Mammoth tobacco, a short-day plant, and *Hyoscyamus niger*, a long-day plant, are grafted so that leafy shoots of both species are available for experimentation. If the grafted plants are exposed to either long-day or short-day conditions, both partners flower. If a graft union is not formed, the flowering stimulus is not translocated from one partner to the other. Such results are interpreted as showing that living cells are essential for the movement of the flowering stimulus within the plant.

Experiments with the short-day plant *Pharbitis* indicate that the synthesis of the flowering stimulus requires about 18 hours and that an additional 4 hours are required for its translocation from the leaf to the apical meristem. Knowing the time necessary for translocation and the distance between the leaf and apical meristem, it is possible to calculate the rate of movement of the flowering stimulus. In *Pharbitis* rates of translocation of the flowering stimulus approach 50 cm h^{-1}. Similar experiments with *Lolium temulentum*, a long-day plant indicate that the velocity of translocation of the flowering stimulus is about 2 cm h^{-1}. These rates are considerably slower than the velocity of translocation of sucrose and other organic solutes in phloem tissue, where velocities up to 150 cm h^{-1} are frequently observed. The fact that the velocities of translocation of the flowering stimulus are so different in short-day and long-day plants suggests that the flowering stimulus might be different in the two kinds of plants.

Despite considerable indirect evidence for the existence of florigen, or a flowering stimulus, limited success has accompanied attempts to isolate such a substance. Most experiments have been carried out on short-day plants—in particular, *Xanthium*, a plant that requires only one inductive photoperiod. Lincoln and associates and Carr found that extracts from leaves of flowering *Xanthium* plants induced flowering when applied to the leaves of *Xanthium* plants maintained under vegetative (long-day) conditions. There is considerable variability in the success of these experiments and it is not possible to say whether the extraction procedure is at fault, whether the active substance or substances are labile, whether the material is not taken up through the leaves of the vegetative plants, or whether other unknown factors are involved.

Hodson and Hamner confirmed that a flower-inducing substance can be extracted from flowering *Xanthium* plants and further demonstrated that the extract induces flowering in vegetative duckweed (*Lemna purpusilla* 6746) plants as well as in vegetative *Xanthium* plants. To obtain flowering in *Xanthium*, it is necessary to supplement the flower-inducing extract with gibberellic acid (GA$_3$). With *Lemna* plants, flowering occurs with just the addition of the *Xanthium* leaf extract. There seems little doubt that leaves of flowering *Xanthium* plants contain an extractable substance or substances capable of bringing into flower plants maintained under vegetative conditions. The material is without effect on long-day plants and a similar substance has not been isolated from flowering long-day plants. Attempts to purify the *Xanthium* extract or identify other active components in the extract have been unsuccessful.

Since the original concept of a flowering hormone was introduced by Chailakhyan in 1937, ideas concerning the nature of the flowering process have undergone considerable change. Rather than evoking a new flowering substance, several investigators showed that the plant growth substance indoleacetic acid (IAA) induces flowering in a number of plants. Following the discovery of new classes of plant growth substances—gibberellins, cytokinins, abscisic acid, ethylene—it was found that these compounds also participated in flowering under certain conditions. It is improbable that a single unique flowering substance—florigen—is involved in the flowering of all plants. Although the sequence of events occurring in leaves during photoperiodic stimulation is not known, it is reasonable to assume that the changes are initiated by phytochrome ($P_{fr} \cdot X$), followed by changes in levels of growth substances, metabolites, and cofactors (ATP, NAD, FAD, nucleotides, etc.). The change in metabolic activity then leads to the formation or activation of a substance(s), which is translocated out of the leaf to the apical meristem. There is no reason to believe that the same sequence of events occurs in all plants or that the same transmissible substance (florigen) is involved. There is considerable evidence supporting the view that different pathways lead to flowering in different plants.

SHOOT APEX

Of the various reactions and partial processes described as being involved in the initiation of flower primordia, none is more crucial than the events taking place at the shoot apex. The initiation of leaves is suddenly switched to the initiation of floral primordia, followed by the development of sepals, petals, anthers, pistils, and so forth. This represents a drastic change in the physiology of the plant and is reflected in changes in internal moisture levels, translocaiton of organic solutes, redistribution of inorganic solutes, and growth rates of various organs.

Morphologists have long been interested in the shoot apex, or apical meristem, and its role in the orderly development of the plant. Wardlaw postulated that the shoot apex is composed of aggregations of cells engaged in the organization and initiation of tissues and organs (Figure 18-6). The meristematic cells of the distal region (D) are responsible for maintaining the continuity and integrity of the growing shoot. Cells in the subdistal region (SD) are similar in appearance to those in the distal region, but because of localized gradients of chemical or biochemical activity, groups of cells are organized as growth centers (GC) that eventually give rise to visible organ primordia in the organogenic region (OR). There primordia develop into recognizable organs in the subapical region (SA). Further maturation of the tissue and organs occurs in the region of maturation (M).

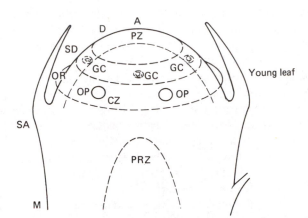

Figure 18-6   An interpretation of the vegetative shoot apex. The apex is considered to be composed of an apical region (A), a peripheral layer of cells in the central zone (CZ), and a zone where the pith and vascular tissue develop, the pith rib meristem (PRZ). It is suggested that within the apex there are localized areas of activity: the distal region (D), where meristematic activity continues to form new cells; the subdistal region (SD), where localized activity results in the formation of growth centers (GC) that, in the organogenic region (OR), develop into organ primordia (OP); the subapical region (SA), where the organs develop and vascular differentiation begins; and the region of maturation (M), where the organs attain their mature condition. (After Wardlaw, 1957; Figure 2.)

It is clear that the vegetative shoot apex is a center of intense activity where cell division, elongation, and differentiation occur. Cells in the distal region are engaged in the synthesis of new cytoplasmic constituents and in the organization of membranes and organelles. Substrates for these synthetic activities are translocated to the apex from the vascular tissues. Furthermore, the organization of the shoot apex follows a rather precise pattern in that the leaves and other lateral organs arise at fixed locations. Leaves may be alternate, opposite, or whorled. They are also arranged in a geometric pattern on the stem so as to produce a characteristic phyllotactic distribution. The orderly initiation and development of a particular leaf pattern indicate that the reactions within the shoot apex are precisely regulated.

The shift from leaf initiation to flower initiation does not denote a major change in the arrangement of cells at the shoot apex inasmuch as flowers are considered to be modified leaves. However, additional structures such as bracts, sepals, petals, stamens, and pistils develop, and such changes probably involve shifts in metabolism. Cytological studies show that an increased mitotic activity is evident within the central zone of the shoot apex (Figure 18-6) within 24 hours after plants are treated with an inductive photoperiod. Histochemical tests reveal that the synthesis of RNA, DNA, and histones accompanies the early mitotic activity. It seems clear that nucleic acid metabolism and the synthesis of new protein occur at the shoot apex following the arrival of some stimulus from the leaves. The altered metabolism within the shoot apex sets up conditions which lead to the formation of floral organs. Further progress in our understanding of floral induction will depend on the isolation and identification of the organ-forming substances produced in the shoot apex as well as the stimulus produced in the leaves.

FLOWER INITIATION: A SUMMARY

Figure 18-1 outlined the general features of the events believed to occur during the initiation of flower primordia. Experimental evidence supports some of the proposed reactions; other reactions are largely hypothetical. There is little question that temperature and duration of light and darkness are the major environmental factors responsible for the initiation of flowering processes. Growth substances also appear to play a role in the initiation of flower primordia, not as the primary initiators of reproductive growth but rather as modifiers of metabolic reactions. These various

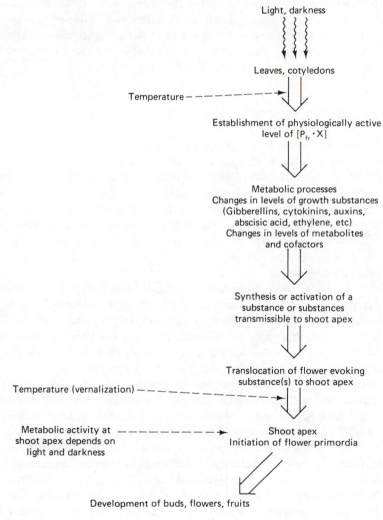

Figure 18-7 Summary of processes participating in the initiation of flower primordia.

reactions are outlined in Figure 18-7, a somewhat more specific version of the reactions originally outlined in Figure 18-1.

It should be noted that flower initiation can be explained by reaction sequences other than those shown in Figure 18-7. Many plants will flower if grown in complete darkness or if all the leaves are removed, for instance. It can be inferred that cells in the shoot apex are capable of initiating flower primordia but are prevented from doing so by an inhibitor produced in the leaf under the influence of light. This inhibitor is translocated to the shoot apex, when it inhibits the initiation of flower primordia. The effect of light and temperature is to initiate metabolic reactions that either prevent the synthesis of the inhibitor or break it down. In either case, the inhibitor is prevented from moving from the leaves to the shoot apex and flower initiation occurs. Inhibitors have been isolated from leaf extracts, but their role in flowering is no better understood than florigen, the flower stimulus. Until the flower stimulus or the flower inhibitor is isolated, identified, and tested on a number of different plants, it is not possible to decide which mechanism is involved in flowering.

TIME MEASUREMENT: THE BIOLOGICAL CLOCK

One of the remarkable features of photoperiodism is the precision of time measurement. In *Xanthium*, for example, the critical day length is 15 hours and 40 minutes. Plants exposed to a day length of 16 hours of light and 8 hours of darkness, a difference of 20 minutes in the length of the light period, remain vegetative. Studies with other plants show that plants can measure time with a precision of about 10 to 15 minutes. In addition to photoperiodism, plants display other processes that indicate some sort of time-measuring system, including such diverse phenomena as sleep movements of leaves, CO_2 evolution by leaves, cell mitosis, enzyme activity, nectary odor, and flower petal movement.

Many of these processes follow a diurnal pattern with a maximum in activity during the light and a minimum during the dark. Other rhythmic processes display maximum activity during the dark and a minimum during the light. Some properties of these diurnal rhythms are demonstrated by the sleep movements of the primary leaves of *Phaseolus multiflorus*, as shown in Figure 18-8. During the day leaves are almost horizontal, whereas during the night the leaves assume a more vertical position. If seeds of *P. multiflorus* are germinated in constant white light and the seedlings maintained under the same conditions, the leaves do not display sleep movements. If a single dark period of 8 to 10 hours is presented to the plant, followed by constant white light, the primary leaves display the pattern of sleep movements shown in the lower part of Figure 18-8. If the plants are placed in constant darkness, instead of being returned to constant white light after sleep movements are initiated, a pattern of sleep movements similar to that shown in Figure 18-8 is displayed but gradually fades out. Note that the periodicity of the sleep movements is not exactly 24 hours but nearer to 27 hours. Observations with other organisms show periodicities in biological activity varying between 22 to 28 hours. Because the periodicities are not exactly 24 hours but

Night (closed)   Day (expanded)

Positions of primary leaves of *Phaseolus multiflorus*

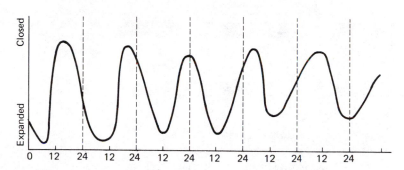

Figure 18-8 Leaf movements of *Phaseolus multiflorus* in continuous dim light, 20° C. High points of the curve refer to closed leaves and low points refer to expanded leaves. (After Bünning, 1967; Figure 1.)

only approximately so, they are referred to as *circadian* (*circa* = about, *diem* = day) rhythms.

The leaf sleep movements demonstrate several properties of biological rhythms. A cue or signal is necessary to initiate the rhythm; but once started, the display continues under constant conditions. With sleep movements in *P. multiflorus*, the cue is the dark period followed by a light period or dark–light. Once initiated by an appropriate signal, the rhythm is "free running" with a periodicity of about 24 hours. Under natural conditions of light and darkness leaves are exposed every 24 hours to a dark–light signal. The free-running rhythm is thought to reflect the presence of an internal fluctuating process or processes, the so-called internal clock.

Bünning (1967) has suggested that there are alternating periods of plant activity sensitive to light (photophile stage) or darkness (scotophile period). These periods are about 12 hours in duration (Figure 18-9). During the photophile phase the plant responds in a particular way to light or darkness; similarly, during the scotophile phase a different response to light and darkness is obtained. Consider the flowering response of short-day (SD) and long-day (LD) plants. In Figure 18-9(b) the SD plant has a critical day length of 9 hours. This 9-hour period always falls within the photophile phase [Figure 18-9(a)] and the scotophile phase is always exposed to darkness. Light

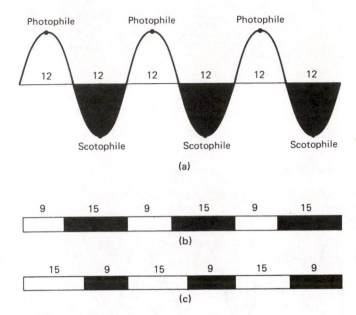

(a)

(b)

(c)

Figure 18-9 Biological activity during alternating periods of light and darkness. (a) There are alternate periods of about 12-hour duration when plants are sensitive to light (photophile) or darkness (scotophile). In (b) there are alternate periods of light (9 h) and darkness (15 h) and the light falls within the photophile and initiates a physiological process. In (c) the light period (14 h) extends into the scotophile segment of the dark rhythm and triggers a different physiological process.

during the scotophile period will negate the photoprocess initiated during the photophile period.

The LD plant [Figure 18-9(c)] has a critical day length of 15 hours and some light always falls in the scotophile period. Under such conditions LD plants flower when light triggers some process during the scotophile phase of plant activity.

The nature of the plant processes (internal clock) taking place during the photophile and scotophile phases is not known, but it appears to be a property of each living cell. Unicellular algae, for example, display biological rhythms, as do cells in callus tissues or in isolated cultures of leaves, roots, and other organs. The precise location of the biological clock within the cell and its chemical or physical nature are not understood, but several possibilities have been proposed. Because only living cells display rhythmic processes, oscillations of enzymic activity may be involved. It is possible to construct a mathematical model composed of a number of enzyme systems that function in a fluctuating pattern. Or a periodic relaxation or contraction of protein molecules may be concerned with the internal clock. In either situation, the periodicities involve metabolic energy.

As noted, plants display a variety of rhythmic processes. They are based on the internal clock and may be phased by external environmental signals. Another example of considerable importance concerns the onset of cold weather in temperate regions. With the approach of winter, plants undergo a series of preparatory changes—alteration in leaf metabolism, translocation of materials out of the leaves, leaf abscission, starch storage in stems—that enable them to survive low temperatures. It might be assumed that these internal changes are triggered by temperature, for the weather is seasonal. Yet there are wide fluctuations in temperature throughout the year, and if plants depended on a particular range of temperature to initiate chemical changes related to leaf abscission, there would be many opportunities for error. Not so with day length. Figure 18-2 shows that the day length changes in a predictable manner at any location and plants with an endogenous time-measuring mechanism are able to couple biochemical changes to day length with great precision.

OTHER ASPECTS OF REPRODUCTIVE DEVELOPMENT

The change from vegetative growth to reproductive growth not only involves switching the shoot apex from leaf formation to flower formation but also involves changes in metabolic patterns of the rest of the plant. These changes include a redistribution of water, growth substances, sugars, nitrogenous compounds, organic acids, and inorganic ions from roots, stems, and leaves to developing flowers. Flowers, and later fruits, function as sinks into which move water and solutes from other parts of the plant. Flower initiation in annual plants usually signals a cessation of stem growth and height increase, although they may be renewed when fruit development is completed.

Following the initiation of floral primordia, further flower development includes the development and maturation of floral parts, such as sepals, petals, stamens, and pistils with their enclosed ovaries. Some flowers may lack certain floral organs—male flowers lack ovaries and so do not develop fruits. Female flowers may lack stamens but are capable of developing fruits. Finally, there are perfect flowers with both stamens and pistils. The degree of completeness of flowers is primarily under genetic control. Environmental factors such as temperature and day length, especially at periods of formation of the male and female gametes, however, may strongly affect the pattern of floral development.

Studies by Steward and associates on floral induction revealed that specific proteins and enzymes developed in the shoot apex prior to the development of floral organs. They studied the shoot apices of *Tulipa*, a bulb plant with a large apical region suitable for chemical and biochemical investigation. The soluble proteins were extracted and separated by acrylamide gel electrophoresis. The separated proteins were then subjected to a variety of tests to ascertain their enzymic characteristics. It was found that certain basic proteins (histones rich in lysine) and several esterases in the apical region underwent marked changes when the growing points were subjected to environmental conditions conducive to floral induction. Also, characteristic protein patterns developed in the regions of the meristem where sepals, petals, and other floral organs developed.

Before mature flowers open (anthesis), the floral organs undergo a series of development changes that prepare the flower for pollination and fertilization. Stamen development includes elongation of the filament and pollen grain formation. In the pistil the style and stigma develop whereas the basal section enlarges as the ovary containing one or more ovules is formed. Morphological and anatomical features of flower development are adequately described for many plants, but details of internal structural and biochemical changes accompanying flower development are less well known.

The opening of the flowers is also a periodic phenomenon. Some open during the early morning hours (morning glory), others during the afternoon (four-o'clock, *Mirabilis jalapa*), and others at night (night-flowering cactus, *Cereus grandiflorus*). Anthesis is frequently accompanied by the release of odoriferous materials that serve to attract insects, birds, and bats to the open flowers, thereby facilitating pollination and fertilization.

Fruit Development

Fruit development is usually considered to start after anthesis, but early aspects of fruit development are initiated soon after flower induction. The early events establish the pattern of development. Fruit growth after anthesis has been studied for many plants. As shown in Figure 18-10, two kinds of growth curves are obtained. The same basic S-shaped growth curve is followed, but in some plants a plateau in size increase is noted

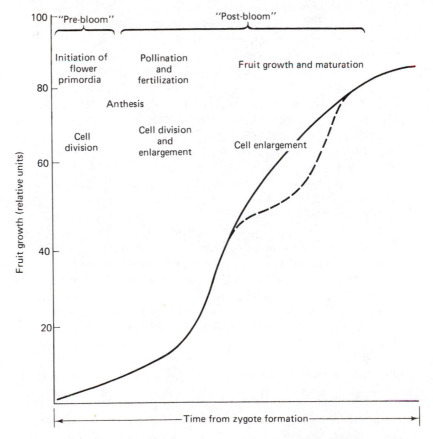

Figure 18-10   Stages in the growth of fruit. Growth may be measured in terms of volume, fresh weight, or dry weight. Some fruits (---) show a plateau in their growth during the cell enlargement stage.

during the cell enlargement phase. Such growth curves are commonly observed in the so-called stone fruits (peach, cherry, plum, etc.), which develop a hard internal pericarp. Nonstone fruits, such as grape, also show a two-phase growth curve, however. Here it is suggested that the plateau in growth is caused by a lack of sugar and other solutes later made available by metabolic activities of the fruit tissue and by translocation from the leafy shoots.

The stages of fruit development (initiation, anthesis, pollination and fertilization, growth, and maturation) are accompanied by changes in cellular and metabolic activity. The earliest stages are characterized by cell division and synthesis of new protoplasm. Growth substances, such as cytokinins and gibberellins, are important during this period. Metabolites for cellular synthesis and the growth substances are translocated to the developing flower (also fruit) from the leaves, stems, and roots during this prebloom period.

The major accomplishment of pollination is the transfer of male gametes to the ovary, where they fuse with the egg nucleus and polar nuclei to form zygote and endosperm nuclei. This process, fertilization, provides the major stimulation to fruit development. Further development of the zygote and endosperm, accomplished through activities of growth substances, such as IAA, cytokinins, and gibberellins, has been described in Chapter 14. Growth of the surrounding ovary and accessory tissues that constitute the mature fruit depends on activities of the zygote and endosperm and also on growth substances and solutes translocated from the rest of the plant. The growing fruit is an active sink that diverts and draws water and solutes from other regions of the plant. Early stages of fruit growth after pollination are characterized by the formation of new cells, but this quickly changes to cell enlargement. Here, again, cytokinins are thought to be essential for cell division whereas IAA and gibberellins function during the cell enlargement phase of fruit growth.

Fruit Ripening

While seed development is taking place, cells in the ovary and accessory tissues are also undergoing a marked change in chemical composition. These chemical changes give rise to the specific combinations of flavor, aroma, color, and consistency characteristic of each kind of fruit. Because of the economic importance of fruits, their chemical composition has been extensively studied. Table 18-3 shows the composition of a number of fruits at maturity and also gives an idea of the nature of the changes in composition taking place during ripening. Note that fruit maturity is defined as the stage when the fruit is normally harvested whereas fruit ripeness denotes the condition when edible. The chemical composition at maturity reflects the presence of materials translocated from other parts of the plant as well as materials formed by metabolic activities of the fruit tissues. Ripening processes are generally degradative in nature.

The data in Table 18-3 show that there are large changes in chemical composition during fruit ripening. Most fruits at maturity are high in either carbohydrate (starch or sugar) or fat but low in protein. During ripening starch is converted to sugar (as in the banana) or (as in the apple) starch decreases while protein and soluble pectins increase.

TABLE 18-3 CHANGES IN CHEMICAL COMPOSITION IN THE EDIBLE
FRACTION OF SOME FRUITS

Constituent	Content at maturity[a] (% fresh weight)	Content on ripening[b] (% of content at maturity)
	Apple	
Starch	2.0	5
Soluble sugars	7.5	99
Organic acid (malic)	1.0	60
Protein	0.2	120
Protopectin	0.7	12
Soluble pectin	0.2	160
	Avocado	
Sugars	0.4	12
Fat	20.0	105
Protein	1.8	110
	Banana	
Starch	20.0	6
Sugars	0.9	2000
Protopectin	0.5	40
Soluble pectin	0.3	150
	Orange	
Sugars	10.0	105
Organic acid (citric)	0.9	85
	Pineapple	
Sugars	15.0	103
Organic acids	0.8	88

[a]"Content at maturity" means composition when the fruit is normally harvested in the mature but not
necessarily the ripe stage.
[b]"Content on ripening" refers to composition at the edible stage.
After Biale (1964).

In the avocado, on the other hand, sugar is converted to fat and protein during
ripening. In many fruits ripening ordinarily occurs while the fruit is still attached to the
tree but may also proceed if the fruit is picked at the mature stage. Ripening then
occurs during storage and may be regulated so as to prolong storage. Ripening of
avocados, however, proceeds only after the fruit is removed from the tree.

The changes in chemical composition of fruits during ripening are accompanied
by alterations in cellular and subcellular organization. Note in Table 18-3 the changes
in protopectin and soluble pectin levels of apple and banana. These constituents are
localized in cell walls and the decrease in protopectin and increase in soluble pectin
denote a softening of the cell wall. Fruit softening and a change in fruit texture are
striking features of ripening in fleshy fruits. Subcellular organelles also undergo

alteration during ripening. Many fruits are green at maturity, but the chlorophyll disappears on ripening. Chloroplasts lose their structural integrity simultaneously with the loss of chlorophyll. Mitochondria also undergo changes during ripening, but in early stages of ripening they appear to be responsible for the maintenance of high respiratory activity.

An especially interesting example of a change in respiratory rate during development occurs in many fruits. When an apple fruit, for example reaches its maximum size but is still green, the respiratory rate declines to a low value. Then there is a sharp increase in the rate of respiration during a relatively short period before ripening (see Figure 18-11). This rise in respiratory rate accompanying the ripening of certain fruits is known as the *climacteric*. The climacteric rise terminates at the climacteric peak, at which time the apple fruit is soft and edible. Storage of apple fruits in atmospheres low in oxygen—about 3% oxygen, 5% carbon dioxide, and 92% nitrogen—has been found to postpone the climacteric and prolong their lifetimes.

As noted in Table 18-3, rather dramatic changes in fruit texture, chemical composition, and color accompany the shift from fruit maturation to ripening. The economic importance of fruits has caused many workers to become interested in trying to regulate maturation and ripening processes in order to prolong fruit storage and improve shipping qualities as well as fruit flavor, texture, and appearance. Many fruits for example, are picked when in the late stages of maturity and are totally unfit for eating. After storage under conditions rather specific for each fruit, the fruit can be induced to enter the ripening stage, whereupon the structural and chemical changes occur that bring the fruit to an edible condition. Storage conditions, depending on the particular fruit handled, may involve low temperatures, high levels of carbon dioxide, or low levels of oxygen.

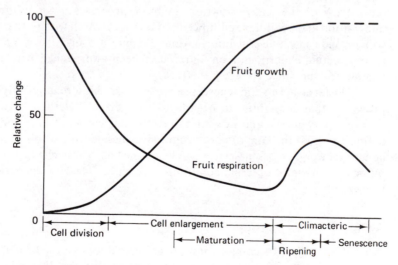

Figure 18-11 Diagrammatic representation of the growth of a climacteric fruit and the course of respiration. In the case of the apple, the total time period on the horizontal axis may be about 5 months. (After Dilley, 1969; Figure 1.)

Because of the large increase in respiratory activity—the climacteric—and other changes associated with ripening, fruit physiologists have been interested in the regulatory mechanisms responsible for controlling ripening. Ethylene (see Chapter 14) has long been known to be involved in the ripening of certain fruits, but the general opinion was that it was necessary to raise the level of ethylene in the environment around the fruit to initiate ripening. With the development of methods capable of measuring low internal concentrations of ethylene, it was found that ethylene production within the fruit tissues increases before other ripening processes take place. There is rather general agreement now that ethylene functions as a fruit-ripening hormone and is responsible, at least in part, for setting in motion all the various structural and biochemical changes characteristic of ripening.

In addition to ethylene, other plant growth substances, such as auxins, cytokinins, gibberellins, and abscisic acid, are present in varying concentrations at different stages of fruit development and it has been suggested that gibberellins and abscisic acid interact in some manner so as to trigger ripening. How these growth substances influence ethylene production is not known.

AGING, SENESCENCE, AND DEATH

Plants grow old and die, a fate shared with other living things. The range of longevity of plants is much greater, however, than that shown by animals. The lifespan of plants ranges from a few weeks in the case of certain annual plants to over 4000 years for one species of pine. There has been interest in the aging processes of animals, particularly humans, for a great many years, but botanists have shown little interest until rather recently. With the realization that the basic processes of aging must ultimately be understood at the cellular and subcellular levels, research with both plants and animals will provide much useful information. Despite a long history of speculation and theory, actual experiments on aging have been approached with modern research techniques only since the early 1950s.

The terms aging and senescence are often used interchangeably in describing the processes that contribute to organismal longevity. Yet these terms refer to quite different phenomena and we shall follow the suggestion of Carr and Pate (1967) and define aging as "the sum total of progressive changes in a whole plant or in one of its constituent organs." Senescence will be defined as "those changes, caused by factors other than harmful external conditions, which are clearly degenerative—ultimately irreversibly so."

Aging

According to the preceding definition, aging includes all the chemical and structural changes occurring in cells, tissues, organs, and whole plants throughout their life. With different life forms—annuals, biennials, perennials, woody, herbaceous, etc.—the patterns of change associated with aging may be quite complex. At any particular time a plant is composed of dead cells, dying cells, new cells, growing cells, and cells in a relatively stable state. Similarly, tissues and organs of a plant will consist of populations of cells of different ages. Because of the complexities of the interactions between these different cell populations, aging of whole plants seldom is studied. Instead the activities of structurally less complex systems, such as single leaves, cotyledons, internodes, roots, flowers, fruits, are investigated. These organs may remain attached to the plant while being investigated or they may be excised and studied in isolation.

Changes in the internal composition of the fifth leaf of field pea (*Pisum arvense*) during development are shown in Figure 18-12. The leaves remained attached to the plant until the time of removal for chemical analysis. Fresh weight, protein content, and soluble nitrogen (total amino acids plus amides) content increases during early growth of the leaf and then decreases as the leaf ages. It is noted, however, that the peak of fresh weight is reached on day 24 whereas the peaks of protein and soluble nitrogen are reached earlier, on days 16 and 11, respectively. Furthermore, within the soluble nitrogen fraction, Figure 18-12(b), there are differences in the times that peak concentrations are reached. The amides, asparagine and glutamine, follow a similar pattern, but the amino acid homoserine reaches a peak concentration about 10 days later.

Figure 18-2(c) shows the changes with time of soluble proteins in the fifth leaf of field pea. These soluble proteins make up the protein fraction shown in Figure 18-12(a). It is seen that some proteins appear to be identical at all ages but differ in concentration. Other proteins are present at some leaf ages but absent at others. The patterns of protein distribution probably reflect differences in metabolic activity. A number of investigations have shown, for example, that embryonic and meristematic tissues utilize carbohydrates by a different metabolic pathway than that followed in mature tissues. Different complements of enzymes must be present in the tissues of varying ages.

Not only are there differences in the kinds and amounts of nitrogenous compounds in aging tissues and organs, but similar differences are to be found in other constituents as well—nucleic acids, nucleotides, carbohydrates, lipids, inorganic ions, growth substances, and so forth. Each leaf, petiole, internode, root, and fruit undergoes shifts in chemical composition similar to those noted for the fifth leaf of field pea. Young, actively growing tissues and organs function as sinks into which move water and solutes from old tissues and organs. Depending on the proximity of source and sink and nature of the vascular connections between the two, there is a ready circulation between old and young parts of the plant. The influence of this circulation of materials on the aging process is not well understood, but it can be assumed to have a profound effect on growth and development.

Not shown in Figure 18-12 is the decline in fresh weight between days 24 and 31, accompanied by a loss of chlorophyll. Between day 31 and abscission of the leaf at day 37 there is a sharp decline in fresh weight and chlorophyll, accompanied by leaf yellowing, a typical symptom of leaf senescence. The overall picture of leaf aging in *Pisum arvense* is as follows. Until maximum leaf area is achieved at day 15, protein

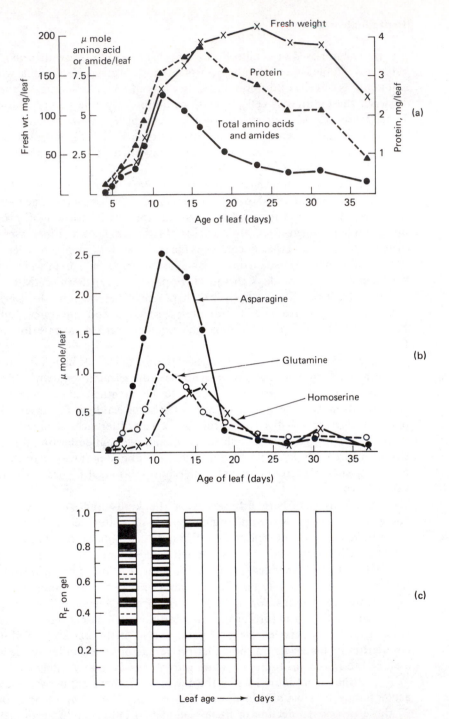

Figure 18-12 Changes with time in the internal composition of the fifth leaf of field pea (*Pisum arvense*): (a) fresh weight, protein, and soluble nitrogen (amino acids plus amides); (b) concentrations of several amino acids; (c) soluble proteins separated by acrylamide gel electrophoresis. (After Carr and Pate, 1967; Figures 6 and 7.)

synthesis and other anabolic activities sustain cell division and cell enlargement. Cell fresh weight continues to increase for an additional 9 to 10 days but then begins to decline. Even before the final leaf area is reached on day 15 the amide fraction (glutamine and asparagine) decreases sharply, but protein synthesis continues for a few more days until it begins to decline around day 17. Both protein synthesis and protein breakdown are occurring. Other important metabolites, such as DNA and RNA, also undergo distinctive changes during the life of the leaf. Studies of cell and organelle structure indicate that the subcellular organization undergoes drastic changes. Starting about the time that fresh weight begins to decline, organelles and membranes become disorganized. Thus there are progressive changes in metabolites and ultracellular organization during the life of the leaf.

At what particular point in the 37-day life of the fifth leaf of *Pisum arvense* would senescence be said to start? There does not appear to be a particular, single characteristic of leaf metabolism or organization that makes an irreversible shift toward senescence. Probably around day 25, when leaf fresh weight starts to decline, the metabolic patterns become irreversible and degenerative processes gradually take over. Until day 25 the senescence events can be reversed in many cases by removing the leaf and treating it so as to promote the initiation of roots. In such rooted leaves the tissue remains green and metabolically active for additional periods of time—weeks or months.

Senescence

Senescence is a phase of the aging process. The major characteristic of senescence is that the metabolic processes are catabolic and eventually become irreversible and terminate in death. In whole plants several general patterns of senescence can be distinguished. Many annual plants display a sequential senescence of leaves, starting with the oldest and continuing up the stem. The pattern of leaf metabolism followed by each leaf is similar to that shown in Figure 18-12. Metabolites, such as amino acids, amides, nucleotides, and carbohydrates, are transported out of the aging leaf to younger leaves. In some plants an abscission layer forms at the base of the petiole before abscission of the leaf. In other cases, the leaves turn yellow or brown and die but remain attached to the plant. The actual number of leaves undergoing senescence and death depends on the particular plant species as well as whether the plant remains vegetative or is induced to reproductive development. In many annuals leaf initiation and growth are halted when reproductive development is initiated. In others leaves continue to be initiated along with reproductive structures and the patterns of leaf senescence are related to fruit development and fruit senescence.

In woody plants, such as deciduous trees in temperate regions, all leaves on the tree enter the senescence phase simultaneously and eventually drop off the plant. The signal for the initiation of senescence is the change in photoperiod, generally long days. The morphogenetic pigment, phytochrome, is the sensor for perception of change in day length. In tropical regions most trees do not display such marked seasonal patterns of senescence and leaf abscission. Instead leaves undergo senescence and abscission continuously along with leaf initiation and growth. Such plants are *evergreen*. In tropical regions where rainfall patterns alternate between wet and dry leaf senescence and abscission may occur during dry seasons.

In another senescence pattern plants may grow for years in a vegetative condition, continuing to produce leaves. When flowering is initiated, leaf development ceases and the organic resources (carbohydrates, proteins, etc.) of the entire plant are mobilized and transported to the developing fruits. Following fruit development and maturation, the entire plant enters senescence and eventually dies. Members of the genus, *Agave*, for example, show such a growth pattern. They may remain vegetative for 35 to 40 years before flowering is initiated. Some palms may grow for 75 to 80 years before the entire plant dies after flower and fruit development.

Changes in biochemical activity of *Pisum* leaves shown in Figure 18-12 encompassed the entire aging process from leaf expansion to abscission. Studies confined more specifically to the senescence phase of leaf development are shown in Figure 18-13. Here changes in physiological activity and chemical composition were followed during leaf senescence of the third pair of leaves attached to intact plants of *Perilla frutescens*. Full leaf expansion is achieved on day 40 and leaf abscission at day 71. Photosynthetic and respiratory activity behave quite differently during the senescence phase [Figure 18-13(a)]. Photosynthesis declined after day 40 even before the loss of chlorophyll starting with day 55 [Figure 18-13(b)]. From about day 55 there was a precipitous decline in both photosynthetic activity and chlorophyll. The rate of respiration, on the other hand, remained unchanged except for a brief burst in activity at day 60, followed by a decline as leaf abscission was approached. Protein concentration increased until about day 55 and then declined sharply in parallel with the decline in chlorophyll content and photosynthetic activity. DNA and RNA concentrations in the leaves did not change drastically until the last few days of senescence, whereupon they decreased sharply at leaf abscission. As noted in the discussion of leaf aging in *Pisum arvense*, there are changes in physiological activity and chemical composition during the senescence phase of aging, but it is difficult to identify any one particular activity as responsible for initiating senescence. The *Perilla frutescens* data shown in Figure 18-13 are in accord with this view. The changes in the protein profile [Figure 18-12(c)] suggest that some enzymes probably disappear or become inactive during senescence. Particular enzymes responsible for senescence, however, have not been identified.

Senescence studies just discussed for *Pisum arvense* and *Perilla frutescens* were carried out on leaves attached to intact plants. Many other studies on senescence have been carried out on excised organs, particularly leaves. Although such systems may be easier to handle experimentally, it is by no means certain that the results observed with excised plant material have a direct bearing on the question of senescence. Excised systems do not participate in the normal circulation of water and solutes from other parts of the plant. Nevertheless, interesting and valuable results have been obtained with excised organs that correlate with results from intact organs. Studies with excised leaves, for example, have shown that senescence is reversible, at least during its early stages. If excised leaves are placed under conditions that permit root formation to take

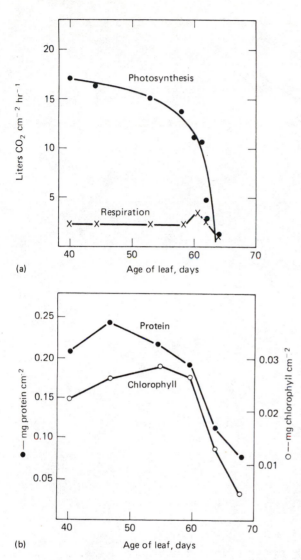

(a)

(b)

Figure 18-13 Changes in physiological activity and internal composition of the attached third pair of leaves of *Perilla frutescens* from the time of completion of leaf expansion on day 40 to abscission at day 64. (a) Photosynthesis and respiration; (b) chlorophyll and protein content. (After Woolhouse, 1967; Figures 2 and 3, Table 2.)

place, the rooted leaves do not display senescence symptoms for long periods of time. Young excised leaves root more easily than mature leaves and very old leaves root with great difficulty. From such studies it has been concluded that the roots are a source of some factor or factors that reverse or inhibit senescence.

The chemical nature of the senescence inhibitor in roots is not known, but several known plant growth substances are likely candidates. Figure 18-14 shows the effect of kinetin (one of the synthetic cytokinins) on the senescence of excised tobacco (*Nicotiana rustica*) leaves. In this study excised tobacco leaves were split in half; one half was then placed with its petiole immersed in water and the other half in a solution of kinetin. At various times the nitrogen and phosphorus contents of the half leaves

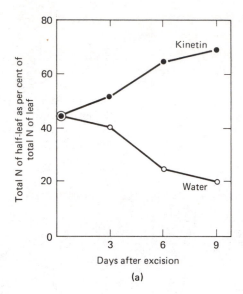

(a)

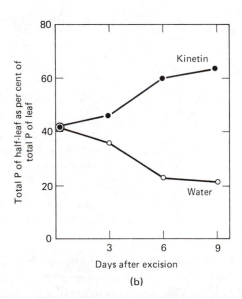

Figure 18-14 Effect of kinetin on the total nitrogen (a) and phosphorus (b) content of excised tobacco (*Nicotiana rustica*) leaves. (After Woolgiehn, 1967; Figure 4.)

(b)

were determined. It is seen that kinetin prevents the decline in nitrogen and phosphorus noted in the water treatment. It has also been shown that kinetin and other synthetic cytokinins prevent the yellowing of attached senescing leaves. In some cases, a single drop of kinetin applied to a senescing leaf is sufficient to prevent the loss of chlorophyll within the treated area whereas the rest of the leaf turns yellow.

Accompanying the physiological and biochemical changes during senescence are changes in cell ultrastructure. Electron microscopic studies of the fine structure of senescing leaves show that subcellular organization is disrupted early in senescence. Mitochondria, chloroplasts, nuclei, and ribosomes are particularly vulnerable and changes in their structure are seen before any visible external changes in the leaves are apparent. It can be imagined that the membranes of the organelles, as well as the plasma membrane, are sites of alteration during senescence. Cellular activities centered in membranes—electron transport, photosynthesis, phosphorylation, ion transport, and so forth—depend on intact membrane structures. Any alterations could lead to those changes associated with senescence.

Theories of Aging and Senescence

From studies with plants and animals a number of theories have been proposed to explain aging processes, particularly those that are degenerative in nature. The theories fall into four general categories.

1. There is an accumulation of toxic or deleterious products within the organism.
2. There is a loss of some essential substance from the organism.
3. The organism undergoes general "wear and tear," a process that damages or destroys the integrity of the subcellular organization or metabolic apparatus.
4. Through mutations, certain cells accumulate deleterious genes (genetic damage).

These theories are not mutually exclusive and different organisms may show aspects of all the aging processes listed.

Attempts to isolate and identify specific toxic substances associated with aging have not been too productive. Many substances, such as the so-called secondary plant substances, accumulate in tissues and organs during aging, but none has been shown to induce aging. Also, during growth and development cells die and the cellular materials are distributed to other parts of the organism, where they may be metabolized by both anabolic and catabolic reactions. As vascular elements (xylem cells) develop, for example, their cellular contents (proteins, nucleic acids, polysaccharides, and the like) are broken down by hydrolytic enzymes and released to adjoining cells. It has been found that indoleacetic acid is synthesized following xylem differentiation, but this process has not been specifically related to aging. Moreover, during growth of the organism cells and subcellular organelles disintegrate and cellular products, such as membrane particles, may accumulate. These particles are rich in lipids, which may be metabolized to peroxides or other reactive molecules of a deleterious nature.

Plants also excrete substances during growth and development. Both inorganic and organic molecules are lost and it is possible that one or more of these molecules may be involved in aging. It has been suggested that a "juvenility factor" is responsible for maintaining a plant in the vegetative phase of development. No such substance, however, has been isolated and identified.

It is unlikely that a single substance will be found responsible for aging and senescence. Multicellular organisms possess complex interacting metabolic processes and interconnecting pathways for the movement of materials throughout the organism. Within such a network, small changes in the rates of important metabolic processes may be propagated throughout the organism. If these processes alter some aspect of nucleic acid metabolism, protein synthesis will be altered. Small alterations in the genetic apparatus may not be noticeably harmful, but if the organism accumulates many genetic alterations during growth and development, eventually irreversible changes may lead ultimately to death.

As noted, aging refers to the sequential changes that occur in a plant as it proceeds from zygote formation to death. Senescence should be regarded as one phase of the normal life cycle of the plant. It is built into the genetic information characteristic of each species of plant. A special theory of senescence is unnecessary. What is needed is a better understanding of all processes and activities of plants during growth and development.

REFERENCES

BARBER, J. T., and F. C. STEWARD. 1968. The proteins of *Tulipa* and their relation to morphogenesis. *Develop. Biol.* 17:326–349.

BERNIER, G., J-M. KINET, and R. M. SACHS. 1981. *The Physiology of Flowering.* Vol. 1, *The Initiation of Flowers*; vol. 2, *Transition to Reproductive Growth.* CRC Press, Boca Raton, Florida.

BIALE, J. B. 1964. Growth, maturation, and senescence in fruits. *Science* 146:880–888.

BORTHWICK H. A., S. B. HENDRICKS, and M. W. PARKER. 1952. The reaction controlling floral initiation. *Proc. Nat. Acad. Sci.* U.S. 38:929–934.

BÜNNING, E. 1967. *The Physiological Clock*, 2nd ed. Academic Press, New York.

BURG, S. P., and E. A. BURG. 1965. Ethylene action and the ripening of fruits. *Science* 148:1190–1196.

CARR, D. J., and J. S. PATE. 1967. Ageing in the whole plant. Pages 559–599 *in* W. H. WOOLHOUSE, ed. *Aspects of the Biology of Ageing.* Society of Experimental Biology Symposium No. 21. Cambridge University Press, Cambridge.

CHAILAKHYAN, M. Kh. 1968 Internal factors of plant flowering. *Ann. Rev. Plant Physiol.* 19:1–36.

DILLEY D. R. 1969. Hormonal control of fruit ripening. *Hort. Sci.* 4:111–114.

EVANS, L. T., ed. 1969. *The Induction of Flowering. Some Case Histories.* Cornell University Press, Ithaca, N.Y.

GARNER, W. W., and H. A. ALLARD. 1920. Effect of the relative length of day and night and other factors of the environment on growth and reproduction in plants. *J. Agr. Res.* 18:553–606.

HENDRICKS, S. B. 1956. Control of growth and reproduction by light and darkness. *Amer. Scient.* 44:229–247.

HILLMAN, W. W. 1961. *The Physiology of Flowering.* Holt, Rinehart and Winston, New York.

HODSON, H. K., and K. C. HAMNER, 1970. Floral inducing extract from *Xanthium. Science* 167:384–385.

LANG, A. 1965. Physiology of flower initiation. Pages 1380–1530 *in Encyclopedia of Plant Physiology*, vol. 15, part 1. Springer-Verlag, Berlin.

LINCOLN, R. G., A. CUNNINGHAM, and K. C. HAMNER. 1964. Evidence for a florigenic acid. *Nature* 202:559–561.

NAYLOR, A. W. 1952. The control of flowering. *Sci. Amer.* 186:49–56.

NAYLOR, A. W. 1961. The photoperiodic control of plant behavior. Pages 331–389 *in Encyclopedia of Plant Physiology*, vol. 16, Springer-Verlag, Berlin.

SALISBURY, F. B. 1963. *The Flowering Process*. Pergamon Press, Elmsford, NY.

SCHWABE, W. W. 1971. Physiology of vegetative reproduction and flowering. Pages 233–411 *in* F. C. STEWARD, ed. *Plant Physiology, a Treatise*, vol. 6A. Academic Press, New York.

SHIBAOKA, H., M. FURUYA, M. KATSUMI, and A. TAKIMOTO, eds, 1978. *Controlling Factors in Plant Development*. Special Issue No. 1. *Botanical Magazine*, Tokyo.

SMITH, H. 1975. *Phytochrome and Photomorphogenesis*. McGraw-Hill Book Company, London.

STEWARD F. C., J. T. BARBER, E. F. BLEICHERT, and W. M. ROCA. 1971. The behavior of shoot apices of *Tulipa* in relation to floral induction. *Develop. Biol.* 25:310–335.

SWEENY, B. M. 1969. *Rhythmic Phenomena in Plants*. Academic Press, New York.

THIMANN, K. V. 1978. Senescence. Pages 19–43 *in* H. SHIBAOKA, M. FURUYA, M. KATSUMI, and A. TAKIMOTO, eds. *Controlling Factors in Plant Development*. Special Issue No. 1, *Botanical Magazine*, Tokyo.

THIMANN, K. V., ed. 1980. *Senescence in Plants*. CRC Press, Boca Raton, Florida.

VINCE-PRUE, D. 1975. *Photoperiodism in Plants*. McGraw-Hill Book Company. New York.

WARDLAW, C. W. 1957. On the organization and reactivity of the shoot apex of vascular plants. *Amer. J. Bot.* 44:176–185.

WOOLGIEHN, R. 1967. Nucleic acid and protein metabolism of excised leaves. Pages 231–246 *in* W. H. WOOLHOUSE, ed. *Aspects of the Biology of Ageing*. Society of Experimental Biology Symposium No. 21. Cambridge University Press, Cambridge University Press, Cambridge.

WOOLHOUSE, H. W. 1967. The nature of senescence in plants. Pages 179–213 *in* W. H. WOOLHOUSE, ed. *Aspects of the Biology of Ageing*. Society of Experimental Biology Symposium No. 21. Cambridge University Press, Cambridge.

Appendix:
International System of Units

The Système International d'Unités (SI) was adopted by the eleventh General Conference on Weights and Measures and endorsed by the International Organization for Standardization in 1960. The system is an extension and refinement of the traditional metric system and is superior to any other in being completely coherent, rational, and comprehensive. In the system there is one, and only one, unit for each physical quantity and the product or quotient of any two SI units yields the unit of the resulting quantity; no numerical factors are involved.

The seven basic units on which the SI is based are listed in Table 1.

TABLE 1 BASIC SI UNITS

Quantity	Name of unit	Unit symbol
Length	meter	m
Mass	kilogram	kg
Time	second	s
Electric current	ampere	A
Thermodynamic temperature	kelvin	K
Luminous intensity	candela	cd
Amount of substance	mole	mol

The basic units are defined as follows.

Meter.   The meter is the length equal to 1650763.73 (exactly) wavelengths in a vacuum of the radiation corresponding to the transition between the energy levels $2p_{10}$ and $5d_5$ of the pure nuclide ^{86}Kr.

Kilogram. The kilogram is the mass of the International Prototype Kilogram, which is in the custody of the Bureau International des Poids et Mesures at Sèvres, France.

Second. The second is the duration of 9192631770 periods of the radiation corresponding to the transition between the two hyperfine levels (F = 4, M_F = 0 and F = 3, M_F = 0) of the ground state of the atom of pure nuclide ^{133}Cs.

Ampere. The ampere is that constant current which, if maintained in two parallel rectilinear conductors, of infinite length, and of negligible circular cross-section, at a distance apart of 1 meter in a vacuum, would produce a force between the conductors equal to 2×10^{-7} newton per meter of length.

Kelvin. The kelvin, or degree Kelvin, is completely defined by the decision of the 1954 Conférence Générale to assign the value 273.16 degrees Kelvin (exactly) to the thermodynamic temperature at the triple point of water.

Candela. The candela, the unit of luminous intensity, is such that the luminance of a blackbody at the freezing point of platinum is 6×10^5 candelas per square meter.

Mole. The mole is an amount of substance of a system which contains as many elementary units as there are carbon atoms in 0.012 kg (exactly) of the pure nuclide ^{12}C. The elementary unit must be specified and may be an atom, a molecule, an ion, an electron, a photon, and so forth, or a specified group of such entities.

Several other units are normally used with the SI system. Although not a part of SI, they remain in common usage. Table 2 lists a few of these units.

TABLE 2 UNITS IN USE WITH THE INTERNATIONAL SYSTEM

Name	Symbol	Value in SI unit
Minute	min	1 min = 60 s
Hour	h	1 h = 60 min = 3600 s
Day	d	1 d = 24 h = 86,400 s
Liter	L	1 L = 10^3 mL = 10^6 μL
Metric ton (tonne)	t	1 t = 10^3 kg

A number of quantities used in plant physiological research are known as derived SI units and are listed in Table 3.

TABLE 3 DERIVED SI UNITS WITH SPECIAL NAMES

Physical quantity	Name of unit	Symbol	Definition of unit
Energy	joule	J	m^2 kg s^{-2}
Force	newton	N	m kg s^{-2}
Pressure	pascal	Pa	m^{-1} kg s^{-2} (N m^{-2})
Power	watt	W	m^2 kg s^{-3} (J s^{-1})

Some of the SI units are of inconvenient size, but the prefixes listed in Table 4 may be used to indicate fractions or multiples of the basic or derived units.

TABLE 4   SOME SI PREFIXES

Multiple or fraction	Prefix	Symbol
10^6	mega	M
10^3	kilo	k
10^2	hecto	h
10^{-1}	deci	d
10^{-2}	centi	c
10^{-3}	milli	m
10^{-6}	micro	μ
10^{-9}	nano	n

Table 5 lists some acceptable exceptions to SI units.

TABLE 5   ACCEPTABLE EXCEPTIONS TO SI UNITS[a]

Concentration	molal = mole per kilogram ($m = \text{mol kg}^{-1}$)
	molar = mole per liter ($M = \text{mol L}^{-1}$)
Pressure	bar = 10^5 Pa
Temperature	degree Celsius (°C)
Energy	calorie (cal) = 4.18 J

[a] These are listed because they are used in this book.

REFERENCES

ANON. 1977. The International System of Units (SI), National Bureau of Standards, U.S. Department of Commerce, NBS Special Publication 330.

INCOLL, L. D., S. P. LONG, and M. R. ASHMORE. 1977. SI units in publications in plant science. Commentaries in Plant Science, No. 28 (April). Pages 331–343 *in Current Advances in Plant Science*.

Index